Introduction to the AdS/CFT Correspondence

Providing a pedagogical introduction to the rapidly developing field of AdS/CFT correspondence, this is one of the first texts to provide an accessible introduction to all the necessary concepts needed to engage with the methods, tools, and applications of AdS/CFT. Without assuming anything beyond an introductory course in quantum field theory, it begins by guiding the reader through the basic concepts of field theory and gauge theory, general relativity, supersymmetry, supergravity, string theory, and conformal field theory, before moving on to give a clear and rigorous account of AdS/CFT correspondence. The final section discusses the more specialized applications, including QCD, quark–gluon plasma, and condensed matter. This book is self-contained and learner-focused, featuring numerous exercises and examples. It is essential reading for both students and researchers across the fields of particle, nuclear, and condensed matter physics.

Horaţiu Năstase is a Researcher at the Institute for Theoretical Physics at the State University of São Paulo, Brazil. To date, his career has spanned four continents: as an undergraduate he studied at the University of Bucharest and at Copenhagen University. He later completed his Ph.D. at the State University of New York at Stony Brook, before moving to the Institute for Advanced Study, Princeton, where his collaboration with David Berenstein and Juan Maldacena defined the pp wave correspondence. He has also held research and teaching positions at Brown University and the Tokyo Institute of Technology.

Introduction to the AdS/CFT Correspondence

HORAŢIU NĂSTASE

Institute for Theoretical Physics at the State University of São Paulo

CAMBRIDGE
UNIVERSITY PRESS

CAMBRIDGE
UNIVERSITY PRESS

University Printing House, Cambridge CB2 8BS, United Kingdom

One Liberty Plaza, 20th Floor, New York, NY 10006, USA

477 Williamstown Road, Port Melbourne, VIC 3207, Australia

314-321, 3rd Floor, Plot 3, Splendor Forum, Jasola District Centre, New Delhi - 110025, India

79 Anson Road, #06-04/06, Singapore 079906

Cambridge University Press is part of the University of Cambridge.

It furthers the University's mission by disseminating knowledge in the pursuit of
education, learning and research at the highest international levels of excellence.

www.cambridge.org
Information on this title: www.cambridge.org/9781107085855

First published 2015

A catalogue record for this publication is available from the British Library

Library of Congress Cataloging in Publication data
Năstase, Horaţiu, 1972– author.
Introduction to the AdS/CFT correspondence / Horaţiu Năstase, Institute for
Theoretical Physics at the State University of São Paulo.
pages cm
ISBN 978-1-107-08585-5
1. Gauge fields (Physics) 2. Supergravity. 3. Supersymmetry. 4. String
models. I. Title.
QC793.3.G38N38 2015
530.14´3–dc23
2015005313

ISBN 978-1-107-08585-5 Hardback

To the memory of my mother,
who inspired me to become a physicist

Contents

Preface

This book is intended as a pedagogical introduction to the rapidly developing field of the AdS/CFT correspondence. This subject has grown to the point where graduate students, as well as researchers, from fields outside string theory or even particle theory, in particular nuclear physics and condensed matter physics, want to learn about it. With this in mind, the book endeavours to introduce AdS/CFT without assuming anything beyond an introductory course in quantum field theory. Some familiarity with the principles of general relativity, supersymmetry or string theory would help the reader follow more easily, but is not necessary, as I introduce all the necessary concepts. I do not overload the book with unnecessary details about these fields, only what I need to give a simple, yet completely rigorous, account of all the basic methods, tools, and applications of AdS/CFT. For more details on these subjects, one can consult a number of good textbooks available for each, which I suggest at the end of the corresponding chapters. When explaining AdS/CFT, I try to give a simple introduction to each method, tool, or application, without aiming for an in-depth or exhaustive treatment. The goal is to introduce most of the AdS/CFT methods, but for an in-depth treatment one should refer to research articles instead. Part I of the book deals with the necessary background material, so someone familiar with this can skip it. Part II describes the basics of AdS/CFT in the context of its best understood example, $\mathcal{N} = 4$ SYM vs. string theory in $AdS_5 \times S^5$. Part III deals with more specialized applications and other dualities, generalizing to the gauge–gravity dualities.

Acknowledgments

It is not possible to write a scientific book without having benefited from learning from teachers, collaborators, and colleagues. So first I would like to thank all of the people who shaped me as a scientist. My mother, through her example as a physicist, first made me realize the beauty of physics and of a scientific career. My high school physics teacher in Romania, Iosif Sever Georgescu, made me realize that I could actually become a physicist, and started my preparation for the physics olympiads, that meant my first contact with international science. My advisor for the undergraduate exchange to NBI in Denmark, Poul Olesen, introduced me to string theory, and all the faculty at NBI made me realize what career I wanted to pursue. My PhD advisor, Peter van Nieuwenhuizen, shaped who I am as a researcher, and many of the subjects described in the first part of this book I learned from him. Of course, I also thank all the other teachers and professors I had along the way. While a postdoc at IAS in Princeton, I learned a lot from Juan Maldacena, who started AdS/CFT, about many of the subjects described in this book. I also thank all my collaborators throughout the years who have helped me understand the concepts presented here.

This book originated from a course I first gave at Tokyo Tech, then at Tokyo Metropolitan University, and finally at the IFT in São Paulo, so I would like to thank all the students that participated in the classes for their input about the material.

In writing this book, I benefited from encouragement and comments on the text from David Berenstein, Aki Hashimoto, and Jeff Murugan. My students Thiago Araujo, Prieslei Goulart, and Renato Costa helped me get rid of errors and typos in an earlier version of the book, and Thiago Araujo and his wife Aline Lima helped me create the figures and make them intelligible. I would like to thank my collaborators, students, and postdocs for their understanding about having less time for interacting with them while I wrote this book.

Introduction

This book gives an introduction to the Anti-de Sitter/Conformal Field Theory correspondence, or AdS/CFT, so it would be useful to first understand what it is about.

From the name, we see that it is a relation between a quantum field theory with conformal invariance (which is a generalization of scaling invariance), living in our flat 4-dimensional space, and string theory, which is a quantum theory of gravity and other fields, living in the background solution of $AdS_5 \times S^5$ (5-dimensional Anti-de Sitter space times a 5-sphere), a curved space with the property that a light signal sent to infinity comes back in a finite time.

The flat 4-dimensional space containing the field theory lives at the boundary (situated at infinity) of the $AdS_5 \times S^5$, thus the correspondence, or equivalence, is said to be an example of *holography*, since it is similar to the way a 2-dimensional hologram encodes the information about a 3-dimensional object. The background $AdS_5 \times S^5$ solution is itself a solution of string theory, as the relevant theory of quantum gravity.

From this description, it is obvious that before we describe AdS/CFT, we must first introduce a number of topics, which is done in Part I of the book. First, we review some relevant notions of quantum field theory, though I assume that the reader has a working knowledge of quantum field theory. Then I describe some basic concepts of general relativity, supersymmetry, and supergravity, since string theory is a supersymmetric theory, whose low energy limit is supergravity. After that, I introduce black holes and p-branes, since the $AdS_5 \times S^5$ string theory background appears as a limit of them. Finally, I introduce string theory, elements of conformal field theory (4-dimensional flat space theories with conformal invariance), and D-branes, which are objects in string theory on which the relevant quantum field theories can be defined.

The AdS/CFT correspondence was put forward by Juan Maldacena in 1997, as a conjectured duality based on a heuristic derivation which will be explained, and until now there is no exact proof for it. However, there is an enormous amount of evidence in its favor in the form of calculations matching on the two sides of the correspondence, turning it into a virtual certainty, so while technically we should append the name "conjecture" to it, this would be a pedantic point, and I shall refrain from doing so.

However, while this is true for all dualities which can be derived in the manner of Maldacena, there are now applications to real-world physics, which I call "phenomenological AdS/CFT," where one uses some general lessons learned from AdS/CFT to engineer a description in terms of quantum field theory that has the right properties to be relevant for systems of interest, but without a microscopic derivation. In this category fall some applications to QCD, quark–gluon plasma, and condensed matter, which are described in detail in Part III of the book. In these cases it is therefore important to realize the conjectural nature of the correspondence.

Another question that we should ask is why is the AdS/CFT correspondence interesting? The reason is that it relates perturbative (weak coupling) string theory calculations in a gravitational theory to nonperturbative (strong coupling) gauge theory calculations, which would otherwise be very difficult to obtain. Of course, the reverse is also true, namely nonperturbative (strong coupling) string theory in a gravitational background is related to perturbative (weak coupling) gauge theory, allowing in principle an (otherwise unknown) definition of the former through the latter, but the rules in this case are much less clear. The strong–weak coupling relation means that AdS/CFT is an example of duality, in the sense of the electric–magnetic duality of Maxwell theory.

The applications to QCD and condensed matter are, however, hampered by the fact that the AdS/CFT duality becomes calculable in the limit of large rank of the gauge group, or "number of colours" on the field theory side, $N_c \to \infty$. Also, the best understood example of $\mathcal{N} = 4$ SYM is very far from the real world, having both supersymmetry and conformal invariance. When we move away from supersymmetry and conformal invariance, the rules are less clear and we can calculate less, as we will see. Nevertheless, AdS/CFT is a developing field, and we have already obtained many useful results and insights, so we can hope that these methods will lead to solving interesting problems that cannot be solved otherwise.

PART I

BACKGROUND

1 Elements of quantum field theory and gauge theory

In this chapter, I review some useful issues about quantum field theory, assuming nevertheless that the reader has seen them before. It will also help to set up the notation and conventions.

1.1 Note on conventions

Throughout this book, I use theorist's conventions, with $\hbar = c = 1$. If we need to reintroduce them, we can use dimensional analysis. In these conventions, there is only one dimensionful unit, $mass = 1/length = energy = 1/time = \ldots$ and when I speak of dimension of a quantity I refer to mass dimension, e.g. the mass dimension of d^4x, $[d^4x]$, is -4.

For the Minkowski metric $\eta^{\mu\nu}$ I use the signature convention mostly plus, thus for instance in 3+1 dimensions the signature will be $(-+++)$, giving $\eta^{\mu\nu} = diag(-1, +1, +1, +1)$. This convention is natural in order to make heavy use of the Euclidean formulation of quantum field theory and to relate to Minkowski space via Wick rotation.

Also, in this book we use *Einstein's summation convention*, i.e. indices that are repeated are summed over. Moreover, the indices summed over will be one up and one down, unless we are in Euclidean space, where up and down indices are the same.

1.2 The Feynman path integral and Feynman diagrams

To exemplify the basic concepts of quantum field theory, and the Feynman diagrammatic expansion, I use the simplest possible example, of a scalar field. A scalar field is a field that under a Lorentz transformation

$$x'^\mu = \Lambda^\mu{}_\nu x^\nu, \tag{1.1}$$

transforms as

$$\phi'(x'^\mu) = \phi(x^\mu). \tag{1.2}$$

We will deal with relativistic field theories, which are also *local*, which means that the action is an integral over functions defined at one point, of the type

$$S = \int L dt = \int d^4x \mathcal{L}(\phi, \partial_\mu \phi). \tag{1.3}$$

Here L is the Lagrangean, whereas $\mathcal{L}(\phi(\vec{x}, t), \partial_\mu \phi(\vec{x}, t))$ is the *Lagrangean density*, that often times by an abuse of notation is also called Lagrangean.

Classically, one varies this action with respect to $\phi(x)$ to give the classical equations of motion for $\phi(x)$,

$$\frac{\partial \mathcal{L}}{\partial \phi} = \partial_\mu \left[\frac{\partial \mathcal{L}}{\partial(\partial_\mu \phi)} \right]. \tag{1.4}$$

Quantum mechanically, the field $\phi(x)$ is not observable anymore, and instead one must use the vacuum expectation value (VEV) of the scalar field quantum operator instead, which is given as a "path integral" in Minkowski space,

$$\langle 0|\hat{\phi}(x_1)|0\rangle = \int \mathcal{D}\phi \, e^{iS[\phi]} \phi(x_1). \tag{1.5}$$

Here the symbol $\int \mathcal{D}\phi$ represents a discretization of a spacetime path $\phi(x_1^\mu) \rightarrow \phi(x_2^\mu)$, followed by integration over the field value at each discrete point:

$$\int \mathcal{D}\phi(x) = \lim_{N \to \infty} \prod_{i=1}^{N} \int d\phi(x_i). \tag{1.6}$$

The action in Minkowski space for a scalar field with only nonderivative self-interactions and a canonical quadratic kinetic term is

$$\begin{aligned} S = \int d^4x \mathcal{L} &= \int d^4x \left[-\frac{1}{2} \partial_\mu \phi \partial^\mu \phi - \frac{1}{2} m^2 \phi^2 - V(\phi) \right] \\ &= \int d^4x \left[\frac{1}{2} \dot{\phi}^2 - \frac{1}{2} |\vec{\nabla}\phi|^2 - \frac{1}{2} m^2 \phi^2 - V(\phi) \right]. \end{aligned} \tag{1.7}$$

A generalization of the scalar field VEV is the *correlation function* or Green's function or *n*-point function,

$$G_n(x_1, \ldots, x_n) = \langle 0|T\{\hat{\phi}(x_1) \ldots \hat{\phi}(x_n)\}|0\rangle = \int \mathcal{D}\phi \, e^{iS[\phi]} \phi(x_1) \ldots \phi(x_n). \tag{1.8}$$

We note, however, that the weight inside the integral, e^{iS}, is highly oscillatory, so the *n*-point functions are hard to define precisely in Minkowski space.

It is much better to Wick rotate to Euclidean space, with signature $+ + \ldots +$, define all objects there, and at the end Wick rotate back to Minkowski space. Both definitions and calculations are then easier. This is also what will happen in the case of AdS/CFT, which will have a natural definition in Euclidean signature, but will be harder to continue back to Minkowski space.

The Wick rotation happens through the relation $t = -it_E$. To rigorously define path integrals, we consider only paths which are periodic in Euclidean time t_E. In the case that the Euclidean time is periodic, a quantum mechanical path integral gives the statistical mechanics partition function at $\beta = 1/kT$ through the Feynman–Kac formula,

$$Z(\beta) = \mathrm{Tr}\{e^{-\beta \hat{H}}\} \left(= \int dq \sum_n |\psi_n(q)|^2 e^{-\beta E_n} = \int dq \langle q, \beta | q, 0 \rangle \right)$$

$$= \int \mathcal{D}q \, e^{-S_E[q]} |_{q(t_E + \beta) = q(t_E)}. \tag{1.9}$$

To obtain the vacuum functional in quantum field theory, we consider the generalization to field theory, for periodic paths with infinite period, i.e. $\lim_{\beta \to \infty} \phi(\vec{x}, t_E + \beta) = \phi(\vec{x}, t_E)$. The Euclidean action is defined through Wick rotation, by the definition

$$iS_M \equiv -S_E. \tag{1.10}$$

This gives for (1.7)

$$S_E[\phi] = \int d^4 x \left[\frac{1}{2} \partial_\mu \phi \partial_\mu \phi + \frac{1}{2} m^2 \phi^2 + V(\phi) \right], \tag{1.11}$$

where, since we are in Euclidean space, $a_\mu b_\mu = a_\mu b^\mu = a_\mu b_\nu \delta^{\mu\nu}$, and time is defined as $t_M \equiv x^0 = -x_0 = -it_E$, $t_E = x_4 = x^4$, and so $x^4 = ix^0$. In this way, the oscillatory factor e^{iS} has been replaced by the highly damped factor e^{-S}, sharply peaked on the minimum of the Euclidean action.

The Euclidean space correlation functions are then defined as

$$G_n^{(E)}(x_1, \dots, x_n) = \int \mathcal{D}\phi \, e^{-S_E[\phi]} \phi(x_1) \dots \phi(x_n). \tag{1.12}$$

We can define a generating functional for the correlation functions, the *partition function*,

$$Z^{(E)}[J] = \int \mathcal{D}\phi \, e^{-S_E[\phi] + J \cdot \phi} \equiv {}_J\langle 0 | 0 \rangle_J, \tag{1.13}$$

where in d dimensions

$$J \cdot \phi \equiv \int d^d x J(x) \phi(x). \tag{1.14}$$

It is so called because at finite periodicity β we have the same relation to statistical mechanics as in the quantum mechanical case,

$$Z^{(E)}[\beta, J] = \mathrm{Tr}\{e^{-\beta \hat{H}_J}\} = \int \mathcal{D}\phi \, e^{-S_E[\phi] + J \cdot \phi} |_{\phi(\vec{x}, t_E + \beta) = \phi(\vec{x}, t_E)}. \tag{1.15}$$

The Euclidean correlation functions are obtained from derivatives of the partition function,

$$G_n^{(E)}(x_1, \dots, x_n) = \frac{\delta}{\delta J(x_1)} \cdots \frac{\delta}{\delta J(x_n)} \int \mathcal{D}\phi \, e^{-S_E + J \cdot \phi} \bigg|_{J=0}$$

$$= \frac{\delta}{\delta J(x_1)} \cdots \frac{\delta}{\delta J(x_n)} Z^{(E)}[J] \bigg|_{J=0}. \tag{1.16}$$

Going back to Minkowski space, we can also define a partition function as a generating functional of the Green's functions,

$$Z[J] = \int \mathcal{D}\phi \, e^{iS[\phi]+i\int d^d x J(x)\phi(x)}, \tag{1.17}$$

that again gives the correlation functions through its derivatives by

$$\begin{aligned}
G_n(x_1, \ldots, x_n) &= \frac{\delta}{i\delta J(x_1)} \cdots \frac{\delta}{i\delta J(x_n)} \int \mathcal{D}\phi \, e^{iS+i\int d^d x J(x)\phi(x)} \Big|_{J=0} \\
&= \frac{\delta}{i\delta J(x_1)} \cdots \frac{\delta}{i\delta J(x_n)} Z[J] \Big|_{J=0}.
\end{aligned} \tag{1.18}$$

The correlation functions can be calculated in perturbation theory in the interaction S_{int}, through the use of Feynman diagrams.

The Feynman theorem relates the correlation functions in the full theory, in the vacuum of the full theory $|\Omega\rangle$, with normalized ratios of correlation functions in the interaction picture, in the vacuum of the free theory $|0\rangle$,

$$\begin{aligned}
&\langle \Omega | T\{\phi_H(x_1) \ldots \phi_H(x_n)\} | \Omega \rangle \\
&= \lim_{T \to \infty(1-i\epsilon)} \frac{\langle 0| T\left\{ \phi_I(x_1) \ldots \phi_I(x_n) \exp\left[-i\int_{-T}^{T} dt H_I(t) \right] \right\} |0\rangle}{\langle 0| T\left\{ \exp\left[-i\int_{-T}^{T} dt H_I(t) \right] \right\} |0\rangle},
\end{aligned} \tag{1.19}$$

where H_I is the interaction Hamiltonian H_i in the interaction picture ($H = H_0 + H_i$), ϕ_I is an interaction picture field, and ϕ_H is a Heisenberg picture field. The denominator cancels vacuum bubbles, which factorize in the calculation, leaving only connected diagrams.

In the path integral formalism and in Euclidean space, we can find correlation functions of the full theory as normalized correlation functions in the interaction picture (divided by the vacuum bubbles), giving again connected diagrams only. For the one-point function and at nonzero source $J(x)$, we obtain the relation

$$\frac{1}{Z[J]} \frac{\delta Z[J]}{\delta J(x)} = \frac{\delta(-W[J])}{\delta J(x)}, \tag{1.20}$$

where $-W[J]$ is defined as the generating functional of connected diagrams, relation solved by

$$Z[J] = \mathcal{N} e^{-W[J]}. \tag{1.21}$$

Here $W[J]$ is called the free energy, again because of the relation with statistical mechanics.

To exemplify the Feynman rules, we use a scalar field action in Euclidean space,

$$S_E[\phi] = \int d^4 x \left[\frac{1}{2}(\partial_\mu \phi)^2 + m^2 \phi^2 + V(\phi) \right]. \tag{1.22}$$

Here I have used the notation

$$(\partial_\mu \phi)^2 = \partial_\mu \phi \partial^\mu \phi = \partial_\mu \phi \partial_\nu \phi \eta^{\mu\nu} = -\dot{\phi}^2 + (\vec{\nabla}\phi)^2. \tag{1.23}$$

Moreover, for concreteness, I use $V = \lambda \phi^4$.

Then, the **Feynman diagram in x space** is obtained as follows. One draws a diagram, in the example in Fig. 1.1a it is the so-called "setting Sun" diagram.

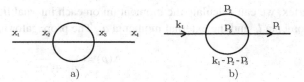

Figure 1.1   a) "Setting Sun" diagram in x-space; b) "Setting Sun" diagram in momentum space.

The Feynman rules are:

0. Draw all the Feynman diagrams for the given correlation function at the given loop order (or given number of vertices).

1. A line between point x and point y represents the Euclidean propagator

$$\Delta(x, y) = [-\partial_\mu \partial^\mu + m^2]^{-1} = \int \frac{d^4 p}{(2\pi)^4} \frac{e^{ip(x-y)}}{p^2 + m^2}, \qquad (1.24)$$

which is a Green's function for the kinetic operator, i.e.

$$[-\partial_\mu \partial^\mu + m^2]_x \Delta(x, y) = \delta(x - y). \qquad (1.25)$$

The analytical continuation (Wick rotation) of the Euclidean propagator to Minkowski space gives the Feynman propagator,

$$D_F(x - y) = \int \frac{d^4 p}{(2\pi)^4} \frac{-i}{p^2 + m^2 - i\epsilon} e^{ip\cdot(x-y)}. \qquad (1.26)$$

2. A 4-vertex at point x represents the vertex factor

$$\int d^4 x (-\lambda). \qquad (1.27)$$

3. Then the value of the Feynman diagram, $F_D^{(N)}(x_1, \ldots, x_n)$ is obtained by multiplying all the above elements, and the value of the n-point function is obtained by summing over diagrams, and over the number of 4-vertices N with a weight factor $1/N!$:

$$G_n(x_1, \ldots, x_n) = \sum_{N \geq 0} \frac{1}{N!} \sum_{\text{diag } D} F_D^{(N)}(x_1, \ldots, x_n). \qquad (1.28)$$

Equivalently, one can use a rescaled potential $\lambda \phi^4 / 4!$ and construct only *topologically inequivalent* diagrams. Then the vertices are still $\int d^4 x (-\lambda)$, but we divide each inequivalent diagram by a statistical weight factor,

$$S = \frac{N! (4!)^N}{\text{\# of equivalent diagrams}}, \qquad (1.29)$$

which equals the number of symmetries of the diagram.

In momentum space, we can use simplified Feynman rules, where we consider as independent momenta the external momenta flowing into the diagram, and integration variables $l_1, \ldots, l_L$ for each independent loop in the diagram. Using momentum conservation at each vertex, we can calculate the momentum on each internal line, p_i, as a function of the loop momenta l_k and the external momenta q_j. The propagator is now

$$\Delta(p) = \frac{1}{p^2 + m^2}, \tag{1.30}$$

and for each internal line (between two internal points) we write $1/(p_i^2 + m^2)$, for each external line (between two points, one of them external) $q/(q_j^2 + m^2)$. The vertex factor is now simply $-\lambda$.

1.3 S matrices vs. correlation functions

We mentioned in the previous section that the VEV of a scalar field is an observable in quantum theory. More precisely, the normalized VEV in the presence of a source $J(x)$,

$$\phi(x; J) = \frac{{}_J\langle 0|\hat{\phi}(x)|0\rangle_J}{{}_J\langle 0|0\rangle_J} = \frac{1}{Z[J]} \int \mathcal{D}\phi \, e^{-S_E[\phi]+J\cdot\phi} \phi(x) = \frac{\delta}{\delta J(x)} \ln Z[J]$$

$$= -\frac{\delta}{\delta J(x)} W[J], \tag{1.31}$$

is called the *classical field ϕ^{cl}* and satisfies a quantum version of the classical field equation. One defines the *quantum effective action* as the Legendre transform of the free energy,

$$\Gamma[\phi^{\mathrm{cl}}] = W[J] + \int d^4 x J(x)\phi^{\mathrm{cl}}(x), \tag{1.32}$$

and finds that it contains the classical action, plus quantum corrections. Then we have the quantum analog of the classical equation of motion with a source $\delta S[\phi]/\delta\phi(x) = J(x)$,

$$\frac{\delta\Gamma[\phi^{\mathrm{cl}}]}{\delta\phi^{\mathrm{cl}}(x)} = J(x). \tag{1.33}$$

The effective action is a generator of the one particle irreducible (1PI) diagrams (except for the 2-point function, where we add an extra term).

To relate to real scatterings, one constructs incoming and outgoing wavefunctions, representing actual states, in terms of the idealized states of fixed (external) momenta $\vec{k}$. These are Schrödinger picture states $\langle\{\vec{p}_i\}|$ and $|\{\vec{k}_j\}\rangle$. We also define Heisenberg picture states whose wavepackets are well isolated at $t = -\infty$, and can be considered noninteracting there (but overlap at other t),

$$|\{\vec{p}_i\} >_{\mathrm{in}}, \tag{1.34}$$

and Heisenberg picture states whose wavepackets are well isolated at $t = +\infty$, and can be considered noninteracting there (but overlap at other t),

$$|\{\vec{p}_i\} >_{\mathrm{out}} . \tag{1.35}$$

Then the S-matrix is defined by

$$\langle\{\vec{p}_i\}|S|\{\vec{k}_j\}\rangle = {}_{\mathrm{out}}\langle\{\vec{p}_i\}|\{\vec{k}_j\}\rangle_{\mathrm{in}}. \tag{1.36}$$

Reduction formula (Lehmann, Symanzik, Zimmermann)

The *LSZ formula* relates S-matrices to correlation functions in momentum space, in Minkowski space, near the physical pole for incoming and outgoing particles.

Define the momentum space Green's functions as

$$\tilde{G}_{n+m}(p_i^\mu, k_j^\mu) = \int \prod_{i=1}^n \int d^4x_i e^{-ip_i \cdot x_i}$$

$$\times \prod_{j=1}^m \int d^4y_j e^{ik_j \cdot y_j} \langle \Omega | T\{\phi(x_1) \cdots \phi(x_n)\phi(y_1) \cdots \phi(y_m)\} | \Omega \rangle. \quad (1.37)$$

Then we have

$$_{\text{in}}\langle \{p_i\}_n | \{k_j\}_m \rangle_{\text{out}}$$

$$= \lim_{p_i^2 \to -m_i^2, k_j^2 \to -m_j^2} \prod_{i=1}^n \frac{(p_i^2 + m^2 - i\epsilon)}{-i\sqrt{Z}} \prod_{j=1}^m \frac{(k_j^2 + m^2 - i\epsilon)}{-i\sqrt{Z}} \tilde{G}_{n+m}(p_i^\mu, k_j^\mu). \quad (1.38)$$

Here Z is the field renormalization factor, and can be defined from the behavior near the physical pole of the full 2-point function,

$$G_2(p) = \int d^4x e^{-ip \cdot x} \langle \Omega | T\{\phi(x)\phi(0)\} | \Omega \rangle \simeq \frac{-iZ}{p^2 + m^2 - i\epsilon}. \quad (1.39)$$

In other words, to find the S-matrix, we put the external lines on a shell, and divide by the full propagators corresponding to all the external lines (but note that Z belongs to two external lines, hence the $\sqrt{Z}$). This implies a diagrammatic procedure called *amputation*: we do not use propagators on the external lines. We also need to consider connected diagrams only, since the S-matrices are normalized objects, and we need to exclude processes where nothing happens and external particles go through without interactions, corresponding to the identity matrix. Therefore we have

$$\langle \{\vec{p}_i\} | S - 1 | \{\vec{k}_j\} \rangle = \left(\sum \text{connected, amputated Feynman diag.} \right) \times (\sqrt{Z})^{n+m}. \quad (1.40)$$

To understand the amputation procedure, consider the setting Sun diagram with external momenta k_1 and p_1 and internal momenta p_2, p_3 and $k_1 - p_2 - p_3$ in Fig. 1.1b. The result for the amputated diagram is in Euclidean space (note that for the S-matrix we must go to Minkowski space instead):

$$(2\pi)^4 \delta^4(k_1 - p_1) \int \frac{d^4p_2}{(2\pi)^4} \frac{d^4p_3}{(2\pi)^4} \lambda^2 \frac{1}{p_2^2 + m^2} \frac{1}{p_3^2 + m^2} \frac{1}{(k_1 - p_2 - p_3)^2 + m_4^2}. \quad (1.41)$$

Feynman path integral with composite operators

Up to now we have considered only correlators of fundamental fields, which are related to external states for the quanta of these fields. But there is no reason to restrict ourselves to this, we can also consider external states corresponding to a composite field $\mathcal{O}(x)$, for instance

$$\mathcal{O}_{\mu\nu}(x) = (\partial_\mu\phi\partial_\nu\phi)(x)(+\cdots). \tag{1.42}$$

We can then define Euclidean space correlation functions for these operators

$$< \mathcal{O}(x_1)\cdots\mathcal{O}(x_n) >_{\text{Eucl}} = \int \mathcal{D}\phi e^{-S_E}\mathcal{O}(x_1)\cdots\mathcal{O}(x_n)$$

$$= \frac{\delta^n}{\delta J(x_1)\cdots\delta J(x_n)}\int \mathcal{D}\phi e^{-S_E+\int d^4x\mathcal{O}(x)J(x)}\big|_{J=0}, \tag{1.43}$$

which can be obtained from the generating functional

$$Z_{\mathcal{O}}[J] = \int \mathcal{D}\phi e^{-S_E+\int d^4x\mathcal{O}(x)J(x)}. \tag{1.44}$$

1.4 Electromagnetism, Yang–Mills fields and gauge groups

Electromagnetism

Up to now we have discussed only scalar fields. Gauge bosons describing forces between particles correspond to vector fields. The simplest example of such a field is the Maxwell field describing the electromagnetic force (the photon),

$$A_\mu(x) = (-\phi(\vec{x},t), \vec{A}(\vec{x},t)), \tag{1.45}$$

where ϕ is the Coulomb potential and $\vec{A}$ is the vector potential.

The field strength is

$$F_{\mu\nu} = \partial_\mu A_\nu - \partial_\nu A_\mu \equiv 2\partial_{[\mu}A_{\nu]}, \tag{1.46}$$

and it contains the electric $\vec{E}$ and magnetic $\vec{B}$ fields as

$$-F_{0i} = F^{0i} = E^i; \quad F_{ij} = \epsilon_{ijk}B^k. \tag{1.47}$$

The observables like $\vec{E}$ and $\vec{B}$ are defined in terms of $F_{\mu\nu}$ (and not A_μ) and as such the theory has a gauge symmetry under a $U(1)$ group, that leaves $F_{\mu\nu}$ invariant,

$$\delta A_\mu = \partial_\mu\lambda; \quad \delta F_{\mu\nu} = 2\partial_{[\mu}\partial_{\nu]}\lambda = 0. \tag{1.48}$$

The Minkowski space action for electromagnetism is

$$S_{\text{Mink}} = -\frac{1}{4}\int d^4xF_{\mu\nu}^2, \tag{1.49}$$

which becomes in Euclidean space (since A_0 and $\partial/\partial x^0 = \partial_t$ rotate in the same way)

$$S_E = \frac{1}{4}\int d^4x(F_{\mu\nu})^2 = \frac{1}{4}\int d^4xF_{\mu\nu}F_{\rho\sigma}\eta^{\mu\rho}\eta^{\nu\sigma}. \tag{1.50}$$

Yang–Mills fields

Yang–Mills fields A_μ^a are self-interacting gauge fields, where a is an index belonging to a nonabelian gauge group. There is thus a 3-point self-interaction of the gauge fields $A_\mu^a, A_\nu^b, A_\rho^c$, that is defined by the *structure constants* of the gauge group $f^a{}_{bc}$, as well as a 4-point self-interaction.

The gauge group G has generators $(T^a)_{ij}$ in the representation R. T^a satisfy the Lie algebra of the group,

$$[T_a, T_b] = f_{ab}{}^c T_c, \tag{1.51}$$

with structure constants $f^a{}_{bc}$. Indices are raised and lowered with the metric g_{ab} on the group space, defined up to a normalization by $\mathrm{Tr}\,_R[T_R^a T_R^b] = t_R g^{ab}$. The group G is most commonly one of the classical groups $SU(N)$, $SO(N)$, $USp(2N)$.

The adjoint representation is defined by the representation for the generators $(T^a)_{bc} = f^a{}_{bc}$. The gauge fields live in the adjoint representation and the field strength is

$$F_{\mu\nu}^a = \partial_\mu A_\nu^a - \partial_\nu A_\mu^a + g f^a{}_{bc} A_\mu^b A_\nu^c. \tag{1.52}$$

One can define $A_\mu = A_\mu^a T_a$ and $F_{\mu\nu} = F_{\mu\nu}^a T_a$ in terms of which we have

$$F_{\mu\nu} = \partial_\mu A_\nu - \partial_\nu A_\mu + g[A_\mu, A_\nu]. \tag{1.53}$$

If one further defines the forms

$$F = \frac{1}{2} F_{\mu\nu} dx^\mu \wedge dx^\nu; \quad A = A_\mu dx^\mu, \tag{1.54}$$

where wedge $\wedge$ denotes antisymmetrization, one has

$$F = dA + g A \wedge A. \tag{1.55}$$

The generators T^a are taken to be antihermitian. Throughout this book, unless otherwise specified, we choose the normalization as defined by (having in mind the application for the trace in the fundamental representation of $SU(N)$)

$$\mathrm{Tr}\, T^a T^b = -\frac{1}{2} \delta^{ab}, \tag{1.56}$$

and here group indices are raised and lowered with δ^{ab}.

The local symmetry under the group G or *gauge symmetry* has now the infinitesimal form

$$\delta A_\mu^a = (D_\mu \epsilon)^a, \tag{1.57}$$

where

$$(D_\mu \epsilon)^a = \partial_\mu \epsilon^a + g f^a{}_{bc} A_\mu^b \epsilon^c. \tag{1.58}$$

The finite form of the transformation is

$$A_\mu^U(x) = U^{-1}(x) A_\mu(x) U(x) + \frac{1}{g} U^{-1} \partial_\mu U(x); \quad U(x) = e^{g \lambda^a(x) T_a} = e^{g \lambda(x)}, \tag{1.59}$$

and if $\lambda^a = \epsilon^a$ is small, we get back $\delta A_\mu^a = (D_\mu \epsilon)^a$.

This transformation leaves invariant the Euclidean action

$$S_E = -\frac{1}{2} \int d^4x \mathrm{Tr} \ (F_{\mu\nu} F^{\mu\nu}) = \frac{1}{4} \int d^4x F^a_{\mu\nu} F^{b\ \mu\nu} \delta_{ab}, \tag{1.60}$$

whereas the field strength transforms *covariantly*, i.e.

$$F'_{\mu\nu} = U^{-1}(x) F_{\mu\nu} U(x). \tag{1.61}$$

1.5 Coupling to fermions and other fields and gauging a symmetry; the Noether theorem

We have analyzed scalars and vectors, but usually matter is fermionic, hence we now show how to deal with matter. Dirac fermions can be thought of as representations of the Clifford algebra,

$$\{\gamma^\mu, \gamma^\nu\} = 2\eta^{\mu\nu} \mathbb{1}, \tag{1.62}$$

and in four Minkowski dimensions they are 4-dimensional complex objects ψ^α on which $(\gamma^\mu)^\alpha{}_\beta$ acts. In a general dimension D, Dirac fermions have $2^{[D/2]}$ complex components. In four dimensions, we introduce the matrix γ_5 as

$$\gamma_5 = -i\gamma^0\gamma^1\gamma^2\gamma^3 = +i\gamma_0\gamma_1\gamma_2\gamma_3, \tag{1.63}$$

which squares to one, $(\gamma_5)^2 = 1$, and in the Weyl representation becomes just

$$\gamma_5 = \begin{pmatrix} \mathbb{1} & \mathbf{0} \\ \mathbf{0} & -\mathbb{1} \end{pmatrix}. \tag{1.64}$$

We also define the *Dirac conjugate*,

$$\bar{\psi} = \psi^\dagger \beta; \quad \beta = i\gamma^0. \tag{1.65}$$

With these conventions, the Dirac action in Minkowski space is written as[1]

$$S_\psi = -\int d^4x \bar{\psi}(\gamma^\mu \partial_\mu + m)\psi. \tag{1.66}$$

In this action, $\bar{\psi}$ is treated as independent of ψ for the purposes of varying, so there are no $1/2$ factors in front. One can consider the *minimal coupling* of the Dirac fermion with the gauge field in Minkowski space,

$$\begin{aligned} \mathcal{L}_\psi &= -\bar{\psi}(\slashed{D} + m)\psi, \\ \slashed{D} &\equiv D_\mu \gamma^\mu; \quad D_\mu \equiv \partial_\mu - ieA_\mu. \end{aligned} \tag{1.67}$$

Going to Euclidean space and considering also a minimal coupling to a complex scalar, obtained by the same substitution $\partial_\mu \to D_\mu$, we have the total action

$$S_E^{\text{total}} = S_{E,A} + \int d^4x[\bar{\psi}(\slashed{D} + m)\psi + (D_\mu\phi)^*D^\mu\phi], \tag{1.68}$$

---

[1] A note on conventions: If instead one uses the metric $\tilde{\eta}_{\mu\nu}$ with signature $(+ - - -)$, then since $\{\tilde{\gamma}^\mu, \tilde{\gamma}^\nu\} = 2\tilde{\eta}^{\mu\nu}$, we have $\tilde{\gamma}^\mu = i\gamma^\mu$, so in Minkowski space one has $\bar{\psi}(i\slashed{\partial} - m)\psi$, with $\bar{\psi} = \psi^\dagger\tilde{\gamma}_0$.

where again there are no 1/2 factors in the scalar action because ϕ and ϕ^* are considered independent for the purposes of varying.

The minimal coupling means that there is a local $U(1)$ symmetry that extends the electromagnetic gauge symmetry (1.48). Now we also have the transformation laws:

$$\psi' = e^{ie\lambda(x)}\psi; \quad \phi' = e^{ie\lambda(x)}\phi, \tag{1.69}$$

under which $D_\mu\psi$ transforms *covariantly* as

$$(D_\mu\psi)' = e^{ie\lambda}D_\mu\psi, \tag{1.70}$$

as does $D_\mu\phi$.

Conversely, we can consider the action of a free Dirac fermion and complex scalar,

$$S_E^{\text{free}} = \int d^4x[\bar{\psi}(\slashed{\partial} + m)\psi + (\partial_\mu\phi)^*\partial^\mu\phi], \tag{1.71}$$

which has a *global $U(1)$* symmetry

$$\psi' = e^{ie\lambda}\psi; \quad \phi' = e^{ie\lambda}\phi. \tag{1.72}$$

Requiring that the global symmetry is promoted to a local one, $\lambda \to \lambda(x)$, requires introducing the minimal coupling to the gauge field transforming as (1.48). This is called *gauging a symmetry*.

The coupling to nonabelian gauge fields is straightforward, one needing to define only the covariant derivative in a representation R, when acting on objects in this representation, as

$$(D_\mu)_{ij} = \delta_{ij}\partial_\mu + g(T_R^a)_{ij}A_\mu^a(x). \tag{1.73}$$

Then, as usual, one replaces $\partial_\mu \to D_\mu$ everywhere, in particular for a fermion $\bar{\psi}\slashed{\partial}\psi \to \bar{\psi}\slashed{D}\psi$.

Irreducible spinor representations

In four Minkowski dimensions, the Dirac representation is reducible as a representation of the Lorentz algebra (or of the Clifford algebra). The irreducible representation (irrep) is found by imposing a constraint on it. The first kind of spinors obtained this way are called Weyl (or chiral) spinors. In the Weyl representation for the gamma matrices, the Dirac spinors split simply as

$$\psi_D = \begin{pmatrix} \psi_L \\ \psi_R \end{pmatrix}, \tag{1.74}$$

which is why the Weyl representation for gamma matrices was chosen. In general, we have

$$\begin{aligned} \psi_L = \frac{\mathbb{1} + \gamma_5}{2}\psi_D &\Rightarrow \frac{\mathbb{1} - \gamma_5}{2}\psi_L = 0, \\ \psi_R = \frac{\mathbb{1} - \gamma_5}{2}\psi_D &\Rightarrow \frac{\mathbb{1} + \gamma_5}{2}\psi_R = 0, \end{aligned} \tag{1.75}$$

and we note that we chose the Weyl representation for gamma matrices such that

$$\frac{1 + \gamma_5}{2} = \begin{pmatrix} 1 & 0 \\ 0 & 0 \end{pmatrix}; \quad \frac{1 - \gamma_5}{2} = \begin{pmatrix} 0 & 0 \\ 0 & 1 \end{pmatrix}. \tag{1.76}$$

Another possible choice for irreducible representation, completely equivalent (in four Minkowski dimensions) to the Weyl representation, is the *Majorana representation*. Majorana spinors satisfy the reality condition

$$\bar{\psi} = \psi^C \equiv \psi^T C, \tag{1.77}$$

where ψ^C is called a Majorana conjugate, thus the reality condition is "Dirac conjugate equals Majorana conjugate," and C is a matrix called a *charge conjugation matrix*. Note that since ψ^T is just another ordering of ψ, whereas $\bar{\psi}$ contains $\psi^\dagger = (\psi^*)^T$, this is indeed a reality condition $\psi^* = (\ldots)\psi$.

The charge conjugation matrix in four Minkowski dimensions satisfies

$$C^T = -C; \quad C\gamma^\mu C^{-1} = -(\gamma^\mu)^T. \tag{1.78}$$

In other dimensions and/or signatures, the definition is more complicated, and it can involve other signs on the right-hand side of the two equations above, as we discuss later in Section 3.3.

For Majorana spinors, we can use the C matrix to raise and lower indices, i.e.

$$\bar{\epsilon}_\alpha = \epsilon^\beta C_{\beta\alpha}; \quad \epsilon^\beta = \bar{\epsilon}_\gamma (C^{-1})^{\gamma\beta}; \quad (C^{-1})^{\alpha\gamma} C_{\gamma\beta} = \delta^\alpha_\beta. \tag{1.79}$$

Note that sometimes we ignore the -1 index on C, and write just C with the indices up, understanding that in the case the indices are up, we mean C^{-1}. Therefore C acts as an antisymmetric metric for spinors, and so the order of contraction is important and is the one defined above. This matters, since for instance $\bar{\epsilon}\chi = \bar{\epsilon}_\alpha \chi^\alpha = -\epsilon^\alpha \bar{\chi}_a$. The matrix C will be used to raise and lower indices on all objects for Majorana spinors, including the gamma matrices themselves, which naturally have the structure $(\gamma^\mu)^\alpha{}_\beta$.

In the Weyl representation, we can choose

$$C = \begin{pmatrix} -\epsilon^{\alpha\beta} & 0 \\ 0 & \epsilon^{\alpha\beta} \end{pmatrix}, \tag{1.80}$$

where we have $\epsilon^{\alpha\beta} = i\sigma^2$, and so[2]

$$C = \begin{pmatrix} -i\sigma^2 & 0 \\ 0 & i\sigma^2 \end{pmatrix} = -i\gamma^0\gamma^2; \quad C^{-1} = \begin{pmatrix} i\sigma^2 & 0 \\ 0 & -i\sigma^2 \end{pmatrix} = -C. \tag{1.81}$$

We can now check explicitly that this C is indeed a representation for the C-matrix, i.e. that it satisfies (1.78).

The action for Majorana fields, with $\bar{\psi}$ related to ψ, is $1/2$ of the action for the Dirac fermion, since we cannot now independently vary ψ and $\bar{\psi}$,

$$S_\psi = -\frac{1}{2} \int d^4x \bar{\psi}(\slashed{\partial} + m)\psi. \tag{1.82}$$

---

[2] In the Weyl representation, $\gamma^\mu = -i \begin{pmatrix} 0 & \sigma^\mu \\ \bar{\sigma}^\mu & 0 \end{pmatrix}$, where $\sigma^\mu = (1, \sigma^i)$ and $\bar{\sigma}^\mu = (1, -\sigma^i)$.

The Noether theorem

We have described local symmetries, and how to gauge (make local) a global symmetry. But for every *global* symmetry of the Lagrangean L, we have a conserved current and its associated conserved charge, which is the statement of the Noether theorem.

The best known examples are the time translation $t \rightarrow t + a$ invariance, corresponding to conserved energy E, and space translation $\vec{x} \rightarrow \vec{x} + \vec{a}$, corresponding to conserved momentum $\vec{p}$. Putting them together we have spacetime translation $x^\mu \rightarrow x^\mu + a^\mu$, corresponding to conserved 4-momentum P^μ. The *currents* corresponding to these charges form the energy-momentum tensor $T_{\mu\nu}$.

Consider the global symmetry $\phi(x) \rightarrow \phi'(x) = \phi(x) + \epsilon \Delta \phi$ that changes the Lagrangean density as

$$\mathcal{L} \rightarrow \mathcal{L} + \epsilon \partial_\mu J^\mu, \tag{1.83}$$

such that the Lagrangean L is invariant, if the fields vanish on the boundary (usually considered at $t = \pm\infty$), since the boundary term

$$\int d^4x \, \partial_\mu J^\mu = \oint_{bd} dS_\mu J^\mu = \int d^3\vec{x} J^0 |_{t=-\infty}^{t=+\infty} \tag{1.84}$$

is then zero. In this case, there exists a conserved current j^μ, i.e.

$$\partial_\mu j^\mu(x) = 0, \tag{1.85}$$

where

$$j^\mu(x) = \frac{\partial \mathcal{L}}{\partial(\partial_\mu \phi)} \Delta \phi - J^\mu. \tag{1.86}$$

For linear symmetries (linear in ϕ), we can define

$$\delta\phi = (\epsilon \Delta \phi)^i \equiv \epsilon^a (T^a)^i_{\ j} \phi^j, \tag{1.87}$$

such that, if $J^\mu = 0$, we have the Noether current

$$j^{a\,\mu} = \frac{\partial \mathcal{L}}{\partial(\partial_\mu \phi)} (T^a)^i_{\ j} \phi^j. \tag{1.88}$$

Applying to translations, $x^\mu \rightarrow x^\mu + a^\mu$, we have for an infinitesimal parameter a^μ

$$\phi(x) \rightarrow \phi(x + a) = \phi(x) + a^\mu \partial_\mu \phi, \tag{1.89}$$

where we have kept only the first terms in the Taylor expansion. The corresponding conserved current is then

$$T^\mu_{\ \nu} \equiv \frac{\partial \mathcal{L}}{\partial(\partial_\mu \phi)} \partial_\nu \phi - \mathcal{L} \delta^\mu_\nu, \tag{1.90}$$

where we have added a term $J^\mu_{(\nu)} = \mathcal{L} \delta^\mu_\nu$, obtaining the conventional definition of the *energy-momentum tensor* or *stress-energy tensor*. The conserved charges are the integrals of the energy-momentum tensor, i.e. P^μ. Note that the above translation gives also the term $J^\mu_{(\nu)}$ from the general formalism, since we can check that for $\epsilon^\nu = a^\nu$, the Lagrangean density changes by $a^\nu \partial_\mu J^\mu_{(\nu)}$.

1.6 Symmetry currents and the current anomaly

We have seen in the last subsection that classically, the current associated with a symmetry of the Lagrangean is conserved, due to the Noether theorem. Quantum mechanically, however, in general we do not need to have conservation, so we can have $\langle \partial_\mu j^\mu \rangle \neq 0$, in which case we say we have an *anomaly*.

If we have quantum conservation of the current, from it we obtain a *Ward identity*. The argument goes as follows. We consider the local version of the symmetry transformation,

$$\phi'^i - \phi^i = \delta\phi^i(x) = \sum_{a,j} \epsilon^a(x)(T^a)_{ij}\phi^j(x). \qquad (1.91)$$

Changing integration variables from ϕ to ϕ' (renaming the variable, really) does nothing, so

$$\int \mathcal{D}\phi' e^{-S[\phi']} = \int \mathcal{D}\phi e^{-S[\phi]}. \qquad (1.92)$$

But now comes a crucial assumption: *if the Jacobian from $\mathcal{D}\phi$ to $\mathcal{D}\phi'$ is 1*, then $\mathcal{D}\phi' = \mathcal{D}\phi$, so

$$0 = \int \mathcal{D}\phi \left[e^{-S[\phi']} - e^{-S[\phi]} \right] = -\int \mathcal{D}\phi \delta S[\phi] e^{-S[\phi]}. \qquad (1.93)$$

This assumption, however, is not true in general, and exactly when it is broken, we obtain a quantum anomaly.

Under the local version of the global symmetry transformation, the *off-shell* action transforms as

$$\delta S = \sum_a \int d^4x (\partial^\mu \epsilon^a(x)) j^a_\mu(x) = -\sum_a \int d^4x \epsilon^a(x)(\partial^\mu j^a_\mu(x)). \qquad (1.94)$$

Using this variation, (1.93) becomes

$$0 = \int d^4x \epsilon^a(x) \int \mathcal{D}\phi e^{-S[\phi]} \partial^\mu j^a_\mu(x). \qquad (1.95)$$

But since the parameters $\epsilon^a(x)$ are arbitrary, we can derive also that

$$\int \mathcal{D}\phi e^{-S[\phi]} \partial^\mu j^a_\mu(x) = 0. \qquad (1.96)$$

As advertised, this is the quantum mechanical version of the classical conservation of the current, namely the averaged version $\langle \partial^\mu j^a_\mu \rangle = 0$. This is one version of the Ward identity, but we can write a relation acting on the partition function and on the correlation functions, which is the usual form of the Ward identities.

Chiral (axial) anomaly

Consider a massless Dirac fermion in 4-dimensional Euclidean space, coupled to an external gauge field A_μ^{external},

$$\mathcal{L} = \bar{\psi} \slashed{D} \psi. \qquad (1.97)$$

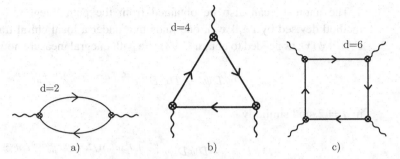

Figure 1.2   a) Anomalous diagram in two dimensions; b) Anomalous diagram (triangle) in four dimensions; c) Anomalous diagram (box) in six dimensions.

The action has the usual $U(1)$ gauge invariance $\psi(x) \to e^{i\alpha}\psi(x)$, $\bar\psi(x) \to \bar\psi(x)e^{-i\alpha}$, but now it also has a *chiral symmetry*,

$$\psi(x) \to e^{i\alpha\gamma_5}\psi(x) \simeq (1 + i\alpha\gamma_5)\psi(x); \quad \bar\psi(x) \to \bar\psi(x)e^{i\alpha\gamma_5} \simeq \bar\psi(x)(1 + i\alpha\gamma_5). \quad (1.98)$$

The action is invariant because $e^{i\alpha\gamma_5}\gamma^\mu = \gamma^\mu e^{-i\alpha\gamma_5}$, but we also easily see that a fermion mass term $m\bar\psi\psi$ would break the symmetry. Using the Noether formula (1.88), we find the conserved current

$$j_\mu^5 = \bar\psi\gamma_\mu\gamma_5\psi. \quad (1.99)$$

One can calculate $\langle \partial^\mu j_\mu^5 \rangle$ in perturbation theory, and find that there is only a one-loop graph that contributes to it.

In $d = 2$ spacetime dimensions, the only diagram that contributes, with a chiral current inserted on one vertex and an external gauge field at the other, is the one in Fig. 1.2a. More precisely, the diagram is for the quantity $h^{\mu\nu}(p)$, giving

$$\langle p^\mu j_\mu^5 \rangle = e p_\mu A_\nu^{\text{external}}(-p) h^{\mu\nu}(p), \quad (1.100)$$

and the result of the calculation is $h^{\mu\nu}(p) = 1/(4\pi)\text{Tr}\,[\gamma^\mu\gamma^\nu\gamma_5]$, giving in Minkowski x-space

$$\langle \partial^\mu j_\mu^5 \rangle = \frac{e}{4\pi}\frac{1}{2}\epsilon^{\mu\nu}F_{\mu\nu}^{\text{external}}. \quad (1.101)$$

A similar calculation in $d = 4$ spacetime dimensions finds that the only diagram that contributes is a triangle, as in Fig. 1.2b, with one vertex coupling to the chiral current and two vertices coupling to the external gauge field, giving the result

$$\langle \partial^\mu j_\mu^5 \rangle = \frac{e^2}{16\pi^2}\frac{1}{2}\epsilon^{\mu\nu\rho\sigma}F_{\mu\nu}^{\text{external}}F_{\rho\sigma}^{\text{external}}. \quad (1.102)$$

One can also prove that the anomaly is *one-loop exact*, meaning that there are no contributions to it from higher loop diagrams.

One can also see that in higher dimensions, the anomaly will be given by higher polygonal one-loop graphs. The anomaly appears only in even dimensions, so the next one will be in $d = 6$, where the diagram is the one-loop box in Fig. 1.2c.

The anomaly can also be obtained from the path integral. Using a regularization method devised by Fujikawa, one finds that under a local chiral transformation $\psi(x) \rightarrow e^{i\alpha(x)\gamma_5}\psi(x)$, as needed to obtain (1.93), the path integral measure now changes according to

$$\mathcal{D}\psi\mathcal{D}\bar{\psi} \rightarrow \mathcal{D}\psi\mathcal{D}\bar{\psi}\, e^{\frac{ie}{4\pi}\int d^2x\alpha(x)1/2\epsilon^{\mu\nu}F^{\text{external}}_{\mu\nu}(x)} \tag{1.103}$$

in $d = 2$, and similarly

$$\mathcal{D}\psi\mathcal{D}\bar{\psi} \rightarrow \mathcal{D}\psi\mathcal{D}\bar{\psi}\, e^{\frac{ie^2}{16\pi}\int d^4x\alpha(x)1/2\epsilon^{\mu\nu\rho\sigma}F^{\text{external}}_{\mu\nu}(x)F^{\text{external}}_{\rho\sigma}(x)} \tag{1.104}$$

in $d = 4$. This corresponds in (1.93) to the same anomaly obtained from Feynman diagrams.

The chiral anomaly can be considered to give an extra contribution to the chiral charge. Indeed, in four dimensions, because $1/2\epsilon^{\mu\nu\rho\sigma}F_{\mu\nu}F_{\rho\sigma} = 4\partial_\mu(\epsilon^{\mu\nu\rho\sigma}A_\nu\partial_\rho A_\sigma)$, we can define a conserved current

$$\tilde{j}^5_\mu = j^5_\mu - \frac{e^2}{4\pi^2}\epsilon_{\mu\nu\rho\sigma}A^\nu\partial^\rho A^\sigma, \tag{1.105}$$

and a conserved charge

$$\tilde{Q}_5 = \int d^3x\tilde{j}^5_0 = Q_5 - \frac{1}{2\pi}S_{CS}[A], \tag{1.106}$$

where the 3-dimensional *Chern–Simons action* is

$$S_{CS}[A_i] = \frac{e^2}{2\pi}\int d^3x\epsilon^{ijk}A_i\partial_jA_k, \tag{1.107}$$

and $i = 1, 2, 3$ are spatial coordinates. Under a gauge transformation $\delta A_i = \partial_i\lambda$, the Chern–Simons action transforms as

$$\delta S_{CS} = \frac{e^2}{\pi}\int d^3x\epsilon^{ijk}\partial_i\lambda\partial_jA_k = \frac{e^2}{\pi}\int dS^iB_i\lambda, \tag{1.108}$$

where B_i is the magnetic field, hence the variation is proportional to the magnetic charge, which is zero in the absence of magnetic monopoles.

Moreover, the anomaly can be obtained from the gauge variation ($\delta A_M = \partial_M\lambda$, $M = 0, 1, \ldots, 4$) of a Chern-Simons term in five dimensions,

$$\delta_g\left[\int d^5x\epsilon^{MNPQR}A_MF_{NP}F_{QR}\right] = \int_M d^5x\partial_M\left[\lambda\epsilon^{MNPQR}F_{NP}F_{QR}\right]$$

$$= \int_{\partial M}d^4x\lambda\epsilon^{\mu\nu\rho\sigma}F_{\mu\nu}F_{\rho\sigma}. \tag{1.109}$$

We will see that this fact is useful in the case of AdS/CFT.

The abelian chiral symmetry can be embedded in a nonabelian theory of massless fermions trivially, by considering the fermions transforming in a representation R of the symmetry group, and using the covariant derivative (1.58). Then the action still has an abelian chiral symmetry, and the chiral current anomaly is just the trace of the previous result (with a (-2) because of our normalization of the trace),

$$\langle \partial^\mu j_\mu^5 \rangle = \frac{(-2)g^2}{16\pi^2} \text{Tr} \left[\frac{1}{2} \epsilon^{\mu\nu\rho\sigma} F_{\mu\nu}^{\text{external}} F_{\rho\sigma}^{\text{external}} \right], \tag{1.110}$$

so we can define a new current

$$\tilde{j}_\mu^5 = j_\mu^5 - \frac{(-2)g^2}{4\pi^2} \epsilon_{\mu\nu\rho\sigma} \text{Tr} \left[A_\nu \partial_\rho A_\sigma + \frac{2}{3} g A_\nu A_\rho A_\sigma \right], \tag{1.111}$$

and a new charge

$$\tilde{Q}_5 = Q_5 - \frac{1}{2\pi} S_{CS}[A]. \tag{1.112}$$

The nonabelian Chern–Simons action is

$$S_{CS}[A_i] = \frac{(-2)g^2}{4\pi} \int d^3x \, \epsilon^{ijk} \text{Tr} \left[A_i \partial_j A_k + \frac{2}{3} g A_i A_j A_k \right], \tag{1.113}$$

and is invariant only under "small" gauge transformations, but can change under "large" gauge transformations that cannot be smoothly connected to the identity,

$$\frac{1}{2\pi} S_{CS}[A] \rightarrow \frac{1}{2\pi} S_{CS}[A] + \frac{1}{4\pi^2} \int d^3x \, \epsilon_{ijk} \text{Tr} \left[\partial^i U U^{-1} \partial^j U U^{-1} \partial^k U U^{-1} \right], \tag{1.114}$$

where the extra term is an integer n called a winding number.

Gauge anomalies

In the case of the chiral anomaly, we have Dirac fermions coupled via a vector current, $\text{Tr} \left[\bar{\psi} \gamma^\mu A_\mu \psi \right] = \text{Tr} \left[j^\mu A_\mu \right]$, and we have a chiral current anomaly, but the vector current is not anomalous.

On the other hand, we can consider a coupling of chiral fermions, $\psi_{R,L}$ to gauge fields, or equivalently a coupling via an axial vector (chiral) current $j_\mu^5 = \bar{\psi} \gamma_\mu \gamma_5 \psi$, in which case there is a potential anomaly in the vector current conservation. But that is a problem, since the vector current corresponds to the local gauge invariance, which should not be broken at the quantum level in order for the quantum theory to be well defined. Indeed, a local symmetry like gauge symmetry changes the number of degrees of freedom (reduces them by one, in this case), so an anomaly in gauge invariance would mean that the number of degrees of freedom at the classical level is different than the number of degrees of freedom at the quantum level (perturbative!), which is clearly impossible.

Consider then the Euclidean action

$$S = \int d^4x \frac{1}{2} \bar{\psi} (1 + \gamma_5) \slashed{D} \psi. \tag{1.115}$$

The Noether current associated with gauge invariance is

$$j^a_\mu = \bar{\psi}\frac{1}{2}(1 + \gamma_5)\gamma_\mu T^a \psi \qquad (1.116)$$

and is classically covariantly conserved,

$$D^\mu j_\mu = 0, \qquad (1.117)$$

and quantum mechanically one again finds an anomaly from the one-loop triangle diagram with a vector current and two external gauge fields,

$$\langle D^\mu j^a_\mu \rangle = \partial_\mu \left(\frac{(-2)}{24\pi^2} \epsilon^{\mu\nu\rho\sigma} \mathrm{Tr} \left[T^a (A_\nu \partial_\rho A_\sigma + \frac{2}{3} A_\nu A_\rho A_\sigma) \right] \right). \qquad (1.118)$$

By symmetry, we see that the anomaly is proportional to $d^{abc} = \mathrm{Tr}\,[T^a(T^b T^c + T^c T^b)]$. This vanishes for $SU(2)$, but not for $SU(N)$, $N > 2$, so in general we need to add several species of chiral fermions such that the total anomaly cancels. These give 't Hooft's consistency conditions for anomaly cancellation.

Other global anomalies

One can similarly analyze other kinds of anomalies, for the conservation of other global currents. Whenever we have a global symmetry, and there are chiral fermions involved, we can analyze its potential anomaly. In fact, this will be the case in theories of interest for AdS/CFT.

For instance, we can consider the massless fermionic Lagrangean in Euclidean space

$$\mathcal{L}_{E,\psi} = \bar{\psi}^i \gamma^\mu D_\mu \psi_i, \qquad (1.119)$$

which is invariant under the global symmetry

$$\delta\psi^i = \epsilon^a (T_a)^i{}_j \psi^j, \qquad (1.120)$$

giving the Noether current

$$j^{a\,\mu} = \bar{\psi}^i \gamma^\mu (T^a)_{ij} \psi^j. \qquad (1.121)$$

It is also invariant under a global symmetry

$$\delta\psi^i = \epsilon^a (T_a)^i{}_j \frac{1 + \gamma_5}{2} \psi^j, \qquad (1.122)$$

giving the Noether current

$$j^{a\,\mu} = \bar{\psi}^i \gamma^\mu (T^a)_{ij} \frac{1 + \gamma_5}{2} \psi^j, \qquad (1.123)$$

which is anomalous, since only chiral fermions run in the loop of the triangle diagram.

Current correlators

The above Noether current for global symmetry is a composite operator, and if the fermion is coupled to gauge fields, the current is also gauge invariant, hence it can represent (create from the vacuum) a physical state.

Therefore one can use the formalism for correlators of composite operators and define the correlators

$$\langle j^{a_1 \; \mu_1}(x_1) \ldots j^{a_n \; \mu_n}(x_n) \rangle = \frac{\delta^n}{\delta A_{\mu_1}^{a_1}(x_1) \ldots \delta A_{\mu_n}^{a_n}(x_n)} \int \mathcal{D}[\text{fields}] e^{-S_E + \int d^4 x j^{\mu,a}(x) A_{\mu}^a(x)}.$$

(1.124)

Then this will be the correlator of some external physical states (i.e. observables). The fact of having an anomaly for j_{μ}^a means that this correlator will in general be nonzero when contracted with the momentum,

$$\partial_{x^{\mu_1}} \langle j^{a_1 \; \mu_1}(x_1) j^{a_2 \; \mu_2}(x_2) \ldots j^{a_n \; \mu_n}(x_n) \rangle \neq 0 \Rightarrow p_{1,\mu_1} \langle j^{a_1 \; \mu_1}(p_1) \ldots j^{a_n \; \mu_n}(p_n) \rangle \neq 0.$$

(1.125)

Important concepts to remember

- Correlation functions and partition functions are easier to define in Euclidean space and then analytically continue.
- The quantum field theory partition function on periodic Euclidean time paths equals a statistical mechanics partition function at finite temperature.
- Correlation functions are given by a Feynman diagram expansion and appear as derivatives of the partition function.
- S matrices defining physical scatterings are obtained via the LSZ formalism from the poles of the correlation functions.
- In the Feynman diagrams for the S-matrices we only have connected and amputated Feynman diagrams.
- Correlation functions of composite operators are obtained from a partition function with sources coupling to the operators.
- Coupling of fields to electromagnetism is done via minimal coupling, replacing the derivatives d with the covariant derivatives $D = d - ieA$.
- Gauging a symmetry, i.e. making local a global one, implies adding a gauge field and making derivatives covariant.
- Yang–Mills fields are self-coupled. Both the covariant derivative and the field strength transform covariantly.
- Classically, the Noether theorem associates every symmetry with a conserved current.
- Quantum mechanically, global symmetries can have an anomaly, i.e the current is not conserved, when inserted inside a quantum average.
- The anomaly comes only from 1-loop polygonal Feynman diagrams. In $d = 4$, it comes from a triangle, thus it only affects the 3-point function.
- In a gauge theory, the current of a global symmetry is gauge invariant, and thus corresponds to some physical state.

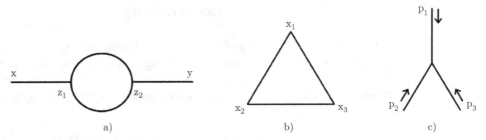

Figure 1.3   a) Setting Sun diagram in x space; b) Triangle diagram in x space; c) Star diagram in p space.

References and further reading

A good introductory course in quantum field theory is Peskin and Schroeder [1], and an advanced level course that has more information, but is more complex, is Weinberg [2]. In this section I have introduced only selected bits of QFT needed in the following.

Exercises

1. If we have the partition function

$$Z_E[J] = \exp\left\{ -\int d^4x\left[\left(\int d^4x_0 K(x,x_0)J(x_0)\right)\left(-\frac{\Box_x}{2}\right)\left(\int d^4y_0 K(x,y_0)J(y_0)\right) \right.\right.$$
$$\left.\left. + \lambda\left(\int d^4x_0 K(x,x_0)(J(x_0))\right)^3 \right] \right\}, \tag{1.126}$$

   write an expression for $G_2(x,y)$ and $G_3(x,y)$.

2. If we have the Euclidean action

$$S_E = \int d^4x\left[\frac{1}{2}(\partial_\mu\phi)^2 + \frac{m^2\phi^2}{2} + \lambda\phi^3 \right], \tag{1.127}$$

   write down the integral for the Feynman diagram in Fig. 1.3a.

3. Show that the Fourier transform of the triangle diagram in x space in Fig. 1.3b is the star diagram in p space in Fig. 1.3c.

4. Derive the Hamiltonian $H(\vec{E}, \vec{B})$ for the electromagnetic field by putting $A_0 = 0$, from the Minkowski space action $S_M = -\int d^4x F_{\mu\nu}^2/4$.

5. Show that $F_{\mu\nu} = [D_\mu, D_\nu]/g$. What is the infinitesimal transformation of $F_{\mu\nu}$? For $SO(d)$ groups, the adjoint representation is antisymmetric, (ab). Calculate $f^{(ab)}{}_{(cd)(ef)}$ and write down $F_{\mu\nu}^{ab}$.

6. Consider the Euclidean action

$$S = \frac{1}{4}\int d^4x F_{\mu\nu}^2 + \int d^4x \bar{\psi}(\slashed{D} + m)\psi + \int d^4x (D_\mu\phi)^* D^\mu\phi \tag{1.128}$$

   and the $U(1)$ electromagnetic transformation. Calculate the Noether current.

Basics of general relativity; Anti-de Sitter space

2.1 Curved spacetime and geometry; the equivalence principle

As the name suggests, general relativity is a generalization of special relativity.

In **special relativity**, one (experimentally) finds that the speed of light is constant in all inertial reference frames, and hence one can fix a system of units where $c = 1$. This becomes one of the postulates of special relativity. As a result, the line element

$$ds^2 = -dt^2 + d\vec{x}^2 = \eta_{\mu\nu}dx^\mu dx^\nu \tag{2.1}$$

is invariant under transformations of coordinates between any inertial reference frames, and is called the invariant distance. Here $\eta_{\mu\nu} = \text{diag}(-1, 1, \ldots, 1)$. Therefore the symmetry group of special relativity is the group that leaves the above line element invariant, namely $SO(1, 3)$, or in general $SO(1, d-1)$. This physically corresponds to transformations between inertial reference frames, and includes as a particular case spatial rotations.

As such, this *Lorentz group* is a generalized rotation group: the rotation group $SO(3)$ is the group of transformations Λ, with $x'^i = \Lambda^i{}_j x^j$ that leaves the 3-dimensional length $d\vec{x}^2$ invariant. The Lorentz transformation is then a generalized rotation

$$x'^\mu = \Lambda^\mu{}_\nu x^\nu; \quad \Lambda^\mu{}_\nu \in SO(1, 3). \tag{2.2}$$

Therefore the statement of special relativity is that physics is Lorentz invariant (invariant under the Lorentz group $SO(1, 3)$ of generalized rotations), just as the statement of Galilean physics is that physics is rotationally invariant. In both cases we start with the statement that the length element is invariant, and generalize to the case of the whole physics being invariant, i.e., physics can be written in the same way in terms of transformed coordinates as in terms of the original coordinates.

In **general relativity**, one considers a more general spacetime, specifically a curved spacetime, defined by the distance between two points, or line element,

$$ds^2 = g_{\mu\nu}(x)dx^\mu dx^\nu, \tag{2.3}$$

where $g_{\mu\nu}(x)$ are arbitrary functions collectively called *the metric* (sometimes one refers to ds^2 as the metric), and x^μ are arbitrary parameterizations of the spacetime, i.e., coordinates on the manifold. This situation is depicted in Fig. 2.1a.

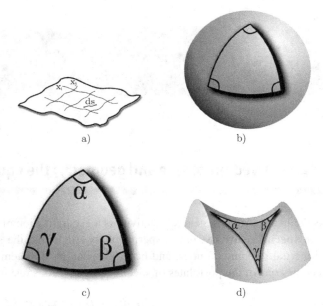

Figure 2.1 a) Curved space: the functional form of the distance between two points depends on local coordinates; b) A triangle on a sphere, made from two meridian lines and a segment of the equator has two angles of 90° ($\pi/2$); c) The same triangle, drawn for a general curved space of positive curvature, emphasizing that the sum of the angles of the triangle exceeds 180° (π); d) In a space of negative curvature, the sum of the angles of the triangle is below 180° (π).

For example for a 2-sphere in angular coordinates θ and ϕ,

$$ds^2 = d\theta^2 + \sin^2\theta\, d\phi^2, \tag{2.4}$$

so $g_{\theta\theta} = 1, g_{\phi\phi} = \sin^2\theta, g_{\theta\phi} = 0$.

As we can see from the definition, the metric $g_{\mu\nu}(x)$ is a symmetric matrix, since it multiplies a symmetric object $dx^\mu dx^\nu$.

To better understand the notion of metric, let us take the example of the sphere, specifically the familiar example of a 2-sphere embedded in 3-dimensional space. Then the metric in the embedding space is the usual Euclidean distance

$$ds^2 = dx_1^2 + dx_2^2 + dx_3^2, \tag{2.5}$$

but if we are on a 2-sphere we have the constraint

$$x_1^2 + x_2^2 + x_3^2 = R^2 \Rightarrow 2(x_1 dx_1 + x_2 dx_2 + x_3 dx_3) = 0$$
$$\Rightarrow dx_3 = -\frac{x_1 dx_1 + x_2 dx_2}{x_3} = -\frac{x_1}{\sqrt{R^2 - x_1^2 - x_2^2}} dx_1 - \frac{x_2}{\sqrt{R^2 - x_1^2 - x_2^2}} dx_2, \tag{2.6}$$

which therefore gives the induced metric (line element) on the sphere

$$ds^2 = dx_1^2 \left(1 + \frac{x_1^2}{R^2 - x_1^2 - x_2^2} \right) + dx_2^2 \left(1 + \frac{x_2^2}{R^2 - x_1^2 - x_2^2} \right) + 2dx_1 dx_2 \frac{x_1 x_2}{R^2 - x_1^2 - x_2^2}$$
$$= g_{ij} dx^i dx^j. \tag{2.7}$$

So this is an example of a curved d-dimensional space which is obtained by embedding it into a flat (Euclidean or Minkowski) $d + 1$ dimensional space. But if the metric $g_{\mu\nu}(x)$ corresponds to arbitrary functions, then one cannot in general embed such a space in flat $d + 1$ dimensional space. Indeed, there are $d(d + 1)/2$ components of $g_{\mu\nu}$, and we can fix d of them to anything (e.g. to 0) by a general coordinate transformation $x'^\mu = x'^\mu(x^\nu)$, where $x'^\mu(x^\nu)$ are d arbitrary functions, so it means that we need to add $d(d - 1)/2$ functions to be able to embed a general metric in a flat space, i.e. we need $d(d - 1)/2$ extra dimensions, with the associated embedding functions $x^a = x^a(x^\mu)$, $a = 1, \ldots, d(d - 1)/2$. In the 3-dimensional example above, $d(d - 1)/2 = 1$, and we need just one embedding function, $x^3(x_1, x_2)$, i.e. we can embed the 2-dimensional surface in three dimensions.

However, even that is not enough, and we need also to make a discrete choice of the signature of the embedding space, *independent of the signature of the embedded space*. For flat spaces, the metric is constant, with $+1$ or -1 on the diagonal, and the signature is given by the $\pm$ values. So 3-dimensional Euclidean means signature $(+1, +1, +1)$, whereas 3-dimensional Minkowski means signature $(-1, +1, +1)$. In three dimensions, these are the only two possible signatures, since we can always redefine the line element by a minus sign, so $(-1, -1, -1)$ is the same as $(+1, +1, +1)$ and $(-1, -1, +1)$ is the same as $(-1, +1, +1)$. Thus, even though a 2-dimensional metric has three components, equal to the three functions available for a 3-dimensional embedding, to embed a metric of *Euclidean* signature in three dimensions one needs to consider both 3-dimensional Euclidean and 3-dimensional Minkowski space, which means that 3-dimensional Euclidean space does not contain all possible *Euclidean* 2-dimensional surfaces.

That means that a general space must be thought of as *intrinsically curved*, defined not by embedding in a given flat space, but by the arbitrary functions $g_{\mu\nu}(x)$ (the metric). In a general space, we define the *geodesic* as the line of shortest distance $\int_a^b ds$ between two points a and b.

In a curved space, the triangle made by three geodesics has an unusual property: the sum of the angles of the triangle, $\alpha + \beta + \gamma$, is not equal to π. For example, if we make a triangle from geodesics on the sphere as in Fig. 2.1b, we can easily convince ourselves that $\alpha + \beta + \gamma > \pi$. In fact, by taking a vertex on the North Pole and two vertices on the Equator, we get $\beta = \gamma = \pi/2$ and $\alpha > 0$. This is the situation for a space with positive curvature, $R > 0$: two parallel geodesics converge to a point as in Fig. 2.1c (by definition, two parallel lines are perpendicular to the same geodesic). In the example given, the two parallel geodesics are the lines between the North Pole and the Equator: both lines are perpendicular to the equator, therefore are parallel by definition, yet they converge at

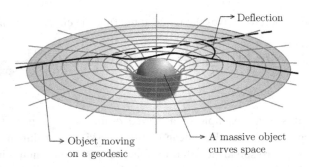

Figure 2.2 Matter curves space (a massive object creates a curvature of spacetime) and then matter (light objects) move on a geodesic, being deflected and creating the effect of gravity.

the North Pole. Because we live in 3-dimensional Euclidean space, and we understand 2-dimensional spaces that can be embedded in it, this case of spaces of positive curvature is the one we can understand easily.

But one can also have a space with negative curvature, $R < 0$, for which $\alpha + \beta + \gamma < \pi$ and two parallel geodesics diverge, as in Fig. 2.1d. Such a space is for instance the so-called *Lobachevski space*, which is a 2-dimensional space of Euclidean signature (like the 2-dimensional sphere), i.e. the diagonalized metric has positive numbers on the diagonal. However, this metric cannot be obtained as an embedding in a Euclidean 3-dimensional space, but rather as an embedding in a Minkowski 3-dimensional space, by

$$ds^2 = dx^2 + dy^2 - dz^2; \quad x^2 + y^2 - z^2 = -R^2. \tag{2.8}$$

Einstein's theory of general relativity

But what has curved spacetime to do with gravity and relativity?

Einstein formulated his theory of general relativity as a way to modify Newton's theory of gravity at strong gravitational field and high velocities to make it consistent with relativity. It was understood that by having only special relativity we cannot do that, and moreover the experimentally observed deflection of light by the Sun would be off by a factor of 2. Hence we are forced to take general relativity instead.

Einstein then proceeded to construct the gravitational theory by making two physical assumptions:

- 1. Gravity is geometry: i.e., matter follows geodesics in a curved space, and the resulting motion (like for instance the deflection of a small object when passing through a localized "dip" of spacetime curvature localized near a point $\vec{r}_0$) appears to us as the effect of gravity, as in Fig. 2.2.

- 2. Matter sources gravity: matter curves space, i.e., the source of spacetime curvature (and thus of gravity) is a matter distribution (in the above, the "dip" is created by the presence of a mass source at $\vec{r}_0$).

We can translate these assumptions into two mathematically well-defined physical principles and an equation for the dynamics of gravity, called the Einstein's equation. The physical principles are:

- a. *Physics is invariant under general coordinate transformations (diffeomorphisms)*

$$x'^{\mu} = x'^{\mu}(x^{\nu}) \Rightarrow ds^2 = g_{\mu\nu}(x)dx^{\mu}dx^{\nu} = ds'^2 = g'_{\mu\nu}(x')dx'^{\mu}dx'^{\nu}. \qquad (2.9)$$

So, further generalizing rotational invariance and Lorentz invariance (special relativity), now not only the line element, but all of physics is invariant under general coordinate transformation, i.e., all the equations of physics take the same form in terms of x^{μ} as in terms of x'^{μ}.

- b. *The equivalence principle*, which can be stated as "there is no difference between acceleration and gravity" OR "if you are in a free falling elevator you cannot distinguish it from being weightless (without gravity)." Einstein imagined a *gedanken experiment* where a person is in a freely falling elevator that falls from a great height towards the surface of the Earth. He cannot determine by any local experiment whether he is freely falling in a gravitational field or is in a weightless situation (until of course he reaches the hard surface of the Earth...). Conversely, imagine the same elevator being now accelerated with a constant acceleration. The person inside it cannot determine by any local experiment whether he is being accelerated or the elevator is fixed and under the influence of a gravitational field.

Note, however, that the equivalence principle is only a *local* statement: for example, if you are falling towards a black hole, tidal forces will pull you apart before you reach it, since gravity acts slightly differently at different points. The quantitative way to write the equivalence principle is

$$m_i = m_g, \quad \text{where } \vec{F} = m_i\vec{a} \text{ (Newton's law) and}$$
$$\vec{F}_g = m_g\vec{g} \text{ (gravitational force)}, \qquad (2.10)$$

i.e., as the equality of the inertial mass (appearing in Newton's force law) with the gravitational mass (appearing in Newton's gravitational law), which are a priori different quantities (there is no a-priori reason for them to be the same, unless there is a principle involved).

In other words, physics is general coordinate (diffeomorphism) invariant, and both gravity and acceleration are manifestations of the curvature of space.

2.2 Kinematics: Christoffel symbols and tensors

Before describing the dynamics of gravity given by the Einstein equations, we must define the kinematics, i.e. the objects used to describe gravity.

As we saw, the metric $g_{\mu\nu}$ changes when we make a coordinate transformation, thus different metrics can describe the same space. More precisely, from (2.9) we obtain the transformation law for the metric

$$g'_{\mu\nu}(x') = \frac{\partial x^\rho}{\partial x'^\mu}\frac{\partial x^\sigma}{\partial x'^\nu}g_{\rho\sigma}(x). \tag{2.11}$$

The infinitesimal form of these general coordinate transformations, for $\delta x^\mu \equiv x'^\mu - x^\mu = \xi^\mu$ small, can be checked to be (see Exercise 4 at the end of the chapter)

$$\delta_\xi g_{\mu\nu}(x) = (\xi^\rho \partial_\rho)g_{\mu\nu} + (\partial_\mu \xi^\rho)g_{\rho\nu} + (\partial_\nu \xi^\rho)g_{\rho\mu}. \tag{2.12}$$

We observe that the first term in this transformation is a translation by ξ^ρ (generalizing the notion of infinitesimal translation in one dimension, or Taylor expansion, $\delta f = (x - x_0)f'(x_0)$), and the other two terms are a kind of tensor generalization of gauge invariance. Indeed, if just μ on $g_{\mu\nu}$ were a gauge field index, we would write $\delta g_{\mu\nu} = \partial_\mu \lambda_\nu$. Since the parameter is ξ^ρ, we need to write $\delta g_{\mu\nu} = (\partial_\mu \xi^\rho)g_{\rho\nu}$ instead, plus another term for the index ν. Because of this, we see that we can think of general coordinate invariance as a kind of local version of translation invariance, or gauge theory of translations. This is not a perfect analogy, but we will see shortly that we can gain much information from it.

Since the metric is symmetric, it has $d(d + 1)/2$ components. But there are d coordinate transformations $x'_\mu(x_\nu)$ one can make that leave the physics invariant, thus we have only $d(d - 1)/2$ degrees of freedom that describe the curvature of space (different physics), but the other d are redundant. Also, by coordinate transformations we can always arrange that $g_{\mu\nu} = \eta_{\mu\nu}$ *around an arbitrary point*, so $g_{\mu\nu}$ is not a good measure for telling whether there is curvature around a point.

To understand the objects that need to be introduced, we first define the notion of general relativistic *tensors*. The notion is the obvious generalization of the notion of special relativity tensors, but some new features appear. The tensors are objects that transform "covariantly" under the general coordinate transformations. We have seen above that general coordinate invariance is a sort of gauge invariance of translations, so transforming covariantly is understood in the same sense as in usual gauge theory, i.e. transformation as the finite transformation of the basic objects of the theory.

A contravariant tensor A^μ is defined as an object that transforms as dx^μ,

$$A'^\mu = \frac{\partial x'^\mu}{\partial x^\nu}A^\nu, \tag{2.13}$$

whereas a covariant tensor B_μ is defined as an object that transforms as the derivative $\partial/\partial x^\mu$ acting on a scalar, i.e., as

$$B'_\mu = \frac{\partial x^\nu}{\partial x'^\mu}B_\nu. \tag{2.14}$$

A general tensor is defined to transform as the product of the transformations of the indices, for instance a tensor T_μ^ν transforms as

$$T'^\mu_\nu = \frac{\partial x'^\mu}{\partial x^\rho}\frac{\partial x^\sigma}{\partial x'^\nu}T^\rho_\sigma. \tag{2.15}$$

Then we see immediately that $g_{\mu\nu}$ is a tensor with two covariant indices, since it transforms as (2.11). But we seem to have a problem, since we can easily check that $\partial_\rho g_{\mu\nu}$ is not a tensor. To remedy this, we look to gauge theory. We see that the solution is to introduce some object that can play the role of gauge field of gravity (or gauge field of local translations, in our analogy above), and to construct a covariant derivative by adding the gauge field to the regular derivative.

The metric itself cannot act as this gauge field of gravity, among other things because it cannot be set to zero by a gauge transformation (i.e., by a general coordinate transformation). Instead, it can be set locally to the flat form, since in a small neighborhood of any given point, a curved space looks flat. Mathematically, that means that we can set the metric fluctuation and its first derivative to zero at that point, i.e. we can write $g_{\mu\nu} = \eta_{\mu\nu} + \mathcal{O}(\delta x^2)$. Locally then, the space looks as if it has an $SO(d-1, 1)$ invariance.

For a covariant derivative involving a gauge field in the adjoint representation $[ab]$ (antisymmetric) of an $SO(d-1, 1)$ group, acting on a fundamental field with index a, we expect something like $D_\mu \phi^a = \partial_\mu \phi^a + (A^a{}_b)_\mu \phi^b$. The difference is that now "gauge" and "coordinate" indices are the same, so we define the covariant derivative acting on a tensor T^μ as

$$D_\mu T^\nu = \partial_\mu T^\nu + (\Gamma^\nu{}_\sigma)_\mu T^\sigma. \tag{2.16}$$

We have separated the indices as in the case of the $SO(d-1, 1)$ gauge field, but now it does not make sense to do so. So we introduce the *Christoffel symbol* $\Gamma^\mu{}_{\nu\rho}$ to act as this "gauge field of gravity". We also define the action on a general tensor with both covariant and contravariant indices, for instance T^ν_μ, as the natural one given the index positions:

$$D_\mu T^\rho_\nu \equiv \partial_\mu T^\rho_\nu + \Gamma^\rho{}_{\sigma\mu} T^\sigma_\nu - \Gamma^\sigma{}_{\mu\nu} T^\rho_\sigma. \tag{2.17}$$

We then impose the requirement that the covariant derivative of the metric tensor vanishes, which is a natural thing to require, since as we saw we can locally put the metric into the Minkowski form (where $\partial_\mu \eta_{\nu\rho} = 0$) by a coordinate (gauge) transformation, and if the Christoffel symbol is truly a gauge field of gravity, it should become zero ("no local gravity") by this gauge transformation. $D_\mu g_{\nu\rho} = 0$ becomes

$$\partial_\mu g_{\nu\rho} - \Gamma^\sigma{}_{\nu\mu} g_{\sigma\rho} - \Gamma^\sigma{}_{\rho\mu} g_{\sigma\nu} = 0, \tag{2.18}$$

which we can verify (left as Exercise 3 to the reader) that is solved by

$$\Gamma^\mu{}_{\nu\rho} = \frac{1}{2} g^{\mu\sigma} (\partial_\rho g_{\nu\sigma} + \partial_\nu g_{\sigma\rho} - \partial_\sigma g_{\nu\rho}). \tag{2.19}$$

Here we have defined the *inverse metric* $g^{\mu\nu}$ as the matrix inverse of the metric, $(g^{-1})_{\mu\nu}$, thus

$$g_{\mu\rho} g^{\rho\sigma} = \delta^\sigma_\mu. \tag{2.20}$$

We can now verify that indeed, as assumed, the Christoffel symbol becomes zero by the coordinate transformation that puts the metric into the locally flat form, since it is linear in first derivatives of the metric, and those are zero. This is as it should be, since a gauge field contains redundancies, and can be set to zero locally (around a point) by a gauge transformation.

We now can finally define an object that characterizes the curvature of space in a covariant manner, called the Riemann tensor. The Riemann tensor is like the "field strength of the gravity gauge field," i.e. of the Christoffel symbol, in that its definition can again be written so as to mimic the definition of the field strength of an $SO(d-1,1)$ gauge group,

$$F_{\mu\nu}^{ab} = \partial_\mu A_\nu^{ab} - \partial_\nu A_\mu^{ab} + A_\mu^{ac} A_\nu^{cb} - A_\nu^{ac} A_\mu^{cb}, \tag{2.21}$$

where a, b, c are fundamental $SO(d-1,1)$ indices, meaning that $[ab]$ (antisymmetric) is an adjoint index. Note that in general, the Yang–Mills field strength is

$$F_{\mu\nu}^A = \partial_\mu A_\nu^A - \partial_\nu A_\mu^A + f^A{}_{BC}(A_\mu^B A_\nu^C - A_\nu^B A_\mu^C), \tag{2.22}$$

and is a *covariant* object under gauge transformations, i.e. it is not yet invariant, but we can construct invariants by simply contracting the indices, for instance by squaring it: $\int (F_{\mu\nu}^{ab})^2$ is a gauge invariant action. Similarly now, the Riemann tensor transforms covariantly under general coordinate transformations, i.e. we can construct invariants by contracting its indices.

As for the construction of the covariant derivative involving $\Gamma^\mu{}_{\nu\rho}$ above, we put brackets in the definition of the Riemann tensor $R^\mu{}_{\nu\rho\sigma}$ only to emphasize the similarity with the $SO(d-1,1)$ gauge field strength:

$$(R^\mu{}_\nu)_{\rho\sigma}(\Gamma) = \partial_\rho (\Gamma^\mu{}_\nu)_\sigma - \partial_\sigma (\Gamma^\mu{}_\nu)_\rho + (\Gamma^\mu{}_\lambda)_\rho (\Gamma^\lambda{}_\nu)_\sigma - (\Gamma^\mu{}_\lambda)_\sigma (\Gamma^\lambda{}_\nu)_\rho, \tag{2.23}$$

the only difference with respect to the Yang–Mills case being that here "gauge" and "spacetime" indices are the same.

From the Riemann tensor we construct by contraction the Ricci tensor

$$R_{\mu\nu} = R^\lambda{}_{\mu\lambda\nu}, \tag{2.24}$$

and the Ricci scalar $R = R_{\mu\nu} g^{\mu\nu}$. The Ricci scalar is coordinate invariant, so it is truly an invariant measure of the curvature of space at a point.

Since the Riemann tensor was constructed like a field strength of a gauge field, it is clear that it should equal the commutator of two covariant derivatives when acting on a tensor, as is the case for Yang–Mills, where $[D_\mu, D_\nu] = F_{\mu\nu}/g$ (see Exercise 5 in Chapter 1). More precisely, we have

$$(D_\mu D_\nu - D_\nu D_\mu) A_\rho = -R^\sigma{}_{\rho\mu\nu} A_\sigma. \tag{2.25}$$

The Riemann tensor also satisfies various symmetry properties, which can be easily checked by direct substitution:

$$R_{\mu\nu\rho\sigma} = -R_{\mu\nu\sigma\rho} = -R_{\nu\mu\rho\sigma} = R_{\rho\sigma\mu\nu}. \tag{2.26}$$

The first equality is obvious when we remember that the first two indices (μ and ν) correspond to gauge indices in the adjoint, antisymmetric representation of $SO(d-1,1)$ in our analogy, and the second is obvious when we remember that ρ and σ correspond to the antisymmetric spatial indices of the field strength $F_{\mu\nu}^{ab}$. Only the last equality is new, due to the identification of gauge and coordinate indices, but can be explicitly checked.

The analogy with the $SO(d-1,1)$ gauge group tells us that the Riemann tensor and its contractions, the Ricci tensor and the Ricci scalar, transform covariantly under general coordinate transformations ("gauge transformations"), i.e. are tensors. We can check this explicitly. We should note that not every object with indices is a tensor. A tensor cannot be set zero by a coordinate transformation (as we can see by its definition above), but the Christoffel symbol can, hence it is not a tensor. The Christoffel symbol involves only first derivatives of the metric, which is why it can be set to zero locally, but the Riemann tensor and its contractions involve two derivatives, and hence cannot be set to zero locally.

To describe physics in curved space, given physics in flat space, we replace the Lorentz metric $\eta_{\mu\nu}$ by the general metric $g_{\mu\nu}$, and Lorentz tensors with general relativity tensors. We should remember that the normal derivative is not a tensor, so it should also be replaced with the covariant derivative.

2.3 Dynamics: Einstein's equations

We now have the tools to turn to the dynamics of gravity, described by Einstein's equations. We proceed to write an action for gravity, and then derive its equations of motion.

However, we first note that the invariant volume of integration over space is not $d^d x$ any more, as in Minkowski or Euclidean space. Indeed, under $x^\mu \to x'^\mu$, we have

$$d^d x = \det\left(\frac{\partial x^\mu}{\partial x'^\nu}\right) d^d x'. \tag{2.27}$$

On the other hand, from (2.11), denoting $\det g_{\mu\nu}$ by g, we get

$$g' = \left[\det\left(\frac{\partial x^\mu}{\partial x'^\nu}\right)\right]^2 g, \tag{2.28}$$

which means that the invariant integration volume is $d^d x \sqrt{-g}$, since

$$\sqrt{-g}\, d^d x = \sqrt{-g'}\, d^d x', \tag{2.29}$$

and the minus sign comes from the Minkowski signature of the metric, which means that $\det g_{\mu\nu} < 0$.

Next we must write a Lagrangean for gravity. The Lagrangean has to be invariant under general coordinate transformations, thus it must be a scalar (a tensor with no indices). There would be several possible choices for such a scalar, but the simplest possible one, the Ricci scalar, turns out to be correct, i.e. compatible with experiment. Thus, one postulates the *Einstein–Hilbert action for gravity:*[1]

$$S_{\text{gravity}} = \frac{1}{16\pi G_N} \int d^d x \sqrt{-g} R. \tag{2.30}$$

In theorists's units ($\hbar = c = 1$), one defines the coefficient of the Einstein action also as $M_{\text{Pl,d}}^{d-2}/2$, where the d-dimensional *Planck mass* is then $M_{\text{Pl,d}} = (8\pi G_N)^{\frac{1}{d-2}}$.

We now want to vary this action with respect to $g_{\mu\nu}$ to obtain the equations of motion of gravity, or equivalently with respect to $g^{\mu\nu}$, since it is simpler.

---

[1] Note on conventions: if we use the $+ - - -$ metric, we get a $-$ in front of the action, since $R = g^{\mu\nu}R_{\mu\nu}$ and $R_{\mu\nu}$ is invariant under constant rescalings of $g_{\mu\nu}$.

The variation of $\sqrt{-g}$ is found as follows. For a general matrix M, we have (using that $0 = \delta(MM^{-1}) = \delta M(M^{-1}) + M(\delta M^{-1}))$

$$\det M = e^{\mathrm{Tr}\,\ln M} \Rightarrow \delta \det M = e^{\mathrm{Tr}\,\ln M}\mathrm{Tr}\,(\delta M M^{-1}) = -\det M \mathrm{Tr}\,((\delta M^{-1})M). \quad (2.31)$$

Applying this for the matrix $g_{\mu\nu}$, we obtain

$$\delta g = -g g_{\mu\nu}\delta g^{\mu\nu}. \quad (2.32)$$

Next, we have the variation of the Ricci scalar

$$\delta R = R_{\mu\nu}\delta g^{\mu\nu} + g^{\mu\nu}\delta R_{\mu\nu}. \quad (2.33)$$

But we now show that the variation of the Ricci tensor is a total derivative, which integrates to zero. To do so, we will use a trick that is common in gravity calculations, so it is worth showing here.

We work in a coordinate system where $\Gamma^{\mu}{}_{\nu\rho}$ is zero in the neighborhood of the point where we calculate. But the derivatives of the Christoffel symbol (containing second derivatives of the metric) can still be nonzero. Then, from the definition of the Riemann tensor (2.23) and the contraction giving the Ricci tensor (2.24), we obtain that

$$\delta R_{\mu\nu} = \delta(\partial_\rho \Gamma^\rho{}_{\mu\nu}) - \delta(\partial_\nu \Gamma^\rho{}_{\mu\rho}). \quad (2.34)$$

On the other hand, we can calculate the variation of the Christoffel symbol. We work in the local system of coordinates where the first derivatives of the metric and the Christoffel symbol are zero. However, the variation of the Christoffel symbol is not zero, and in fact must be a tensor. We can write a covariant expression for it, as

$$\delta \Gamma^{\mu}{}_{\nu\rho} = \frac{1}{2}g^{\mu\lambda}(D_\rho \delta g_{\lambda\nu} + D_\nu \delta g_{\lambda\rho} - D_\lambda \delta g_{\nu\rho}). \quad (2.35)$$

Here we have turned the ∂s into Ds adding for free a Γ, which is zero in our coordinate system, and cancelled the term proportional to $\delta g^{\mu\lambda}$, since it contains only first derivatives of the metric, which are zero. Finally, at the end, noticing that we have a covariant equation (written only in terms of tensors), it must be valid in every system of coordinates, not just in our special one. Of course, we can also verify that (2.35) is correct by direct substitution, but it is considerably longer.

Finally, given that $\delta \Gamma^{\mu}{}_{\nu\rho}$ is a tensor, we can turn the normal derivatives in (2.34) into covariant derivatives for free in our system where Γs are zero, but then again the right-hand side is written in terms of tensors only, so is valid not only in our particular system of coordinates, but in an arbitrary one. Then, multiplying the relation by $g^{\mu\nu}$ and using the fact that the covariant derivative of the metric is zero, $D_\rho g_{\mu\nu} = 0$, we obtain

$$g^{\mu\nu}\delta R_{\mu\nu} = D_\mu(g^{\nu\rho}\delta \Gamma^{\mu}{}_{\nu\rho}) - D_\rho(g^{\nu\rho}\delta \Gamma^{\mu}{}_{\nu\mu}) \equiv D_\mu U^\mu, \quad (2.36)$$

i.e. that this variation is written as a total covariant derivative, which integrates to a boundary term (zero in the bulk), $\int d^d x \sqrt{-g} D_\mu U^\mu$.

Putting all the pieces together, we have that the variation of the Einstein–Hilbert action for gravity (2.30) is

$$\delta S_{\mathrm{gravity}} = \frac{1}{16\pi G_N} \int d^d x \sqrt{-g}\ \delta g^{\mu\nu}\left[R_{\mu\nu} - \frac{1}{2}g_{\mu\nu}R\right]. \quad (2.37)$$

Then finally we obtain the equations of motion of pure gravity, or Einstein's equations without matter,

$$R_{\mu\nu} - \frac{1}{2}g_{\mu\nu}R = 0. \tag{2.38}$$

As we mentioned, the action and the equations of motion above are not fixed by theory, they just happen to agree well with experiments. In fact, in quantum gravity/string theory, the gravitational action S_g could have quantum corrections of different functional form (e.g., $\int d^d x\sqrt{-g}R^2$, $\int d^d x\sqrt{-g}R_{\mu\nu\rho\sigma}R^{\mu\nu\rho\sigma}$, etc.).

The next step is to put matter in curved space, since one of the physical principles that defined general relativity was that matter sources gravity. The way to do this follows from the rules described at the end of the last subsection. For instance, the kinetic term for a scalar field in Minkowski space was

$$S_{M,\phi} = -\frac{1}{2}\int d^4 x (\partial_\mu\phi)(\partial_\nu\phi)\eta^{\mu\nu}, \tag{2.39}$$

and it becomes now

$$-\frac{1}{2}\int d^4 x\sqrt{-g}(D_\mu\phi)(D_\nu\phi)g^{\mu\nu} = -\frac{1}{2}\int d^4 x\sqrt{-g}(\partial_\mu\phi)(\partial_\nu\phi)g^{\mu\nu}, \tag{2.40}$$

where the last equality, of the partial derivative with the covariant derivative, is only valid for a scalar field. In general, we have covariant derivatives in the action.

The variation of the matter action gives the energy-momentum tensor, known from electromagnetism, though perhaps not by this general definition. By definition, we have (if we were to use the $+ - - -$ metric, it would be natural to define it with a $+$)

$$T_{\mu\nu} = -\frac{2}{\sqrt{-g}}\frac{\delta S_{\text{matter}}}{\delta g^{\mu\nu}}. \tag{2.41}$$

Then the sum of the gravity and matter action

$$S_{\text{total}} = \frac{1}{16\pi G_N}\int d^d x\sqrt{-g}R + S_{\text{matter}}, \tag{2.42}$$

with the variation

$$\delta S_{\text{total}} = \frac{1}{16\pi G_N}\int d^d x\sqrt{-g}\delta g^{\mu\nu}\left[R_{\mu\nu} - \frac{1}{2}g_{\mu\nu}R\right] - \int d^d x\sqrt{-g}\frac{\delta g^{\mu\nu}}{2}T_{\mu\nu}, \tag{2.43}$$

gives the equations of motion

$$R_{\mu\nu} - \frac{1}{2}g_{\mu\nu}R = 8\pi G_N T_{\mu\nu}, \tag{2.44}$$

known as Einstein's equations. As an example, for a scalar field with the action (2.40), we have the energy-momentum tensor

$$T^\phi_{\mu\nu} = \partial_\mu\phi\partial_\nu\phi - \frac{1}{2}g_{\mu\nu}(\partial_\rho\phi)^2. \tag{2.45}$$

2.4 Global structure: Penrose diagrams

Spaces of interest are infinite in extent, but have complicated topological and causal structures, that are difficult to visualize and understand. To make sense of them, we use the so-called Penrose diagrams. These are diagrams that preserve the causal and topological structure of space, but have infinity at a finite distance on the diagram.

To construct a Penrose diagram, we note that light propagation, defining the causal structure, is along $ds^2 = 0$, thus an overall factor (known as a "conformal factor") in ds^2 is irrelevant for this. So we make coordinate transformations that bring infinity to a finite distance, and drop the conformal factors. For convenience, we usually get some type of flat space at the end of the calculation. Then, in the diagram, light rays are at 45 degrees ($\delta x = \delta t$ for light, in the final coordinates).

Example 1 As a simple first example, we calculate and then draw the Penrose diagram of 2-dimensional Minkowski space,

$$ds^2 = -dt^2 + dx^2, \tag{2.46}$$

where $-\infty < t, x < +\infty$.

We first make a transformation to "lightcone coordinates"

$$u_\pm = t \pm x \Rightarrow ds^2 = -du_+ du_-, \tag{2.47}$$

followed by a transformation of the lightcone coordinates that makes them finite,

$$u_\pm = \tan \tilde{u}_\pm; \quad \tilde{u}_\pm = \frac{\tau \pm \theta}{2}, \tag{2.48}$$

where the last transformation goes back to space-like and time-like coordinates θ and τ. Now the metric is

$$ds^2 = \frac{1}{4 \cos^2 \tilde{u}_+ \cos^2 \tilde{u}_-}(-d\tau^2 + d\theta^2). \tag{2.49}$$

By dropping the overall (conformal) factor we get back a flat 2-dimensional space, but now of finite extent. Indeed, we have that $|\tilde{u}_\pm| \leq \pi/2$, thus $|\tau \pm \theta| \leq \pi$, so the Penrose diagram is a diamond (a rotated square), as in Fig. 2.3a.

Example 2 For 3-dimensional Minkowski space, the metric is again

$$ds^2 = -dt^2 + dr^2(+r^2 d\theta^2). \tag{2.50}$$

By dropping the angular dependence (the $r^2 d\theta^2$ term) in order to again draw a 2-dimensional diagram, we get the same metric as before, just that now $r > 0$. Therefore everything follows in the same way, just that $\theta > 0$ in the final form. Thus

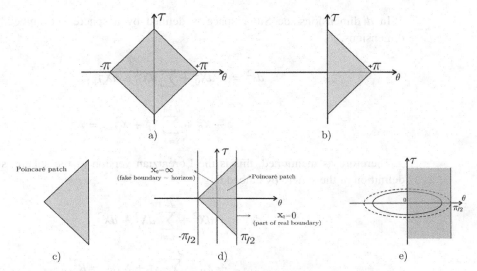

Figure 2.3 Penrose diagrams: a) Penrose diagram of 2-dimensional Minkowski space; b) Penrose diagram of 3-dimensional Minkowski space; c) Penrose diagram of the Poincaré patch of Anti-de Sitter space; d) Penrose diagram of global AdS_2 (2-dimensional Anti-de Sitter), with the Poincaré patch emphasized; $x_0 = 0$ is part of the boundary, but $x_0 = \infty$ is a fake boundary (horizon); e) Penrose diagram of global AdS_d for $d \geq 2$, it is half the Penrose diagram of AdS_2 rotated around the $\theta = 0$ axis.

for 3-dimensional, and higher dimensional, Minkowski space, the Penrose diagram is a triangle (the $\tau > 0$ half of the 2-dimensional Penrose diagram), as in Fig. 2.3b.

More precisely, the full metric after the same transformations and rescalings as in the 2-dimensional case gives in general

$$ds^2 = -d\tau^2 + d\theta^2 + \sin^2\theta \, d\Omega_{d-2}^2. \tag{2.51}$$

It turns out that these simple examples contain all the information needed in order to understand the Penrose diagrams relevant for AdS space, to be described next, and for pp waves, described in Part II of the book.

2.5 Anti-de Sitter space: definition, metrics, Penrose diagram

Anti-de Sitter space is a space of Lorentzian signature $(-++\ldots+)$, but of constant *negative* curvature. Thus it is a Lorentzian signature analog of the Lobachevski space, discussed in the first section of this chapter, which was a space of Euclidean signature and of constant negative curvature.

The anti- in Anti-de Sitter is because *de Sitter* space is defined as a space of Lorentzian signature and of constant *positive* curvature, thus a Lorentzian signature analog of the sphere. Indeed, we know that the sphere is the unique space of Euclidean signature and constant positive curvature.

In d dimensions, de Sitter space is defined by a sphere-like embedding in $d + 1$ dimensions,

$$ds^2 = -dX_0^2 + \sum_{i=1}^{d-1} dX_i^2 + dX_{d+1}^2$$

$$- X_0^2 + \sum_{i=1}^{d-1} X_i^2 + X_{d+1}^2 = R^2. \tag{2.52}$$

Therefore as mentioned, this is the Lorentzian version of the sphere, since from the definition of the sphere by embedding,

$$ds^2 = + dX_0^2 + \sum_{i=1}^{d-1} dX_i^2 + dX_{d+1}^2$$

$$+ X_0^2 + \sum_{i=1}^{d-1} X_i^2 + X_{d+1}^2 = R^2, \tag{2.53}$$

we have changed the minus signs in front of X_0^2 and dX_0^2.

From the definition (2.52), de Sitter space in d dimensions is explicitly invariant under the group $SO(1, d)$, which in fact is defined as the group of transformations $X'^\mu = \Lambda^\mu_{\ \nu} X^\nu$ ($\mu = 0, 1, \ldots, d - 1, d + 1$) that leaves invariant the $d + 1$ dimensional Minkowski metric. The definition involves the $d + 1$-dimensional Minkowski metric and an $SO(d, 1)$-invariant constraint. Note also that the d-dimensional sphere (2.53) is explicitly invariant under $SO(d + 1)$ rotations of the $d + 1$ embedding coordinates, $X'^\mu = \Omega^\mu_{\ \nu} X^\nu$, for the same reason: both the embedding constraint and the embedding metric are explicitly $SO(d + 1)$-invariant.

Similarly, in d dimensions, Anti-de Sitter space (or AdS space) is defined by a Lobachevski-like embedding in $d + 1$ dimensions,

$$ds^2 = - dX_0^2 + \sum_{i=1}^{d-1} dX_i^2 - dX_{d+1}^2$$

$$- X_0^2 + \sum_{i=1}^{d-1} X_i^2 - X_{d+1}^2 = -R^2, \tag{2.54}$$

with the only difference being the same sign change in front of X_0^2 and dX_0^2, and is therefore the *Lorentzian version of Lobachevski space*.

It is explicitly invariant under the group $SO(2, d - 1)$ that rotates the coordinates $X_\mu = (X_0, X_{d+1}, X_1, \ldots, X_{d-1})$ by $X'^\mu = \Lambda^\mu_{\ \nu} X^\nu$, as both the embedding metric and embedding equations are invariant under this transformation.

The metric of this space can be written in different forms, corresponding to different coordinate systems. To write them, as in the case of the sphere described in the first subsection of this chapter, we must find solutions of the embedding Equation (2.54).

Consider the following solution,

$$X_0 = \frac{1}{2u}\left(1 + u^2(R^2 + \vec{x}^2 - t^2)\right),$$
$$X^{d+1} = Rut,$$
$$X^i = Rux^i, \quad i = 1, \ldots, d-2$$
$$X^{d-1} = \frac{1}{2u}\left(1 - u^2(R^2 - \vec{x}^2 + t^2)\right). \tag{2.55}$$

Substituting in (2.54), we obtain

$$ds^2 = R^2\left(u^2(-dt^2 + \sum_{i=1}^{d-2} dx_i^2) + \frac{du^2}{u^2}\right), \tag{2.56}$$

where we see that $0 < u < +\infty$. This is AdS space in *Poincaré coordinates*. This form is explicitly invariant under $ISO(1, d-2)$, the Poincaré group of rotations and translations on $(t, \vec{x})$ and $SO(1, 1)$, a scaling symmetry acting by $(t, \vec{x}, u) \to (\lambda t, \lambda \vec{x}, \lambda^{-1} u)$.

We can also change variables to $u = 1/x_0$, obtaining another form of the Poincaré coordinates,

$$ds^2 = \frac{R^2}{x_0^2}\left(-dt^2 + \sum_{i=1}^{d-2} dx_i^2 + dx_0^2\right), \tag{2.57}$$

where $-\infty < t, x_i < +\infty$, but $0 < x_0 < +\infty$.

Therefore, up to a conformal factor, this is just flat d-dimensional Minkowski space, thus its Penrose diagram is the same, the one of 3-dimensional Minkowski space, a triangle, as in Fig. 2.3c. But this sounds like an odd situation: could it be that two really different spaces have the same Penrose diagram?

The answer is no, because it turns out that the Poincaré coordinates do not cover the whole of AdS space, defined by the embedding (2.54), so it is rather that the triangle is the Penrose diagram for a patch of AdS space, the *Poincaré patch*.

To see this, first change coordinates as $x_0/R = e^{-y}$, obtaining

$$ds^2 = e^{+2y}\left(-dt^2 + \sum_{i=1}^{d-2} dx_i^2\right) + R^2 dy^2. \tag{2.58}$$

However, one now discovers that despite the coordinates being infinite in extent, one does not cover all of space in these coordinates! If we send a light ray to infinity in y coordinates ($x_0 = 0$), which is a boundary of the space, we have $ds^2 = 0$, and we consider it also at constant x_i, obtaining

$$t = \int dt = R \int^\infty e^{-y} dy < \infty, \tag{2.59}$$

so it takes a finite amount of time t for light to reach the boundary, but since t is not finite, light can in principle go further: it can "reflect" from the boundary and travel back to another region of AdS space. In fact, as mentioned, we find that the Poincaré coordinates only cover a patch, the "Poincaré patch" of the AdS space, and we can extend to the full AdS space, finding coordinates that cover it all.

In the Poincaré coordinates, we can understand Anti-de Sitter space as a $d-1$ dimensional Minkowski space in $(t, x_1, \ldots, x_{d-2})$ coordinates, with a "warp factor" (gravitational potential) that depends only on the additional coordinate x_0.

A coordinate system that does cover the whole of space is called the (system of) global coordinates, and is found by the following solution of (2.54):

$$
\begin{aligned}
X_0 &= R \cosh \rho \cos \tau, \\
X_i &= R \sinh \rho \Omega_i, \quad i = 1, \ldots, d-1 \\
X_{d+1} &= R \cosh \rho \sin \tau,
\end{aligned}
\tag{2.60}
$$

where Ω_i are Euclidean coordinates for the unit sphere ($\Omega_i \Omega_i = 1$). Substituting into (2.54), we find the metric

$$
ds_d^2 = R^2(-\cosh^2 \rho \, d\tau^2 + d\rho^2 + \sinh^2 \rho \, d\vec{\Omega}_{d-2}^2),
\tag{2.61}
$$

where $d\vec{\Omega}_{d-2}^2$ is the metric on the unit $d-2$-dimensional sphere.

This metric is written in a suggestive form, since the metric on the d-dimensional sphere can be written in a similar way in terms of the metric on the $d-2$-dimensional sphere,

$$
ds_d^2 = R^2(\cos^2 \rho \, dw^2 + d\rho^2 + \sin^2 \rho \, d\vec{\Omega}_{d-2}^2),
\tag{2.62}
$$

therefore, the AdS global coordinates metric is given by the analytical continuation in θ, $\cosh(i\theta) = \cos \theta$, $\sinh(i\theta) = i \sin \theta$ from the sphere.

As we see from (2.60), the coordinate τ appears naturally to belong to $[0, 2\pi]$, and in fact we can check that this, together with $\rho \geq 0$, covers the embedding hyberboloid (2.54) once. Near $\rho = 0$ (the "center" of AdS space), the metric is $ds^2 \simeq R^2(-d\tau^2 + d\rho^2 + \rho^2 d\vec{\Omega}_{d-2}^2)$, which means the space has the topology of $S^1 \times \mathbb{R}^{d-1}$, with S^1 being the periodic time, giving acausal closed timelike curves. The solution to this problem is simple though: we just need to "unwrap" the circle S^1, i.e. consider $-\infty < \tau < +\infty$ with no identifications, thus obtaining a causal spacetime, known as the *universal cover of AdS space*. In this book, we always consider the universal cover of AdS space whenever we talk about global AdS space, and we shall cease saying "universal cover."

Finally, the change of coordinates $\tan \theta = \sinh \rho$ gives a form of the global coordinate AdS metric that can be used to write the Penrose diagram of the full space,

$$
ds_d^2 = \frac{R^2}{\cos^2 \theta}(-d\tau^2 + d\theta^2 + \sin^2 \theta \, d\vec{\Omega}_{d-2}^2),
\tag{2.63}
$$

where $0 \leq \theta \leq \pi/2$ in all dimensions except two, where $-\pi/2 \leq \theta \leq \pi/2$, and τ is arbitrary.

Penrose diagram and boundary of space

From this metric we infer the Penrose diagram of global AdS_d space (Anti-de Sitter space in d dimensions) by just dropping the conformal factor $R^2/\cos^2 \theta$. In general dimension d, we obtain the metric

$$
ds^2 = -d\tau^2 + d\theta^2 + \sin^2 \theta \, d\vec{\Omega}_{d-2}^2,
\tag{2.64}
$$

known as the "Einstein static Universe," which is formally of the same form as (2.51), except with an infinite range for τ.

Therefore, the Penrose diagram of AdS_2 is an infinite strip between $\theta = -\pi/2$ and $\theta = +\pi/2$. The "Poincaré patch" covered by the Poincaré coordinates, is a triangle region of the diagram, with its vertical boundary being a segment of the infinite vertical boundary of the global Penrose diagram, as in Fig. 2.3d.

The boundary of AdS_2 space is given in Poincaré coordinates by the place where the conformal factor in (2.57) becomes singular, $x_0 = 0$ and $x_0 = \infty$. But $x_0 = \infty$ corresponds in global coordinates to the "center" of AdS_2, $\theta = 0$, and so is a fake boundary through which we can analytically continue the space. Whereas $x_0 = 0$ corresponds to the real boundary of AdS_2, though it can be reached in finite time, as we saw above. In global coordinates, the full boundary of AdS_2 is obtained, namely the $\theta = \pm\pi/2, \tau$ arbitrary lines, for which the conformal factor is again infinite.

The Penrose diagram of AdS_d is similar, but it is a cylinder obtained by the revolution of the infinite strip between $\theta = 0$ and $\theta = \pi/2$ around the $\theta = 0$ axis, as in Fig. 2.3e. The "circle" of the revolution represents in fact a $d - 2$ dimensional sphere. Therefore, the boundary of AdS_d (d-dimensional Anti-de Sitter space) is $\mathbb{R}_\tau \times S_{d-2}$, the infinite vertical line of time τ times a $d - 2$-dimensional sphere. This is important in defining AdS/CFT correctly.

The metric on the correct boundary of the Poincaré patch is the metric on the $x_0 = 0$ slice (or $x_0 = \epsilon$) without the conformal factor,

$$ds^2 = -dt^2 + \sum_{i=1}^{d-2} dx_i^2, \tag{2.65}$$

whereas the metric on the boundary of the global AdS space is the metric on the $\theta = \pi/2$ slice (or $\theta = \pi/2 - \epsilon$) without the conformal factor,

$$ds^2 = -d\tau^2 + d\vec{\Omega}_{d-2}^2. \tag{2.66}$$

Analytical continuation to Euclidean signature

It will turn out that, as for the case of a general quantum field theory reviewed in Chapter 1, it is easier to define AdS/CFT for a Euclidean version of AdS space. We have already mentioned the fact that the embedding equation for AdS space is the analytically continued version of Euclidean-signature Lobachevski space, just as the embedding equation for dS space is the Euclidean version of the sphere. Therefore we refer to the d-dimensional sphere as "Euclidean de Sitter in d dimensions, EdS_d", and to the d-dimensional version of Lobachevski space as "Euclidean Anti-de Sitter space in d dimensions, $EAdS_d$."

But we now want to understand better $EAdS_d$ from the analytical continuation. The Wick rotation $X_{d+1}^{(E)} = -iX_{d+1}$ turns the embedding relation of AdS_d, (2.54), into the embedding relation for d-dimensional Lobachevski space $EAdS_d$,

$$ds_E^2 = - dX_0^2 + \sum_{i=1}^{d-1} dX_i^2 + (dX_{d+1}^{(E)})^2$$

$$- X_0^2 + \sum_{i=1}^{d-1} X_i^2 + (X_{d+1}^{(E)})^2 = -R^2. \tag{2.67}$$

But we also see that the corresponding Wick rotation for global coordinates is $\tau_E = -i\tau$. Since $\cos\tau = \cosh\tau_E$ and $\sin\tau = i\sinh\tau_E$, the solution to the embedding Equation (2.60) turns into

$$\begin{aligned} X_0 &= R\cosh\rho\cosh\tau_E, \\ X_i &= R\sinh\rho\,\Omega_i, \quad i = 1,\dots,d-1 \\ X_{d+1}^{(E)} &= R\cosh\rho\sinh\tau_E, \end{aligned} \tag{2.68}$$

leading to the global Euclidean AdS metric

$$\begin{aligned} ds_{E,d}^2 &= R^2(\cosh^2\rho\,d\tau_E^2 + d\rho^2 + \sinh^2\rho\,d\vec{\Omega}_{d-2}^2) \\ &= \frac{R^2}{\cos^2\theta}(d\tau_E^2 + d\theta^2 + \sin^2\theta\,d\vec{\Omega}_{d-2}^2). \end{aligned} \tag{2.69}$$

On the other hand, the same analytical continuation gives the analytical continuation of the Poincaré patch. The Wick rotation $t_E = -it$ turns the solution of the embedding equation (2.55) to

$$\begin{aligned} X_0 &= \frac{1}{2u}\left(1 + u^2(R^2 + \vec{x}^2 + t_E^2)\right), \\ X_{d+1}^{(E)} &= Rut_E, \\ X^i &= Rux^i, \quad i = 1,\dots,d-2 \\ X^{d-1} &= \frac{1}{2u}\left(1 - u^2(R^2 - \vec{x}^2 - t_E^2)\right), \end{aligned} \tag{2.70}$$

leading to the Euclidean Poincaré metric

$$\begin{aligned} ds_E^2 &= R^2\left(u^2(dt_E^2 + \sum_{i=1}^{d-2} dx_i^2) + \frac{du^2}{u^2}\right) \\ &= \frac{R^2}{x_0^2}\left(dt_E^2 + \sum_{i=1}^{d-2} dx_i^2 + dx_0^2\right). \end{aligned} \tag{2.71}$$

Now the metric on the boundary of the global Euclidean AdS metric,

$$ds_E^2 = d\tau_E^2 + d\vec{\Omega}_{d-2}^2, \tag{2.72}$$

can be related by dropping a conformal factor to the metric on the boundary of the Euclidean Poincaré metric,

$$ds_E^2 = dt_E^2 + \sum_{i=1}^{d-2} dx_i^2 = d\tilde{\rho}^2 + \tilde{\rho}^2 d\vec{\Omega}_{d-2}^2 = e^{2\tau_E}(d\tau_E^2 + d\vec{\Omega}_{d-2}^2), \tag{2.73}$$

where we have first written flat space in spherical coordinates with radius $\tilde{\rho}$, and then redefined $\tilde{\rho} = e^{\tau_E}$.

We see that the boundary of AdS_d space is a flat $d-1$-dimensional space, up to a possible conformal factor (in the case of the global coordinates metric). However, we also note an important subtlety. The Wick rotation on the boundary in global coordinates, $d\tau_E = -d\tau^2$, is different than the Wick rotation on the boundary in Poincaré metric, $dt_E^2 = -dt^2$, as we can easily check! While the latter is the usual Wick rotation on the plane, the former is a Wick rotation for radial quantization (for "radial time"). This is the reason why Wick rotation of AdS/CFT from Euclidean signature is difficult to do, and there are still unknowns about it.

Cosmological constant

Finally, let me mention another important property of Anti-de Sitter space: it is a solution of the Einstein equation with a constant energy-momentum tensor, known as a *cosmological constant*, thus $T_{\mu\nu} = -\Lambda g_{\mu\nu}$, coming from a constant term in the action, $-\int d^4x \sqrt{-g}\Lambda$, so the Einstein equation is

$$\mathcal{R}_{\mu\nu} - \frac{1}{2} g_{\mu\nu} \mathcal{R} = 8\pi G_N (-\Lambda) g_{\mu\nu}, \qquad (2.74)$$

where $\Lambda < 0$ for AdS and we have written the Ricci tensor and scalar with $\mathcal{R}$ in order not to confuse it with the AdS radius R. Multiplying by $g^{\mu\nu}$, we obtain

$$\mathcal{R} = \frac{2d}{d-2} 8\pi G_N \Lambda, \qquad (2.75)$$

which shows that indeed, we obtain a space of constant negative curvature. For the case of positive curvature space with Euclidean signature we obtain the sphere S_d, corresponding to $\Lambda > 0, \mathcal{R} > 0, g > 0$; for the case of positive curvature space with Minkowski signature we obtain de Sitter space dS_d, with $\Lambda > 0, \mathcal{R} > 0, g < 0$; for the case of negative curvature space with Minkowski signature we obtain Anti-de Sitter space AdS_d, with $\Lambda < 0, \mathcal{R} < 0, g < 0$.

On the AdS solution in Poincaré coordinates (2.57), we can calculate the components of the Ricci scalar, using the definition of the Christoffel symbols and the Riemann tensor in terms of the metric:[2]

$$\mathcal{R}_{00} = -\frac{d-1}{x_0^2}; \quad \mathcal{R}_{ab} = -\frac{d-1}{x_0^2}\eta_{ab}; \quad \mathcal{R}_{0a} = 0; \quad a = (t,i)$$

$$\Rightarrow \mathcal{R}_{\mu\nu} = -\frac{d-1}{R^2} g_{\mu\nu}, \qquad (2.76)$$

so that the Ricci scalar is

$$\mathcal{R}_{AdS} = -\frac{d(d-1)}{R^2} \qquad (2.77)$$

---

[2] The only nonzero components of the Christoffel symbol are found to be $\Gamma^0{}_{ab} = \frac{1}{x_0}\eta_{ab}; \Gamma^0{}_{00} = -\frac{1}{x_0}; \Gamma^a{}_{0b} = \Gamma^a{}_{b0} = -\frac{1}{x_0}\delta^a_b$ and the only nonzero components of the Riemann tensor are found to be $R^0{}_{a0b} = -R^0{}_{ab0} = -\frac{1}{x_0^2}\eta_{ab}, R^a{}_{bcd} = \frac{1}{x_0^2}(\delta^a_d\eta_{bc} - \delta^a_c\eta_{bd})$ and $R^a{}_{0b0} = -R^a{}_{00b} = -\frac{1}{x_0^2}\delta^a_b$, which together can be written as $R^\mu{}_{\nu\rho\sigma} = \frac{1}{R^2}(\delta^\mu_\sigma g_{\nu\rho} - \delta^\mu_\rho g_{\nu\sigma})$.

and the cosmological constant is, from (2.75),

$$\Lambda_{\text{AdS}} = -\frac{(d-1)(d-2)}{16\pi G_N R^2} = -\frac{(d-1)(d-2)M_{\text{Pl,d}}^{d-2}}{2R^2}. \tag{2.78}$$

Consider the transformation

$$\sinh \rho = \frac{r}{R}; \quad t = \frac{\bar{t}}{R}, \tag{2.79}$$

on the global coordinate metric (2.61). We obtain the new metric

$$ds^2 = -\left(1 + \frac{r^2}{R^2}\right)d\bar{t}^2 + \frac{dr^2}{1 + \frac{r^2}{R^2}} + r^2 d\vec{\Omega}_{d-2}^2. \tag{2.80}$$

This form of the metric is useful because if we write it in terms of the cosmological constant Λ as

$$ds^2 = -\left(1 - \frac{2\Lambda}{(d-1)(d-2)M_{\text{Pl,d}}^{d-2}}r^2\right)d\bar{t}^2 + \frac{dr^2}{1 - \frac{2\Lambda}{(d-1)(d-2)M_{\text{Pl,d}}^{d-2}}r^2} + r^2 d\vec{\Omega}_{d-2}^2, \tag{2.81}$$

it is valid both in the AdS case $\Lambda < 0$ and in the dS case $\Lambda > 0$.

We can now write a transformation that takes us between the *Euclidean version* (with $+d\bar{t}^2$ signature) of this new form of the global coordinates metric and the Poincaré form,

$$\frac{r}{R}\vec{\Omega} = \frac{\vec{x}}{x_0}; \quad e^{\frac{\bar{t}}{R}} = \sqrt{x_0^2 + \vec{x}^2} = x_0\sqrt{1 + \frac{r^2}{R^2}}. \tag{2.82}$$

Important concepts to remember

- In general relativity, space is intrinsically curved.
- In general relativity, physics is invariant under general coordinate transformations.
- Gravity is the same as curvature of space, or gravity = local acceleration.
- The Christoffel symbol acts like a gauge field of gravity, giving the covariant derivative.
- Its field strength is the Riemann tensor, whose scalar contraction, the Ricci scalar, is an invariant measure of curvature.
- One postulates the action for gravity as $(1/(16\pi G_N)) \int \sqrt{-g}R$, giving Einstein's equations.
- To understand the causal and topological structure of curved spaces, we draw Penrose diagrams, which bring infinity to a finite distance in a controlled way.
- de Sitter space is the Lorentzian signature version of the sphere; Anti-de Sitter space is the Lorentzian version of Lobachevski space, a space of constant negative curvature.
- Anti-de Sitter space in d dimensions has $SO(2, d-1)$ invariance.
- The Poincaré coordinates only cover part of Anti-de Sitter space, despite having maximum possible range (over the whole real line), related to the fact that one can send a light ray to infinity in a finite time.
- Global coordinates cover the whole (universal cover of) AdS space.
- The Penrose diagram of (the global) AdS space is a cylinder.

- Its boundary is $\mathbb{R}_t \times S^{d-2}$, conformal to $\mathbb{R}^{d-1}$.
- Anti-de Sitter space has a cosmological constant.

References and further reading

For a very basic (but not too detailed) introduction to general relativity you can try the general relativity chapter in Peebles [3]. A good and comprehensive treatment is given in [4], which has a very good index, and detailed information, but you need to be selective in reading only the parts you are interested in. An advanced treatment, with an elegance and concision that a theoretical physicist should appreciate, is found in the general relativity section of Landau and Lifshitz [5], though it might not be the best introductory book. A more advanced and thorough book for the theoretical physicist is Wald [6].

Exercises

1. Parallel the derivation in the text to find the metric on the 2-sphere in its usual form,

$$ds^2 = R^2(d\theta^2 + \sin^2\theta \, d\phi^2), \tag{2.83}$$

   from the 3-dimensional Euclidean metric.
2. Show that on-shell, the graviton has degrees of freedom corresponding to a transverse $(d-2$ indices$)$ symmetric traceless tensor.
3. Show that the metric $g_{\mu\nu}$ is covariantly constant $(D_\mu g_{\nu\rho} = 0)$ by substituting the Christoffel symbols.
4. Prove that the general coordinate transformation on $g_{\mu\nu}$,

$$g'_{\mu\nu}(x') = g_{\rho\sigma}(x)\frac{\partial x^\rho}{\partial x'^\mu}\frac{\partial x^\sigma}{\partial x'^\nu}, \tag{2.84}$$

   reduces for infinitesimal tranformations to

$$\partial_\xi g_{\mu\nu}(x) = (\xi^\rho \partial_\rho)g_{\mu\nu} + (\partial_\mu \xi^\rho)g_{\rho\nu} + (\partial_\nu \xi^\rho)g_{\rho\mu}. \tag{2.85}$$

5. Prove that the commutator of two covariant derivatives when acting on a covariant vector gives the action of the Riemann tensor on it, Equation (2.25).
6. Parallel the calculation in 2-dimensions to show that the Penrose diagram of 3-dimensional Minkwoski space, with an angle $(0 \le \phi \le 2\pi)$ supressed, is a triangle.
7. Substitute the coordinate transformation

$$X_0 = R\cosh\rho\cos\tau; \quad X_i = R\sinh\rho\Omega_i; \quad X_{d+1} = R\cosh\rho\sin\tau, \tag{2.86}$$

   to find the global metric of AdS space from the embedding $(2, d-1)$ signature flat space.

3 Basics of supersymmetry

3.1 Lie algebras; the Coleman–Mandula theorem

In the 1960s people were asking: "what kind of symmetries are possible in particle physics?"

We know the Poincaré symmetry $ISO(3, 1)$ defined by the Lorentz generators $J_{\mu\nu}$ of the $SO(3, 1)$ Lorentz group and the generators of 3+1-dimensional translation symmetries, P_a,

$$
\begin{aligned}
[J_{\mu\nu}, J_{\rho\sigma}] &= -(\eta_{\mu\rho}J_{\nu\sigma} + \eta_{\nu\sigma}J_{\mu\rho} - \eta_{\mu\sigma}J_{\nu\rho} - \eta_{\nu\rho}J_{\mu\sigma}), \\
[P_\mu, J_{\nu\rho}] &= (\eta_{\mu\nu}P_\rho - \eta_{\mu\rho}P_\nu), \\
[P_\mu, P_\nu] &= 0 \ .
\end{aligned}
\tag{3.1}
$$

We also know that there are possible internal symmetries T_r of particle physics, such as the local $U(1)$ of electromagnetism, the local $SU(3)_c$ (color) of QCD or the (approximate) global $SU(2)$ symmetry of isospin. These generators form a Lie algebra

$$
[T_r, T_s] = f_{rs}{}^t T_t \ .
\tag{3.2}
$$

So the question arose: can they be combined, i.e. $[T_s, P_\mu] \neq 0, [T_s, J_{\mu\nu}] \neq 0$, such that maybe we could embed, say the $SU(2)$ of isospin together with the $SU(2)$ of spin into a larger group?

The answer turned out to be NO, in the form of the Coleman–Mandula theorem, which says that if the Poincaré and internal symmetries were to combine, the S matrices for all processes would be zero.

Note that we can have $ISO(3, 1)$ as part of a larger group, but this group will not involve an internal symmetry, but rather other transformations of spacetime. The group in question is called a conformal group, which in four dimensions is $SO(4, 2)$, and the study of conformal field theories (theories invariant under this group) is the main subject of AdS/CFT. We explain conformal field theories in a later chapter.

3.2 Supersymmetry: a symmetry between bosons and fermions

Like all theorems, the Coleman–Mandula theorem is only as strong as its assumptions, and one of them is that the final algebra is a Lie algebra.

But people realized that one can generalize the notion of Lie algebra to a *graded Lie algebra* and thus evade the theorem. A graded Lie algebra is an algebra that has some generators Q^i_α that satisfy not a commuting law: but an anticommuting law:

$$\{Q^i_\alpha, Q^j_\beta\} = \text{other generators.} \tag{3.3}$$

Then the generators $P_\mu, J_{\mu\nu}$, and T_r are called "even generators" and the Q^i_α are called "odd" generators. The graded Lie algebra is then of the type

$$[\text{even, even}] = \text{even}; \quad \{\text{odd, odd}\} = \text{even}; \quad [\text{even, odd}] = \text{odd}, \tag{3.4}$$

and the commutation and anticommutation relations correspond to what we expect if "even = boson", "odd = fermion". The graded Lie algebra satisfies generalized Jacobi identities which follow from it (consider that bosons commute, fermions anticommute and boson and fermion commute),

$$[[B_1, B_2], B_3] + [[B_3, B_1], B_2] + [[B_2, B_3], B_1] = 0,$$
$$[[B_1, B_2], F_3] + [[F_3, B_1], B_2] + [[B_2, F_3], B_1] = 0,$$
$$\{[B_1, F_2], F_3\} + \{[B_1, F_3], F_2\} + [\{F_2, F_3\}, B_1] = 0,$$
$$[\{F_1, F_2\}, F_3] + [\{F_1, F_3\}, F_2] + [\{F_2, F_3\}, F_1] = 0. \tag{3.5}$$

So such a graded Lie algebra generalization of the Poincaré + internal symmetries is possible. But what kind of symmetry would a Q^i_α generator describe? The equation

$$[Q^i_\alpha, J_{\mu\nu}] = (\ldots)Q^i_\beta \tag{3.6}$$

means that Q^i_α must be in a representation of $J_{\mu\nu}$ (the Lorentz group), since $[(\Phi), J_{\mu\nu}] = (\ldots)\Phi$ means by definition Φ is in a representation of $J_{\mu\nu}$. Because of the anticommuting nature of Q^i_α ($\{Q_\alpha, Q_\beta\} = $ others), we choose the spinor representation, which in particular means that

$$[Q^i_\alpha, J_{\mu\nu}] = \frac{1}{2}(\gamma_{\mu\nu})_\alpha{}^\beta Q^i_\beta. \tag{3.7}$$

The spinors we will work with are Majorana spinors, which as we saw satisfy the reality condition

$$\bar{Q}^i_\beta = Q^{i\alpha} C_{\alpha\beta}, \tag{3.8}$$

so strictly speaking we should write instead of Q^i_α, $\bar{Q}^i_\alpha$, but since Majorana spinor indices are raised and lowered with C, it is never ambiguous, so we can be sloppy about including the bar. Raising and lowering of anticommuting Majorana spinor indices is done as

$$\psi_\beta = \psi^\alpha C_{\alpha\beta}; \quad \psi^\beta = \psi_\alpha C^{-1\,\alpha\beta}, \tag{3.9}$$

where $C_{\alpha\beta}$ is antisymmetric and $C^{-1\,\alpha\beta} \equiv C^{\alpha\beta}$.

The reason we use Majorana spinors is convenience, since it is easier to prove various supersymmetry identities, and then in the Lagrangean we can always go from a Majorana to a Weyl spinor and vice versa.

But a spinor field times a boson field gives a spinor field. Therefore, when acting with Q^i_α (spinor) on a boson field, we get a spinor field. More precisely Q^i_α is a spinor, with α a spinor index and i a label, thus the parameter of the transformation law, ϵ^i_α is a spinor also.

Therefore a Q_α^i *gives a symmetry between bosons and fermions, called* **supersymmetry**! Thus,

$$\delta \ boson = fermion; \quad \delta \ fermion = boson. \tag{3.10}$$

If $i = 1, \ldots, \mathcal{N}$ we say we have $\mathcal{N}$ supersymmetries.

$\{Q_\alpha, Q_\beta\}$ is called the supersymmetry algebra, and the above graded Lie algebra is called the superalgebra.

Since $\{Q_\alpha^i, Q_\beta^j\}$ are symmetric in the exchange of $(i\alpha)$ with $(j\beta)$, we can have structures symmetric in $(\alpha\beta)$ and in (ij) and also structures antisymmetric in both $(\alpha\beta)$ and (ij). Since using (1.78) we can easily verify that[1]

$$(C\gamma^\mu)_{\alpha\beta} = (C\gamma^\mu)_{\beta\alpha}; \quad (C\gamma^{\mu\nu})_{\alpha\beta} = (C\gamma^{\mu\nu})_{\beta\alpha}; \quad (C\gamma_5)_{\alpha\beta} = -(C\gamma_5)_{\beta\alpha}, \tag{3.11}$$

it follows that the supersymmetry algebra can be in principle of the type

$$\{Q_\alpha^i, Q_\beta^j\} = m(C\gamma^\mu)_{\alpha\beta}P_\mu\delta^{ij} + n(C\gamma^{\mu\nu})_{\alpha\beta}J_{\mu\nu}\delta^{ij} + C_{\alpha\beta}U^{ij} + (C\gamma_5)_{\alpha\beta}V^{ij}, \tag{3.12}$$

where $U^{ij} = -U^{ji}$ and $V^{ij} = -V^{ji}$ are antisymmetric matrices called *central charges* since they commute with the rest of the algebra. However, from the Jacobi indentities of the superalgebra (3.5) one can prove that $n = 0$, and we can normalize the generators such that $m = 2$.

Then the most general form of the N-extended superalgebra in four dimensions with central charges is

$$\{Q_\alpha^i, Q_\beta^j\} = 2(C\gamma^\mu)_{\alpha\beta}P_\mu\delta^{ij} + C_{\alpha\beta}U^{ij} + (C\gamma_5)_{\alpha\beta}V^{ij} \ . \tag{3.13}$$

Note that in higher dimensions we could have more central charges, but in four dimensions this is the most general possibility. Using the Jacobi indentities (3.5) one finds also that

$$[Q_\alpha^i, P_\mu] = 0 \ . \tag{3.14}$$

Finally, there is an internal (global) symmetry that can act on the index i of Q_α^i and rotates them according to

$$[Q_\alpha^i, T_r] = (V_r)^i{}_j Q_\alpha^j, \tag{3.15}$$

i.e., $(V_r)^i{}_j$ is the representative of T_r in the representation of Q_α^i for the Lie algebra. In general (and in the absence of central charges) we can have complex representations acting on the N objects $i = 1, \ldots, N$, which means that in general for N-extended supersymmetry we can have $U(N)$ internal symmetry.

---

[1] As a matter of notation, observe that for us

$$(C\gamma^\mu)_{\alpha\beta} \equiv C_{\alpha\gamma}(\gamma^\mu)^\gamma{}_\beta = -(\gamma^\mu)_{\alpha\beta},$$

where in the last expression we have simply lowered the index α with our rules for raising and lowering spinor indices.

3.3 Spinors in various dimensions

In this book we deal with various spacetime dimensions, not just $d = 4$, so it is important to realize how to define a spinor in general.

For the Lorentz group $SO(1, d - 1)$ there is always a representation called the spinor representation χ_α defined by the fact that there exist gamma matrices $(\gamma_\mu)^\alpha{}_\beta$ satisfying the Clifford algebra

$$\{\gamma_\mu, \gamma_\nu\} = 2g_{\mu\nu}, \tag{3.16}$$

where $g_{\mu\nu}$ is the $SO(1, d - 1)$ invariant metric, i.e. d-dimensional Minkowski, that takes spinors into spinors $(\gamma_\mu)^\alpha{}_\beta \chi^\beta = \tilde{\chi}^\alpha$. In d spacetime dimensions, these *Dirac spinors* have $2^{[d/2]}$ complex components, but this representation is not irreducible.

For an irreducible representation, we must impose either the *Weyl* (chirality) condition, or the *Majorana* (reality) condition, or in the case of $d = 2$ and 10 Minkowski dimensions both, obtaining Weyl, Majorana, or Majorana–Weyl spinors.

The Weyl spinor condition exists only in $d = 2n$ (even) dimensions, but the Majorana condition (or sometimes the *modified Majorana* condition, involving another matrix besides $C)^2$ can always be defined.

The proof of the fact that Dirac spinors have $2^{[d/2]}$ components is constructive. In $d = 2$ Euclidean dimensions, the Pauli matrices σ^i, satisfying

$$\sigma^i \sigma^j = \delta^{ij} + i\epsilon^{ijk}\sigma^k, \tag{3.17}$$

form a representation of the Clifford algebra, since we get $\{\sigma^i, \sigma^j\} = 2\delta^{ij}$. We can choose for instance $\sigma_1 = \gamma_1$ and $\sigma_2 = \gamma_2$. Then we can define the analog of γ_5,

$$\gamma_{d+1}(= \text{``}\gamma_3\text{''}) = \sigma_3 = -i\sigma_1\sigma_2 \tag{3.18}$$

that satisfies $\{\gamma_i, \gamma_3\} = 0$, $(\gamma_3)^2 = 1$. In $d = 3$ Euclidean dimensions then, we can choose $\gamma_1, \gamma_2, \gamma_3$ as the objects satisfying the Clifford algebra. In general, we will have that in $d = 2n + 1$, γ_{2n+1} (the analog of γ_5), becomes the last gamma matrix, but otherwise does not change the dimension of the spinor representation. That is, both $d = 2n$ and $d = 2n + 1$ have the same dimension of the spinor representation.

Then in $d = 4 = 2n$ we can choose various representations, but for instance one that is possible is constructed by tensor products as follows

$$\Gamma^a = \gamma^a \otimes \sigma_3, \ a = 1, \ldots, 2n,$$
$$\Gamma^{2n+i} = 1 \otimes \sigma^i, \ i = 1, 2, \tag{3.19}$$

which we have written in a way immediately generalizable to constructing $d = 2n + 2$-dimensional gamma matrices from $d = 2n$-dimensional gamma matrices. This is a representation of the Clifford algebra, since

$$\{\Gamma^a, \Gamma^b\} = \{\gamma^a, \gamma^b\} \otimes (\sigma_3)^2 = 2\delta^{ab}1 \otimes 1_2,$$
$$\{\Gamma^i, \Gamma^j\} = 1 \otimes \{\sigma^i, \sigma^j\} = 2\delta^{ij}1 \otimes 1_2,$$
$$\{\Gamma^a, \Gamma^i\} = \gamma^a \otimes \{\sigma^3, \sigma^i\} = 0. \tag{3.20}$$

---

[2] The resulting spinors are sometimes also called symplectic Majorana spinors.

We can then also define Γ_{d+1} in $d = 2n + 2$ dimensions as $(\gamma_{2n+1} = i\gamma_1 \cdots \gamma_{2n})$,

$$\Gamma_{2n+3} = \Gamma_1 \Gamma_2 \cdots \Gamma_{2n+2} = -i\gamma_{2n+1} \otimes \sigma_1 \sigma_2 = \gamma_{2n+1} \otimes \sigma_3, \tag{3.21}$$

and satisfies $(\Gamma_{2n+3})^2 = 1$ and $\{\Gamma_{2n+3}, \Gamma^M\} = 0$, $M = (a, i)$.

In a general even dimension $d = 2n$, we then define the chiral projectors as

$$P_L = \frac{\mathbb{1} + \Gamma_{2n+1}}{2}; \quad P_R = \frac{\mathbb{1} - \Gamma_{2n+1}}{2}, \tag{3.22}$$

and then the Weyl condition is

$$P_L \psi_R = 0 \quad \text{or} \quad P_R \psi_L = 0, \tag{3.23}$$

defining ψ_R, ψ_L.

To define Majorana spinors, we need to define the general charge conjugation matrix. For more details, see [13].

The Majorana reality condition is defined as usual by

$$\bar{\chi}^D \equiv \chi^\dagger \tilde{\gamma}_t = \bar{\chi}^C \equiv \chi^T C, \tag{3.24}$$

where $\tilde{\gamma}_t$ is the gamma matrix in the time direction for Minkowski signature and $\mathbb{1}$ for Euclidean. The condition that $\bar{\chi}^D$ and $\bar{\chi}^C$ both satisfy the same Dirac equation gives a condition on C. Indeed, the Dirac equation $0 = (\gamma^\mu \partial_\mu + M)\chi$ gives, when taking the dagger,

$$0 = [(\gamma^\mu \partial_\mu + M)\chi]^\dagger = \chi^\dagger (\gamma^{\mu\dagger} \overleftarrow{\partial}_\mu + M) \Rightarrow$$
$$0 = \chi^\dagger (\gamma^{\mu\dagger} \overleftarrow{\partial}_\mu + M)\tilde{\gamma}_t = \chi^\dagger \tilde{\gamma}_t (\pm \gamma^\mu \overleftarrow{\partial}_\mu + M), \tag{3.25}$$

where we have used that $(\gamma^\mu)^\dagger = +\gamma^\mu$ for spacelike μ and $(\gamma^\mu)^\dagger = -\gamma^\mu$ for timelike μ, implying $(\gamma^\mu)^\dagger \tilde{\gamma}_t \partial_\mu = \pm \tilde{\gamma}_t \gamma^\mu \partial_\mu$. On the other hand, we can take the transpose of the Dirac equation and get

$$0 = [(\gamma^\mu \partial_\mu + M)\chi]^T C = \chi^T C C^{-1} (\gamma^{\mu T} \overleftarrow{\partial}_\mu + M)C. \tag{3.26}$$

Since $\bar{\chi}^D = \chi^\dagger \tilde{\gamma}_t$ and $\bar{\chi}^C = \chi^T C$ must obey the same equation, by comparing the two results we obtain

$$C\gamma^\mu C^{-1} = \sigma \gamma^{\mu T}, \tag{3.27}$$

where $\sigma^2 = 1$, i.e., $\sigma = \pm 1$. We call C_+ the matrix satisfying the relation with $\sigma = +1$ and C_- the matrix satisfying the relation with $\sigma = -1$. In even dimensions, both C_+ and C_- exist, but in odd dimensions only one of them can exist, not both.

The matrix C must be symmetric or antisymmetric, as can be seen by applying twice the above relation for $(\gamma^\mu)^T$, and using that $(\gamma^{\mu T})^T = \gamma^\mu$, which gives

$$(C^{-1} C^T)^{-1} \gamma^\mu (C^{-1} C^T) = \gamma^\mu \Rightarrow$$
$$C^{-1} C^T = a \mathbb{1}, \quad a^2 = 1 \Rightarrow a = \pm 1. \tag{3.28}$$

From (3.27) and (3.28), $C\gamma_\mu$ is symmetric or antisymmetric $((C\gamma_\mu)^T = \sigma a C\gamma_\mu)$. But for the existence of an action for anticommuting spinors, we must have $C\gamma_\mu$ symmetric, since

$$\int \bar{\chi}\,\partial\!\!\!/\chi = \int \chi^T C\gamma^\mu \partial_\mu \chi = \int (\chi^T C\gamma^\mu \partial_\mu \chi)^T = -\int (\partial_\mu \chi^T)(C\gamma^\mu)^T \chi$$
$$= +\int \chi^T (C\gamma^\mu)^T \partial_\mu \chi. \tag{3.29}$$

In the second equality we use the fact that the transpose of a number is the same number, in the third we use the fact that the spinors are anticommuting, and in the last we use partial integration. By comparing the first and last form, we see that we need $C\gamma^\mu$ to be symmetric. If one uses commuting spinors instead, the same argument finds that we need $C\gamma^\mu$ to be antisymmetric. In Minkowski space we generally want to use anticommuting (usual) spinors, whereas in Euclidean space, having in mind the application to Kaluza–Klein compactifications (to be studied later), we generally want to use commuting spinors.

The condition of symmetry of $C\gamma^\mu$ in Minkowski space means that we need $\sigma a = +1$, as we can easily check; therefore, in its absence we cannot define spinors.

A summary of the possibilities in various Minkowski dimensions is as follows. A more complete list (including Euclidean spinors) is found in [13].

- $d = 2$: We have both C_+, C_-, satisfying $C_+^T = C_+, C_-^T = -C_-$, and we can construct Majorana spinors with both. We can also construct Majorana–Weyl spinors using both.
- $d = 3$: We have a C_-, satisfying $C_-^T = -C_-$, and we can construct Majorana spinors with it.
- $d = 4$: We have both C_+ and C_-, satisfying $C_-^T = -C_-$ and $C_+^T = -C_+$, but we can use only C_- to construct Majorana spinors (C_+ has $\sigma a = -1$).
- $d = 5$: We have only a C_+, with $C_+^T = -C_+$, but we cannot use it to construct Majorana spinors, since $\sigma a = -1$.
- $d = 6$: We have both a C_+ and a C_-, satisfying $C_+^T = -C_+$ and $C_-^T = C_-$, but neither can be used to define Majorana spinors, since $\sigma a = -1$.
- $d = 7$: We have only a C_-, with $C_-^T = C_-$, but we cannot use it to construct Majorana spinors, since $\sigma a = -1$.
- $d = 8$. We have both a C_+ and a C_-, satisfying $C_+^T = C_+$, $C_-^T = C_-$, but only C_+ can be used to construct Majorana spinors, since C_- has $\sigma a = -1$.
- $d = 9$. We have only a C_+, with $C_+^T = C_+$, and we can use it to construct Majorana spinors.
- $d = 10$. We have both a C_+ and a C_-, satisfying $C_+^T = C_+$, $C_-^T = -C_-$, and we can use both to construct Majorana spinors, as well as Majorana–Weyl spinors.
- $d = 11$. We have only a C_-, with $C_-^T = -C_-$, which can be used to construct Majorana spinors.

We see that in $d = 5, 6, 7$ Minkowski dimensions it seems that we cannot define Majorana spinors. However, if we have more than one spinor, in particular an even number of spinors, we can define a *modified Majorana condition*, with Majorana conjugate

$$\bar{\chi}_C^i \equiv \chi_j^T \Omega^{ji} C, \quad i = 1, 2, \tag{3.30}$$

if Ω is antisymmetric, $\Omega^T = -\Omega$, and unitary, which implies $\Omega\Omega^* = -\mathbb{1}$ (since $\Omega^\dagger = \Omega^{T*} = -\Omega^*$). The simplest choice is if Ω is the symplectic matrix,

$$\Omega = \begin{pmatrix} \mathbf{0} & \mathbb{1} \\ -\mathbb{1} & \mathbf{0} \end{pmatrix}, \tag{3.31}$$

which means that the spinors χ^i are in the fundamental representation of $USp(2N)$.

3.4 The 2-dimensional Wess–Zumino model: on-shell supersymmetry

We will start by explaining supersymmetry for the simplest possible models, which occur in two dimensions.

As we saw, a general (Dirac) fermion in d dimensions has $2^{[d/2]}$ complex components, therefore in two dimensions it has two complex dimensions, and thus a Majorana fermion will have two real components. An on-shell Majorana fermion (that satisfies the Dirac equation, its equation of motion) will then have a single component (since the Dirac equation is a matrix equation that relates half of the components to the other half).

Since we have a symmetry between bosons and fermions, the number of degrees of freedom of the bosons must match the number of degrees of freedom of the fermions (the symmetry will map a degree of freedom to another degree of freedom). This matching can be:

- on-shell, in which case we have *on-shell supersymmetry*; OR
- off-shell, in which case we have *off-shell supersymmetry*.

Thus, in two dimensions, the simplest possible model has one Majorana fermion ψ (which has one degree of freedom on-shell), and one real scalar ϕ (also one on-shell degree of freedom). We can then obtain **on-shell supersymmetry** and get the *Wess–Zumino model in two dimensions*.

Free Wess–Zumino model

The action of a free boson and a free fermion in two Minkowski dimensions is[3]

$$S = -\frac{1}{2} \int d^2x [(\partial_\mu \phi)^2 + \bar{\psi} \partial\!\!\!/ \psi], \tag{3.32}$$

and this is actually the action of the free Wess–Zumino model. From the action, the mass dimension of the scalar is $[\phi] = 0$, and of the fermion is $[\psi] = 1/2$ (the mass dimension of $\int d^2x$ is -2 and of ∂_μ is $+1$, and the action is dimensionless).

---

[3] Note that the Majorana reality condition implies that $\bar{\psi} = \psi^T C$ is not independent from ψ, thus we have a 1/2 factor in the fermionic action.

To write down the supersymmetry transformation between the boson and the fermion, we start by varying the boson into fermion times ϵ, i.e.

$$\delta\phi = \bar{\epsilon}\psi \equiv \bar{\epsilon}_\alpha \psi^\alpha \equiv \epsilon^\beta C_{\beta\alpha} \psi^\alpha. \tag{3.33}$$

This is a definition, but it is also the simplest thing we can have (we need both ϵ and ψ on the right-hand side). From this we infer that the mass dimension of ϵ is $[\epsilon] = -1/2$. This also defines the order of indices in contractions $\bar{\chi}\psi$ ($\bar{\chi}\psi = \bar{\chi}_\alpha \psi^\alpha$ and $\bar{\chi}_\alpha = \chi^\beta C_{\beta\alpha}$). By dimensional reasoning, for the reverse transformation we must add an object of mass dimension one with no free vector indices, and the only one such object available to us is $\slashed{\partial}$, thus

$$\delta\psi = \slashed{\partial}\phi\epsilon. \tag{3.34}$$

We can check that the above free action is indeed invariant on-shell under this symmetry.

Majorana spinor identities for fermion bilinears

For this, we must use the Majorana spinor identities for fermion bilinears. We start with two identities valid in both two and four dimensions. In both dimensions we use the C-matrix C_-, which obeys the same relations (1.78) in both dimensions. The identities are

$$1)\ \ \bar{\epsilon}\chi = +\bar{\chi}\epsilon; \quad 2)\ \ \bar{\epsilon}\gamma_\mu\chi = -\bar{\chi}\gamma_\mu\epsilon. \tag{3.35}$$

To prove the first identity, we write $\bar{\epsilon}\chi = \epsilon^\alpha C_{\alpha\beta}\chi^\beta$, but $C_{\alpha\beta}$ is antisymmetric and ϵ and χ anticommute, being spinors, thus we get $-\chi^\beta C_{\alpha\beta}\epsilon^\alpha = +\chi^\beta C_{\beta\alpha}\epsilon^\alpha$. To prove the second, we use the fact that, from (1.78), $C\gamma_\mu = -\gamma_\mu^T C = \gamma_\mu^T C^T = (C\gamma_\mu)^T$, thus now $(C\gamma_\mu)$ is symmetric (as we also mentioned in the previous subsection) and the rest is the same.

We can write two more relations, which now, however, depend on dimension. In two Minkowski dimensions we define

$$\gamma_3 = i\gamma_0\gamma_1, \tag{3.36}$$

and in four Minkowski dimensions we defined (see (1.63)) $\gamma_5 = i\gamma_0\gamma_1\gamma_2\gamma_3$. We then get

$$3)\ \ \bar{\epsilon}\gamma_3\chi = -\bar{\chi}\gamma_3\epsilon; \quad \bar{\epsilon}\gamma_5\chi = +\bar{\chi}\gamma_5\epsilon,$$
$$4)\ \bar{\epsilon}\gamma_\mu\gamma_3\chi = -\bar{\chi}\gamma_\mu\gamma_3\epsilon; \quad \bar{\epsilon}\gamma_\mu\gamma_5\chi = +\bar{\chi}\gamma_\mu\gamma_5\epsilon. \tag{3.37}$$

To prove these, we need also that $C\gamma_3 = +i\gamma_0^T\gamma_1^T C = -i(C\gamma_1\gamma_0)^T = +(C\gamma_3)^T$ (also as claimed in the last section), whereas $C\gamma_5 = +i\gamma_0^T\gamma_1^T\gamma_2^T\gamma_3^T C = -i(C\gamma_3\gamma_2\gamma_1\gamma_0)^T = -(C\gamma_5)^T$, as well as $\{\gamma_\mu, \gamma_3\} = \{\gamma_\mu\gamma_5\} = 0$ and $\{\gamma_\mu^T, \gamma_3^T\} = -\{\gamma_\mu^T, \gamma_5^T\} = 0$.

Then the variation of the action gives

$$\delta S = -\int d^2x \left[-\phi\Box\delta\phi + \frac{1}{2}\delta\bar{\psi}\slashed{\partial}\psi + \frac{1}{2}\bar{\psi}\slashed{\partial}\delta\psi \right] = -\int d^2x[-\phi\Box\delta\phi + \bar{\psi}\slashed{\partial}\delta\psi], \tag{3.38}$$

where in the second equality we have used partial integration together with identity 2) above. Then substituting the transformation law we get

$$\delta S = -\int d^2x[-\phi\Box\bar{\epsilon}\psi + \bar{\psi}\slashed{\partial}\slashed{\partial}\phi\epsilon]. \tag{3.39}$$

But we have

$$\partial\!\!\!/\,\partial\!\!\!/ = \partial_\mu \partial_\nu \gamma^\mu \gamma^\nu = \partial_\mu \partial_\nu \frac{1}{2}\{\gamma_\mu, \gamma_\nu\} = \partial_\mu \partial_\nu g^{\mu\nu} = \Box, \tag{3.40}$$

and by using this identity, together with two partial integrations, we obtain that $\delta S = 0$.

Therefore the action is invariant without the need for the equations of motion, so it would seem that this is an off-shell supersymmetry. However, the invariance of the action is not enough, since we have not proven that the above transformation law closes on the fields, i.e. that by acting twice on every field and forming the Lie algebra of the symmetry, we get back to the same field, or that we have a *representation of the Lie algebra* on the fields.

As we saw, the graded Lie algebra of supersymmetry is generically of the type

$$\{Q_\alpha^i, Q_\beta^j\} = 2(C\gamma^\mu)_{\alpha\beta} P_\mu \delta^{ij} + \dots, \tag{3.41}$$

and in the case of a single supersymmetry, as for the 2-dimensional Wess–Zumino model, we do not have any $+\dots$ and the above algebra is complete. In order to represent it on the fields, we note that in general, for a symmetry, $\delta_\epsilon = \epsilon^a T_a$, i.e. the symmetry variation is understood as the variation parameter times the generator. In the case of susy (supersymmetry), we then have $\delta_\epsilon = \epsilon^\alpha Q_\alpha$, so multiplying the algebra with ϵ_1^α from the left and ϵ_2^β from the right, we get on the left-hand side

$$\epsilon_1^\alpha Q_\alpha Q_\beta \epsilon_2^\beta + \epsilon_1^\alpha Q_\beta Q_\alpha \epsilon_2^\beta = \epsilon_1^\alpha Q_\alpha Q_\beta \epsilon_2^\beta - \epsilon_2^\beta Q_\beta Q_\alpha \epsilon_1^\alpha = -[\delta_{\epsilon_1}, \delta_{\epsilon_2}], \tag{3.42}$$

and on the right-hand side we get, using that P_μ is a translation, so is represented on the fields by ∂_μ,

$$2\bar\epsilon_1 \gamma^\mu \epsilon_2 \partial_\mu = -(2\bar\epsilon_2 \gamma^\mu \epsilon_1)\partial_\mu. \tag{3.43}$$

All in all, the algebra we need to represent is

$$[\delta_{\epsilon_1}, \delta_{\epsilon_2}] = 2\bar\epsilon_2 \gamma^\mu \epsilon_1 \partial_\mu. \tag{3.44}$$

In other words, we need to find

$$[\delta_{\epsilon_{1\alpha}}, \delta_{\epsilon_{2\beta}}] \begin{pmatrix} \phi \\ \psi \end{pmatrix} = 2\bar\epsilon_2 \gamma^\mu \epsilon_1 \partial_\mu \begin{pmatrix} \phi \\ \psi \end{pmatrix}. \tag{3.45}$$

We get

$$[\delta_{\epsilon_1}, \delta_{\epsilon_2}]\phi = \delta_{\epsilon_1}(\bar\epsilon_2 \psi) - (1 \leftrightarrow 2) = \bar\epsilon_2(\partial\!\!\!/\phi)\epsilon_1 - (1 \leftrightarrow 2) = 2\bar\epsilon_2 \gamma^\rho \epsilon_1 \partial_\rho \phi, \tag{3.46}$$

where in the last equality we have used the Majorana spinor relation 2) above. Thus the algebra is indeed realized on the scalar, without the use of the equations of motion. On the spinor, we have

$$[\delta_{\epsilon_1}, \delta_{\epsilon_2}]\psi = \delta_{\epsilon_1}(\partial\!\!\!/\phi)\epsilon_2 - (1 \leftrightarrow 2) = (\bar\epsilon_1 \partial_\mu \psi)\gamma^\mu \epsilon_2 - (1 \leftrightarrow 2). \tag{3.47}$$

Fierz identities

To proceed further, we need to use the so-called "Fierz identities" (or "Fierz recoupling"). In two Minkowski dimensions, these read

$$M\chi(\bar{\psi}N\phi) = -\sum_j \frac{1}{2}MO_jN\phi(\bar{\psi}O_j\chi), \tag{3.48}$$

(the minus is a consequence of changing the order of two fermions) where M and N are arbitrary matrices, χ, ψ, ϕ are arbitrary spinors and the set of matrices $\{O_j\}$ is $= \{\mathbb{1}, \gamma_\mu, \gamma_5\}$ and is a complete set on the space of 2×2 matrices (we have four independent matrices for four components). The identity follows from the completeness relation for the matrices $\{O_i\}$,

$$\delta_\alpha^\beta \delta_\gamma^\delta = \frac{1}{2}(O_i)_\alpha^\delta (O_i)_\gamma^\beta . \tag{3.49}$$

This is a completeness relation since by multiplying with an M_β^γ, we obtain the decomposition of an arbitrary matrix M into O_i,

$$M^\delta{}_\alpha = \frac{1}{2}\mathrm{Tr} \ (MO_i)(O_i)^\delta{}_\alpha . \tag{3.50}$$

We note that the factor $1/2$ is related to the normalization $\mathrm{Tr} \ (O_iO_j) = 2\delta_{ij}$.

In four Minkowski dimensions, we have

$$M\chi(\bar{\psi}N\phi) = -\sum_j \frac{1}{4}MO_jN\phi(\bar{\psi}O_j\chi) \tag{3.51}$$

instead, since now $\mathrm{Tr} \ (O_iO_j) = 4\delta_{ij}$, and the $\{O_i\}$ is now a complete set of 4×4 matrices, given by $O_i = \{\mathbb{1}, \gamma_\mu, \gamma_5, i\gamma_\mu\gamma_5, i\gamma_{\mu\nu}\}$. Here as usual $\gamma_{\mu\nu} = 1/2[\gamma_\mu, \gamma_\nu]$ (six matrices), so in total we have 16 independent matrices for 16 components.

Using the Fierz relation (3.48) for $M = \gamma_\mu, N = \partial_\mu$, we have for (3.47),

$$
\begin{aligned}
\gamma^\mu &\epsilon_2(\bar{\epsilon}_1\partial_\mu\psi) - 1 \leftrightarrow 2 \\
&= -\frac{1}{2}[\gamma^\mu\mathbb{1}\partial_\mu\psi(\bar{\epsilon}_1\mathbb{1}\epsilon_2) + \gamma^\mu\gamma_\nu\partial_\mu\psi(\bar{\epsilon}_1\gamma^\nu\epsilon_2) + \gamma^\mu\gamma_3\partial_\mu\psi(\bar{\epsilon}_1\gamma_3\epsilon_2)] - 1 \leftrightarrow 2 \\
&= +\gamma^\mu\gamma_\nu\partial_\mu\psi(\bar{\epsilon}_2\gamma_\nu\epsilon_1) + \gamma^\mu\gamma_3\partial_\mu\psi(\bar{\epsilon}_2\gamma_3\epsilon_2) \\
&= 2(\bar{\epsilon}_2\gamma^\mu\epsilon_1)\partial_\mu\psi - \gamma^\nu(\partial\!\!\!/\psi)(\bar{\epsilon}_2\gamma_\nu\epsilon_1) - \gamma_3(\partial\!\!\!/\psi)(\bar{\epsilon}_2\gamma_3\epsilon_1),
\end{aligned} \tag{3.52}
$$

where in the second line we used Majorana relations 1), 2), and 3) above.

Thus now we do not obtain a representation of the susy algebra on ψ in general, since we have the last two extra terms. But these extra terms vanish on-shell, when $\partial\!\!\!/\psi = 0$, hence now we have a realization of *on-shell supersymmetry*.

3.5 The 2-dimensional Wess–Zumino model: off-shell supersymmetry

In two dimensions, an off-shell Majorana fermion has two degrees of freedom, but a scalar has only one. Thus to close the algebra of the Wess–Zumino model off-shell, we need one extra scalar field F. But on-shell, we must get back the previous model, thus the extra scalar F needs to be auxiliary (non-dynamical, with no propagating degree of freedom).

That means that its action is $\int F^2/2$, thus the action of the off-shell free Wess–Zumino model is

$$S = -\frac{1}{2} \int d^2x [(\partial_\mu \phi)^2 + \bar{\psi} \slashed{\partial} \psi - F^2]. \tag{3.53}$$

From the action we see that F has mass dimension $[F] = 1$, and the equation of motion of F is $F = 0$. The off-shell Wess–Zumino model algebra does not close on ψ, thus we need to add to $\delta \psi$ a term proportional to the equation of motion of F. By dimensional analysis, $F\epsilon$ has the right dimension. Since $F(= 0)$ itself is an equation (bosonic) of motion, its variation δF should be the fermionic equation of motion, and by dimensional analysis $\bar{\epsilon} \slashed{\partial} \psi$ is OK. Thus the transformations laws are

$$\delta \phi = \bar{\epsilon} \psi; \quad \delta \psi = \slashed{\partial} \phi \epsilon + F \epsilon; \quad \delta F = \bar{\epsilon} \slashed{\partial} \psi. \tag{3.54}$$

Then we have

$$\delta_{\epsilon_1} \delta_{\epsilon_2} \phi = \delta_{\epsilon_1}(\bar{\epsilon}_2 \psi) = \bar{\epsilon}_2 \slashed{\partial} \phi \epsilon_1 + \bar{\epsilon}_2 \epsilon_1 F, \tag{3.55}$$

and using Majorana spinor relations 1) and 2) (3.35), we get

$$[\delta_{\epsilon_1}, \delta_{\epsilon_2}] \phi = 2(\bar{\epsilon}_2 \gamma^\mu \epsilon_1) \partial_\mu \phi. \tag{3.56}$$

Here we have no modification with respect to the on-shell case, and the algebra is still represented on ϕ. On the other hand,

$$\delta_{\epsilon_1} \delta_{\epsilon_2} \psi = \delta_{\epsilon_1}(\slashed{\partial} \phi \epsilon_2 + F \epsilon_2) = \gamma^\mu \epsilon_2 (\bar{\epsilon}_1 \partial_\mu \psi) + (\bar{\epsilon}_1 \slashed{\partial} \psi) \epsilon_2, \tag{3.57}$$

so in the commutator on ψ we get the extra term

$$\begin{aligned}
(\bar{\epsilon}_1 \slashed{\partial} \psi) \epsilon_2 &= -\frac{1}{2}[1 \cdot \slashed{\partial} \psi (\bar{\epsilon}_1 1 \epsilon_2) + \gamma^\mu \slashed{\partial} \psi (\bar{\epsilon}_1 \gamma_\mu \epsilon_2) + \gamma_3 \slashed{\partial} \psi (\bar{\epsilon}_1 \gamma_3 \epsilon_2)] - 1 \leftrightarrow 2 \\
&= -(\bar{\epsilon}_1 \gamma_\mu \epsilon_2) \gamma^\mu \slashed{\partial} \psi - (\bar{\epsilon}_1 \gamma_3 \epsilon_2) \gamma_3 \slashed{\partial} \psi \\
&= (\bar{\epsilon}_2 \gamma_\mu \epsilon_1) \gamma^\mu \slashed{\partial} \psi + (\bar{\epsilon}_2 \gamma_3 \epsilon_1) \gamma_3 \slashed{\partial} \psi,
\end{aligned} \tag{3.58}$$

where we have used the Fierz identity with $M = 1, N = \slashed{\partial}$, and we have again used Majorana spinor relations 1), 2), 3) (3.35, 3.37). These extra terms exactly cancel the extra terms in (3.52), and we get a representation of the algebra on ψ as well,

$$[\delta_{\epsilon_1}, \delta_{\epsilon_2}] \psi = 2(\bar{\epsilon}_2 \gamma^\mu \epsilon_1) \partial_\mu \psi. \tag{3.59}$$

It is left as an exercise (Exercise 4) to check that the algebra also closes on F.

3.6 The 4-dimensional Wess–Zumino model

Free on-shell Wess–Zumino model in four dimensions

Similarly to the 2-dimensional case, in four dimensions the on-shell Wess–Zumino model has one Majorana fermion, which, however, now has two real on-shell degrees of freedom, thus we need two real scalars, A and B, to form an on-shell supersymmetry invariant Wess–Zumino model. Therefore the action for the *free on-shell Wess–Zumino model* is

$$S_0 = -\frac{1}{2} \int d^4x \left[(\partial_\mu A)^2 + (\partial_\mu B)^2 + \bar{\psi} \slashed{\partial} \psi \right], \tag{3.60}$$

and the transformation laws are almost the same as in two dimensions, except that now B acquires an $i\gamma_5$ to distinguish it from A, thus

$$\delta A = \bar{\epsilon} \psi; \quad \delta B = \bar{\epsilon} i\gamma_5 \psi; \quad \delta \psi = \slashed{\partial}(A + i\gamma_5 B)\epsilon. \tag{3.61}$$

The proof of invariance of the action under these rules exactly follows the 2-dimensional case, so will not be repeated here, but is left as an exercise (Exercise 2). We can also show that the algebra (3.44) is represented on the fields *on-shell* only.

Free off-shell Wess–Zumino model in four dimensions

Also as in two dimensions, off-shell the Majorana fermion has four degrees of freedom, so one needs to introduce one auxiliary scalar for each propagating scalar. Therefore the action of the off-shell Wess–Zumino model is

$$S = S_0 + \int d^4x \left[\frac{F^2}{2} + \frac{G^2}{2} \right], \tag{3.62}$$

and the transformation rules are

$$\delta A = \bar{\epsilon} \psi; \quad \delta B = \bar{\epsilon} i\gamma_5 \psi; \quad \delta \psi = \slashed{\partial}(A + i\gamma_5 B)\epsilon + (F + i\gamma_5 G)\epsilon;$$
$$\delta F = \bar{\epsilon} \slashed{\partial} \psi; \quad \delta G = \bar{\epsilon} i\gamma_5 \slashed{\partial} \psi. \tag{3.63}$$

The invariance of the action and the representation of the algebra (3.44) on the fields is slightly more involved than in the 2-dimensional case, but follows in a similar manner.

One can form a complex field $\phi = A + iB$ and one complex auxiliary field $M = F + iG$, thus the Wess–Zumino multiplet in four dimensions is (ϕ, ψ, M).

We have written the free Wess–Zumino model in two dimensions and four dimensions, but one can write down interactions for them as well, that preserve the supersymmetry. We will not write them now, though we will do so when we introduce the notion of superspace, with which it is easier to organize them.

3.7 Two-component notation, extended supersymmetry algebra, and multiplets; R-symmetry

In four Minkowski dimensions, it is useful to use the 2-component notation, using dotted and undotted indices. In this section we use $A, B = 1, \ldots, 4$ for 4-component spinor indices, in order to reserve $\alpha, \dot{\alpha}$ for the 2-component indices, as is conventional. A general Dirac spinor is written as

$$\psi^A = \begin{pmatrix} \psi_\alpha \\ \bar{\chi}^{\dot{\alpha}} \end{pmatrix}, \tag{3.64}$$

where $\alpha, \dot\alpha = 1, 2$ and the relation between dotted and undotted indices for the same 2-component spinor is given by $\bar\chi^{\dot\alpha} = \epsilon^{\dot\alpha\dot\alpha}(\chi_\alpha)^*$. We use the representation for the C-matrix

$$C_{AB} = \begin{pmatrix} \epsilon^{\alpha\beta} & 0 \\ 0 & \epsilon_{\dot\alpha\dot\beta} \end{pmatrix}, \tag{3.65}$$

where $\epsilon^{12} = \epsilon^{\dot1\dot2} = +1$, $\epsilon_{\dot\alpha\dot\beta} = -\epsilon^{\dot\alpha\dot\beta} = (\epsilon^{\dot\alpha\dot\beta})^{-1}$, and for the gamma matrices

$$\gamma^\mu = \begin{pmatrix} 0 & \sigma^\mu \\ \bar\sigma^\mu & 0 \end{pmatrix}, \tag{3.66}$$

where $(\sigma^\mu)_{\alpha\dot\alpha} = (\mathbb{1}, \vec\sigma)_{\alpha\dot\alpha}$ and $(\bar\sigma^\mu)^{\alpha\dot\alpha} = \epsilon^{\alpha\beta}\epsilon^{\dot\alpha\dot\beta}(\sigma^\mu)_{\beta\dot\beta} = (\mathbb{1}, -\vec\sigma)^{\alpha\dot\alpha}$.

A Majorana spinor has $\psi_\alpha = \chi_\alpha$ (a single independent 2-component spinor), i.e. it is

$$\begin{pmatrix} \psi_\alpha \\ \bar\psi^{\dot\alpha} \end{pmatrix}. \tag{3.67}$$

Finally, we use the notation $\psi\chi \equiv \psi^\alpha\chi_\alpha$ and $\bar\psi\bar\chi \equiv \bar\psi_{\dot\alpha}\bar\chi^{\dot\alpha}$, which is derived from the 4-dimensional one,

$$\bar\psi\chi = \psi^A C_{AB}\chi^B = \psi^\beta\chi_\beta + \bar\psi_{\dot\beta}\bar\chi^{\dot\beta} = \psi\chi + \bar\psi\bar\chi, \tag{3.68}$$

together with the obvious definitions (coming from our definitions of raising and lowering indices with C_{AB} and C^{AB}) $\psi^\beta = \psi_\alpha\epsilon^{\alpha\beta}$, $\bar\psi_{\dot\beta} = \bar\psi^{\dot\alpha}\epsilon_{\dot\alpha\dot\beta}$.

Then in 2-component spinor notation, the $\mathcal{N} = 1$ supersymmetry algebra

$$\{Q_A, Q_B\} = 2(C\gamma^\mu)_{AB}P_\mu, \tag{3.69}$$

becomes (note that $Q_A = C_{AB}Q^B$, where Q^B is of the form (3.64))

$$\begin{aligned} \{Q_\alpha, \bar{Q}_{\dot\alpha}\} &= -2(\sigma^\mu)_{\alpha\dot\alpha}P_\mu, \\ \{Q_\alpha, Q_\beta\} &= 0; \quad \{\bar{Q}_{\dot\alpha}, \bar{Q}_{\dot\beta}\} = 0. \end{aligned} \tag{3.70}$$

More generally, the $\mathcal{N}$-extended supersymmetry algebra *without central charges*,

$$\{Q_A^i, Q_B^j\} = +2(C\gamma^\mu)_{AB}\delta^{ij}P_\mu, \tag{3.71}$$

where $i, j = 1, \ldots, \mathcal{N}$, in 2-component Majorana spinor notation, with $\bar{Q}_{i\dot\alpha} = (Q_\alpha^i)^*$, becomes

$$\begin{aligned} \{Q_\alpha^i, \bar{Q}_{j\dot\alpha}\} &= -2(\sigma^\mu)_{\alpha\dot\alpha}\delta_j^i P_\mu, \\ \{Q_\alpha^i, Q_\beta^j\} &= 0; \quad \{\bar{Q}_{i\dot\alpha}, \bar{Q}_{j\dot\beta}\} = 0. \end{aligned} \tag{3.72}$$

Massless irreducible representations

For massless states we can find a frame where $P^\mu = p(1, 0, 0, 1)$, so $P_\mu = p(-1, 0, 0, 1)$, and $\sigma^\mu P_\mu = p(-\mathbb{1} + \sigma_3)$. Therefore the algebra reduces to

$$\{Q_\alpha^i, \bar{Q}_{j\dot\alpha}\} = 2p(\mathbb{1} - \sigma_3)\delta_j^i = 4p\begin{pmatrix} 0 & 0 \\ 0 & 1 \end{pmatrix}\delta_j^i. \tag{3.73}$$

Therefore $\{Q_1^i, Q_1^j\} = 0$, and since they are conjugate to each other, we must impose $Q_1^i|\psi\rangle = \bar{Q}_1^i|\psi\rangle = 0$ for $|\psi\rangle$ a physical state. On the other hand, since $\{Q_2^i, \bar{Q}_2^j\} = 4p\delta^{ij}$, we can define fermionic creation and annihilation operators,

$$a^i = \frac{1}{2\sqrt{p}}Q_2^i; \quad a^{\dagger i} = \frac{1}{2\sqrt{p}}\bar{Q}_{i2}, \tag{3.74}$$

satisfying the usual algebra

$$\{a^i, a^{\dagger j}\} = \delta^{ij}; \quad \{a^i, a^j\} = \{a^{\dagger i}, a^{\dagger j}\} = 0. \tag{3.75}$$

We can thus use Wigner's method to find the irreducible representations (irreps), by acting on a "vacuum" state of given helicity λ, $|\Omega_\lambda\rangle$, with the $\mathcal{N}$ creation operators $a^{\dagger i}$, that will lower the helicity. Helicity is defined by the Lorentz generator $J_3 = J_{12}$, so $J_3|\Omega_\lambda\rangle = \lambda|\Omega_\lambda\rangle$. But the commutation relation (3.7), written now as

$$[\bar{Q}_2^i, J_3] = +\frac{1}{2}\bar{Q}_2^i, \tag{3.76}$$

means that $a^{\dagger i}$ lowers helicity, since

$$a^{\dagger i}J_3|j\rangle - J_3 a^{\dagger i}|j\rangle = \frac{1}{2}a^{\dagger i}|j\rangle \Rightarrow J_3(a^{\dagger i}|j\rangle) = \left(j - \frac{1}{2}\right)(a^{\dagger i}|j\rangle). \tag{3.77}$$

The algebra of fermionic creation and annihilation operators has irreducible representations (depending on a given vacuum) of dimension $2^{\mathcal{N}}$, obtained by acting or not with each of the $\mathcal{N}$ $a^{\dagger i}$. Relevant examples are then

$$\mathcal{N} = 1 : |\lambda\rangle, \quad |\lambda - 1/2\rangle = a^\dagger|\lambda\rangle,$$
$$\mathcal{N} = 2 : |\lambda\rangle, \quad 2|\lambda - 1/2\rangle = (a^{\dagger 1}|\lambda\rangle, a^{\dagger 2}|\lambda\rangle)), \quad |\lambda - 1\rangle = a^{\dagger 1}a^{\dagger 2}|\lambda\rangle,$$
$$\mathcal{N} = 4 : |\lambda\rangle, \quad 4|\lambda - 1/2\rangle = a^{\dagger i}|\lambda\rangle, \quad 6|\lambda - 1\rangle = (a^{\dagger i}a^{\dagger j})|\lambda\rangle, i \neq j;$$
$$4|\lambda - 3/2\rangle = (a^{\dagger i}a^{\dagger j}a^{\dagger k})|\lambda\rangle, i \neq j \neq k; \quad |\lambda - 2\rangle = a^{\dagger 1}a^{\dagger 2}a^{\dagger 3}a^{\dagger 4}|\lambda\rangle. \tag{3.78}$$

But the irreducible representations are not necessarily CPT invariant. CPT invariance in particular reverses the sign of the helicity, and if the representation is not invariant under CPT, we must add the CPT-conjugate representation to it. The relevant cases we obtain are

$$\mathcal{N} = 1, \quad \lambda = 1/2 : |1/2\rangle, |0\rangle; \; |-1/2\rangle, |0\rangle,$$
$$\lambda = 1 : |1\rangle, |1/2\rangle; \; |-1\rangle, |-1/2\rangle,$$
$$\mathcal{N} = 2, \quad \lambda = 1/2 : |1/2\rangle, 2|0\rangle, |-1/2\rangle; \; |-1/2\rangle, 2|0\rangle, |1/2\rangle,$$
$$\lambda = 1 : |1\rangle, 2|1/2\rangle, |0\rangle; \; |-1\rangle, 2|-1/2\rangle, |0\rangle,$$
$$\mathcal{N} = 4, \quad \lambda = 1 : |1\rangle, 4|1/2\rangle, 6|0\rangle, 4|-1/2\rangle, |-1\rangle. \tag{3.79}$$

Note that only the $\mathcal{N} = 4$ representation is self-conjugate under CPT, the rest are two CPT conjugate representations added up.

Also note that if we start with helicity if we want to have a theory with at most vectors, but no spins greater than 1, the initial helicity λ of the vacuum cannot be greater than 1, and after acting with all the $\mathcal{N}$ creation operators we should get back to at most -1 helicity (the other helicity of a vector), but not smaller. That requires $\mathcal{N} \leq 2 \times (1 - (-1)) = 4$, so $\mathcal{N} = 4$ is the maximum allowed supersymmetry such that we have spins ≤ 1 only.

Multiplets

The representations above correspond to the following field content:

- $\mathcal{N} = 1$ multiplets.

  -The $\mathcal{N} = 1, \lambda = 1/2$ representation corresponds to a Majorana spinor and two real scalars, i.e. the Wess–Zumino, chiral or scalar multiplet studied before, since a spinor has two on-shell states with helicities $\pm 1/2$.

  -The $\mathcal{N} = 1, \lambda = 1$ representation corresponds to a Majorana spinor and a vector, since a vector has two on-shell states with helicities ± 1, forming a *vector multiplet*. The vector multiplet is (λ^a, A^a_μ), where a is an adjoint index. The vector A_μ in four dimensions has two on-shell degrees of freedom, since it has four components, minus one gauge invariance symmetry parameterized by an arbitrary ϵ^a, $\delta A^a_\mu = \partial_\mu \epsilon^a$ giving three off-shell components. In the covariant gauge $\partial^\mu A_\mu = 0$, the equation of motion $k^2 = 0$ is supplemented with the constraint $k^\mu \epsilon^a_\mu(k) = 0$ ($\epsilon^a_\mu(k)$ =polarization), which has only two independent solutions. The two degrees of freedom of the gauge field match the two degrees of freedom of the on-shell fermion.

- $\mathcal{N} = 2$ multiplets.

  -The $\mathcal{N} = 2, \lambda = 1/2$ representation has a field content of two $\mathcal{N} = 1$ WZ multiplets (ψ_1, ϕ_1) and (ψ_2, ϕ_2), together forming the *hypermultiplet*.

  -The $\mathcal{N} = 2, \lambda = 1$ representation has the field content of an $\mathcal{N} = 1$ WZ multiplet (ψ, ϕ), plus an $\mathcal{N} = 1$ vector multiplet (A_μ, λ), forming together the $\mathcal{N} = 2$ vector multiplet.

- $\mathcal{N} = 4$ multiplets.

  Finally, the $\mathcal{N} = 4, \lambda = 1$ representation has the field content of an $\mathcal{N} = 2$ vector and an $\mathcal{N} = 2$ hypermultiplet, or three $\mathcal{N} = 1$ WZ multiplets (ψ_i, ϕ_i), $i = 1, 2, 3$, and one $\mathcal{N} = 1$ vector multiplet (A_μ, ψ_4), forming together the $\mathcal{N} = 4$ vector multiplet. They can be rearranged into $(A^a_\mu, \psi^{ai}, \phi_{[ij]})$, where $i = 1, .., 4$ is an $SU(4) = SO(6)$ index, $[ij]$ is the six dimensional antisymmetric representation of $SU(4)$ or the fundamental representation of $SO(6)$, and i is the fundamental representation of $SU(4)$ or the spinor representation of $SO(6)$. The field $\phi_{[ij]}$ has complex entries but satisfies a reality condition,

$$\phi^\dagger_{ij} = \phi^{ij} \equiv \frac{1}{2} \epsilon^{ijkl} \phi_{kl}. \tag{3.80}$$

There are possible generalizations where the vacuum itself has a spin j, giving an extra multiplicity of $2j + 1$ to the representations, but we do not consider them here.

Massive representations of the algebra without central charges

In this case we can write in the rest frame $P^\mu = M(1, 0, 0, 0)$, so $\sigma^\mu P_\mu = -M$, and the algebra reduces to

$$\{Q^i_\alpha, \bar{Q}_{j\dot\alpha}\} = 2M \mathbb{1} \delta^i_j = 2M \begin{pmatrix} \mathbb{1} & 0 \\ 0 & \mathbb{1} \end{pmatrix}. \tag{3.81}$$

Now we have twice as many creation and annihilation operators,

$$a_\alpha^i = \frac{1}{\sqrt{2M}} Q_\alpha^i; \quad a_\alpha^{\dagger i} = \frac{1}{\sqrt{2M}} \bar{Q}_{i\dot\alpha}, \; \alpha = 1, 2, \tag{3.82}$$

satisfying the usual algebra

$$\{a_\alpha^i, a_\beta^{\dagger j}\} = \delta^{ij}\delta_{\alpha\beta}; \quad \{a_\alpha^i, a_\beta^j\} = \{a_\alpha^{\dagger i}, a_\beta^{\dagger j}\} = 0. \tag{3.83}$$

The number of states in the irreducible representation is now $2^{2\mathcal{N}}$.

Massive representations of the algebra with central charges

The algebra with central charges (3.13) reduces in component notation to

$$\begin{aligned}
\{Q_\alpha^i, \bar{Q}_{j\dot\beta}\} &= -2(\sigma^\mu)_{\alpha\dot\beta}\delta_j^i P_\mu, \\
\{Q_\alpha^i, Q_\beta^j\} &= 2\epsilon_{\alpha\beta}Z^{ij}, \\
\{\bar{Q}_{i\dot\alpha}, \bar{Q}_{j\dot\beta}\} &= 2\epsilon_{\dot\alpha\dot\beta}Z_{ij}^*.
\end{aligned} \tag{3.84}$$

For simplicity we will focus on the $\mathcal{N} = 2$ case. We can go in the rest frame, where $-2\sigma^\mu P_\mu = 2M$. We can also diagonalize the antisymmetric matrix Z^{ij} by a global $SU(2)$ transformation (part of the internal symmetry group) acting on the Q^is, obtaining $Z^{ij} = Z\epsilon^{ij}$. Moreover, by a $U(1)$ rotation (also part of the internal symmetry group) acting on the Q^is, we can make Z real, i.e. $Z^{ij} = Z\epsilon^{ij}$. Then the algebra becomes

$$\begin{aligned}
\{Q_\alpha^i, \bar{Q}_{j\dot\beta}\} &= 2M\delta_{\alpha\dot\beta}\delta_j^i, \\
\{Q_\alpha^i, Q_\beta^j\} &= 2Z\epsilon_{\alpha\beta}\epsilon^{ij}, \\
\{\bar{Q}_{i\dot\alpha}, \bar{Q}_{j\dot\beta}\} &= 2Z\epsilon_{\dot\alpha\dot\beta}\epsilon_{ij},
\end{aligned} \tag{3.85}$$

where $\bar{Q}_i^{\dot\alpha} = (Q_\alpha^i)^\dagger$ and $\bar{Q}_{i\dot\alpha} = \bar{Q}_i^{\dot\beta}\epsilon_{\dot\beta\dot\alpha}$. Then by defining

$$\begin{aligned}
a_\alpha &= \frac{1}{\sqrt{2}}[Q_\alpha^1 + \epsilon_{\alpha\dot\beta}\bar{Q}_{2\dot\beta}] & a_\alpha^\dagger &= \frac{1}{\sqrt{2}}[\bar{Q}_{1\dot\alpha} + \epsilon_{\alpha\beta}Q_\beta^2], \\
b_\alpha &= \frac{1}{\sqrt{2}}[Q_\alpha^1 - \epsilon_{\alpha\dot\beta}\bar{Q}_{2\dot\beta}] & a_\alpha^\dagger &= \frac{1}{\sqrt{2}}[\bar{Q}_{1\dot\alpha} - \epsilon_{\alpha\beta}Q_\beta^2],
\end{aligned} \tag{3.86}$$

we obtain the algebra

$$\{a_\alpha, a_\beta^\dagger\} = 2(M - Z)\delta_{\alpha\beta}; \quad \{b_\alpha, b_\beta^\dagger\} = 2(M + Z)\delta_{\alpha\beta}, \tag{3.87}$$

and the rest zero. This in turn means that we have the inequality

$$M \geq |Z|, \tag{3.88}$$

which is called the *Bogomolnyi–Prasad–Sommerfield, or BPS bound*, as it is related to the similar bound on the mass of solitons which is examined in a later chapter.

We see then that in the case of saturation of the bound, $M = |Z|$, we are back to having only $\mathcal{N}$ creation operators, and so to having a representation which is $2^\mathcal{N}$ dimensional, as in the massless case. Thus we obtain *short multiplets*.

R-symmetry

We have already described the fact that there is an internal symmetry that rotates the supercharges $Q_{i\alpha}$. When representing the algebra on fields, it turns into a global symmetry of the action, rotating the fields, called R-symmetry. For $\mathcal{N}$-extended supersymmetry, the maximum allowed is $U(N)$, as mentioned, but in specific cases it is smaller.

We are mostly interested in four dimensions, where the spinors can be chosen to be Weyl (complex). In particular, $\mathcal{N} = 4$ supersymmetry has only the $\mathcal{N} = 4$ vector multiplet, which as we saw, has an $SU(4)$ R-symmetry. For $\mathcal{N} = 2$ supersymmetry, we have generically $SU(2)$ R-symmetry. For hypermultiplets, it rotates the $2\,\mathcal{N} = 1$ multiplets, and for the vectors it rotates the two fermions in the multiplet.

We will also be interested in three dimensions, where the spinors are necessarily Majorana (real) and the natural action on N spinors is $SO(N)$. The maximal supersymmetry with spins ≤ 1 is $\mathcal{N} = 8$, which has $SO(8)$ R-symmetry. The case of $\mathcal{N} = 6$ is of interest, when the generic R-symmetry is $SO(6) = SU(4)$.

3.8 $\mathcal{N} = 1$ superspace in four dimensions

We have seen how we can have on-shell supersymmetry, when the susy algebra closes only on-shell, or off-shell supersymmetry, when the susy algebra closes off-shell, but we need to introduce auxiliary fields (which have no propagating degrees of freedom) to realize it. In these cases, the actions and susy rules were guessed, though we had a semi-systematic way of doing it.

However, it would be more useful if we had a formalism with *manifest* supersymmetry, i.e. the supersymmetry is built into the formalism, and we do not need to guess or check anything. Such a formalism is known as the *superspace* formalism. Instead of fields which are functions of the (bosonic) position $\phi(x)$ only, we consider a more general space called superspace, involving a fermionic coordinate θ^A as well, besides the usual x^μ, i.e. we consider fields that are functions on superspace, $\phi(x, \theta)$, in such a way that supersymmetry is manifest.

But for a fermionic variable θ, $\{\theta, \theta\} = \theta^2 = 0$, so a general function of a single fermionic variable can be Taylor expanded as $f(\theta) = a + b\theta$ only. Since in four dimensions, θ^A has four components, we can have functions which have at most one of each of the θs, i.e. up to θ^4.

The $\mathcal{N} = 1$ supersymmetry algebra in 2-component spinor notation (3.70) can be represented on superfields $\phi(z^M) = \phi(x, \theta) = \phi(x^\mu, \theta_\alpha, \bar{\theta}^{\dot\alpha})$ in terms of derivative operators by

$$\begin{aligned}
Q_\alpha &= \partial_\alpha - i(\sigma^\mu)_{\alpha\dot\alpha}\bar{\theta}^{\dot\alpha}\partial_\mu, \\
\bar{Q}_{\dot\alpha} &= -\partial_{\dot\alpha} + i(\sigma^\mu)_{\alpha\dot\alpha}\theta^\alpha\partial_\mu, \\
P_\mu &= -i\partial_\mu.
\end{aligned}$$

(3.89)

When checking the algebra, we should note that $\partial_\alpha \equiv \partial/\partial\theta^\alpha$ and $\partial_{\bar\alpha} \equiv \partial/\partial\bar\theta^{\bar\alpha}$ are also fermions, so anticommute (instead of commuting) among themselves and with different θs.

Then by definition, the variation under supersymmetry (with parameters ξ_α, $\bar\xi^{\bar\alpha}$) of the superspace coordinates z^M is $\delta z^M = (\xi Q + Q\bar\xi)z^M$, giving explicitly

$$
\begin{aligned}
x^\mu &\to x'^\mu = x^\mu + i\theta\sigma^\mu\bar\xi - i\xi\sigma^\mu\bar\theta,\\
\theta &\to \theta' = \theta + \xi,\\
\bar\theta &\to \bar\theta' = \bar\theta + \bar\xi.
\end{aligned}
\tag{3.90}
$$

Now we can also define another representation of the supersymmetry algebra, just with the opposite sign in the nontrivial anticommutator,

$$
\begin{aligned}
D_\alpha &= \partial_\alpha + i(\sigma^\mu)_{\alpha\dot\alpha}\bar\theta^{\dot\alpha}\partial_\mu,\\
\bar D_{\dot\alpha} &= -\partial_{\dot\alpha} - i(\sigma^\mu)_{\alpha\dot\alpha}\theta^\alpha\partial_\mu,
\end{aligned}
\tag{3.91}
$$

i.e. giving

$$
\{D_\alpha, \bar D_{\dot\alpha}\} = -2i(\sigma^\mu)_{\alpha\dot\alpha}\partial_\mu,
\tag{3.92}
$$

which then anticommute with the Qs, as we can easily check.

If we write general superfields of some Lorentz spin, we in general obtain reducible representations of supersymmetry. In order to obtain irreducible representations of supersymmetry, we must further constrain the superfields, without breaking the supersymmetry. In order for that to happen, the constraints must anticommute with the supersymmetry generators. Since we already know that the Ds anticommute with the Qs, the constraints that we write will be made up of Ds.

We now consider the simplest superfield, namely a scalar superfield $\Phi(x, \theta)$. To obtain an irreducible representation, we try the simplest possible constraint, namely

$$
\bar D_{\dot\alpha}\Phi = 0,
\tag{3.93}
$$

which is called a *chiral constraint*, thus obtaining a *chiral superfield*, which is in fact an irreducible representation of supersymmetry. Then the complex conjugate constraint, $D_\alpha\Phi = 0$ results in an *antichiral superfield*.

In order to solve the constraint, we find objects which solve it, made up of the $x^\mu, \theta^\alpha, \bar\theta^{\dot\alpha}$. We first construct

$$
y^\mu = x^\mu + i\theta\sigma^\mu\bar\theta,
\tag{3.94}
$$

and then we can check that

$$
\bar D_{\dot\alpha}y^\mu = 0; \quad \bar D_{\dot\alpha}\theta^\beta = 0,
\tag{3.95}
$$

which means that an arbitrary function of y and θ is a chiral superfield. Since $\bar\theta$ does not solve the constraint, we can also reversely say that we can write a chiral superfield as a function of y and θ. We can now write the expansion in θ of the chiral superfield as

$$
\Phi = \Phi(y, \theta) = \phi(y) + \sqrt{2}\theta\psi(y) + \theta\theta F(y),
\tag{3.96}
$$

where by definition we write $\theta^2 = \theta\theta = \theta^\alpha\theta_\alpha$, $\bar\theta^2 = \bar\theta\bar\theta = \bar\theta_{\dot\alpha}\bar\theta^{\dot\alpha}$. Note then that

$$
\epsilon^{\alpha\beta}\frac{\partial}{\partial\theta^\alpha}\frac{\partial}{\partial\theta^\beta}\theta\theta = -4.
\tag{3.97}
$$

Here ϕ is a complex scalar, ψ_α can be extended to a Majorana spinor, and F is a complex auxiliary scalar field. All in all, we see that we obtain the same multiplet as the off-shell WZ multiplet, (ϕ, ψ, F).

The fields of the multiplet are found in terms of covariant derivatives of the superfield as (note that $D_\alpha \theta^\beta = \delta_\alpha^\beta$, $D_\alpha y^\mu = 2i(\sigma^\mu)_{\alpha\dot\alpha}\bar\theta^{\dot\alpha}$)

$$\phi(x) = \Phi|_{\theta=\bar\theta=0},$$

$$\psi_\alpha(x) = \frac{1}{\sqrt{2}}D_\alpha\Phi|_{\theta=\bar\theta=0},$$

$$F(x) = -\frac{D^2\Phi|_{\theta=\bar\theta=0}}{4}. \tag{3.98}$$

Note that $D^2 = \epsilon^{\alpha\beta}D_\alpha D_\beta$ so, as observed above, $D^2\theta^2|_{\theta=\bar\theta=0} = -4$.

We can also expand the ys in Φ in terms of the θs, and obtain

$$\Phi = \phi(x) + \sqrt{2}\theta\psi(x) + \theta^2 F(x)$$
$$+ i\theta\sigma^\mu\bar\theta\partial_\mu\phi(x) - \frac{i}{\sqrt{2}}\theta^2(\partial_\mu\psi\sigma^\mu\bar\theta) + \theta^2\bar\theta^2\partial^2\phi(x). \tag{3.99}$$

We next turn to writing actions in terms of superfields. Note that fermionic integration is the same as the derivative, being *defined* by

$$\int d\theta\, 1 = 0; \qquad \int d\theta\,\theta = 1, \tag{3.100}$$

so we can write $\int d\theta = d/d\theta$. In terms of the 4-dimensional θ and $\bar\theta$, we define

$$d^2\theta = -\frac{1}{4}d\theta^\alpha d\theta^\beta \epsilon_{\alpha\beta}, \tag{3.101}$$

such that $\int d^2\theta\,\theta\theta = 1$.

Then we can also derive the following identities (defining also $d^2\bar\theta = -1/4 d\bar\theta^{\dot\alpha}d\bar\theta^{\dot\beta}\epsilon_{\dot\alpha\dot\beta}$ and $\bar D^2 = \bar D^\alpha \bar D_\alpha$):

$$\int d^4x \int d^2\theta = -\frac{1}{4}\int d^4x D^2|_{\theta\bar\theta=0} = -\frac{1}{4}\int d^4x D^\alpha D_\alpha|_{\theta=\bar\theta=0},$$
$$\int d^4x \int d^2\bar\theta = -\frac{1}{4}\int d^4x \bar D^2|_{\theta\bar\theta=0} = -\frac{1}{4}\int d^4x \bar D^\alpha \bar D_\alpha|_{\theta=\bar\theta=0}. \tag{3.102}$$

We could in principle apply the same procedure for $\int d^4\theta \equiv \int d^2\theta d^2\bar\theta$, but now we have to be careful, since D and $\bar D$ do not anticommute, so their order matters.

We can now write the most general action for a chiral superfield. We can write an arbitrary function K of Φ and $\Phi^\dagger$, which then we must integrate over the whole superspace, i.e. over $\int d^4\theta$, and a function W of Φ only, which will be a function only of y and θ, but not $\bar\theta$. Since we can shift the y integration to x integration only, thus leaving no need for integration over $\bar\theta$, W must be integrated only over $d^2\theta$. We can then write the most general action for a chiral superfield as

$$\mathcal{L} = \int d^4\theta K(\Phi, \Phi^\dagger) + \int d^2\theta W(\Phi) + \int d^2\bar\theta \bar W(\Phi^\dagger). \tag{3.103}$$

Here K is called the *Kähler potential*, giving kinetic terms, and W is called the *superpotential*, giving interactions.

If the supersymmetric theory we have is not fundamental, but is an effective theory embedded into a more fundamental one, i.e. is valid only below a certain UV scale, as for instance in the case of the effective $\mathcal{N} = 1$ supersymmetric low energy theory coming from a string compactification, then K and W can be anything. But if the supersymmetric theory is supposed to be fundamental, being valid until very large energies, then we need to have a renormalizable theory.

For a renormalizable theory, we have

$$
K = \Phi^\dagger \Phi,
$$
$$
W = \lambda\Phi + \frac{m}{2}\Phi^2 + \frac{g}{3}\Phi^3. \tag{3.104}
$$

Indeed, a renormalizable theory needs to have couplings of mass dimension ≥ 0, since if we have a coupling λ of negative mass dimension, we can form an effective dimensionless coupling $\lambda E^\#$ that grows to infinity with the energy, which is related to its power counting nonrenormalizability. We can check that Φ has dimension 1, since its first component is the scalar ϕ, of dimension 1, whereas $\int d\theta$ is like $\partial/\partial\theta$, which has mass dimension $+1/2$ (ψ has dimension 3/2, and Φ has dimension 1, thus θ has dimension$-1/2$). Therefore K has dimension 2, and W has dimension 3. That singles out only the terms we wrote as being renormalizable.

Also, in components, the only renormalizable terms are mass terms, Yukawa terms $\psi\psi\phi$ (of dimension 4, thus with massless coupling), and scalar self-interactions of at most ϕ^4, since $\lambda_n\phi^n$ needs to have dimension 4, giving $[\lambda_n] = 4 - n \geq 0$. We now calculate the action in components, and we obtain only the above terms. We first write for the superpotential terms

$$
\int d^4x \int d^2\theta \left(\lambda\Phi + \frac{m}{2}\Phi^2 + \frac{g}{3}\Phi^3\right) = -\frac{1}{4}\int d^4x D^2 \left(\lambda\Phi + \frac{m}{2}\Phi^2 + \frac{g}{3}\Phi^3\right)|_{\theta=\bar{\theta}=0}, \tag{3.105}
$$

and from

$$
D^2(\Phi^2)|_{\theta=\bar{\theta}=0} = 2(D^2\Phi)|_{\theta=\bar{\theta}=0}\Phi|_{\theta=\bar{\theta}=0} + 2(D^\alpha\Phi)|_{\theta=\bar{\theta}=0}(D_\alpha\Phi)|_{\theta=\bar{\theta}=0},
$$
$$
D^2(\Phi^3)|_{\theta=\bar{\theta}=0} = 3(D^2\Phi)|_{\theta=\bar{\theta}=0}\Phi|_{\theta=\bar{\theta}=0}\Phi|_{\theta=\bar{\theta}=0}
$$
$$
+ 6(D^\alpha\Phi)|_{\theta=\bar{\theta}=0}(D_\alpha\Phi)|_{\theta=\bar{\theta}=0}\Phi|_{\theta=\bar{\theta}=0}, \tag{3.106}
$$

and the definitions (3.98), we obtain

$$
\int d^4x \int d^2\theta\, W(\Phi) = -\frac{1}{4}\int d^4x[2m\bar{\psi}\psi + 4g\phi\bar{\psi}\psi - 4F(\lambda + m\phi + g\phi^2)]. \tag{3.107}
$$

For the Kähler potential term, we have to use the fact (left as Exercise 6) that for a chiral superfield,

$$
\bar{D}^2D^2\Phi = 16\Box\Phi \Rightarrow D^2\bar{D}^2\Phi^\dagger = 16\Box\Phi^\dagger, \tag{3.108}
$$

and to remember the commutation relation (3.92), which implies

$$
D^2\bar{D}^2 = \bar{D}^2D^2 - 8i(\sigma^\mu)_{\alpha\dot{\alpha}}\partial_\mu\bar{D}^{\dot{\alpha}}D^\alpha + 16\Box. \tag{3.109}
$$

Then, for the Kähler potential term we obtain

$$
\frac{1}{16} \int d^4 x D^2 \bar{D}^2 (\Phi^\dagger \Phi)|_{\theta=\bar{\theta}=0} = \frac{1}{16} \int d^4 x [(D^2 \bar{D}^2 \Phi^\dagger)|_{\theta=\bar{\theta}=0} \Phi|_{\theta=\bar{\theta}=0}
$$
$$
+ (\bar{D}^2 \Phi^\dagger)|_{\theta=\bar{\theta}=0} (D^2 \Phi)|_{\theta=\bar{\theta}=0}
$$
$$
- 8i(\sigma^\mu)^{\alpha\dot{\alpha}} (\partial_\mu \bar{D}_{\dot{\alpha}} \Phi^\dagger)|_{\theta=\bar{\theta}=0} (D_\alpha \Phi)|_{\theta=\bar{\theta}=0}], \quad (3.110)
$$

giving finally the kinetic terms

$$
\int d^4 x [\phi^* \Box \phi + F^* F - i(\partial_\mu \bar{\psi}^{\dot{\alpha}})(\sigma^\mu)_{\alpha\dot{\alpha}} \psi^\alpha]. \quad (3.111)
$$

We can eliminate the F auxiliary field, obtaining

$$
F^* = -(\lambda + m\phi + g\phi^2), \quad (3.112)
$$

and replacing it in the action we get the potential term

$$
- \int d^4 x |\lambda + m\phi + g\phi^2|^2. \quad (3.113)
$$

All in all, the action (3.103) for the renormalizable on-shell WZ model in components is

$$
S = \int d^4 x \left[\phi^* \Box \phi - i(\partial_\mu \bar{\psi})(\sigma^\mu)^T \psi - m\bar{\psi}\psi - 2\mathrm{Re}[g\phi \bar{\psi}\psi] - |\lambda + m\phi + g\phi^2|^2 \right].
$$
$$
(3.114)
$$

More generally, an important quantity in supersymmetric theories is the scalar potential. For the case of several WZ multiplets, the potential will be the sum of the *F-terms*, in the same way as above for a single F,

$$
V = \sum_i |F_i|^2, \quad (3.115)
$$

i.e., we replace the auxiliary scalar fields F_i by their equations of motion,

$$
F_i = \frac{\partial W}{\partial \Phi^i} \bigg|_{\Phi_j = \phi_j(\theta=0)}. \quad (3.116)
$$

3.9 The $\mathcal{N} = 1$ Super Yang–Mills (SYM) action

We have seen that, besides the chiral or WZ multiplet, for $\mathcal{N} = 1$ supersymmetry with spins ≤ 1 we can also construct a vector multiplet, composed on-shell of a vector A_μ^a and a spinor λ^a, which must live in the same adjoint representation as the vector, i.e. a $(1, 1/2)$ multiplet. The action for a spinor interacting with a vector is easy to write down:

$$
S = (-2) \int d^4 x \mathrm{Tr} \left[-\frac{1}{4} F_{\mu\nu}^2 - \frac{1}{2} \bar{\lambda} \not{D} \lambda \right], \quad (3.117)
$$

where the -2 comes from the trace normalization, $\mathrm{Tr}\ (T^a T^b) = -1/2\delta^{ab}$. Here $\lambda = \lambda^a T^a$, the same as for the gauge field. Off-shell, a vector has three degrees of freedom, and a

Majorana spinor has four (they reduce off-shell to half, i.e. two), which means that we need to introduce an auxiliary field, also in the adjoint representation of the gauge group, D^a, with $D = D^a T^a$. The off-shell action is then

$$S_{\mathcal{N}=1\ SYM} = (-2) \int d^4 x \mathrm{Tr} \left[-\frac{1}{4} F_{\mu\nu}^2 - \frac{1}{2} \bar{\lambda} \not{D} \lambda + \frac{D^2}{2} \right]. \tag{3.118}$$

We can easily write the supersymmetry rules. The transformation of the boson A_μ^a should be something like $\sim \bar{\epsilon} \lambda$, but one needs to get the indices to match, which can be done by introducing the constant matrix γ_μ. The transformation of λ should be of the type $\sim \partial A \epsilon + D\epsilon$, where $D = 0$ on-shell. However, λ transforms covariantly under a gauge transformation, which means we should get ∂A to transform covariantly also, uniquely fixing it to be $F_{\mu\nu}$. But because of Lorentz invariance, we should actually multiply it with the constant matrix $\gamma^{\mu\nu}$. We find that Lorentz invariance and gauge invariance uniquely fix the structure of the possible terms, though not the coefficients (which have to be found by checking the invariance of the action). For D^a, the transformation law should be $\bar{\epsilon}$ times a possible constant, times the spinor equation of motion, since $D^a = 0$ is also an equation of motion. This again fixes the structure of δD^a. All in all, we obtain

$$\delta A_\mu^a = \bar{\epsilon} \gamma_\mu \lambda^a,$$
$$\delta \lambda^a = \left[-\frac{1}{2} \gamma^{\mu\nu} F_{\mu\nu}^a + i \gamma_5 D^a \right] \epsilon,$$
$$\delta D^a = i \bar{\epsilon} \gamma_5 \not{D} \lambda^a. \tag{3.119}$$

We will not check the invariance of the action under these susy rules, it is left as an exercise (Exercise 10). Instead, we explain the construction of the Yang–Mills multiplet from superspace, where supersymmetry is manifest from the start.

We consider a general (arbitrary function of superspace) *real* scalar superfield V, i.e. one that satisfies $V = V^\dagger$. Therefore we have a $V(x, \theta, \bar{\theta})$ that can be expanded in the fermionic coordinates, the Taylor expansion terminating when we write all the spinors, i.e. at order $\theta^2 \bar{\theta}^2$. The coefficient of the $\theta \bar{\theta}$ term contains a vector, as $\theta \sigma^\mu \bar{\theta} A_\mu$, therefore if we want to describe the gauge multiplet, we need to impose the existence of a gauge invariance on V. This is done in the abelian case as follows:

$$V \to V + i\Lambda - i\Lambda^\dagger, \tag{3.120}$$

where Λ is a *chiral* superfield (so $\Lambda^\dagger$ is antichiral). This allows us to set to zero a Majorana spinor, corresponding to the coefficient of θ and $\bar{\theta}$ in the fermionic Taylor expansion, a complex scalar, corresponding to the coefficient of θ^2 and $\bar{\theta}^2$, and an auxiliary scalar field, corresponding to the 0th order term. All in all, in this *Wess–Zumino gauge*, we are left with the *vector multiplet* V,

$$V = -\theta \sigma^\mu \bar{\theta} A_\mu + i\theta^2 (\bar{\theta} \bar{\lambda}) - i\bar{\theta}^2 (\theta \lambda) + \frac{1}{2} \theta^2 \bar{\theta}^2 D. \tag{3.121}$$

By imposing the Wess–Zumino gauge, we lose manifest (superspace) supersymmetry acting on superfields, though we still have the usual component field supersymmetry defined above. We are also left with a remnant of the super-gauge invariance, the usual gauge invariance $A_\mu \rightarrow A_\mu + \partial_\mu \Lambda(x)$. We can define a super-field strength for the gauge superfield,

$$W_\alpha = -\frac{1}{4}\bar{D}^2 D_\alpha V, \tag{3.122}$$

which is gauge invariant since $\bar{D}_{\dot{\alpha}}\Lambda = 0 = D_\alpha \Lambda^\dagger$. It is a chiral superfield, $\bar{D}_{\dot{\alpha}}W_\alpha = 0$, Majorana $(W_\alpha)^\dagger = W_{\dot{\alpha}}$, and satisfies the reality constraint

$$D^\alpha W_\alpha = D^{\dot{\alpha}} W_{\dot{\alpha}} \leftrightarrow \operatorname{Im}(D^\alpha W_\alpha) = 0. \tag{3.123}$$

In the nonabelian case, the gauge transformation is

$$e^{-2V} \rightarrow e^{i\Lambda^\dagger} e^{-2V} e^{-i\Lambda}, \tag{3.124}$$

where $\Lambda = \Lambda^a T_a$, $V = V^a T_a$ and the super-field strength is

$$W_\alpha = \frac{1}{8}\bar{D}^2(e^{2V}D_\alpha e^{-2V}) = W_\alpha^a T_a. \tag{3.125}$$

It transforms covariantly, i.e. as

$$W_\alpha \rightarrow e^{i\Lambda} W_\alpha e^{-i\Lambda}. \tag{3.126}$$

In terms of W_α, the $\mathcal{N} = 1$ SYM action can be written as

$$S_{\mathcal{N}=1\,SYM} = -\frac{(-2)}{4}\int d^4x d^2\theta \operatorname{Tr}\, W_\alpha W^\alpha + h.c. \tag{3.127}$$

When $\mathcal{N} = 1$ SYM is coupled to chiral multiplets (and not only in that case), an important quantity is the scalar potential. We saw in the previous subsection that for just the chiral multiplets it was given by the F-terms, and we saw above that for $\mathcal{N} = 1$ SYM a priori there was a potential $g^2 D^2$, where D is the auxiliary field in the multiplet, just that for the pure SYM, $D = 0$ on-shell. When we couple the SYM to chiral multiplets Φ_i, we obtain a nontrivial *D-term* D^a given by

$$D^a = \Phi^\dagger T^a \Phi \Big|_{\Phi_i = \phi_i(\theta=0)} = \phi^{\dagger i}(T^a)_{ij}\phi^j, \tag{3.128}$$

where $(T^a)_{ij}$ is the generator in the fundamental representation. The scalar potential is then the sum of F-terms and D-terms,

$$V = \sum_i |F_i|^2 + g^2 D^a D^a, \tag{3.129}$$

where we replace F_i and D^a with their on-shell values.

In an abelian theory, we can add to the action a supersymmetric term called the *Fayet–Iliopoulos term* (FI term),

$$\mathcal{L}_{FI} = \int d^2\theta d^2\bar{\theta}\xi^a V^a, \tag{3.130}$$

which just shifts the D-term,

$$D^a = \xi^a + \phi^{\dagger i}(T^a)_{ij}\phi^j. \tag{3.131}$$

3.10 The $\mathcal{N} = 4$ Super Yang–Mills (SYM) action

We are finally ready to write down the $\mathcal{N} = 4$ SYM action, which will be the main field theory to be discussed via the AdS/CFT correspondence. To do so, it is useful to write down the action via "dimensional reduction" from ten dimensions. We discuss in a later chapter all the details of dimensional reduction, but we can give at this point an abbreviated version. In its simplest form, we consider a theory in a higher dimension, and then restrict the dependence of the fields to be only on four dimensions. We must also decompose the ten dimensional fields according to 4-dimensional spin.

In ten dimensions, the minimal spinor is Majorana–Weyl, as we explained. That means that from the $2^{[10/2]} = 32$ complex components of a Dirac spinor, only 16 real components remain. On-shell we lose half the components, for eight real fermionic degrees of freedom. This matches with the $10 - 2 = 8$ bosonic degrees of freedom of an on-shell vector. Therefore the fields of $\mathcal{N} = 1$ Super Yang–Mills in ten dimensions are the vector A_M, $M = 0, \ldots, 9$ and the spinor Ψ_Π, $\Pi = 1, \ldots, 16$, satisfying

$$\Gamma_{11}\Psi = \Psi, \quad \bar{\Psi} = \Psi^T C_{10}. \tag{3.132}$$

The action is the same one as in four dimensions, namely

$$S_{10d, \mathcal{N}=1 SYM} = (-2) \int d^{10}x \, \mathrm{Tr} \left[-\frac{1}{4} F^{MN} F_{MN} - \frac{1}{2} \bar{\lambda} \Gamma^M D_M \lambda \right]. \tag{3.133}$$

As we mentioned, the first step in dimensional reduction is to restrict the dependence of fields to only four dimensions, in particular replacing $d^{10}x$ with d^4x. The second step is the decomposition of the fields according to their 4-dimensional Lorentz spin. We first decompose the gamma matrices and the C-matrix as

$$\Gamma_M = (\gamma_\mu \otimes \mathbb{1}, \gamma_5 \otimes \gamma_m), \text{ i.e. } \Gamma_\mu = \gamma_\mu \otimes \mathbb{1}; \quad \Gamma_m = \gamma_5 \otimes \gamma_m; \quad \Gamma_{11} = \gamma_5 \otimes \gamma_7,$$
$$C_{10} = C_4 \otimes C_6. \tag{3.134}$$

We can check that this decomposition is consistent with the properties of the gamma matrices and C-matrices. Since in ten Minkowski dimensions and in four Minkowski dimensions $C\Gamma$ is symmetric, as we saw in Section 3.3, and from the above we have $C_{10}\Gamma^\mu = C_4\gamma^\mu \otimes C_6$, C_6 must be symmetric, which restricts it to be C_{6-}. Then, as $C_{10}^T = C_{4-}^T \otimes C_{6-}^T = -C_{10}$, C_{10} must be C_{10-}. The 10-dimensional Majorana conjugate in 4-dimensional notation is now

$$\bar{\psi}_M \equiv \psi^T C_4 \otimes C_6, \tag{3.135}$$

and should equal the 10-dimensional Dirac conjugate, and then the 10-dimensional Weyl condition restricts the 10-dimensional spinors to decompose into four 4-dimensional spinors, $\Psi_\Pi = \psi_{\alpha i}$, $i = 1, .., 4$. The 10-dimensional vector A_M decomposes into a 4-dimensional vector A_μ and six scalars ϕ_m, $m = 1, \ldots, 6$. Moreover, the scalars can be reorganised:

$$A_M = (A_\mu, \phi_m), \quad \phi_{[ij]} \equiv \phi_m \tilde{\gamma}^m_{[ij]}, \tag{3.136}$$

where the Clebsh-Gordan coefficients for the transition from $SO(6)$ indices m to $SU(4)$ indices i, $i = 1, \ldots, 4$ (where the Weyl spinor representation of $SO(6)$, with $+1$ eigenvalue for γ_7, $\gamma_7 \psi = +\psi$, is identified with the fundamental representation of $SU(4)$) with antisymmetric six representation $[ij]$ are normalized,[4]

$$\tilde{\gamma}^m_{[ij]} \equiv \frac{1}{2}(C_6 \gamma_m \gamma_7)_{[ij]}; \quad \tilde{\gamma}^m_{[ij]} \tilde{\gamma}_n^{[ij]} = \delta_n^m. \tag{3.137}$$

The fields $\phi_{[ij]}$ have complex entries but satisfy a reality condition,

$$\phi_{ij}^\dagger = \phi^{ij} \equiv \frac{1}{2}\epsilon^{ijkl}\phi_{kl}. \tag{3.138}$$

Substituting the reduction ansatz in the 10-dimensional action, we obtain

$$\begin{aligned}
S_{4d, \mathcal{N}=4 \text{ SYM}} &= (-2)\int d^4x \, \text{Tr} \, \left[-\frac{1}{4}F_{\mu\nu}^2 - \frac{1}{2}\bar{\psi}_i \slashed{D} \psi^i - \frac{1}{2}D_\mu \phi_{ij} D^\mu \phi^{ij} \right. \\
&\qquad\qquad\qquad \left. - g\bar{\psi}^i[\phi_{ij}, \psi^j] - \frac{g^2}{4}[\phi_{ij}, \phi_{kl}][\phi^{ij}, \phi^{kl}] \right] \\
&= (-2)\int d^4x \, \text{Tr} \, \left[-\frac{1}{4}F_{\mu\nu}^2 - \frac{1}{2}\bar{\psi}_i \slashed{D} \psi^i - \frac{1}{2}D_\mu \phi_m D^\mu \phi^m \right. \\
&\qquad\qquad\qquad \left. - g\bar{\psi}^i[\phi_n, \psi^j]\tilde{\gamma}^n_{[ij]} - \frac{g^2}{4}[\phi_m, \phi_n][\phi^m, \phi^n] \right], \quad (3.139)
\end{aligned}$$

where $D_\mu = \partial_\mu + g[A_\mu, \;]$.

The action can be rewritten in $\mathcal{N} = 1$ superfield form. The multiplet is composed of three chiral superfields Φ^l, $l = 1, 2, 3$, and a vector multiplet W_α. The action has the unique renormalizable Kähler potential $K = \Phi_l^\dagger \Phi_l$, and superpotential

$$W = \text{Tr} \, (\Phi_1[\Phi_2, \Phi_3]) = \frac{1}{3}\epsilon_{lmn}\text{Tr} \, [\Phi^l \Phi^m \Phi^n]; \tag{3.140}$$

Both in the original 10-dimensional action and in the final 4-dimensional version, only this on-shell version is known, there is no fully off-shell formulation with Lorentz invariance.

Some details on the reduction are as follows. The field strength F_{MN} splits as $(F_{\mu\nu}, D_\mu\phi_m, g[\phi_m, \phi_n])$ for $(F_{\mu\nu}, F_{\mu m}, F_{mn})$, which immediately gives the bosonic terms in the 4-dimensional action from the $F_{MN}F^{MN}$ term. The covariant derivative of the spinor, $D_M\lambda$, splits into $D_\mu\psi_i$ and $g[\phi_m, \psi_i]$, which gives rise to the terms involving fermions.

The 10-dimensional supersymmetry transformations are, as in four dimensions,

$$\begin{aligned}
\delta A_M^a &= \bar{\epsilon}\Gamma_M \lambda^a, \\
\delta \lambda^a &= -\frac{1}{2}\Gamma^{MN}F_{MN}^a \epsilon.
\end{aligned} \tag{3.141}$$

---

[4] Note that $\tilde{\gamma}_n^{[ij]} = (C_6^{-1} \tilde{\gamma}_n C_6^{-1})^{[ij]}$ and C_6 is symmetric.

Plugging in the reduction ansatz, we obtain the 4-dimensional transformation laws,

$$\delta A_\mu^a = \bar{\epsilon}_i \gamma_\mu \psi^{ai}$$
$$\delta \phi_a^{[ij]} = 2\bar{\epsilon}^{[i} \psi^{j]a}$$
$$\delta \lambda^{ai} = -\frac{\gamma^{\mu\nu}}{2} F_{\mu\nu}^a \epsilon^i - 2\gamma^\mu D_\mu \phi^{a,[ij]} \epsilon_j + 2g f^a{}_{bc} (\phi^b \phi^c)^{[ij]} \epsilon_j;$$
$$(\phi^a \phi^b)^i{}_j \equiv \phi^{a,i}{}_k \phi^{b,k}{}_j.$$

$$(3.142)$$

The $\mathcal{N} = 4$ Super Yang–Mills action has, as we can see, an $SO(6) = SU(4)$ global symmetry, which is an R-symmetry. In the above construction, it appears as a remnant of the six reduced dimensions, with the scalars organized as ϕ_m in the fundamental representation of $SO(6)$, or as we wrote them, as $\phi_{[ij]}$, in the antisymmetric representation of $SU(4)$. The spinors ψ^i are in the spinor representation of $SO(6)$, or the fundamental representation of $SU(4)$. The $\mathcal{N} = 4$ supersymmetry multiplet $(A_\mu^a, \psi^{ai}, \phi_{[ij]})$ splits into an $\mathcal{N} = 2$ vector multiplet and an $\mathcal{N} = 2$ hypermultiplet, or one $\mathcal{N} = 1$ vector multiplet (A_μ, ψ_4) and three $\mathcal{N} = 1$ chiral multiplets $(\psi_q, \phi_q), q = 1, 2, 3$.

Important concepts to remember

- A graded Lie algebra can contain the Poincaré algebra, internal algebra and supersymmetry.
- The supersymmetry Q_α relates bosons and fermions.
- If the on-shell numbers of degrees of freedom of bosons and fermions match we have on-shell supersymmetry, if the off-shell numbers match we have off-shell supersymmetry.
- N-extended supersymmetry can have central charges, and an $\subseteq U(N)$ global symmetry.
- For off-shell supersymmetry, the supersymmetry algebra must be realized on the fields.
- The prototype for all (linear) supersymmetry is the 2-dimensional Wess–Zumino model, with $\delta\phi = \bar{\epsilon}\psi, \delta\psi = \not{\partial}\phi\epsilon$. Off-shell, there is a real auxiliary scalar as well.
- The Wess–Zumino model in four dimensions has a fermion and a complex scalar on-shell. Off-shell there is also an auxiliary complex scalar.
- The susy algebra can be written in terms of $a, a^\dagger$ operators and represented in terms of $a^\dagger$s acting on a vacuum $|\Omega\rangle$.
- We have a BPS bound, $M \geq |Z|$.
- BPS-saturated representations are short, the same as massless ones.
- Superspace is made up of the usual space x^μ and a spinorial coordinate θ^α.
- Superfields are fields in superspace and can be expanded up to linear order in θ components, $f(\theta) = a + b\theta$, since $\theta^2 = 0$.
- Irreducible representations of susy are obtained by imposing constraints in terms of the covariant derivatives D on superfields, since the Ds commute with the susy generators' Qs, and thus preserve susy.
- A chiral superfield is an arbitrary function $\Phi(y, \theta)$ of $y^\mu = x^\mu + i\theta\sigma^\mu\bar{\theta}$ and θ.
- Fermionic integrals and derivatives are the same.

- The action for a chiral superfield has a function $K(\Phi, \bar{\Phi})$ called the Kähler potential, giving kinetic terms, and a function $W(\Phi)$, called the superpotential, giving potentials and Yukawas.
- To derive the component Lagrangean from the superfield one, we can either do the full θ expansion, or (simpler) use the fact that $\int d^4x \int d^2\theta = -1/4 \int d^4x D^2|_{\theta=0}$ (and its c.c.) and the definitions $\phi(x) = \Phi|_{\theta=0}$, $\psi(x) = 1/\sqrt{2}D_\alpha \Phi|_{\theta=0}$, etc., but we need to be careful with the Kähler potential.
- The on-shell vector multiplet has a gauge field and a fermion.
- The $\mathcal{N} = 2$ vector multiplet is made up of an $\mathcal{N} = 1$ vector multiplet and an $\mathcal{N} = 1$ chiral multiplet; an $\mathcal{N} = 2$ hypermultiplet is made up of two $\mathcal{N} = 1$ chiral multiplets.
- The $\mathcal{N} = 2$ vector and hypermultiplets together make up the unique $\mathcal{N} = 4$ supermultiplet of spin ≤ 1, the vector.
- The $\mathcal{N} = 4$ supersymmetric vector multiplet ($\mathcal{N} = 4$ SYM) has one gauge field, four fermions and six scalars, all in the adjoint of the gauge field.

References and further reading

For a very basic introduction to supersymmetry, see the introductory parts of [7] and [8]. Good introductory books are West [9] and Wess and Bagger [10]. An advanced book that is more complex but contains a lot of useful information is [11]. An advanced student might want to try also volume 3 of Weinberg [2], which is more recent than the above, but it is more demanding to read and mostly uses approaches seldom adopted in string theory. A book with a modern approach but emphasizing phenomenology is [12]. For a good treatment of spinors in various dimensions, and spinor identities (symmetries and Fierz rearrangements) see [13]. For an earlier but less detailed acount, see [14].

Exercises

1. Prove that the matrix

$$C_{AB} = \begin{pmatrix} \epsilon^{\alpha\beta} & 0 \\ 0 & \epsilon_{\dot{\alpha}\dot{\beta}} \end{pmatrix}; \epsilon^{\alpha\beta} = \epsilon^{\dot{\alpha}\dot{\beta}} = \begin{pmatrix} 0 & 1 \\ -1 & 0 \end{pmatrix} \tag{3.143}$$

   is a representation of the 4-dimensional C matrix, i.e. $C^T = -C, C\gamma^\mu C^{-1} = -(\gamma^\mu)^T$, if γ^μ is represented by

$$\gamma^\mu = \begin{pmatrix} 0 & \sigma^\mu \\ \bar{\sigma}^\mu & 0 \end{pmatrix}; \quad (\sigma^\mu)_{\alpha\dot{\alpha}} = (1, \vec{\sigma})_{\alpha\dot{\alpha}}; \quad (\bar{\sigma}^\mu)^{\alpha\dot{\alpha}} = (1, -\vec{\sigma})^{\alpha\dot{\alpha}}. \tag{3.144}$$

2. Show that the susy variation of the 4-dimensional on-shell Wess–Zumino model is zero, paralleling the 2-dimensional WZ model.
3. Using the general form of the Fierz identities, check that in four dimensions we have

$$(\bar{\lambda}^a \gamma^\mu \lambda^c)(\bar{\epsilon}\gamma_\mu \lambda^b) f_{abc} = 0, \tag{3.145}$$

   using the fact that f_{abc} is totally antisymmetric, and the identities $\gamma_\mu \gamma_\rho \gamma^\mu = -2\gamma_\rho$, $\gamma_\mu \gamma_{\rho\sigma} \gamma^\mu = 0$ (prove these as well).

4. Prove that if ϵ, χ are 4-dimensional Majorana spinors, we have

$$\bar{\epsilon}\gamma_\mu\gamma_5\chi = +\bar{\chi}\gamma_\mu\gamma_5\epsilon. \tag{3.146}$$

5. Prove that, for

$$S = -\frac{1}{2}\int d^4x[(\partial_\mu\phi)^2 + \bar{\psi}\not{\partial}\psi], \tag{3.147}$$

we have

$$[\delta_{\epsilon_1}, \delta_{\epsilon_2}]\phi = 2\bar{\epsilon}_2\not{\partial}\epsilon_1\phi,$$
$$[\delta_{\epsilon_1}, \delta_{\epsilon_2}]\psi = 2(\bar{\epsilon}_2\gamma^\rho\epsilon_1)\partial_\rho\psi - (\bar{\epsilon}_2\gamma^\rho\epsilon_1)\gamma_\rho\not{\partial}\psi. \tag{3.148}$$

6. For the off-shell WZ model in two dimensions,

$$S = -\frac{1}{2}\int d^2x[(\partial_\mu\phi)^2 + \bar{\psi}\not{\partial}\psi - F^2]. \tag{3.149}$$

Check that

$$[\delta_{\epsilon_1}, \delta_{\epsilon_2}]F = 2(\bar{\epsilon}_2\gamma^\mu\epsilon_1)\partial_\mu F. \tag{3.150}$$

7. Check explicitly (without the use of y^μ) that $\bar{D}_{\dot{\alpha}}\Phi = 0$, where

$$\Phi = \phi(x) + \sqrt{2}\psi(x) + \theta^2 F(x) + i\theta\sigma^\mu\bar{\theta}\partial_\mu\phi(x) - \frac{i}{\sqrt{2}}\theta^2(\partial_\mu\psi\sigma^\mu\bar{\theta}) + \theta^2\bar{\theta}^2\partial^2\phi(x). \tag{3.151}$$

8. Prove that for a chiral super-field

$$D^2\bar{D}^2\Phi = 16\Box\Phi. \tag{3.152}$$

9. Consider the Lagrangean

$$\mathcal{L} = \int d^2\theta d^2\bar{\theta}\,\Phi_i^\dagger\Phi_i + \left(\int d^2\theta W(\Phi_i) + h.c.\right). \tag{3.153}$$

Do the θ integrals to obtain in components

$$\mathcal{L} = (\partial_\mu\phi_i)^\dagger\partial^\mu\phi_i - i\bar{\psi}_i\bar{\sigma}^\mu\partial_\mu\psi_i + F_i^\dagger F_i$$
$$+ \frac{\partial W}{\partial\phi_i}F_i + \frac{\partial\bar{W}}{\partial\phi_i^\dagger}F_i^\dagger - \frac{1}{2}\frac{\partial^2 W}{\partial\phi_i\partial\phi_j}\psi_i\psi_j - \frac{1}{2}\frac{\partial^2\bar{W}}{\partial\phi_i^\dagger\partial\phi_j^\dagger}\bar{\psi}_i\bar{\psi}_j. \tag{3.154}$$

10. Check the invariance of the $\mathcal{N} = 1$ off-shell SYM action

$$S = \int d^4x\left[-\frac{1}{4}(F_{\mu\nu}^a)^2 - \frac{1}{2}\bar{\psi}^a\not{D}\psi_a + \frac{1}{2}D_a^2\right] \tag{3.155}$$

under the supersymmetry transformations

$$\delta A_\mu^a = \bar{\epsilon}\gamma_\mu\psi^a; \quad \delta\psi^a = \left(-\frac{1}{2}\gamma^{\mu\nu}F_{\mu\nu}^a + i\gamma_5 D^a\right)\epsilon; \quad \delta D^a = i\bar{\epsilon}\gamma_5\not{D}\psi^a. \tag{3.156}$$

11. Calculate the number of off-shell degrees of freedom of the on-shell $\mathcal{N} = 4$ SYM action. Propose a set of bosonic + fermionic auxiliary fields that could make the number of degrees of freedom match. Are they likely to give an off-shell formulation, and why?

Basics of supergravity

In this chapter we examine supergravity, which is a supersymmetric theory of gravity. However, since the construction involves fermions, we have to learn how to couple fermions to gravity. Also, the theory is constructed as a gauge theory of (local) supersymmetry. Therefore we need first to describe another formalism for general relativity.

4.1 The vielbein and spin connection formulation of general relativity

In Chapter 2, we saw that gravity is defined by the metric $g_{\mu\nu}$, which in turn defines the Christoffel symbol $\Gamma^\mu{}_{\nu\rho}(g)$, which is like a gauge field of gravity, with the Riemann tensor $R^\mu{}_{\nu\rho\sigma}(\Gamma)$ playing the role of its field strength.

But there is a formulation that makes the gauge theory analogy more manifest, namely in terms of the "vielbein" e^a_μ and the "spin connection" ω^{ab}_μ. The word "vielbein" comes from the german "viel" which means many and "bein" which means leg. It was introduced in four dimensions, where it is known as "vierbein", since "vier" means four. In various dimensions one uses "einbein, zweibein, dreibein, ..." $(1, 2, 3 = \text{ein, zwei, drei})$, or generically "vielbein", as we do here.

Any curved space is locally flat, if we look at a scale much smaller than the scale of the curvature. That means that locally, we have the Lorentz invariance of special relativity. The vielbein is an object that makes that local Lorentz invariance manifest. It is a sort of square root of the metric, i.e.

$$g_{\mu\nu}(x) = e^a_\mu(x)e^b_\nu(x)\eta_{ab} , \qquad (4.1)$$

so in $e^a_\mu(x)$, μ is a "curved" index, acted upon by a general coordinate transformation (so that e^a_μ is a covariant vector of general coordinate transformations, like a gauge field), and a is a newly introduced "flat" index, acted upon by a local Lorentz gauge invariance. That is, around each point we define a small flat neighborhood called "tangent space," and a is a tensor index living in that local Minkowski space, acted upon by Lorentz transformations. The Lorentz transformation is *local*, since the tangent space on which it acts changes at each point on the curved manifold, i.e. it is local.

Note that the description in terms of $g_{\mu\nu}$ or e^a_μ is equivalent, since both contain the same number of degrees of freedom. At first sight, we might think that while $g_{\mu\nu}$ has $d(d+1)/2$ components (a symmetric matrix), e^a_μ has d^2; but on e^a_μ we act with another local symmetry not present in the metric, local Lorentz invariance,

$$e^a_\mu \to \Lambda^a{}_b e^b_\mu \,, \tag{4.2}$$

so we can set $d(d-1)/2$ components to zero using this (the number of components of the antisymmetric matrix $\Lambda^a{}_b$), so it has in fact also $d^2 - d(d-1)/2 = d(d+1)/2$ components.

On both the metric and the vielbein we also have general coordinate transformations. As we saw in Section 2.2, an infinitesimal general coordinate transformation ("Einstein" transformation) $\delta x^\mu = \xi^\mu$ acting on the metric gives

$$\delta_\xi g_{\mu\nu}(x) = (\xi^\rho \partial_\rho) g_{\mu\nu} + (\partial_\mu \xi^\rho) g_{\rho\nu} + (\partial_\nu \xi^\rho) g_{\rho\nu} \,, \tag{4.3}$$

and the first term corresponds to a translation, thus the general coordinate transformations are the general relativity version, i.e. the *local* version, of the (global) P_μ translations in special relativity (in special relativity we have a global parameter ξ^μ, but now we have a local $\xi^\mu(x)$).

On the vielbein e^a_μ, the infinitesimal coordinate transformation gives

$$\delta_\xi e^a_\mu(x) = (\xi^\rho \partial_\rho) e^a_\mu + (\partial_\mu \xi^\rho) e^a_\rho \,, \tag{4.4}$$

thus it acts only on the curved index μ. On the other hand, the local Lorentz transformation

$$\delta_{\text{l.L.}} e^a_\mu(x) = \lambda^a{}_b(x) e^b_\mu(x) \tag{4.5}$$

acts in the usual manner, except now the parameter is local.

Thus the vielbein is like a sort of gauge field, with one covariant vector index and a gauge group index, though not quite, since the group index a is in the fundamental instead of the adjoint of the Lorentz group.

But there is one more "gauge field," ω^{ab}_μ, the "spin connection," which is defined as the "connection" (mathematical name for a gauge field) for the action of the Lorentz group on spinors. Now $[ab]$ is an index in the adjoint of the Lorentz group $SO(1, d-1)$ (the antisymmetric representation), and at least the covariant derivative on the spinors will have the standard form in a gauge theory.

That is, while we have already defined the action of the covariant derivative on tensors (bosons), we have yet to define it on spinors (fermions). The curved space covariant derivative acting on spinors acts as the gauge field covariant derivative on a spinor, by (here $1/4\Gamma_{ab} \equiv 1/2[\Gamma_a, \Gamma_b]$ is the generator of the Lorentz group in the spinor representation, so we have the usual formula $D_\mu \phi = \partial_\mu \phi + A^a_\mu T_a \phi$)

$$D_\mu \psi = \partial_\mu \psi + \frac{1}{4} \omega^{ab}_\mu \Gamma_{ab} \psi. \tag{4.6}$$

This definition means that $D_\mu \psi$ is the object that transforms as a tensor under general coordinate transformations and it implies that ω^{ab}_μ acts as a gauge field on any local Lorentz index a.

But now we seem to have too many degrees of freedom for gravity. We have seen that the vielbein alone has the same degrees of freedom as the metric, so for a formulation of gravity completely equivalent to Einstein's we need to fix ω in terms of e. If there are no *dynamical* fermions (i.e. fermions that have a kinetic term in the action) then this constraint is given by $\omega^{ab}_\mu = \omega^{ab}_\mu(e)$, a fixed function defined through the "vielbein postulate" or "no torsion constraint" (the antisymmetrization below is with "strength one," which means that when multiplying with an antisymmetric tensor, we can drop the antisymmetrization, as we always do unless noted):

$$T^a_{[\mu\nu]} \equiv 2D_{[\mu}e^a_{\nu]} = 2\partial_{[\mu}e^a_{\nu]} + 2\omega^{ab}_{[\mu}e^b_{\nu]} = 0. \tag{4.7}$$

Note that we can also start by imposing the alternative condition

$$D_\mu e^a_\nu \equiv \partial_\mu e^a_\nu + \omega^{ab}_\mu e^b_\nu - \Gamma^\rho{}_{\mu\nu}e^a_\rho = 0, \tag{4.8}$$

and antisymmetrize it to obtain the same as the one above, since $\Gamma^\rho{}_{\mu\nu}$ is symmetric. This latter condition is also sometimes called the vielbein postulate, though it fixes not only $\omega(e)$ but also the Christoffel symbol in terms of e, $\Gamma(e)$. We will also explain shortly that in the presence of fermions, the vielbein postulate is modified, and $\omega = \omega(e, \psi)$.

The solution to the vielbein postulate is

$$\omega^{ab}_\mu(e) = \frac{1}{2}e^{a\nu}(\partial_\mu e^b_\nu - \partial_\nu e^b_\mu) - \frac{1}{2}e^{b\nu}(\partial_\mu e^a_\nu - \partial_\nu e^a_\mu) - \frac{1}{2}e^{a\rho}e^{b\sigma}(\partial_\rho e_{c\sigma} - \partial_\sigma e_{c\rho})e^c_\mu, \tag{4.9}$$

which is left for the reader to verify as Exercise 1.

Here T^a is called the "torsion", and as we can see it is a sort of field strength of e^a_μ, and the vielbein postulate says that the torsion (field strength of vielbein) is zero.

But we can also construct an object that is a field strength of ω^{ab}_μ,

$$R^{ab}_{\mu\nu}(\omega) = \partial_\mu \omega^{ab}_\nu - \partial_\nu \omega^{ab}_\mu + \omega^{ab}_\mu \omega^{bc}_\nu - \omega^{ab}_\nu \omega^{bc}_\mu, \tag{4.10}$$

and this time the definition is exactly the definition of the field strength of a gauge field of the local Lorentz group $SO(1, d - 1)$ (though there are still subtleties in trying to make the identification of ω^{ab}_μ with a gauge field of the Lorentz group). So, unlike the case of the formula for the Riemann tensor as a function of Γ, where gauge and spatial indices were of the same type, we now have a clear definition.

From the fact that the two objects ($R(\omega)$ and $R(\Gamma)$) have formally the same formula, we can guess the relation between them, and we can check that this guess is actually correct. We have

$$R^{ab}_{\rho\sigma}(\omega(e)) = e^a_\mu e^{-1,\nu b} R^\mu{}_{\nu\rho\sigma}(\Gamma(e)). \tag{4.11}$$

That means that the $R^{ab}_{\mu\nu}$ is actually just the Riemann tensor with two indices flattened (turned from curved to flat using the vielbein). That in turn implies that we can define the Ricci scalar in terms of $R^{ab}_{\mu\nu}$ as

$$R = R^{ab}_{\mu\nu}e^{-1\,\mu}_a e^{-1\,\nu}_b, \tag{4.12}$$

and since as matrices, $g = e\eta e$, we have $-\det g = (\det e)^2$, so the Einstein–Hilbert action in d dimensions is

$$S_{\mathrm{EH}} = \frac{1}{16\pi G_N} \int d^d x (\det e) R^{ab}_{\mu\nu}(\omega(e)) e^{-1,\mu}_a e^{-1,\nu}_b. \tag{4.13}$$

The formulation just described of gravity in terms of e and ω is the *second order formulation*, so called because ω is not independent, but is a function of e. In general, we call first order a formulation involving an auxiliary field, which usually means that the action becomes first order in derivatives, or in propagating fields (a standard example would be going from the usual Maxwell action $-\int(\partial_{[\mu}A_{\nu]})^2$ to a new form for it $\int[-2F^{\mu\nu}(\partial_{[\mu}A_{\nu]}) + F^2_{\mu\nu}]$ where $F_{\mu\nu}$ is an independent auxiliary field). The second order formulation is obtained by eliminating the auxiliary field, and is usually second order in derivatives and/or propagating fields.

But notice that if we make ω an independent variable in the above Einstein–Hilbert action, the ω equation of motion gives exactly $T^a_{\mu\nu} = 0$, i.e. the "vielbein postulate" that we needed to postulate before. Thus we might as well make ω independent without changing the classical theory (only possibly the quantum version). This is then the *first order formulation* of gravity, also known as Palatini formalism, in terms of independent fields $(e^a_\mu, \omega^{ab}_\mu)$.

To prove that the equation of motion for ω^{ab}_μ is $T^a = 0$, we first use the relation (valid in four dimensions only, though with an obvious generalization to higher dimensions)

$$(\det e)e^{-1\,[\mu}_a e^{-1\,\nu]}_b = \frac{1}{4}\epsilon^{\mu\nu\rho\sigma}\epsilon_{abcd}e^c_\rho e^d_\sigma, \tag{4.14}$$

(which follows from the definition of the determinant, $\det e^a_\mu = \frac{1}{4!}\epsilon_{abcd}\epsilon^{\mu\nu\rho\sigma}e^a_\mu e^b_\nu e^c_\rho e^d_\sigma$), to write the Einstein–Hilbert action in four dimensions as

$$S_{\text{EH}} = \frac{1}{16\pi G_N}\frac{1}{4}\int d^4x\,\epsilon^{\mu\nu\rho\sigma}\epsilon_{abcd}R^{ab}_{\mu\nu}(\omega)e^c_\rho e^d_\sigma \left(= \frac{1}{16\pi G_N}\int \epsilon_{abcd}R^{ab}(\omega) \wedge e^c \wedge e^d\right). \tag{4.15}$$

This action's variation with respect to ω gives

$$\epsilon_{abcd}\epsilon^{\mu\nu\rho\sigma}(D_\nu e^c_\rho)e^d_\sigma = 0, \tag{4.16}$$

which implies

$$T^a_{[\mu\nu]} \equiv 2D_{[\mu}e^a_{\nu]} = 0. \tag{4.17}$$

We now also note that if there are fundamental fermions, i.e. fermions present in the action of the theory, their kinetic term will contain the covariant derivative, via $\bar{\psi}\!\!\not{\!D}\psi$, hence in the equation of motion of ω we will get new terms, involving fermions. Therefore we will have $\omega = \omega(e) + \psi\psi$ terms, and we get a nonzero fermionic torsion. The function will still be a fixed function, but we see that in that case it is more useful to start with the first order formulation, with ω coupled to fermions through the covariant derivative, and then find the equation of motion for ω in order to find $\omega(e, \psi)$. After that we can move to the second order formulation by defining $\omega = \omega(e, \psi)$. It would be mistaken to start with a "second order formulation" with $\omega = \omega(e)$ in that case, since it would lead to contradictions, as we can easily understand from the first order formulation.

4.2 Counting degrees of freedom of on-shell and off-shell fields

We saw that for supersymmetric theories, we need to match the bosonic and fermionic degrees of freedom. Therefore, before continuing, we study systematically how to count degrees of freedom on-shell and off-shell.

Off-shell degrees of freedom

- Scalar, either propagating (with a kinetic action with derivatives), or auxiliary (with algebraic equation of motion), it is always one degree of freedom.
- Gauge field A_μ, with transformation law $\delta A_\mu = D_\mu \lambda$. A_μ has d components, but we can use the gauge transformation, with one parameter $\lambda(x)$, to fix one component of A_μ to whatever we like, therefore we have $d - 1$ independent degrees of freedom ("dofs").
- Graviton. In the $g_{\mu\nu}$ formulation, we have a symmetric matrix, with $d(d + 1)/2$ components, but we have a "gauge invariance" = general coordinate tranformations, with parameter $\xi^\mu(x)$, which can be used to fix d components, therefore we have $d(d + 1)/2 - d = d(d - 1)/2$ independent degrees of freedom. Equivalently, in the vielbein formulation, e_μ^a has d^2 components. We subtract the "gauge invariance" of general coordinate transformations with $\xi^\mu(x)$, but now we also have local Lorentz invariance with parameter $\lambda^{ab}(x)$, giving $d^2 - d - d(d - 1)/2 = d(d - 1)/2$ independent degrees of freedom again.
- For a spinor of spin 1/2, ψ^α, we saw that in the Majorana spinor case we have $n \equiv 2^{[d/2]}$ real components.
- For a *gravitino* ψ_μ^α, which is a vector-spinor field describing the propagation of a mode of spin 3/2, we have nd components. But now again we have a "gauge invariance" as we will explain, namely local supersymmetry, also called *supergravity*, acting by $\delta\psi = D_\mu \epsilon$, so again we can use it to fix n components corresponding to the arbitrary spinor ϵ. That means that we have $n(d - 1)$ independent degrees of freedom.
- Antisymmetric tensor $A_{\mu_1...\mu_r}$, with field strength $F_{\mu_1...\mu_{r+1}} = (r+1)! \partial_{[\mu_1} A_{\mu_2...\mu_{r+1}]}$ and gauge invariance $\delta A_{\mu_1...\mu_r} = \partial_{[\mu_1} \Lambda_{\mu_2...\mu_r]}$, $\Lambda_{\mu_1...\mu_{r-1}} \neq \partial_{[\mu_1} \lambda_{\mu_2...\mu_{r-1}]}$. By subtracting the gauge invariances, we obtain

$$\binom{d}{r} - \binom{d-1}{r-1} = \frac{(d-1)\ldots(d-r)}{1 \cdot 2 \cdot \ldots r} = \binom{d-1}{r}, \qquad (4.18)$$

i.e., an $A_{\mu_1...\mu_r}$ where the indices run over $d - 1$ values instead of d values.

On-shell degrees of freedom

- Scalar: the Klein–Gordon equation of motion for the scalar does not constrain away the scalar, just the functional form of the degree of freedom (restricted to $k^2 = 0$ in momentum space), so the propagating scalar still has one degree of freedom. The auxiliary degree of freedom of course has nothing on-shell.
- Gauge field A_μ. The equation of motion is $\partial^\mu(\partial_\mu A_\nu - \partial_\nu A_\mu) = 0$ and in principle we should analyze the restrictions it makes on components. But it is easier to use a trick: Consider the equation of motion in covariant (Lorentz) gauge, $\partial^\mu A_\mu = 0$. Then the equation becomes just the Klein–Gordon equation $\Box A_\nu = 0$, which as we said, does not constrain anything. But now the covariant gauge condition we used imposes one constraint on the degrees of freedom, specifically on the polarization vectors. If $A_\mu \propto \epsilon_\mu(k)$, then we get $k^\mu \epsilon_\mu(k) = 0$. Since off-shell we had $d - 1$ degrees of freedom, now we have $d - 1 - 1 = d - 2$. These degrees of freedom correspond to *transverse*

components for the gauge field. As is well known, the *longitudinal* components, A_0 (time direction) and A_z (in the direction of propagation) are not propagating, and only the transverse ones are. The condition $k^\mu \epsilon_\mu(k)$ is a transversality condition, since it says that the polarization vector $\epsilon_\mu(k)$ is perpendicular to the momentum k^μ (direction of propagation).

- Graviton $g_{\mu\nu}$. The equation of motion for the linearized graviton ($\kappa_N h_{\mu\nu} \equiv g_{\mu\nu} - \eta_{\mu\nu}$) follows from the Fierz–Pauli action

$$\mathcal{L} = \frac{1}{2} h_{\mu\nu,\rho}^2 + h_\mu^2 - h^\mu h_{,\mu} + \frac{1}{2} h_{,\mu}^2; \quad h_\mu \equiv \partial^\nu h_{\nu\mu}; \quad h \equiv h^\mu{}_\mu \,, \qquad (4.19)$$

which is the linearized part of the Einstein–Hilbert action, and where comma denotes derivative (e.g. $h_{,\rho} \equiv \partial_\rho h$). We leave the checking of this fact as an exercise (Exercise 7). Again in principle we should analyze the restrictions this complicated equation of motion makes on components, but we have the same trick: If we impose the de Donder gauge condition

$$\partial^\nu \bar{h}_{\mu\nu} = 0; \quad \bar{h}_{\mu\nu} \equiv h_{\mu\nu} - \eta_{\mu\nu} \frac{h}{2} \,, \qquad (4.20)$$

the equation of motion again becomes just the Klein–Gordon equation $\Box \bar{h}_{\mu\nu} = 0$, which just restricts the functional form through $k^2 = 0$, but not the degrees of freedom. But again the gauge condition now imposes d constraints on the polarization tensors $\bar{h}_{\mu\nu} \propto \epsilon_{\mu\nu}(k)$, namely $k^\mu \epsilon_{\mu\nu}(k) = 0$, so on-shell we lose d degrees of freedom, and are left with

$$\frac{d(d-1)}{2} - d = \frac{(d-1)(d-2)}{2} - 1. \qquad (4.21)$$

These correspond to the graviton fluctuations $\delta h_{\mu\nu}$ being transverse (μ, ν run only over the $d-2$ transverse direction) and traceless. Indeed, now again $k^\mu \epsilon_{\mu\nu}(k) = 0$ is a transversality condition, since it states that the polarization tensor of the graviton is perpendicular to the direction of propagation.

- Spinor of spin 1/2. The Dirac equation in momentum space,

$$(\not{p} - m)u(p) = 0, \qquad (4.22)$$

relates 1/2 of the components in $u(p)$ to the other half, thus we are left with only $n/2$ degrees of freedom on-shell.

- Gravitino ψ_μ^α. Naively, we would say that it is a spinor $\times$ a gauge field, so $n/2(d-2)$ degrees of freedom. But there is a subtlety. The component that is an irreducible representation is not the full ψ_μ^α, but only the gamma-traceless part. Indeed, we have the decomposition in terms of Lorentz spin, $1 \otimes 1/2 = 3/2 \oplus 1/2$, where the 1/2 component is $\gamma^\mu \psi_\mu$, since we can see that it transforms to itself, thus is a sub-representation. So we need to first impose the condition $\gamma^\mu \psi_\mu = 0$ (eliminating the 1/2 representation), and then we can use the vector times spinor ($1 \otimes 1/2$) counting. All in all we get $n/2(d-2) - n/2 = n/2(d-3)$ degrees of freedom.

- Antisymmetric tensor $A_{\mu_1 \dots \mu_r}$. Again this object is a generalization of the gauge field, and so imposing the covariant gauge condition

$$\partial^{\mu_1} A_{\mu_1 \dots \mu_r} = 0 \,, \qquad (4.23)$$

we get the Klein–Gordon equation $\Box A_{\mu_1...\mu_r} = 0$, so we have transversality constraints on the polarization tensors:

$$k^{\mu_1} \epsilon_{\mu_1...\mu_r}(k) = 0. \tag{4.24}$$

We again obtain only transverse components for the antisymmetric tensor, i.e.

$$\binom{d-2}{r} = \frac{(d-2)\ldots(d-1-r)}{1 \cdot 2 \cdot \ldots r} \tag{4.25}$$

independent degrees of freedom.

4.3 Local supersymmetry: supergravity

Supergravity can be defined in two independent ways that give the same result. It is a supersymmetric theory of gravity; and it is also a theory of local supersymmetry. Thus we could either take Einstein gravity and supersymmetrize it, or we can take a supersymmetric model and make the supersymmetry local. In practice we use a combination of the two.

We want a theory of local supersymmetry, which means that we need to make the rigid ϵ^α transformation local. We know from gauge theory that if we want to make a global symmetry local we need to introduce a gauge field for the symmetry. For example, for the globally $U(1)$-invariant complex scalar with action $-\int |\partial_\mu \phi|^2$, $\phi \to e^{i\alpha}\phi$ invariant, if we make $\alpha = \alpha(x)$ (local), we need to add the $U(1)$ gauge field A_μ that transforms by $\delta A_\mu = \partial_\mu \alpha$ and write covariant derivatives $D_\mu = \partial_\mu - iA_\mu$ everywhere.

Now, the gauge field would be "A_μ^α" (since the supersymmetry acts on the index α), which we denote in fact by $\psi_{\mu\alpha}$ and call the gravitino.

Here μ is a curved space index ("curved") and α is a local Lorentz spinor index ("flat"). In flat space, an object $\psi_{\mu\alpha}$ would have the same kind of indices (curved = flat) and we can then show that $\mu\alpha$ forms a spin 3/2 field – though on-shell we need to remove the gamma-trace, as we saw in the previous subsection – therefore the same is true in curved space.

The fact that we have a supersymmetric theory of gravity means that the gravitino must be transformed by supersymmetry into some gravity variable, thus $\psi_{\mu\alpha} = Q_\alpha(gravity)$. But the index structure tells us that the gravity variable cannot be the metric, but something with only one curved index, namely the vielbein. Thus the gravitino is the superpartner of the vielbein. In conclusion, the gravitino is at the same time the superpartner of the vielbein, and the "gauge field of local supersymmetry." We also see that supergravity needs the vielbein–spin connection formulation of gravity, introduced earlier in this chapter.

We can now count the degrees of freedom for the $\mathcal{N} = 1$ multiplets in three dimensions and four dimensions and check that we have an equal number. As we saw, the supermultiplets need to have at least e_μ^a and ψ_μ^α, and to match bosonic with fermionic degrees of freedom.

Three dimensions

We try starting with three dimensions, in the hope of finding a simpler system. On-shell, e_μ^a has $1 \cdot 2/2 - 1 = 0$ degrees of freedom, and ψ_μ^α has $2/2(3-3) = 0$ degrees of freedom, thus $\{e_\mu^a, \psi_\mu^\alpha\}$ do form a trivial multiplet by themselves, the on-shell $\mathcal{N} = 1$ supergravity multiplet, with no propagating degrees of freedom. Off-shell, e_μ^a has $2 \cdot 3/2 = 3$ degrees of freedom, and ψ_μ^α has $2(3-1) = 4$ degrees of freedom, so we need one bosonic auxiliary degree of freedom. This is a scalar, that we will call S, giving the off-shell $\mathcal{N} = 1$ supergravity multiplet $\{e_\mu^a, S, \psi_\mu^\alpha\}$. We see that to obtain a nontrivial on-shell theory, we need to look at four dimensions.

Four dimensions

On-shell, e_μ^a has $3 \cdot 2/2 - 1 = 2$ degrees of freedom, and ψ_μ^α has $4/2(4-3) = 2$ degrees of freedom, so again $\{e_\mu^a, \psi_\mu^\alpha\}$ form the $\mathcal{N} = 1$ supergravity multiplet by themselves, but now it is a nontrivial one. Off-shell, e_μ^a has $4 \cdot 3/2 = 6$ degrees of freedom, whereas ψ_μ^α has $4(4-1) = 12$ degrees of freedom. We see that now the minimal choice would involve six bosonic auxiliary degrees of freedom. But other choices are possible. We could for instance use ten bosonic auxiliary degrees of freedom and four fermionic ones (one auxiliary Majorana spin 1/2 spinor), etc. There are thus several possible choices for auxiliary fields that have been used in the literature. A useful set is the *minimal set* (A_μ, S, P) (two real scalars and one vector). We can also write $M = S + iP$.

4.4 $\mathcal{N} = 1$ on-shell supergravity in four dimensions

We now turn to the construction of the $\mathcal{N} = 1$ on-shell supergravity model in four dimensions. To write down the supersymmetry transformations, we start with the vielbein. In analogy with the Wess–Zumino model, where $\delta\phi = \bar{\epsilon}\phi$, or the vector multiplet, where the gauge field variation is $\delta A_\mu^a = \bar{\epsilon}\gamma_\mu\psi^a$, it is easy to see that the vielbein variation has to be

$$\delta e_\mu^a = \frac{k_N}{2}\bar{\epsilon}\gamma^a\psi_\mu \,, \tag{4.26}$$

where k_N is the Newton constant and appears for dimensional reasons.

Since ψ is like a gauge field of local supersymmetry, for its transformation law we expect something like $\delta A_\mu = D_\mu\epsilon$. Therefore we must have

$$\delta\psi_\mu = \frac{1}{k_N}D_\mu\epsilon; \quad D_\mu\epsilon = \partial_\mu\epsilon + \frac{1}{4}\omega_\mu^{ab}\gamma_{ab}\epsilon \,, \tag{4.27}$$

plus maybe more terms. For the $\mathcal{N} = 1$ supergravity there are in fact no other terms, but for $\mathcal{N} > 1$ there are.

We now turn to writing the action. For gravity, we need to write the Einstein–Hilbert action. But in which formulation? In principle we could write the form

$$S = \frac{1}{2k_N^2} \int d^4x \sqrt{-g} R(\Gamma) , \tag{4.28}$$

where $\Gamma = \Gamma(g)$ (the second order formulation, the usual one used by Einstein). Or we could consider the original formulation of Palatini, the first order formulation with an independent Γ and the metric $g_{\mu\nu}$. But as we have already mentioned, due to the fact that we have spinors in the theory, the spin connection appears in the covariant derivative, and so we need the vielbein–spin connection formulation of gravity.

Thus we write the Einstein–Hilbert action in the form

$$\begin{aligned} S_{\text{EH}} &= \frac{1}{2k_N^2} \int d^4x (\det e) R_{\mu\nu}^{ab}(\omega) (e^{-1})_a^\mu (e^{-1})_b^\nu \\ &= \frac{1}{4} \frac{1}{2k_N^2} \int d^4x \epsilon_{abcd} \epsilon^{\mu\nu\rho\sigma} e_\mu^a e_\nu^b R_{\rho\sigma}^{cd} \\ &\equiv \frac{1}{2k_N^2} \int \epsilon_{abcd} e^a \wedge e^b \wedge R^{cd}(\omega) , \end{aligned} \tag{4.29}$$

where the first line is valid in any dimension, in the second line we use (4.14) to write (4.15) valid only in four dimensions, and in the last line we use form language. Also, from now on we will drop the -1 power on the inverse vielbein, understanding whether we have the vielbein or the inverse vielbein by the position of the curved index (index down is vielbein, index up is inverse vielbein). Again in form language,

$$R^{ab} = d\omega^{ab} + \omega^{ac} \wedge \omega^{cb}. \tag{4.30}$$

Next we consider the action for a gravitino. The action for a free spin 3/2 field in flat space is the Rarita–Schwinger action, which is

$$\begin{aligned} S_{\text{RS}} &= -\frac{1}{2} \int d^dx \bar\psi_\mu \gamma^{\mu\nu\rho} \partial_\nu \psi_\rho \\ &= +\frac{i}{2} \int d^4x \epsilon^{\mu\nu\rho\sigma} \bar\psi_\mu \gamma_5 \gamma_\nu \partial_\rho \psi_\sigma , \end{aligned} \tag{4.31}$$

where the first form is valid in all dimensions and the second form is only valid in four dimensions ($i\epsilon^{\mu\nu\rho\sigma} \gamma_5 \gamma_\nu = -\gamma^{\mu\rho\sigma}$ in four dimensions, $\gamma_5 = i\gamma_0\gamma_1\gamma_2\gamma_3$). In curved space, this becomes

$$\begin{aligned} S_{\text{RS}} &= -\frac{1}{2} \int d^dx (\det e) \bar\psi_\mu \gamma^{\mu\nu\rho} D_\nu \psi_\rho \\ &= +\frac{i}{2} \int d^4x \epsilon^{\mu\nu\rho\sigma} \bar\psi_\mu \gamma_5 \gamma_\nu D_\rho \psi_\sigma . \end{aligned} \tag{4.32}$$

We can now write the action of $\mathcal{N} = 1$ on-shell supergravity in four dimensions as just the sum of the Einstein–Hilbert action and the Rarita–Schwinger action:

$$S_{\mathcal{N}=1} = S_{\text{EH}}(\omega, e) + S_{\text{RS}}(\psi_\mu), \tag{4.33}$$

and the supersymmetry transformations rules are just the ones defined previously,

$$\delta e_\mu^a = \frac{k_N}{2}\bar{\epsilon}\gamma^a\psi_\mu; \quad \delta\psi_\mu = \frac{1}{k_N}D_\mu\epsilon. \tag{4.34}$$

However, this is not yet enough to specify the theory. We must specify the formalism and various quantities:

- Second order formalism: The independent fields are e_μ^a and ψ_μ, and ω is not an independent field. But now there is a dynamical fermion (ψ_μ), so the torsion $T_{\mu\nu}^a$ is not zero anymore, thus $\omega \neq \omega(e)$! In fact,

$$\omega_\mu^{ab} = \omega_\mu^{ab}(e,\psi) = \omega_\mu^{ab}(e) + \psi\psi \text{ terms} \tag{4.35}$$

is found by varying the action with respect to ω, as in the $\psi = 0$ case:

$$\frac{\delta S_{\mathcal{N}=1}}{\delta\omega_\mu^{ab}} = 0 \Rightarrow \omega_\mu^{ab}(e,\psi). \tag{4.36}$$

It is left to the reader (as Exercise 2) to find the explicit form of the $\psi\psi$ terms.

- First order formalism: All fields, ψ, e, ω, are independent. But now we must supplement the action with a transformation law for ω. It is

$$\delta\omega_\mu^{ab}(\text{first order}) = -\frac{1}{4}\bar{\epsilon}\gamma_5\gamma_\mu\tilde{\psi}^{ab} + \frac{1}{8}\bar{\epsilon}\gamma_5(\gamma^\lambda\tilde{\psi}_\lambda^b e_\mu^a - \gamma^\lambda\tilde{\psi}_\lambda^a e_\mu^b),$$
$$\tilde{\psi}^{ab} \equiv \epsilon^{abcd}\psi_{cd}; \quad \psi_{ab} \equiv e_a^\mu e_b^\nu(D_\mu\psi_\nu - D_\nu\psi_\mu). \tag{4.37}$$

In this first order formalism, on-shell the variation of ω should reduce to the one of the second order formalism, where we use the chain rule for $\omega(e,\psi)$ and we substitute δe and $\delta\psi$. We can indeed find that this is so, using the fact (easily checked) that the equation of motion for ψ_μ is

$$\sim \gamma^\lambda\tilde{\psi}_{\lambda\mu} = 0. \tag{4.38}$$

- 1.5 order formalism: The 1.5 order formalism is a simple but powerful observation which simplifies calculations, so is the most useful. We use second order formalism, but in the action $S(e,\psi,\omega(e,\psi))$ whenever we vary it, we do not vary $\omega(e,\psi)$ by the chain rule, since it is multiplied by $\delta S/\delta\omega$, which is equal to zero in the second order formalism:

$$\delta S = \frac{\delta S}{\delta e}\delta e + \frac{\delta S}{\delta\psi}\delta\psi + \frac{\delta S}{\delta\omega}\left(\frac{\delta\omega}{\delta e}\delta e + \frac{\delta\omega}{\delta\psi}\delta\psi\right). \tag{4.39}$$

Of course, that means that when we write the action, we have to write $\omega(e,\psi)$ without substituting the explicit form in terms of e and ψ.

One can check the invariance of the action (4.33) under the local susy (4.34), but we will not do it here, since it is somewhat involved.

For completeness, we can write also the other transformation laws for the supergravity fields. For the Einstein transformations, we have

$$\delta_E e_\mu^a = \xi^\nu\partial_\nu e_\mu^a + (\partial_\mu\xi^\nu)e_\nu^a,$$
$$\delta_E\omega_\mu^{ab} = \xi^\nu\partial_\nu\omega_\mu^{ab} + (\partial_\mu\xi^\nu)\omega_\nu^{ab},$$
$$\delta_E\psi_\mu = \xi^\nu\partial_\nu\psi_\mu + (\partial_\mu\xi^\nu)\psi_\nu, \tag{4.40}$$

whereas for the local Lorentz transformations, we have

$$\delta_{lL} e_\mu^a = \lambda^{ab} e_\mu^b,$$
$$\delta_{lL} \omega_\mu^{ab} = D_\mu \lambda^{ab} = \partial_\mu \lambda^{ab} + \omega_\mu^{ac} \lambda^{cb} - \omega_\mu^{bc} \lambda^{ca},$$
$$\delta_{lL} \psi_\mu = -\lambda^{ab} \frac{1}{4} \gamma_{ab} \psi_\mu. \tag{4.41}$$

4.5 Generic features of supergravity theories

The $\mathcal{N} = 1$ supergravity multiplet in four dimensions is $(e_\mu^a, \psi_{\mu\alpha})$, as we saw, and has spins $(2, 3/2)$.

It can also couple with other $\mathcal{N} = 1$ supersymmetric multiplets of lower spin: the chiral multiplet of spins $(1/2, 0)$ and the gauge multiplet of spins $(1, 1/2)$ that have already been described, as well as the so-called *gravitino multiplet*, composed of a gravitino and a vector, thus spins $(3/2, 1)$.

By adding appropriate numbers of such multiplets we obtain the $\mathcal{N} = 2, 3, 4, 5, 6, 8$ supergravity multiplets. Here $\mathcal{N}$ is the number of supersymmetries, and since it acts on the graviton, there should be exactly $\mathcal{N}$ gravitini in the multiplet, so that each supersymmetry maps the graviton to a different gravitino.

The limit for supergravity is $\mathcal{N} = 8$ supersymmetries, for the same reason that the limit for vector fields is $\mathcal{N} = 4$, as seen in Chapter 3.7. Namely, if we want to have only spins ≤ 2, we can at most start with a vacuum with helicity $\lambda = 2$. By the successive action of all the $\mathcal{N}$ supersymmetry operators, which lower the helicity by 1/2 each, we can at most reach helicity -2. That means that the maximum number of supersymmetries for supergravity is $2 \times (2 - (-2)) = 8$. The only thing left to understand is why we would only allow spins ≤ 2? The reason is that there seems to be no good way to write an interacting theory of a finite number of fields with spins > 2.

Coupling to supergravity of a supersymmetric multiplet is a generalization of coupling to gravity, which means putting fields in curved space. Now we put fields in curved space and also introduce a few more couplings.

We will denote the $\mathcal{N} = 1$ supersymmetry multiplets by brackets, e.g. $(1, 1/2)$, $(1/2, 0)$, etc. Then $\mathcal{N} = 2$ supergravity is obtained by coupling the $\mathcal{N} = 1$ supergravity multiplet $(2, 3/2)$ (graviton plus gravitino) to the $\mathcal{N} = 1$ gravitino multiplet $(3/2, 1)$, i.e. gravitino plus (abelian) vector, for a total of graviton, two gravitini, and an abelian scalar.

If we minimally couple the gravitinos in the 4-dimensional $\mathcal{N} = 2$ multiplet to an abelian gauge field, we obtain *gauged supergravity*. In fact, the gauged supergravity is only a deformation by the coupling constant g of the ungauged model, so the abelian gauge field is in fact the one already in the $\mathcal{N} = 2$ supergravity multiplet. The new gravitino transformation law is

$$\delta \psi_\mu^i = D_\mu(\omega(e, \psi)) \epsilon^i + g \gamma_\mu \epsilon^i + g A_\mu \epsilon^i. \tag{4.42}$$

Thus we have a constant term $(g \gamma_\mu \epsilon^i)$ in this transformation law, so it is natural to find that we must add a constant term in the action as well, namely a cosmological constant term, $\int e\Lambda$. This cosmological constant is negative, leading to the fact that the simplest background of gauged supergravity is Anti-de Sitter, AdS. Unlike the ungauged supergravity, it does not admit a Minkowski background. Thus, in fact, gauged supergravity is AdS supergravity.

In the usual, "ungauged models," the photons are not coupled to the fermions, i.e. the gauge coupling $g = 0$. But these models have *global* symmetries. One can couple the gauge fields to the fermions, thus "gauging" (making local) some subset of the global symmetry. In general (except the $\mathcal{N} = 2$ case above) abelian fields become nonabelian (Yang–Mills), i.e. self-coupled. Another way to obtain the gauged models is by adding a cosmological constant and requiring invariance, since we saw that gauged supergravity needs AdS space.

Next we consider the $\mathcal{N} = 3$ supergravity multiplet, which is composed of the supergravity multiplet $(2, 3/2)$, two gravitino multiplets $(3/2, 1)$, and a vector multiplet $(1, 1/2)$. Together, they correspond to the fields $\{e^a_\mu, \psi^i_\mu, A^i_\mu, \lambda\}$, for $i = 1, 2, 3$.

We can also minimally couple the $\mathcal{N} = 3$ multiplet with gauge fields, and as before, the gauge fields have to be the same three gauge fields in the ungauged multiplet. We also find that we must add a negative cosmological constant, and find again that gauged supergravity is AdS supergravity. The difference is that now, under the gauge coupling deformation, the gauge fields become nonabelian (the ungauged model had abelian vector fields).

The next possibility is the $\mathcal{N} = 4$ supergravity multiplet, which is the first to also contain scalars. It is composed of the $\mathcal{N} = 1$ multiplets $(2, 3/2), 3 \times (3/2, 1), 3 \times (1, 1/2), (1, 0)$, together making $\{e^a_\mu, \psi^i_\mu, A^k_\mu, B^k_\mu, \lambda^i, \phi, B\}$, where $i = 1, \dots, 4$, $k = 1, 2, 3$, A^k_μ are vectors, B^k_μ are axial vectors, ϕ is scalar, and B is pseudoscalar. The same comments as above apply for the gauging of this model. But in general, we can gauge a subset of the vectors, so there are various gaugings possible.

The $\mathcal{N} = 5$ supergravity multiplet is composed of the $\mathcal{N} = 1$ multiplets $(2, 3/2), 4 \times (3/2, 1), 6 \times (1, 1/2), 5 \times (1/2, 0)$, together making the graviton, five gravitini, ten vectors, 11 spin 1/2 fermions, and ten real scalars. The $\mathcal{N} = 6$ supergravity multiplet is composed of the $\mathcal{N} = 1$ multiplets $(2, 3/2), 5 \times (3/2, 1), 11 \times (1, 1/2), 15 \times (1/2, 0)$, together making the graviton, six gravitini, 16 vectors, 26 spin 1/2 fermions, and 30 real scalars.

We could imagine that we could have $\mathcal{N} = 7$ supergravity, but if we impose this susy, we obtain $\mathcal{N} = 8$ as well, so the next model is in fact $\mathcal{N} = 8$ supergravity. As mentioned, it is the maximal possible model in four dimensions. It is composed of the supergravity multiplet $(2, 3/2) + 7$ gravitino multiplets $(3/2, 1) + 21$ vector multiplets $(1, 1/2) + 35$ chiral multiplets $(1/2, 0)$. The fields are $\{e^a_\mu, \psi^i_\mu, A^{IJ}_\mu, \chi_{ijk}, v\}$ which are: one graviton, eight gravitinos ψ^i_μ, 28 photons A^{IJ}_μ, 56 spin 1/2 fermions χ_{ijk}, and 70 scalars in the matrix v.

Higher dimensions

The $\mathcal{N} = 8$ supergravity multiplet can be obtained by dimensional reduction of $\mathcal{N} = 1$ supergravity in 11 dimensions. In fact, 11 dimensions is the maximal dimension from which we can reduce to obtain $\mathcal{N} = 8$ in four dimensions, since the eight 4-dimensional gravitini make up a single gravitino in 11 dimensions, but would make less than one gravitino in higher dimensions. The field content of $\mathcal{N} = 8$ supergravity is a graviton e_μ^a, a gravitino $\psi_{\mu\alpha}$, and a three index antisymmetric tensor $A_{\mu\nu\rho}$.

We can have other supergravity theories in all dimensions and with various super-symmetries, such that when reducing to four dimensions, we would get at most eight supersymmetries.

One new feature in higher dimensions is that it is possible also to have antisymmetric tensor fields $A_{\mu_1,\ldots,\mu_n}$, which are just an extension of abelian vector fields. We have seen in Section 4.2 how to treat them in flat space. In curved space the generalization is trivial. The field strength is still

$$F_{\mu_1,\ldots,\mu_{n+1}} = (n+1)\partial_{[\mu_1} A_{\mu_2,\ldots,\mu_{n+1}]}\,, \tag{4.43}$$

with the gauge invariance

$$\delta A_{\mu_1,\ldots,\mu_n} = \partial_{[\mu_1} \Lambda_{\mu_2,\ldots,\mu_n]}. \tag{4.44}$$

Putting the action in curved space is obvious, having the curved space volume element and contracting with the metric,

$$-\frac{1}{2(n+1)!} \int d^d x (\det e) F^2_{\mu_1,\ldots,\mu_{n+1}}. \tag{4.45}$$

The antisymmetric tensor fields appear in the supersymmetry transformation law for the gravitini $\delta\psi_\mu^i$ generically through $\Gamma \cdot F$ terms, like

$$\Gamma^{\mu_1\ldots\mu_n} F_{\mu\mu_1\ldots\mu_n} \epsilon^i. \tag{4.46}$$

Important concepts to remember

- Vielbeins are defined by $g_{\mu\nu}(x) = e_\mu^a(x) e_\nu^b(x) \eta_{ab}$, by introducing a Minkowski space in the neighborhood of a point x, giving local Lorentz invariance.
- The spin connection is the gauge field needed to define covariant derivatives acting on spinors. In the absence of dynamical fermions, it is determined as $\omega = \omega(e)$ by the vielbein postulate: the torsion is zero.
- The field strength of this gauge field is related to the Riemann tensor.
- In the first order formulation (Palatini), the spin connection is independent, and is determined from its equation of motion.
- Supergravity is a supersymmetric theory of gravity and a theory of local supersymmetry.
- The gauge field of local supersymmetry and superpartner of the vielbein (graviton) is the gravitino ψ_μ.
- In three dimensions on-shell, there are no degrees of freedom (dof) for the $\mathcal{N} = 1$ super-gravity, whereas in four dimensions there are two bosonic and two fermionic degrees of freedom.
- In four dimensions off-shell, we need six bosonic auxiliary dofs more than the fermionic auxiliary dofs. Choosing just the bosonic auxiliary fields (A_μ, S, P) is the minimal set.
- 4-dimensional on-shell supergravity is the first nontrivial case (with propagating degrees of freedom) and is composed of e_μ^a and ψ_μ.

- Supergravity (local supersymmetry) is of the type $\delta e_\mu^a = (k_N/2)\bar\epsilon \gamma^a \psi_\mu + \ldots, \delta\psi_\mu = (D_\mu \epsilon)/k_N + \ldots$
- The action for gravity in supergravity is the Einstein–Hilbert action in the vielbein–spin connection formulation.
- The action for the gravitino is the Rarita–Schwinger action.
- The most useful formulation is the 1.5 order formalism: second order formalism, but do not vary $\omega(e,\psi)$ by the chain rule.
- For each supersymmetry we have a gravitino. The maximal supersymmetry in $d = 4$ is $\mathcal{N} = 8$.
- Gauged supergravity is AdS supergravity, and is an extension by a gauge coupling parameter of the ungauged models.
- Supergravity theories in higher dimensions can contain antisymmetric tensor fields.
- The maximal dimension for a supergravity theory is $d = 11$, with a unique model composed of $e_\mu^a, \psi_\mu, A_{\mu\nu\rho}$.

References and further reading

The vielbein and spin connection formalism for general relativity is hard to find in standard general relativity books, but one can find some information, for instance, in the supergravity review [14]. An introduction to supergravity, but one which might be hard to follow for the beginning student, is found in West [9] and Wess and Bagger [10]. A good supergravity course, that starts at an introductory level and reaches quite far, is [14]. In this chapter, I followed mostly [14] (you can find more details in Sections 1.2–1.6 of the reference). A good and complete recent book is [15].

Exercises

1. Check that

$$\omega_\mu^{ab}(e) = \frac{1}{2}e^{av}(\partial_\mu e_\nu^b - \partial_\nu e_\mu^b) - \frac{1}{2}e^{bv}(\partial_\mu e_\nu^a - \partial_\nu e_\mu^a) - \frac{1}{2}e^{a\rho}e^{b\sigma}(\partial_\rho e_{c\sigma} - \partial_\sigma e_{c\rho})e_\mu^c \quad (4.47)$$

    satisfies the no-torsion (vielbein) constraint, $T_{\mu\nu}^a = 2D_{[\mu}e_{\nu]}^a = 0$.
2. Find $\omega_\mu^{ab}(e,\psi) - \omega_\mu^{ab}(e)$ in the second order formalism for $\mathcal{N} = 1$ supergravity.
3. Write down the free gravitino equation of motion in curved space.
4. Calculate the number of off-shell bosonic and fermionic degrees of freedom of $\mathcal{N} = 8$ on-shell supergravity in four dimensions, with field content $\{(2, 3/2) + 7 \times (3/2, 1) + 21 \times (1, 1/2) + 35 \times (1/2, 0)\}$, specifically $\{e_\mu^a, \psi_\mu^i, A_\mu^{[IJ]}, \chi_{[IJK]}, v\}$, where $i, j, k = 1, \ldots, 8; I, J = 1, \ldots, 8$; and $v =$ matrix of 70 real scalars (the scalar in the WZ multiplet $(1/2, 0)$ is complex).
5. Consider the spinors η^I satisfying the "Killing spinor equation"

$$D_\mu \eta^I = \pm \frac{i}{2}\gamma_\mu \eta^I. \quad (4.48)$$

Prove that they live on a space of constant positive curvature (a sphere), by computing the curvature of the space.

6. Write down explicitly the variation of the $\mathcal{N} = 1$ 4-dimensional supergravity action in 1.5 order formalism, as a function of δe and $\delta \psi$.

7. Prove that the linearized action for the graviton perturbation ($\kappa_N h_{\mu\nu} \equiv g_{\mu\nu} - \eta_{\mu\nu}$) is the Fierz–Pauli action

$$\mathcal{L} = \frac{1}{2} h_{\mu\nu,\rho}^2 + h_\mu^2 - h^\mu h_{,\mu} + \frac{1}{2} h_{,\mu}^2; \quad h_\mu \equiv \partial^\nu h_{\nu\mu}; \quad h \equiv h^\mu{}_\mu , \qquad (4.49)$$

by expanding the Einstein–Hilbert action ($\kappa_N^2 \equiv 8\pi G$).

5

Kaluza–Klein dimensional reduction

We have already briefly encountered dimensional reduction, when we wrote the $\mathcal{N} = 4$ SYM model as a dimensional reduction of the 10-dimensional $\mathcal{N} = 1$ SYM model. However, in that case it seemed like just a useful trick to obtain the right action. In this chapter, however, we study the possibility that spacetime really has more dimensions than $3 + 1$, and we want to describe physics from the point of view of $3 + 1$ dimensions.

The idea of extra dimensions is an old one, going back to Theodor Kaluza (1921) and Oskar Klein (1926). One considers that space is a direct product space, $M_D = M_4 \times K_n$, where K_n is a compact space, like for instance a sphere S^n or torus $T^n = S^1 \times \ldots \times S^1$, as considered by Klein. The reason why we feel only four dimensions is that the size of K_n is very small, comparable with the Planck scale, so we cannot probe it. Kaluza used this construction to unify forces (in particular, gravitation and electromagnetism), or more precisely the fields associated with them. This is also the way in which we use it nowadays, as several fields have a common higher dimensional origin. The resulting theory is generally known as the Kaluza–Klein (KK) theory.

One considers in every point in spacetime x^μ a space K_n that can depend on x^μ, $K_n(x^\mu)$ with coordinates y^m. The total spacetime is (x^μ, y^m), and functions of the total spacetime are now $\phi(x^\mu, y^m)$ instead of $\phi(x^\mu)$.

For example, in the simplest case of a circle S^1, one can write functions $\phi(x^\mu, y)$ and Fourier expand them

$$\phi(x^\mu, y) = \sum_{n \geq 0} e^{\frac{iny}{R}} \phi_n(x^\mu) \,, \tag{5.1}$$

which means that we have a sum of fields $\phi_n(x^\mu)$ from the point of view of $3 + 1$ dimensions. The KK theory is a generalization of this analysis.

Note that there is also another way to have extra dimensions, the "braneworld" scenario, where our $3 + 1$ dimensional world lives on a wall, or "brane". This possibility is related to D-branes, relevant for AdS/CFT, which are studied in a later chapter.

5.1 The KK background, KK expansion, and KK reduction

There are three metrics that sometimes go by the name of KK metric, so we should distinguish between them:

- **The KK background metric**. The fact that space is $M_4 \times K_n$ means that the *background* is a solution of the equations of motion which is of direct product type,

$$g_{MN}(\vec{x}, \vec{y}) = \begin{pmatrix} g^{(0)}_{\mu\nu}(\vec{x}) & 0 \\ 0 & g^{(0)}_{mn}(\vec{y}) \end{pmatrix}, \quad M = (\mu, m). \tag{5.2}$$

Note that the metric is itself one of the fields of the theory, so it is a variable, so when we write $M_4 \times K_n$ we only mean the background, not the full fluctuating metric. Also note that in general, the background has to be a solution of the supergravity equations of motion; however, sometimes one considers the case when it is not. Here $g^{(0)}_{\mu\nu}$ is a background metric in four dimensions, usually Minkowski, de Sitter or Anti-de Sitter, and $g^{(0)}_{mn}$ is the metric on the compact space K_n.

- **The KK expansion**. This is an *exact* decomposition, the generalization of the Fourier expansion on a circle, or the spherical harmonic expansion on the 2-sphere. In the case of the Fourier expansion, as mentioned, the Fourier theorem says we can always expand

$$\phi(\vec{x}, y) = \sum_n \phi_n(\vec{x}) e^{\frac{iny}{R}}, \tag{5.3}$$

if y is on a circle of radius R. On a 2-sphere, we can similarly always write

$$\phi(\vec{x}, \theta, \phi) = \sum_{lm} \phi_{lm}(\vec{x}) Y_{lm}(\theta, \phi). \tag{5.4}$$

Here the functions in which we expand are eigenfunctions of the Laplacean, since

$$\partial_y^2 e^{\frac{iny}{R}} = -\left(\frac{n}{R}\right)^2 e^{\frac{iny}{R}},$$
$$\Delta_2 Y_{lm}(\theta, \phi) = -\frac{l(l+1)}{R^2} Y_{lm}(\theta, \phi), \tag{5.5}$$

where in the second line we include an R^2 for a 2-sphere of radius R, though of course the Y_{lm}s correspond to $R = 1$.

Similarly, in a general case, we can always write

$$\phi(\vec{x}, \vec{y}) = \sum_{q, I_q} \phi_q^{I_q}(\vec{x}) Y_q^{I_q}(\vec{y}), \tag{5.6}$$

where $Y_q^{I_q}(\vec{y})$ is also called *spherical harmonic*, as in the 2-sphere case. Here q is an index that measures the eigenvalue of the Laplacean, like l for S^2, and I_q is an index in some representation of the symmetry group (like m for S^2 which takes values in a representation of the $SO(3) = SU(2)$ invariance group of S^2, namely a spin l representation). The $Y_q^{I_q}$ are also eigenfunctions of the Laplacean on K_n, i.e.

$$\Delta_n Y_q^{I_q}(\vec{y}) = -m_q^2 Y_q^{I_q}(\vec{y}). \tag{5.7}$$

From the 4-dimensional point of view, we get for $\phi(\vec{x}, \vec{y}) = \phi_q^{I_q}(\vec{x}) Y_q^{I_q}(\vec{y})$,

$$\Box_D \phi(\vec{x}, \vec{y}) = (\Box_4 + \Delta_n)\phi(\vec{x}, \vec{y}) = (\Box_4 - m_q^2)\phi(\vec{x}, \vec{y}), \tag{5.8}$$

and so if $\phi(\vec{x}, \vec{y})$ is a D-dimensional massless field, $\Box_D \phi(\vec{x}, \vec{y}) = 0$, it looks like a 4-dimensional massive field with mass m_q,

$$\left[(\Box - m_q^2) \phi_q^{I_q}(\vec{x}) \right] Y_q^{I_q}(\vec{y}) = 0. \tag{5.9}$$

This is the statement that in order to see structure on K_n, we must use some energy, at least equal to m_q, if we want to see information at the level of the $Y_q^{I_q}$ spherical harmonic.

Thus the KK expansion is a mathematical equality, and contains no information other than the metric of the background we expand around.

- **The KK reduction ansatz.** This is an *ansatz*, which means it is a guess, it is not guaranteed to work. Since we want to say that the compact space has a very small size, and we cannot probe it, we must find an effective 4-dimensional description which does not see the K_n. This is the dimensional reduction ansatz, which is: we keep only fields in the $n = 0$ representation, i.e. "independent of y", though in general there is a given y-dependence, namely of $Y_0(\vec{y})$, but this dependence is the simplest we can have.

  Also, in general it is not necessarily the first representation that is kept for all fields, but rather it could be $n = 1$ or $n = 2$ for some fields. In the case of supergravity, the relevant factor is that we need to keep a 4-dimensional supermultiplet. Also note that for M_4 being AdS for example, m_q is not necessarily zero, we could have fields that are a bit tachyonic, namely $m_q^2 < 0$, but still above a bound (the so-called Breitenlohner–Freedman bound, that is explained later on in the book), or massive. The relevant fact is still that we keep the lowest supermultiplet.

  Thus in the KK dimensional reduction ansatz we keep generically speaking

$$\phi(\vec{x}, \vec{y}) = \phi_0(\vec{x}) Y_0(\vec{y}). \tag{5.10}$$

In summary, the KK background metric is a solution, the KK expansion is a parameterization, and the KK reduction ansatz is an ansatz.

5.2 The KK dimensional reduction

We saw that KK dimensional reduction corresponds to keeping only one of the fields in the KK expansion in spherical harmonics, usually the $n = 0$ one.

Fields with spin

One of the most important properties of KK reduction, which is why it was originally introduced by Kaluza, and why it is useful nowadays, is the property of unifying various fields with spin into a single higher dimensional field.

The point is that various components of a field with spin (with indices) act as different fields in the lower dimension (e.g., in four dimensions).

For instance, consider electromagnetism in a higher dimension, with a Maxwell field $A_M(\vec{x}, \vec{y})$; $\mu = 0, 1, 2, 3$; $m = 4, \ldots, 4 + m$; $M = (\mu, m)$. It splits as

$$A_M(\vec{x}, \vec{y}) = (A_\mu(\vec{x}, \vec{y}), A_m(\vec{x}, \vec{y})). \tag{5.11}$$

Then A_μ behaves as a 4-dimensional gauge field: if $\Lambda^\mu{}_\nu \in SO(1, 3)$ is a Lorentz transformation in $3 + 1$ dimensions, it acts on it as a vector, $A'_\mu(\vec{x}', \vec{y}) = \Lambda_\mu{}^\nu A_\nu(\vec{x}, \vec{y})$. On the other hand, A_m behaves as a scalar: $\Lambda^\mu{}_\nu$ acts on it trivially, $A'_m(\vec{x}') = A_m(\vec{x})$.

We assume that the theory in $D = 4 + n$ dimensions is Lorentz invariant, i.e. $SO(1, 3 + n)$ invariant. This is broken to $SO(1, 3) \times SO(n)$ by the background. Then $SO(n)$ acts only on A_m, and on $\vec{y}$ instead of $\vec{x}$, meaning it becomes an *internal* symmetry.

The KK expansions for A_μ and A_m are

$$A_\mu(\vec{x}, \vec{y}) = \sum_{q, I_q} A_\mu^{q, I_q}(\vec{x}) Y_q^{I_q}(\vec{y}),$$
$$A_m(\vec{x}, \vec{y}) = \sum_{q, I_q} A^{q, I_q}(\vec{x}) Y_m^{q, I_q}(\vec{y}). \tag{5.12}$$

The $SO(n)$ acts on the spherical harmonics Y_m^{q, I_q} only, and thus $A^{q, I_q}(\vec{x})$ are 4-dimensional scalars.

The KK reduction ansatz would be keeping only one mode, usually $q = 0$.

For gravity, similarly, we decompose the metric into the lower dimensional tensors as

$$g_{MN}(\vec{x}, \vec{y}) = \begin{pmatrix} g_{\mu\nu}(\vec{x}, \vec{y}) & g_{\mu m}(\vec{x}, \vec{y}) \\ g_{m\mu}(\vec{x}, \vec{y}) & g_{mn}(\vec{x}, \vec{y}) \end{pmatrix}, \tag{5.13}$$

and then $g_{\mu\nu}(\vec{x}, \vec{y})$ is an $SO(1, 3)$ symmetric tensor, i.e. a metric, $g_{\mu m} = g_{m\mu}$ is an $SO(1, 3)$ vector, i.e. $g'_{\mu m}(\vec{x}', \vec{y}) = \Lambda_\mu{}^\nu g_{\nu m}(\vec{x}, \vec{y})$, and $g_{mn}(\vec{x}, \vec{y})$ are $SO(1, 3)$ scalars. Under the split $SO(1, 3 + n) \to SO(1, 3) \times SO(n)$, as before $SO(n)$ acts on the index m, which belongs to spherical harmonics. And as before, the KK reduction ansatz corresponds to keeping only one mode, the $q = 0$ mode, with $Y_0(\vec{y}) = 1$, since the lower dimensional graviton $g_{\mu\nu}$ cannot have indices in a nontrivial representation I_q.

5.3 The expansion of various fields on tori

The torus $T^n = (S^1)^n$ is obtained by periodic identifications in $\mathbb{R}^n$, and as such the metric on it is flat, $g_{mn}^{(0)} = \delta_{mn}$. Therefore the KK background metric is

$$g_{MN} = \begin{pmatrix} g_{\mu\nu}^{(0)} & 0 \\ 0 & \delta_{mn} \end{pmatrix}. \tag{5.14}$$

The KK expansion for the metric fluctuation is just a product of the Fourier expansions on the circles in T^n, i.e.

$$g_{MN} = \begin{pmatrix} g_{\mu\nu}(\vec{x},\vec{y}) = g^{(0)}_{\mu\nu}(\vec{x}) + \sum_{\{n_i\}} h^{\{n_i\}}_{\mu\nu}(\vec{x}) e^{\frac{in_i y_i}{R_i}}; & g_{\mu m}(\vec{x},\vec{y}) = \sum_{\{n_i\}} B^{m,\{n_i\}}_{\mu}(\vec{x}) e^{\frac{in_i y_i}{R_i}} \\ g_{m\mu}(\vec{x},\vec{y}) = g_{\mu m}(\vec{x},\vec{y}); & g_{mn} = \delta_{mn} + \sum_{\{n_i\}} h^{\{n_i\}}_{mn}(\vec{x}) e^{\frac{in_i y_i}{R_i}} \end{pmatrix}.$$

$$(5.15)$$

Thus here the spherical harmonics are just products of Fourier modes,

$$Y_{\{n_i\}}(\vec{y}) = \prod_i e^{\frac{in_i y_i}{R_i}}. \tag{5.16}$$

The KK reduction ansatz for the metric is

$$g_{MN} = \begin{pmatrix} g^{(0)}_{\mu\nu}(\vec{x}) + h^{\{0\}}_{\mu\nu}(\vec{x}); & g_{\mu m}(\vec{x}) = B^{m\{0\}}_{\mu}(\vec{x}) \\ g_{m\mu}(\vec{x}) = g_{\mu m}(\vec{x}); & g_{mn}(\vec{x}) = \delta_{mn} + h^{\{0\}}_{mn}(\vec{x}) \end{pmatrix}. \tag{5.17}$$

Here obviously $g_{\mu\nu}(\vec{x}) = g^{(0)}_{\mu\nu}(\vec{x}) + h^{\{0\}}_{\mu\nu}(\vec{x})$ is the 4-dimensional metric, $g_{\mu m}(\vec{x})$ are vectors from the point of view of four dimensions, since they have a single 4-dimensional vector index μ, more precisely we have n vectors $B^{m\{0\}}_{\mu}(\vec{x})$, and $g_{mn}(\vec{x})$ are 4-dimensional scalars.

As we saw, a gauge field $A_M(\vec{x},\vec{y})$ splits into $A_\mu(\vec{x},\vec{y})$ which is a vector from the 4-dimensional point of view, and $A_m(\vec{x},\vec{y})$ which are scalars in four dimensions. Under KK dimensional reduction, we obtain the 4-dimensional vector $A_\mu(\vec{x})$ and the 4-dimensional scalars $A_m(\vec{x})$.

We can also have antisymmetric tensors, $p+1$-forms $A_{M_1...M_{p+1}}$. Under KK dimensional reduction, $A_{M_1...M_{p+1}}$ splits into $A_{\mu_1...\mu_{p+1}}$, which is again an antisymmetric tensor ($p+1$-form) and $A_{\mu_1...\mu_k m_{k+1}...m_{p+1}}$, which are k-forms, up to $A_{\mu_1...\mu_{p-n+1} m_{p-n+2}...m_{p+1}}$, which are $p - n + 1$-forms.

Considering next fermions, a D-dimensional spinor ($D = 4 + n$) on $M_4 \times K_n$ splits under the KK reduction ansatz into many spinors in the lower dimension,

$$\lambda_A(\vec{x},\vec{y}) = \lambda^i_\tau(\vec{x}), \tag{5.18}$$

i.e., we obtain many spinors on M_4, with spinor index τ and labelled by the i index for the torus.

5.4 Consistent truncation and nonlinear ansatz

As we mentioned, the KK expansion is always valid, since it is just a generalized Fourier theorem. But the KK reduction ansatz is not, except in the case of the torus T^n, when it is always valid, as we will shortly see. In general the KK reduction ansatz is not consistent (i.e. valid), except *at the linearized level*, i.e. for terms quadratic in the action.

Indeed, making a truncation to just the lowest mode ϕ_0, and setting the rest to zero ($\phi_q = 0$) is in general not a solution of the higher dimensional (D-dimensional) equations of motion. If it is a solution to the higher dimensional equations of motion, we say we have a *consistent truncation*.

What can go wrong? To see that, consider a ϕ^3 coupling in the higher dimension, and focus on a single term in the action resulting from its KK expansion, namely on

$$(\dots) \int d^d \vec{x} \sqrt{\det g^{(0)}_{\mu\nu}} \phi_q^{I_q}(\vec{x}) \phi_0^{I_0}(\vec{x}) \phi_0^{J_0}(\vec{x}) \times \int d^n \vec{y} \sqrt{\det g^{(0)}_{mn}} Y_q^{I_q}(\vec{y}) Y_0^{I_0}(\vec{y}) Y_0^{J_0}(\vec{y}). \quad (5.19)$$

This term acts as a source for the equations of motion of ϕ_n, in general of the type

$$(\Box - m_q^2) \phi_q^{I_q}(\vec{x}) = (\dots) \phi_0^{I_0}(\vec{x}) \phi_0^{J_0}(\vec{x}). \quad (5.20)$$

It is therefore inconsistent (not a solution of the equations of motion for ϕ_q) to set ϕ_q to zero, while keeping ϕ_0. But there is a way to avoid this problem: the above equation of motion is the equation of motion of ϕ_q, which only appears after integrating over $\vec{y}$ and writing the reduced action in d dimensions for $\phi_q(\vec{x})$ and $\phi_0(\vec{x})$. But in integrating, it can happen that

$$\int d^n \vec{y} \sqrt{\det g^{(0)}_{mn}} Y_q^{I_q}(\vec{y}) Y_0^{I_0}(\vec{y}) Y_0^{J_0}(\vec{y}) \quad (5.21)$$

could be zero, and in that case the truncation is consistent, and we have a consistent dimensional reduction ansatz.

This is indeed what happens for the torus, since there $Y_0^{I_0}(\vec{y}) = 1$, and we obtain

$$\int dy Y^{I_n} = \int dy e^{\frac{iny}{R}} = 0 \quad (5.22)$$

for $n \neq 0$. Therefore for the torus we always have a consistent truncation, as we claimed.

We can also have a generalization of this case, namely if we have some global symmetry G for the fields in the KK expansion, and under the dimensional reduction ansatz we keep ALL the singlets of G (fields that do not transform under G), then we obtain the same result. Indeed, if $Y_0^{I_0}$ and $Y_0^{J_0}$ are singlets, then $Y_0^{I_0} Y_0^{J_0}$ is also a singlet, whereas $Y_q^{I_q}$ is not, since we assumed that we kept all the singlets. Then by spherical harmonic orthogonality, or rather by the need for G-group invariance, we have

$$\int Y_q^{I_q}(Y_0^{I_0} Y_0^{J_0}) = 0. \quad (5.23)$$

If we have an inconsistent truncation, we can sometimes make it consistent by making a nonlinear redefinition of the fields, i.e. something of the type

$$\phi_q' = \phi_q + a\phi_0^2 + \dots,$$
$$\phi_0' = \phi_0 + \sum_{pq(\text{including } 0)} c_{pq} \phi_p \phi_q. \quad (5.24)$$

Equivalently, we can make a *nonlinear KK ansatz* from the beginning. This would then only come from the KK expansion after the nonlinear redefinition, but otherwise needs to be considered on its own.

The simplest example of a nonlinear KK ansatz is the one needed to get the correct d-dimensional Einstein action (in an Einstein frame) from the D-dimensional Einstein action. Namely, we need to write

$$g_{\mu\nu}(\vec{x}, \vec{y}) = g_{\mu\nu}(\vec{x}) \left[\frac{\det g_{mn}(\vec{x}, \vec{y})}{\det g_{mn}^{(0)}(\vec{y})} \right]^{-\frac{1}{d-2}} . \tag{5.25}$$

To check this formula completely would take some calculation, but we can make a simple check. As we know, $\Gamma \sim g^{-1}\partial g$ and $R_{\mu\nu} \sim \partial\Gamma + \Gamma\Gamma$, which means that under a constant scale transformation $g_{\mu\nu} \to \lambda g_{\mu\nu}$, $R_{\mu\nu}$ will be invariant. In the D-dimensional Einstein action we have $\int \sqrt{g^{(D)}} R^{(D)}$, and from $R^{(D)} = R_{MN}^{(D)} g^{MN}$ we only look at the sum over d-dimensional indices $\mu\nu$, $\tilde{R} = R_{\mu\nu}^{(D)} g^{\mu\nu}$, that contain the d-dimensional Einstein action (the other terms contain gauge fields and scalars). Then under $g_{\mu\nu} \to g_{\mu\nu}\lambda$ (with g_{mn} untouched),

$$\sqrt{g^{(D)}} \tilde{R} = \sqrt{g^{(d)}} \sqrt{\det g_{mn}} \lambda^{\frac{d}{2}-1} \tilde{R} , \tag{5.26}$$

which means that indeed we need to take $\lambda = [\det g_{mn} / \det g_{mn}^{(0)}]^{-1/(d-2)}$.

5.5 Example: original Kaluza–Klein reduction

The idea of Kaluza and Klein was to unify gravity ($g_{\mu\nu}$) and electromagnetism (B_{μ}) into a 5-dimensional metric g_{MN}. The linearized KK reduction ansatz for the 5-dimensional metric g_{MN} would then be, according to our general analysis,

$$g_{MN} = \begin{pmatrix} g_{\mu\nu}(\vec{x}) & g_{\mu 5} = B_{\mu}(\vec{x}) \\ g_{5\mu} = B_{\mu}(\vec{x}) & g_{55} = \phi(\vec{x}) \end{pmatrix} . \tag{5.27}$$

This ansatz is always consistent, since we are on a circle, and we have kept all the zero modes. But since experimentally we do not observe a massless scalar ϕ, Kaluza and Klein wanted to choose the background value $\phi = 1$ (as we said, for tori $g_{mn}^{(0)} = \delta_{mn}$), i.e. to put the fluctuation in ϕ to zero. But this further truncation is inconsistent, i.e. it does not satisfy the equations of motion! So we cannot unify gravity and electromagnetism in this simple way. Therefore we need to keep ϕ, in which case we do have a consistent ansatz, i.e. theoretically valid, just that it does not agree with experiments, since we do not see ϕ. But in this case, even though the reduction ansatz is consistent, we still need to write nonlinear modifications in order to get both the action for gravity in the standard Einstein form, and the action for electromagnetism in the standard Maxwell form. Finally, the nonlinear KK reduction ansatz is

$$g_{MN} = \begin{pmatrix} g_{\mu\nu}(\vec{x})\phi^{-1/2}(\vec{x}) & B_{\mu}(\vec{x})\phi(\vec{x}) \\ B_{\mu}(\vec{x})\phi(\vec{x}) & \phi(\vec{x}) \end{pmatrix} , \tag{5.28}$$

which we can rewrite as

$$g_{MN} = \Phi^{-1/3}(\vec{x}) \begin{pmatrix} g_{\mu\nu}(\vec{x}) & B_{\mu}(\vec{x})\Phi(\vec{x}) \\ B_{\mu}(\vec{x})\Phi(\vec{x}) & \Phi(\vec{x}) \end{pmatrix} , \tag{5.29}$$

redefining the scalar field by $\phi = \Phi^{2/3}$.

5.6 General properties; symmetries

On a general compact space, the linearized KK ansatz for the off-diagonal components of the metric is

$$g_{\mu m}(\vec{x}, \vec{y}) = B_\mu^{AB}(\vec{x}) V_m^{AB}(\vec{y}) , \qquad (5.30)$$

where $V_m^{AB}(\vec{y})$ is called a Killing vector, and it has an index in an adjoint of the gauge group of symmetries of the compact space; A, B are fundamental indices, and B_μ^{AB} is a gauge field. That means that in general, for each independent Killing vector we get one corresponding gauge field.

However, since in supergravity we deal with vielbeins instead of metrics, it means we need to explain what happens to them as well. For vielbeins we also have the local Lorentz transformations, which we can use to fix part of the vielbein, which has otherwise more components than the metric. We denote the flat indices with α for noncompact and a for compact space. We can use the off-diagonal part of the local Lorentz transformations to fix the gauge $E_m^\alpha = 0$, thus breaking down the total local Lorentz invariance $SO(1, D-1)$ to just $SO(1, d-1) \times SO(D-d)$. Then we have an ansatz for E_μ^α compatible with the ansatz for $g_{\mu\nu}$, namely

$$E_\mu^\alpha(\vec{x}, \vec{y}) = e_\mu^\alpha(\vec{x}) \left[\frac{\det E_m^a(\vec{x}, \vec{y})}{\det e_m^{(0)a}(\vec{y})} \right]^{-\frac{1}{d-2}} , \qquad (5.31)$$

whereas for the remaining off-diagonal vielbein we write an ansatz in terms of gauge fields,

$$E_\mu^a(\vec{x}, \vec{y}) = B_\mu^m(\vec{x}, \vec{y}) E_m^a(\vec{x}, \vec{y}), \qquad (5.32)$$

$$B_\mu^m(\vec{x}, \vec{y}) = B_\mu^{AB}(\vec{x}) V_m^{AB}(\vec{y}). \qquad (5.33)$$

Note that the multiplication by E_m^a was needed in order to curve the index on V^{AB}, as it should be. Finally, for the scalars in E_m^a, there is no general recipe, and we must write an ansatz on a case by case basis.

On a general product space $M_4 \times K_n$, a spinor splits into a spinor on M_4 times a spinor on K_n,

$$\lambda_A(\vec{x}, \vec{y}) = \lambda_\tau^I(\vec{x}) \eta_i^I(\vec{y}) , \qquad (5.34)$$

where the index I is an index in a spinor representation of the symmetry group G of the compact space, $A = \{\tau, i\}$ is an $SO(1, D-1)$ spinor index that splits into τ, a spinor index on M_4 and i, a spinor index on K_n.

Since both in D dimensions and in d dimensions we have the spin-statistics theorem, it means that both the spinors λ_A and the spinors λ_τ^I must be anticommuting. But that in turn means that necessarily $\eta_i^I(\vec{y})$ must be *commuting spinors*.

On spaces with symmetries, the $\eta_i^I(\vec{y})$ are so-called "Killing spinors", which are a sort of square root of the Killing vectors.

The Killing vectors (so named after Wilhelm Killing) satisfy the equation

$$D_{(\mu} V_{\nu)}^{AB} = 0 , \qquad (5.35)$$

where the covariant derivative uses the background metric on the compact space K_n.

The Killing spinors on a sphere satisfy

$$D_\mu \eta_i^I = c(\gamma_\mu \eta^I)_i \equiv c e_\mu^\alpha (\gamma_\alpha \eta^I)_i \,, \tag{5.36}$$

where c is a constant.

Moreover, on a sphere, the Killing vectors and the Killing spinors are related by

$$V_\mu^{AB} = \bar{\eta}^I \gamma_\mu \eta^J (\gamma^{AB})_{IJ} \,, \tag{5.37}$$

since the gamma matrices $(\gamma^A)_{IJ}$ relate vector (A) and spinor (I) indices.

But how do we define Killing spinors more generally? To do so, we note that for 4-dimensional $\mathcal{N} = 1$ supergravity, we had $\delta_{\text{susy}} \psi_\mu = D_\mu \epsilon$, and moreover, the γ_μ term is also present in the supersymmetry transformation law of the gravitino in certain cases of reduction of higher dimensional supergravities.

It then follows that the more general definition of the Killing spinor is of a spinor that preserves some supersymmetry,

$$\delta_{\text{susy}} \lambda_A(\vec{x}, \vec{y}) = 0. \tag{5.38}$$

This condition in general will imply a condition of the type

$$D_\mu \eta^I = (\text{fields} \times \gamma \text{ matrices})_\mu |_{\text{bgr}} \eta^I \,, \tag{5.39}$$

and in turn that means that we will use a KK reduction ansatz of the type (5.34). That is, we keep only as many spinors as there are Killing spinors. The reason is that, since they will preserve supersymmetry, by the susy algebra they will be massless because $\{Q, Q\} \sim H$, thus $Q|0\rangle = 0 \Rightarrow H|0\rangle = 0$, whereas other states will be massive. Since in the KK reduction we are supposed to keep all the massless modes, the above ansatz follows.

We will see that in general we can construct all the "massless spherical harmonics" from Killing spinors, therefore the Killing spinors are good basis objects for describing the compact space.

Symmetries

On a torus, all the fields of the same spin (scalars, vectors, antisymmetric tensors) group into multiplets of some global symmetry group G (which symmetry group is, however, not obvious a priori, without knowing the theory we are KK reducing).

If we now compactify the same theory on a nontrivial space K_n of the same dimensionality as the torus above, for instance the sphere S^n, the abelian Killing spinors (i.e. trivial, $V_m = 1$) of the torus will change into nonabelian Killing spinors of some gauge group $H \subset G$, therefore the abelian vector fields in a representation of the global group G from the torus case now re-group as nonabelian fields of part or all of G, that is, we are *gauging the global symmetry* (making it local).

Important concepts to remember

- In KK reduction, we consider a product space $M_D = M_d \times K_n$.
- There are three KK metrics: the background metric, the KK expansion, and the KK reduction ansatz.

- The background metric is a solution of the product space type, the KK expansion is a generalization of the Fourier expansion, which is always valid, and the KK reduction ansatz means truncation to only one mode, and is a priori valid only at the linearized level.
- The KK expansion is in terms of spherical harmonics, which are eigenfunctions of the Laplacean on the compact space.
- On the torus, the spherical harmonics are just products of Fourier mode exponentials, and the fields split into fields of different d-dimensional spin, according to the split $M = (\mu, m)$.
- The truncation to the zero modes (KK reduction) is a priori inconsistent at the nonlinear level, i.e. it could happen not to satisfy the D-dimensional equations of motion.
- On a torus, or if we have some global symmetry group G, and keep ALL the singlets under the symmetry, the linear KK reduction is consistent.
- Sometimes, a nonlinear redefinition of fields, or equivalently a nonlinear KK reduction ansatz from the beginning, will turn a reduction ansatz consistent.
- In the original KK ansatz, the truncation $\phi = 1$ is inconsistent.
- To get the EH action in d dimensions, we redefine $g_{\mu\nu}$ by $[\det g_{mn} / \det g_{mn}^{(0)}]^{-1/(d-2)}$, and for the vielbein E_μ^α by $[\det e_m^a / \det e_m^{(0)a}]^{-1/(d-2)}$.
- The off-diagonal metric gives a gauge field for each Killing vector, $g_{\mu m} = B_\mu^{AB} V_m^{AB}$, or $E_\mu^a = B_\mu^{AB} V_m^{AB} E_m^a$.
- Spinors are expanded into d-dimensional spinors times Killing spinors $\lambda_A = \lambda_\tau^I \eta_i^I$.
- Killing spinors preserve some supersymmetry.

References and further reading

For the Kaluza–Klein approach to supergravity, see [16]. For more details, see for instance [36] and references therein.

Exercises

1. For a 4-sphere, the Euclidean embedding coordinates Y^A are scalar spherical harmonics, satisfying $Y^A Y_A = 1$ (and so $Y^A D_\mu^{(0)} Y^A = 0$, $D_\mu^{(0)} Y^A D_\nu^{(0)} Y^A = g_{\mu\nu}^{(0)}$.) Prove then that

$$\epsilon_{A_1 \dots A_5} dY^{A_1} \wedge dY^{A_2} = 3\sqrt{g^{(0)}} \epsilon_{\mu\nu\rho\sigma} dx^\mu \wedge dx^\nu \partial^\rho Y^{[A_3} \partial^\sigma Y^{A_4} Y^{A_5]}. \tag{5.40}$$

2. For the original KK metric

$$g_{MN} = \phi^{-1/3} \begin{pmatrix} g_{\mu\nu} & B_\mu \phi \\ B_\mu \phi & \phi \end{pmatrix}, \tag{5.41}$$

prove that $g_{\mu\nu}$ is the metric in an Einstein frame.

3. Prove that if $g_{\mu m}(x, y) = B_\mu^{AB}(x) V_m^{AB}(y)$ and we choose the general coordinate transformation with parameter

$$\xi_m(x, y) = \lambda^{AB}(x) V_m^{AB}(y) , \qquad (5.42)$$

then the transformation with parameter $\lambda^{AB}(x)$ is the nonabelian gauge transformation of B_μ^{AB}. Note: Use the fact that $V^{AB} = V^{mAB} \partial_m$ satisfies the nonabelian algebra.

4. Let Y^A be six Cartesian coordinates for the 5-sphere S^5. Then Y^A are vector spherical harmonics and $Y^{A_1 \ldots A_n} = Y^{(A_1} \ldots Y^{A_n)} - traces$ is a totally symmetric traceless spherical harmonic (i.e. $Y^{A_1 \ldots A_n} \delta_{A_m A_p} = 0$, $\forall \ 1 \leq m, p \leq n$). Check that, as polynomials in 6d, $Y^{A_1 \ldots A_n}$ satisfy $\Box_{6d} Y^{A_1 \ldots A_n} = 0$. Expressing $\Box_{6d}$ in terms of $\Box_{S^5}$ and ∂_r (where $Y^A Y^A \equiv r^2$), check that $Y^{A_1 \ldots A_n}$ are eigenfunctions with eigenvalues $-k(k+5-1)/r^2$.

5. Dimensionally reduce an 11-dimensional Maxwell-type 3-form gauge field A_{MNP} on a 7-torus T^7, down to four dimensions.

6. Show that starting with a Euclidean Lagrangean

$$\mathcal{L} = \frac{1}{2}(\partial_\mu \phi)^2 + \frac{1}{2}m^2\phi^2 + \lambda\phi^3 , \qquad (5.43)$$

for which the bulk 3-point function is equal to λ, and doing a transformation $\phi = \tilde{\phi} + a\tilde{\phi}^2$, we obtain the same bulk 3-point function, but also get a boundary 3-point function contribution.

6.1 The Schwarzschild solution: metric, horizon, black holes

The Schwarzschild solution (1916)

The Schwarzschild solution is a solution to the Einstein equation without matter ($T_{\mu\nu} = 0$), namely

$$R_{\mu\nu} - \frac{1}{2} g_{\mu\nu} R = 0. \tag{6.1}$$

It is in fact the most general solution of Einstein's equation with $T_{\mu\nu} = 0$ and spherical symmetry, which is a theorem of Birkhoff from 1923. It means that by general coordinate transformations we can always bring the metric, in the case of spherical symmetry and no matter, to this form. In this subsection, we deal only with four spacetime dimensions.

To find a solution, without using the full power of general relativity, we will take a detour to understand its Newtonian limit. For Newtonian gravity, the local form of Gauss' law is

$$\vec{\nabla}^2 U_{\text{Newton}} = 4\pi G_N \rho_m, \tag{6.2}$$

where ρ_m is the density of matter. In Exercise 7 of Chapter 4 we saw that the quadratic action for perturbations coming from the Einstein–Hilbert action is the Fierz–Pauli action. In the de Donder gauge, the equation of motion is $\partial_\rho^2 \bar{h}_{\mu\nu} = 0$, which means that the gauge-fixed form of the action is (remember that $g_{\mu\nu} = \eta_{\mu\nu} + \kappa_N h_{\mu\nu}$)

$$\mathcal{L}_{\text{FP,g.f.}} = \frac{1}{2} (\partial_\rho \bar{h}_{\mu\nu})^2. \tag{6.3}$$

The nonrelativistic energy-momentum tensor (dust matter) is $T_{\mu\nu} = \text{diag}(\rho, p, p, p)$ with $p \simeq 0$. For a static, spherically symmetric solution, we expect only h_{00} and $h_{11} = h_{22} = h_{33} \equiv h_{ii}$ to be nonzero. On the other hand, $\bar{h}_{\mu\nu} = h_{\mu\nu} - h\eta_{\mu\nu}/2$ ($h = \eta^{\mu\nu} h_{\mu\nu}$), so $\bar{h}_{00} = (h_{00} + 3h_{ii})/2$ and $\bar{h}_{ii} = 3/2(h_{00} - h_{ii})$. Then the Einstein equation with matter, component ii, gives $\Box h_{00} = \Box h_{ii}$, and then the Einstein equation component 00, becomes in the static case

$$\vec{\nabla}^2 (\kappa_N h_{00}) = -8\pi G_N \rho, \tag{6.4}$$

which means that in the Newtonian limit we can identify

$$\kappa_N h_{00} = -2U_{\text{Newton}} = \kappa_N h_{ii}. \tag{6.5}$$

Therefore for weak fields $\kappa_N h_{\mu\nu} \ll 1$ and nonrelativistic, $v \ll 1$, we can always put the metric in the form

$$ds^2 \simeq -(1+2U_N)dt^2 + (1-2U_N)d\vec{x}^2 = -(1+2U_N)dt^2 + (1-2U_N)(dr^2 + r^2 d\Omega_2^2). \tag{6.6}$$

Consider now the case of a pointlike source, $\rho = M\delta^3(x)$, so

$$\Delta U_{\text{Newton}} = 4\pi G_N M\delta^3(x) \Rightarrow \Delta \kappa_N h_{00} = -8\pi G_N M\delta^3(x). \tag{6.7}$$

The solution is

$$\kappa_N h_{00} = \kappa_N h_{ii} = -2U_N = +\frac{2MG_N}{r}, \tag{6.8}$$

which means that $g_{00} = -(1 - 2MG_N/r)$, which becomes 0 at $r = 2MG_N$, meaning it is an apparent singularity. But then we should see this behavior also in the spatial part of the metric. As it stands, $g_{ii} = 1 + 2MG_N/r$ does not have it, but $1/(1 - 2MG_N/r)$ does, and both have the same weak field limit.

In fact, the correct Schwarzschild metric is

$$ds^2 = -\left(1 - \frac{2MG_N}{r}\right)dt^2 + \frac{dr^2}{1 - \frac{2MG_N}{r}} + R^2 d\Omega_2^2. \tag{6.9}$$

Of course, this metric has a source at $r = 0$, which is how we find the solution from the Newtonian approximation. But the point is that if the space is empty at $r \geq r_0$, with r_0 some arbitrary value, and is spherically symmetric, Birkhoff's theorem says that we should obtain the Schwarzschild metric for $r \geq r_0$ (and maybe a modified solution at $r \leq r_0$).

But the solution apparently becomes singular at $r_H = 2MG > 0$, so it would seem that it cannot reach its source at $r = 0$? This would be a paradoxical situation, since then what would be the role of the source? It would seem as if we do not really need a point mass to create this metric.

If the Schwarzschild solution is valid all the way down to $r = r_H$ (not just to some $r_0 > r_H$ which is the case for, let us say, the gravitational field of the Earth, in which case r_0 is the Earth's radius), then we call that solution a *Schwarzschild black hole*. We call $r = r_H$ the *event horizon* of the black hole. In general a black hole is a solution that has an event horizon, whose properties we now study.

First, let us investigate the propagation of light, which is the fastest possible signal. If light propagates radially (along $d\theta = d\phi = 0$), its propagation on $ds^2 = 0$ implies

$$dt = \frac{dr}{1 - \frac{2MG_N}{r}}. \tag{6.10}$$

This means that near r_H we have

$$dt \simeq 2MG_N \frac{dr}{r - 2MG_N} \Rightarrow t \simeq 2MG_N \ln(r - 2MG_N) \to \infty. \tag{6.11}$$

In other words, from the point of view of an asymptotic observer, who measures coordinates r, t (since at large r, $ds^2 \simeq -dt^2 + dr^2 + r^2 d\Omega_2^2$), it takes an infinite time for light to reach r_H. And conversely, it takes an infinite time for a light signal from $r = r_H$ to reach the observer at large r. That means that $r = r_H$ is cut-off from causal communication with $r = r_H$. For this reason, $r = r_H$ is called an "event horizon." Nothing can reach, or escape from the event horizon in finite time.

Observation: However, quantum mechanically, Hawking proved that black holes radiate thermally, thus thermal radiation does escape the event horizon of the black hole. We will understand this better in a later chapter.

But is the event horizon of the black hole singular or not?

The answer is actually NO. In gravity, the metric is not gauge invariant, it changes under coordinate transformations. The appropriate gauge invariant (general coordinate transformations invariant) quantity that measures the curvature of space is the Ricci scalar R, as we saw. One can calculate it for the Schwarzschild solution (we will not do it here) and one obtains that at the event horizon

$$R \sim \frac{1}{r_H^2} = \frac{1}{(2MG_N)^2} = \text{finite!} \qquad (6.12)$$

Since the curvature of space at the horizon is finite, an observer falling into a black hole does not feel anything special at $r = r_H$, other than a finite curvature of space creating some tidal force pulling him apart with finite strength.

So for an observer at large r, the event horizon looks singular, but for an observer falling into the black hole it does not seem remarkable at all. This shows that in general relativity, more than in special relativity, different observers see apparently different events. For instance, in special relativity, synchronicity of two events is relative, which is still true in general relativity, but now there are more examples of relativity.

An observer at fixed r close to the horizon sees an apparently singular behavior: if $dr = 0, d\Omega = 0$, then

$$ds^2 = -\frac{dt^2}{1 - \frac{2MG_N}{r}} = -d\tau^2 \Rightarrow d\tau = \sqrt{-g_{00}}dt = \frac{dt}{\sqrt{1 - \frac{2MG_N}{r}}}, \qquad (6.13)$$

thus the time measured by that observer becomes infinite as $r \to r_H$, and we get an infinite time dilation: an observer fixed at the horizon is "frozen in time" from the point of view of the observer at infinity. Of course, a freely falling observer sees that he falls through the event horizon in a finite time (nothing special happens for him as he falls through the event horizon), but from the point of view of the observer at infinity, it takes an infinite time for the freely falling observer to pass through the event horizon.

6.2 Continuation inside the horizon; global structure

Since there is no singularity at the event horizon, it means that there must exist coordinates that continue inside the horizon, and there are indeed. The first such coordinates were found by Eddington (in 1924) and Finkelstein (in 1958). So Finkelstein rediscovered them 34 years later, without being aware of Eddington's work, which shows that the subject of black holes was not so popular back then. In fact, it was only in the 1960s it started becoming popular. The Eddington–Finkelstein coordinates, however, do not cover all the geometry.

The first set of coordinates that cover all the geometry was found by Kruskal and Szekeres in 1960, and they give maximum insight into the physics, so we will describe them here.

One first introduces the "tortoise" coordinates r_* by imposing

$$\frac{dr}{1 - \frac{2MG_N}{r}} = dr_* \Rightarrow r_* = r + 2MG_N \ln\left(\frac{r}{2MG_N} - 1\right), \tag{6.14}$$

which gives the metric

$$ds^2 = \left(1 - \frac{2MG_N}{r}\right)(-dt^2 + dr_*^2) + r^2(r_*)d\Omega_2^2. \tag{6.15}$$

Next one introduces the null (lightcone) coordinates considered by Eddington and Finkelstein,

$$u = t - r_*; \quad v = t + r_*, \tag{6.16}$$

such that light ($ds^2 = 0$) travels at $u = $ constant or $v = $ constant. Finally, one introduces Kruskal coordinates,

$$\bar{u} = -4MG_N e^{-\frac{u}{4MG_N}}; \quad \bar{v} = +4MG_N e^{\frac{v}{4MG_N}}. \tag{6.17}$$

Then the region $r \geq 2MG_N$ becomes $-\infty < r_* < +\infty$, thus $-\infty < \bar{u} \leq 0, 0 \leq \bar{v} < +\infty$. But the metric in Kruskal coordinates is

$$ds^2 = -\frac{2MG_N}{r} e^{-\frac{r}{2MG_N}} d\bar{u}d\bar{v} + r^2 d\Omega_2^2, \tag{6.18}$$

where r stands for the implicit $r(\bar{u}, \bar{v})$. This metric is nonsingular at the horizon $r = 2MG_N$, and thus can be analytically continued for general values of $\bar{u}, \bar{v}$, covering all the real line $\bar{u}, \bar{v} \in \mathbb{R}$, having four quadrants (I–IV) instead of one (I)! The only obstruction to the $(\bar{u}, \bar{v})$ covering all $\mathbb{R}^2$ is the $r = 0$ singularity.

We can find the relation between $\bar{u}, \bar{v}$, and r that is valid generally, since from the definitions of $\bar{u}, \bar{v}$, we have

$$-\frac{\bar{u}\bar{v}}{(4MG_N)^2} = e^{\frac{v-u}{4MG_N}} = e^{\frac{r_*}{2MG_N}} = e^{\frac{r}{2MG_N}}\left(\frac{r}{2MG_N} - 1\right). \tag{6.19}$$

This means that the $r = 0$ singularity (restricting the $(\bar{u}, \bar{v})$ space) corresponds to $\bar{u}\bar{v} = (4MG_N)^2$.

The resulting *Kruskal diagram*, or diagram in Kruskal coordinates $\bar{t}, \bar{r}$, where $\bar{u} = \bar{t} - \bar{r}$, $\bar{v} = \bar{t} + \bar{r}$, is given in Fig. 6.1. The diagonal lines $\bar{u} = 0$ and $\bar{v} = 0$ separate the four quadrants, with the original quadrant (I) being $\bar{u} \leq 0, \bar{v} \geq 0$. The $r = 0$ singularity corresponds to $\bar{u}\bar{v} = (4MG_N)^2$, i.e. $\bar{t}^2 - \bar{r}^2 = (4MG_N)^2$.

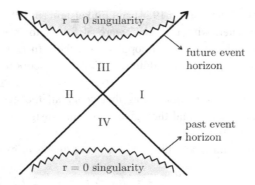

Kruskal diagram of the Schwarzschild black hole.

The Penrose diagram can be obtained from (6.18). Since modulo the conformal factor, we just have the flat space $ds^2 = d\bar{u}d\bar{v}$, the Penrose diagram is a subset of the diagram of 2-dimensional flat space, which as we saw in Chapter 2 is a diamond, restricted by the two $\bar{t}^2 - \bar{r}^2 = (4MG_N)^2$ lines. We make the transformations (2.48), modified by a $4MG_N$ factor, i.e.

$$\bar{u} = 4MG_N \tan \tilde{u}_+; \quad \bar{v} = 4MG_N \tan \tilde{u}_-; \quad \tilde{u}_\pm = \frac{\tau \pm \theta}{2}. \tag{6.20}$$

Then the lines corresponding to the $r = 0$ singularity are

$$1 = \tan \frac{\tau + \theta}{2} \tan \frac{\tau - \theta}{2} = \frac{\sin^2(\tau/2) - \sin^2(\theta/2)}{1 - \sin^2(\tau/2) - \sin^2(\theta/2)}, \tag{6.21}$$

i.e. $\sin^2(\tau/2) = 1/2$, thus $\tau = \pm\pi/2$, whereas the maximum value for τ (obtained for $\theta = 0$) is $|\tau| = \pi$. The resulting Penrose diagram is given in Fig. 6.2a. It has two causally disconnected (by two event horizons) asymptotic regions (left and right), and an $r = 0$ singularity shielded by an event horizon in the future (up) and past (down). The Penrose diagram of a physical black hole, obtained from a collapsing star, is given in Fig. 6.2b. It has no singularity in the past, and otherwise has only one asymptotic region (right) and one horizon.

6.3 Solutions with charge; solutions inside AdS space

One can add also a point electric charge to the Schwarzschild black hole at the same point as the mass, thus obtaining the *Reissner–Nordstrom black hole*. We thus require the electric field

$$F_{rt} = \frac{Q}{4\pi \epsilon_0 r^2} \Rightarrow A_t = -\frac{Q}{4\pi \epsilon_0 r}. \tag{6.22}$$

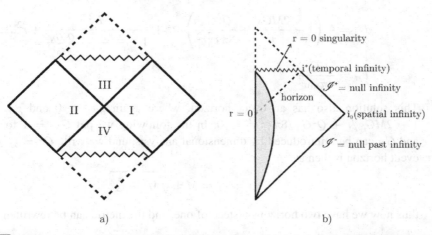

Figure 6.2 a) Penrose diagram of the eternal Schwarzschild black hole (time independent solution). The dotted line gives the completion of the Penrose diagram of flat 2-dimensional (Minkowski) space; b) Penrose diagram of a physical black hole, obtained from a collapsing star (the curved line). The dotted line gives the completion of the Penrose diagram of flat $d > 2$-dimensional (Minkowski) space.

Then the energy-momentum tensor is

$$T_{\mu\nu} = F_{\mu\rho}F_\nu{}^\rho - \frac{1}{4}g_{\mu\nu}F_{\rho\sigma}F^{\rho\sigma} \Rightarrow$$

$$T_{tt} = (F_{rt})^2 g^{rr} - \frac{1}{2}(F_{rt})^2 g^{rr} \simeq \frac{1}{2}\frac{Q^2}{(4\pi\epsilon_0)^2 r^4}, \tag{6.23}$$

where in the last equality for T_{tt} we use the background $g_{\mu\nu} \simeq \eta_{\mu\nu}$, valid for the first order solution. Then the equation of motion in the Newtonian approximation, (6.7), gets modified to

$$\Delta U_N = 4\pi G_N \left(M\delta^3(x) + \frac{Q^2}{(4\pi\epsilon_0)^2 2r^4}\right). \tag{6.24}$$

But since

$$\vec{\nabla}^2 \frac{1}{r^2} = \left(\frac{d^2}{dr^2} + \frac{2}{r}\frac{d}{dr}\right)\frac{1}{r^2} = \frac{2}{r^4},$$

$$\vec{\nabla}^2 \frac{1}{r} = -4\pi\delta^3(x), \tag{6.25}$$

we have the solution

$$U_N = -\frac{MG_N}{r} + \frac{Q^2 G_N}{4\pi\epsilon_0^2 4r^2}. \tag{6.26}$$

But we saw that $2U_{\text{Newton}} = -\kappa_N h_{tt}$ in the Newtonian approximation, and in the full solution the same appears in the denominator in g_{rr}. Therefore the full Reissner–Nordstrom solution is

$$ds^2 = -\left(1 - \frac{2MG_N}{r} + \frac{Q^2 G_N}{8\pi \epsilon_0^2 r^2}\right) dt^2 + \frac{dr^2}{1 - \frac{2MG_N}{r} + \frac{Q^2 G_N}{8\pi \epsilon_0^2 r^2}} + r^2 d\Omega_2^2,$$

$$F_{rt} = \frac{Q}{4\pi \epsilon_0 r^2}. \tag{6.27}$$

This solution also has an event horizon, where again $g_{tt} = 0$ and $g_{rr} = \infty$, so $1 - 2MG_N/r + Q^2 G_N/8\pi \epsilon_0^2 r^2 = 0$. In the following we put $G_N = 1$ for simplicity, and G_N can be reintroduced by dimensional analysis, and we write $\tilde{Q}^2 = Q^2/8\pi \epsilon_0^2$. The event horizon is then at

$$r = r_\pm = M \pm \sqrt{M^2 - \tilde{Q}^2}, \tag{6.28}$$

thus now we have two horizons instead of one, and the metric can be rewritten as

$$ds^2 = -\Delta dt^2 + \frac{dr^2}{\Delta} + r^2 d\Omega_2^2; \quad \Delta = \left(1 - \frac{r_+}{r}\right)\left(1 - \frac{r_-}{r}\right),$$

$$\tilde{Q} = \sqrt{r_+ r_-}; \quad M = \frac{r_+ + r_-}{2}. \tag{6.29}$$

However, if $M < \tilde{Q}$, there is no horizon at all, as we can see from (6.28), just a "naked singularity" at $r = 0$, i.e. the singularity is not covered by a horizon. This is believed to be excluded on physics grounds: there are a number of theorems saying that naked singularities should not occur under certain very reasonable assumptions. Therefore we must have $M \geq \tilde{Q}$, which is called the Bogomolnyi–Prasad–Sommerfield bound, or BPS bound. In supersymmetric theories (with the Reissner–Nordstrom solution embedded in them), this is the same bound that is obtained from the susy algebra, for the mass to be greater than the central charge of the algebra.

The case of saturation of the bound, $M = \tilde{Q}$, is special as we saw in the case of the susy algebra. In this "extremal black hole" case, we obtain the solution

$$ds^2 = -\left(1 - \frac{M}{r}\right)^2 dt^2 + \left(\frac{dr}{1 - \frac{M}{r}}\right)^2 + r^2 d\Omega_2^2, \tag{6.30}$$

and by a change of coordinates $r = M + \bar{r}$ we get

$$ds^2 = -\frac{1}{\left(1 + \frac{M}{\bar{r}}\right)^2} dt^2 + \left(1 + \frac{M}{\bar{r}}\right)^2 (d\bar{r}^2 + \bar{r}^2 d\Omega_2^2). \tag{6.31}$$

Here

$$H = 1 + \frac{M}{\bar{r}} \tag{6.32}$$

is a harmonic function, i.e. it satisfies

$$\Delta_{(3)} H = -4\pi M \delta^3(r). \tag{6.33}$$

We see that the extremal solutions are defined by a harmonic function in three dimensions.

There is a simple generalization to solutions with both electric and magnetic charge,

$$F_{rt} = \frac{\tilde{Q}_e}{\sqrt{2\pi} r^2}; \quad F_{\theta\phi} = Q_m \sin\theta,$$ (6.34)

where in the metric we replace $\tilde{Q}_e^2$ by

$$Q^2 = \tilde{Q}_e^2 + Q_m^2.$$ (6.35)

One can put the Reissner–Nordstrom black hole inside an Anti-de Sitter space as well as follows. The Anti-de Sitter space metric can be written as (2.81) in a form parameterized in the same way as the black hole metric, in terms of a function Δ. Therefore the AdS-Reissner–Nordstrom solution is obtained by combining the Δ of the two, giving

$$ds^2 = -\Delta dt^2 + \frac{dr^2}{\Delta} + r^2 d\Omega_2^2; \quad \Delta \equiv 1 - \frac{2MG_N}{r} + \frac{\tilde{Q}^2 G_N}{r^2} - \frac{8\pi G_N \Lambda r^2}{3}.$$ (6.36)

The only other parameter one can add to a black hole is the angular momentum J, in which case, however, the metric is quite complicated. There are so-called "no hair theorems" stating that black holes are characterized only by Q, M, and J (any other charge or parameter would be called "hair" of the black hole).

6.4 Black holes in higher dimensions

Electromagnetism in D dimensions

In three spatial dimensions, the local form of Gauss' law is

$$\vec{\nabla} \cdot \vec{E} = \frac{\rho}{\epsilon_0}.$$ (6.37)

The electric field of a point charge is obtained by integrating it over a ball B^3 with a sphere S^2 at its boundary and using the Stokes theorem,

$$\int_{B^3} dV \vec{\nabla} \cdot \vec{E} = \int_{B^3} dV \frac{\rho}{\epsilon_0} = \frac{Q}{\epsilon_0}$$
$$= \int_{S^2(R)} d\vec{S} \cdot \vec{E} = \text{vol}(S^2) R^2 E(R),$$ (6.38)

where $\text{vol}(S^2)$ is the area of the 2-sphere, 4π. Therefore the electric field is

$$E(R) = \frac{Q}{\text{vol}(S^2)\epsilon_0 R^2} = \frac{Q}{4\pi \epsilon_0 R^2}.$$ (6.39)

We can easily generalize this to d spatial dimensions ($D = d+1$ spacetime dimensions). The local Gauss' law is the same (6.37). Integrating it, we obtain

$$\int_{B^d} dV \vec{\nabla} \cdot \vec{E} = \int_{B^d} dV \frac{\rho}{\epsilon_0} = \frac{Q}{\epsilon_0}$$

$$= \int_{\partial B^d = S^{d-1}(R)} d\vec{S} \cdot \vec{E} = \text{vol}(S^{d-1}) R^{d-1} E(R), \tag{6.40}$$

which gives the electric field

$$E(R) = \frac{Q}{R^{d-1} \text{vol}(S^{d-1}) \epsilon_0}. \tag{6.41}$$

But since the volume of the sphere is

$$\text{vol}(S^{d-1}) \equiv \Omega_{d-1} = \frac{2\pi^{d/2}}{\Gamma(d/2)}, \tag{6.42}$$

we obtain

$$E(R) = \frac{Q}{R^{d-1}} \frac{\Gamma(d/2)}{2\pi^{d/2} \epsilon_0}. \tag{6.43}$$

Newtonian gravity in D dimensions

For Newtonian gravity, the local Gauss' law in D spacetime dimensions is

$$\vec{\nabla}^2 U_N = 4\pi G_N^{(D)} \rho_m. \tag{6.44}$$

The integrated version in 3+1 dimensions is

$$\int_{B^3} dV \vec{\nabla}^2 U_N = 4\pi G_N^{(4)} \int_{B^3} dV \rho_m = 4\pi G_N^{(4)} M$$

$$= \int_{S^2(R)} d\vec{S} \cdot \vec{\nabla} U_N = \text{vol}(S^2) R^2 |\vec{g}(R)|. \tag{6.45}$$

We have for the gravitational force $\vec{F} = -m\vec{\nabla} U_N$, which gives for the gravitational acceleration (the equivalent of the electric field)

$$|\vec{g}(R)| = \frac{G_N^{(4)} M}{r^2}. \tag{6.46}$$

Before generalizing, we relate $G_N^{(D)}$ with $G_N^{(4)}$ under KK compactification. For KK compactification, with the product metric on $M_4 \times K^n$ ($D = 4 + n$), we have $\sqrt{g^{(D)}} = \sqrt{g^{(4)}}\sqrt{g^{(n)}}$. Substituting in the D-dimensional Einstein–Hilbert action,

$$S = \frac{1}{16\pi G_N^{(D)}} \int d^D x \sqrt{-g^{(D)}} R^{(D)}, \tag{6.47}$$

and integrating over K_n, we obtain

$$\frac{1}{G_N^{(4)}} = \int d^n x \frac{\sqrt{-g^{(n)}}}{G_N^{(D)}} = \frac{V^{(n)}}{G_N^{(D)}}, \tag{6.48}$$

which gives

$$G_N^{(D)} = G_N^{(4)} V^{(n)}. \tag{6.49}$$

Then $G_N^{(D)}$ has dimensions of $[G_N^{(D)}] = L^{D-2}$ and we have $1/(8\pi G_N^{(D)}) = [M_{\text{Pl}}^{(D)}]^{D-2}$ and $\kappa_N = \sqrt{8\pi G_N^{(D)}}$.

The integrated version of Gauss' law for gravity in d space dimensions ($D = d + 1$) is now

$$\int_{B^d} dV \vec{\nabla}^2 U_N = 4\pi G_N^{(D)} \int_{B^d} dV \rho_m = 4\pi G_N^{(D)} M$$

$$= \int_{\partial B^d = S^{d-1}(R)} d\vec{S} \cdot \vec{\nabla} U_N = \text{vol}(S^{d-1}) R^{d-1} |\vec{g}(R)|. \tag{6.50}$$

Then the gravitational acceleration is

$$|\vec{g}(R)| = \frac{4\pi G_N^{(D)} \Gamma((D-1)/2) M}{2\pi^{\frac{D-1}{2}} R^{D-2}} = \frac{\Gamma((D-1)/2)}{2[M_{\text{Pl}}^{(D)}]^{D-2} 2\pi^{\frac{D-1}{2}}} \frac{M}{R^{D-2}}, \tag{6.51}$$

and the gravitational potential is

$$U_N(R) = -\frac{4\pi G_N^{(D)}}{(D-3)\text{vol}(S^{D-2})} \frac{M}{R^{D-3}} = -\frac{2\pi^{\frac{3-D}{2}} \Gamma((D-1)/2)}{D-3} \frac{M G_N^{(D)}}{R^{D-3}}$$

$$= -\frac{\Gamma((D-1)/2)}{4\pi^{\frac{D-1}{2}} (D-3)[M_{\text{Pl}}^{(D)}]^{D-2}} \frac{M}{R^{D-3}} \equiv -\frac{C^{(D)} G_N^{(D)} M}{R^{D-3}}. \tag{6.52}$$

Black hole solutions

The Schwarzschild solution is then

$$ds^2 = -(1 + 2U_N)dt^2 + \frac{dr^2}{1 + 2U_N} + r^2 d\Omega_{D-2}^2$$

$$= -\left(1 - \frac{2C^{(D)} G_N^{(D)} M}{r^{D-3}}\right) dt^2 + \frac{dr^2}{1 - \frac{2C^{(D)} G_N^{(D)} M}{r^{D-3}}} + r^2 d\Omega_{D-2}^2. \tag{6.53}$$

To obtain the Reissner–Nordstrom solution, we use $F_{tr}(r) = E(r)$ in (6.43) in the energy momentum tensor:

$$T_{\mu\nu} = -\frac{2}{\sqrt{-g}} \frac{\delta}{\delta g^{\mu\nu}} \int d^D x \sqrt{-g} \left(-\frac{1}{4} F_{\mu\nu} F^{\mu\nu}\right) = F_{\mu\rho} F_\nu{}^\rho - \frac{1}{4} g_{\mu\nu} F_{\rho\sigma} F^{\rho\sigma}. \tag{6.54}$$

The tt component is

$$T_{tt} = (F_{rt})^2 g^{rr} - \frac{1}{2}(F_{rt})^2 g^{rr} \simeq \frac{1}{2}(F_{tr})^2 = \frac{Q^2}{2(\Omega_{D-2}\epsilon_0)^2 r^{2(D-2)}}, \tag{6.55}$$

giving the equation

$$\Delta U_N = 4\pi G_N^{(D)} \left[M\delta^{D-1}(x) + \frac{Q^2}{2(\Omega_{D-2}\epsilon_0)^2 r^{2(D-2)}}\right]. \tag{6.56}$$

From

$$\vec{\nabla}^2 \frac{1}{r^{2(D-3)}} = \left(\frac{d^2}{dr^2} + \frac{D-2}{r}\frac{d}{dr}\right)\frac{1}{r^{2(D-3)}} = \frac{2(D-3)^2}{r^{2(D-2)}},$$

$$\vec{\nabla}^2 \frac{1}{r^{D-3}} = -(D-3)\Omega_{D-2}\delta^{D-1}(x), \tag{6.57}$$

we get the potential

$$U_N = -\frac{4\pi}{(D-3)\Omega_{D-2}}\frac{MG_N^{(D)}}{r^{D-3}} + \frac{4\pi}{4(D-3)^2\Omega_{D-2}^2\epsilon_0^2}\frac{Q^2 G_N^{(D)}}{r^{2(D-2)}}$$

$$\equiv -\frac{C^{(D)}MG_N^{(D)}}{r^{D-3}} + \frac{C'^{(D)}Q^2 G_N^{(D)}}{r^{2(D-3)}}. \tag{6.58}$$

Then we write the metric from $-g_{tt} \equiv F(r) = 1 + 2U_N(r)$, as

$$ds^2 = -F(r)dt^2 + \frac{dr^2}{F(r)} + r^2 d\Omega_{D-2}^2. \tag{6.59}$$

The solution has two horizons at $F(r) = 0$, i.e.

$$1 - 2\frac{C^{(D)}MG_N^{(D)}}{r^{D-3}} + 2\frac{C'^{(D)}Q^2 G_N^{(D)}}{r^{2(D-3)}} = 0, \tag{6.60}$$

with solutions

$$(r_\pm)^{D-3} = C^{(D)}MG_N^{(D)} \pm \sqrt{(C^{(D)}MG_N^{(D)})^2 - 2C'^{(D)}Q^2 G_N^{(D)}}. \tag{6.61}$$

The BPS bound is now

$$M^2 \geq \frac{2C'^{(D)}}{C^{(D)2}G_N^{(D)}}Q^2. \tag{6.62}$$

When the bound is saturated, we have an extremal solution, with

$$r_+ = r_- = r_H = \left[C^{(D)}MG_N^{(D)}\right]^{\frac{1}{D-3}} = \left[Q\sqrt{2C'^{(D)}}G_N^{(D)}\right]^{\frac{1}{D-3}}, \tag{6.63}$$

and $F(r)$ becomes

$$F(r) = \left[1 - \left(\frac{r_H}{r}\right)^{D-3}\right]^2. \tag{6.64}$$

Then we can define new coordinates $\bar{r}$ by

$$r^{D-3} = \bar{r}^{D-3} + r_H^{D-3}, \tag{6.65}$$

which gives

$$1 - \left(\frac{r_H}{r}\right)^{D-3} = \frac{1}{1 + \left(\frac{r_H}{\bar{r}}\right)^{D-3}} \equiv \frac{1}{f(\bar{r})}. \tag{6.66}$$

We obtain the extremal black hole metric

$$ds^2 = -f(\bar{r})^{-2}dt^2 + f(\bar{r})^{\frac{2}{D-3}}(d\bar{r}^2 + \bar{r}^2 d\Omega_{D-2}^2). \tag{6.67}$$

The details are left as an exercise (Exercise 6).

We see that for this extremal black hole, there is a horizon only at $\bar{r} = 0$, which coincides with the singularity, and is written in terms of the harmonic function

$$f(\bar{r}) = 1 + \left(\frac{r_H}{\bar{r}}\right)^{D-3},\tag{6.68}$$

satisfying

$$\Delta f(\bar{r}) = -4\pi G_N^{(D)} M \delta^{D-1}(x).\tag{6.69}$$

6.5 Black holes extended in p spatial dimensions: "p-brane solutions"

We can have other generalizations of black holes as well, called "black p-branes." These are black holes that extend in p spatial dimensions. The terminology comes from the word mem-brane which is now called a 2-brane, that is, it extends in two spatial dimensions.

In the absence of charges, the generalization is trivial, and we obtain black Schwarzschild p-branes. By KK reduction on a p-dimensional torus T^p (flat space with identifications), the solution should be the Schwarzschild solution. Therefore we have

$$ds^2 = -\left(1 - \frac{2C^{(D-p)}G_N^{(D-p)}M}{r^{D-3-p}}\right)dt^2 + d\vec{x}_p^2 + \frac{dr^2}{1 - \frac{2C^{(D-p)}G_N^{(D-p)}M}{r^{D-3-p}}} + r^2 d\Omega_{D-2-p}^2.\tag{6.70}$$

But we have a more interesting generalization, charged extremal p-branes, but not charged under electromagnetic fields, but rather under antisymmetric tensor fields $A_{\mu_1...\mu_{p+1}}$.

Note that in four dimensions, the only localized extremal p-branes are the black holes. An extended object can be either a cosmic string (one spatial extension) or a domain wall (two spatial extensions). However, we will shortly see that, like the extremal Reissner–Nordstrom black holes, the extremal p-branes are defined by harmonic functions in $D - p - 1$ dimensions (the black hole, with $p = 0$, in $D = 4$ is defined by a harmonic function in three dimensions). Thus for a cosmic string, the harmonic function would be in two dimensions, which is $H = \ln|z|$ ($z = x_1 + ix_2$), whereas for a domain wall, the harmonic function would be in one dimension, which is $H = 1 + a|x|$. In both cases, the harmonic function increases away from its source, so both the cosmic string and the domain wall extremal p-brane solutions would affect the whole space. They are therefore quite unlike black holes, and not quite physical.

But in dimensions higher than four, we can have black hole-like objects extended in p spatial dimensions that are localized in space (i.e., don't grow at infinity). These are the "black p-branes," charged under a $p + 1$-form antisymmetric tensor field, and have complicated metrics described by *two* harmonic functions. The extremal case is easier to explain, so we will start with it, and it is described by a single harmonic function.

In electromagnetism, a static electric charge, i.e. an electron, has only A_0 nonzero (the 0 component was called t previously), therefore it is a solution to the action

$$S = \int d^4x \left[-\frac{F_{\mu\nu}^2}{4} + j^\mu A_\mu \right], \qquad (6.71)$$

with j^μ a delta function source, $j^0 = Q\delta^3(x)$, and the rest zero. Therefore the source term is $\int d^4x j^\mu A_\mu = \int d^4x j^0 A_0$, and the A_μ equation of motion gives the electric field of the electron.

Similarly, we find that an electric p-brane in D dimensions carries electric charge Q_p with respect to the $p+1$-form field $A_{\mu_1...\mu_{p+1}}$. By analogy with the above, there should be a source coupling

$$\int d^D x j^{\mu_1...\mu_{p+1}} A_{\mu_1...\mu_{p+1}} \rightarrow \int d^D x j^{01...p} A_{01...p}. \qquad (6.72)$$

It therefore follows that a source for the $A_{01...p}$ field will be of the type $j^{01...p} = Q_p \delta^{(D-p-1)}(x)$, which is an object extended in p spatial dimensions plus time. The solution of the source coupling is an object with nonzero $A_{01...p}$, and indeed the p-brane has such a nonzero field.

The electric p-brane in the absence of gravity would be a solution to

$$S = \int d^D x \left[-\frac{1}{2(p+2)!} F_{\mu_1...\mu_{p+2}}^2 + j^{\mu_1...\mu_{p+1}} A_{\mu_1...\mu_{p+1}} \right], \qquad (6.73)$$

given by

$$A_{01...p} = -\frac{C_p Q_p}{r^{D-p-3}}, \qquad (6.74)$$

where C_p is a constant. Using (6.57), this will satisfy

$$\Delta_{(D-p-1)} A_{01...p} = [(D-p-3)\Omega_{D-p-2} C_p] Q_p \delta^{(D-p-1)}(x), \qquad (6.75)$$

which allows, for instance, normalizing of the coefficient of Q_p to 1, giving

$$C_p = \frac{1}{(D-p-3)\Omega_{D-p-2}}. \qquad (6.76)$$

When coupling to gravity, this will be a solution to

$$S = S_D(p+1) + S_{p+1}$$
$$S_D(p+1) = \frac{1}{2\kappa_N^2} \int d^D x \sqrt{-g} \left[R - \frac{1}{2(p+2)!} e^{-a(p)\phi} F_{(p+2)}^2 - \frac{1}{2}(\partial_\mu \phi)^2 \right], \qquad (6.77)$$

where, defining $\tilde{p} = D - p - 4$,

$$a(p) = \sqrt{4 - \frac{2(p+1)(\tilde{p}+1)}{p+\tilde{p}+2}}. \qquad (6.78)$$

Note that in this action we also have a scalar ϕ, called the dilaton. The source term in D dimensions must include the coupling to gravity of the source term above, i.e.

$$S_{\text{source}} = \int d^D x \sqrt{-g} j^{\mu_1 \ldots \mu_{p+1}} A_{\mu_1 \ldots \mu_{p+1}}, \tag{6.79}$$

with $j^{\mu_1 \ldots \mu_{p+1}}$ a certain delta function localizing it onto $p+1$ dimensions, but in fact it contains a source for the dilaton, as well as for the metric. It is

$$S_{p+1} = T_{p+1} \int d^{p+1}\xi \left[-\frac{1}{2} \sqrt{-\gamma} \gamma^{ij} \partial_i X^M \partial_j X^N G_{MN} e^{a(p)\phi/(p+1)} + \frac{p-1}{2} \sqrt{-\gamma} \right.$$
$$\left. -\frac{1}{(p+1)!} \epsilon^{i_1 \ldots i_{p+1}} \partial_{i_1} X^{M_1} \ldots \partial_{i_{p+1}} X^{M_{p+1}} A_{M_1 \ldots M_{p+1}} \right]. \tag{6.80}$$

We come back to describing this source term in the next chapter, after describing the fundamental string action.

There is a BPS bound for the *tension* $T_p = M/V_p$ (mass per unit p-dimensional volume), $T_p \geq c_p Q_p$, where c_p is a numerical constant. At saturation of the bound, we have the *extremal p-brane solution*.

The easiest to describe are these extremal solutions, which as we will see are relevant for string theory. Here, we simply write them down, since deriving them is quite complicated. Then we want also to describe solutions of $D = 10$ supergravity theory (a part of which can be written as (6.80)), that approximates string theory at moderate energies. Their extremal p-brane solutions are

$$ds^2_{\text{string}} = H_p^{-1/2}(-dt^2 + d\vec{x}_p^2) + H_p^{1/2}(d\bar{r}^2 + \bar{r}^2 d\Omega_{8-p}^2)$$
$$= H_p^{-1/2}(-dt^2 + d\vec{x}_p^2) + H_p^{1/2} d\vec{x}_{9-p}^2,$$
$$e^{-2\phi} = H_p^{\frac{p-3}{2}},$$
$$A_{01\ldots p} = -\frac{1}{2}(H_p^{-1} - 1), \tag{6.81}$$

where H_p is a harmonic function of $\vec{x}_{9-p}$, i.e.

$$\Delta_{(9-p)} H_p = -[(7-p)\Omega_{8-p} 2C_p] Q_p \delta^{(9-p)}(x^i); \quad \Rightarrow H_p = 1 + \frac{2C_p Q_p}{\bar{r}^{7-p}}, \tag{6.82}$$

and the coordinate $\bar{r}$ is related to r in (6.74) by

$$r^{7-p} = \bar{r}^{7-p} + 2C_p Q_p, \tag{6.83}$$

similarly to relation (6.65) for the Reissner–Nordstrom case.

Here ds^2_{string} is known as the "string metric" and is related to the usual "Einstein metric" defined until now by a conformal factor

$$ds^2_{\text{Einstein}} = e^{-\phi/2} ds^2_{\text{string}}, \tag{6.84}$$

and $A_{01\ldots p}$ is some antisymmetric tensor ("gauge") field present in the 10-dimensional supergravity theory (there are several), and ϕ is the "dilaton" field, which is a scalar field that is related to the string theory coupling constant by $g_s = e^{-\phi}$.

The 10-dimensional gravitational action in the string metric is

$$\int d^{10}x \sqrt{-g_{\text{string}}} e^{-2\phi} R(g_{\text{string}}), \tag{6.85}$$

and the rescaling by $e^{-\phi/2}$ of the metric is such that it removes the $e^{-2\phi}$ factor in front of R ($\sqrt{-g_{\text{string}}} \to e^{5\phi/2} \sqrt{-g_{\text{Einstein}}}$, $g_{\text{string}}^{\mu\nu} R_{\mu\nu,\text{string}} \to e^{-\phi/2} g_{\text{Einstein}}^{\mu\nu} R_{\mu\nu,\text{Einstein}} + \ldots$).

The black p-brane solutions away from extremality are parameterized by $\mu_p = T_p - c_p Q_p > 0$. With respect to the extremal solution, only the metric is modified to

$$ds^2 = H_p^{-1/2}(-f(r)dt^2 + d\vec{x}_p^2) + H_p^{1/2}\left(\frac{dr^2}{f(r)} + r^2 d\Omega_{8-p}^2\right), \tag{6.86}$$

where

$$f(r) = 1 - \frac{\tilde{c}_p \mu_p}{r^{7-p}}. \tag{6.87}$$

Important concepts to remember

- The Newtonian limit of general relativity gives $ds^2 \simeq -(1+2U_N)dt^2 + (1-2U_N)(dr^2 + r^2 d\Omega^2)$.
- The Schwarzschild solution is the most general solution with spherical symmetry and no sources. Its source is located behind the event horizon.
- The exact solution is $ds^2 = -(1+2U_N)dt^2 + dr^2/(1+2U_N) + r^2 d\Omega^2$.
- If the solution is valid down to the horizon, it is called a black hole.
- Light takes an infinite time to reach the horizon, from the point of view of the far away observer, and one has an infinite time dilation at the horizon ("frozen in time").
- Classically, nothing escapes the horizon (quantum mechanically, we have Hawking radiation).
- The horizon is not singular, and one can analytically continue inside it via the Kruskal coordinates.
- Black hole solutions with charge (Reissner–Nordstrom) have $M \geq \tilde{Q}$ and can also be written in terms of U_N only.
- The $M = \tilde{Q}$ solutions (extremal) are defined by a harmonic function and have a collapsed horizon coinciding with the singularity.
- p-brane solutions without charge are just Schwarzschild black holes extended on a flat torus.
- p-brane solutions with electric charge are (extremal or not) black hole solutions that extend in p spatial dimensions. They also carry charge under an antisymmetric tensor field $A_{\mu_1 \ldots \mu_{p+1}}$, and are determined by a harmonic function.
- They are obtained from supergravity with a source term for the spacetime fields.

References and further reading

For an introduction to black holes, the relevant chapters in [4] are probably a good starting point. A very advanced treatment of the topological properties of black holes can be found in Hawking and Ellis [17]. They also have a good treatment of Penrose diagrams, so one can read that selectively. For an introduction to p-brane solutions of supergravity, see the review [18]. To understand the usefulness of p-branes, one can look at Tseytlin's "harmonic function rule" developed in [19]. To understand the meaning of extremal p-branes, one can look at the rule for making an extremal solution non-extremal, found in [20].

Exercises

1. Check the transformation from Schwarzschild coordinates to Kruskal coordinates.
2. Verify that the Penrose diagram for an astrophysical black hole (from a collapsing star) is the one in Fig. 6.2b.
3. Consider the *ingoing Eddington–Finkelstein coordinates* v and r, with u defined in (6.16). Show that the metric becomes

$$ds^2 = -\left(1 - \frac{2MG}{r}\right)dv^2 + 2dvdr + r^2d\Omega_2^2. \qquad (6.88)$$

Similarly, consider the *outgoing Eddington–Finkelstein coordinates* u and r, with v defined in (6.16). Show that now the metric becomes

$$ds^2 = -\left(1 - \frac{2MG}{r}\right)du^2 - 2dudr + r^2d\Omega_2^2. \qquad (6.89)$$

4. Check that $H = 1 + a/r^{7-p}$ is a good harmonic function for a p-brane. Check that $r = 0$ is an event horizon (it traps light).
5. The electric current of a point charge is $j^\mu = Q\frac{dx^\mu}{d\tau}\delta^{d-1}(x^\mu(\tau))$. Write an expression for the $p+1$-form current of a p-brane, $j^{\mu_1\cdots\mu_{p+1}}$.
6. Prove that the change of coordinates

$$r^{D-3} = \bar{r}^{D-3} + r_H^{D-3} \qquad (6.90)$$

takes the extremal black hole metric to

$$ds^2 = -f(\bar{r})^{-2}dt^2 + f(\bar{r})^{\frac{2}{D-3}}(d\bar{r}^2 + \bar{r}^2d\Omega_{D-2}^2). \qquad (6.91)$$

String theory is the theory of relativistic strings. That is, not strings like violin strings, but strings that move at the speed of light. They do not have a compression mode: the energy density along a string is not a Lorentz invariant, so cannot appear as a physical variable in a relativistic theory. They only have a vibration mode, unlike, e.g. a massive cosmic string or a violin string.

However, they can have tension, i.e. energy per unit length, which resists forces pulling the string apart. The point is that if one stretches the string the energy density stays the same, just the length increases, thus *energy = tension × length*.

Because they have tension, the only possible action for a string is the one that minimizes the area traversed by the string, i.e. the "worldsheet". We will see that this expectation is correct. However, before we write the action, we will review what happens for particles, which is a simpler version of the case of strings.

7.1 Worldline particle action and worldline construction of Feynman diagrams

The action for a nonrelativistic particle is

$$S = \int dt L = \int dt \frac{m\dot{\vec{x}}^2}{2}. \tag{7.1}$$

The action for a relativistic particle is an action that reduces to the above at small speeds, namely (here τ is proper time, $ds^2 = -d\tau^2$)

$$S = -m(c^2) \int d\tau; \quad d\tau^2 = -\eta_{\mu\nu} dx^\mu dx^\nu, \tag{7.2}$$

that is, the mass times the invariant element on the worldline of the particle, i.e. the proper time along the path of the particle. The action can be rewritten in terms of the position of the particle, $X^\mu(\tau)$, more precisely in terms of $\dot{X}^\mu \equiv dX^\mu/d\tau$, as

$$S = -m \int d\tau \sqrt{-\dot{X}^\mu \dot{X}^\nu \eta_{\mu\nu}}. \tag{7.3}$$

Note that in the nonrelativistic limit we have $d\tau^2 = dt^2 - d\vec{x}^2/c^2 = dt^2(1 - v^2/c^2)$, so the action becomes

$$S = -mc^2 \int dt \sqrt{1 - \frac{v^2}{c^2}} \simeq \int dt \left[-mc^2 + \frac{mv^2}{2} \right], \tag{7.4}$$

as it should.

This action has *reparameterization invariance*. That is, we can use any parameter τ, not just the proper time. Under $\tau' = \tau'(\tau)$, $dx^\mu/d\tau = (dx^\mu/d\tau')d\tau'/d\tau$, so the action becomes

$$S = -mc^2 \int d\tau \frac{d\tau'}{d\tau} \sqrt{-\eta_{\mu\nu} \frac{dX^\mu}{d\tau'} \frac{dX^\nu}{d\tau'}}. \tag{7.5}$$

That means the action is indeed reparameterization invariant. Note that the paths are the same, $X'^\mu(\tau'(\tau)) = X^\mu(\tau)$.

The equations of motion of the action in (7.3) are obtained by varying the action,

$$\delta S = -m \int d\tau \delta \left(\sqrt{-\dot{X}^\mu \dot{X}_\mu} \right) = +m \int d\tau \frac{d}{d\tau} \left[-\frac{\eta_{\mu\nu}\dot{X}^\mu}{\sqrt{-\dot{X}^\rho \dot{X}_\rho}} \right] \delta X^\nu + \delta X^\mu m \frac{dX_\mu}{d\tau} \Bigg|_{\tau_i}^{\tau_f}. \tag{7.6}$$

Here we have used $-(\dot{X}^\mu)^2 = -ds^2/d\tau^2 \equiv 1$. The 4-momentum is

$$p^\mu = mu^\mu = m \frac{dX^\mu}{d\tau}, \tag{7.7}$$

and in terms of this momentum the equation of motion is the equation of the free particle,

$$\frac{dp^\mu}{d\tau} = 0. \tag{7.8}$$

This of course looks rather trivial, we obtain just the free motion in a straight line. However, if we write the same action in curved space instead, replacing $\eta_{\mu\nu} \to g_{\mu\nu}$, we will get the free motion along a geodesic in spacetime. The geodesic equation is then nontrivial, and can be understood as the interaction of the particle with the gravitational field. In more general terms, we can say that background fields (like the metric) appearing in the particle or string actions will give interaction effects.

Coupling the particle to a background charge is done by adding the $A_\mu j^\mu$ to the action, i.e.

$$\int d\tau A_\mu(X^\rho(\tau)) \left(q \frac{dX^\mu}{d\tau} \right) = \int d^4 x A_\mu(X^\rho(\tau)) q \frac{dX^\mu}{d\tau} \delta^3(X^\rho(\tau))$$

$$\equiv \int d^4 x A_\mu(X^\rho(\tau)) j^\mu(X^\rho(\tau)). \tag{7.9}$$

But what is the usefulness of the particle action for quantum field theory?

Let us suppose that we do not know how to do quantum field theory and/or the precise theory we have. We can then still *construct Feynman diagrams*, considered as describing particles propagating in spacetime, for instance as in Fig. 7.1.

To construct such a Feynman diagram, we need:

- The propagator from x to y.
- The vertex factor at x and y: this contains the coupling g, thus it defines a particular theory.

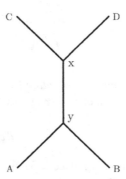

Feynman diagram in x space: from x to y we have the particle propagator.

- Rules about how to integrate (in this case, $\int d^4x \int d^4y$). For particles, this is obvious, but for strings, we need to carefully define a path integral construction. There are subtleties due to the possibility of overcounting if we use naive integration.

The propagator from x to y for a massless particle can be written as (here $\Box$ is the kinetic operator)

$$\langle x|\Box^{-1}|y\rangle = \int_0^\infty d\tau \, \langle y|e^{-\tau\Box}|x\rangle. \qquad (7.10)$$

But now we can use a trick: a massive nonrelativistic particle has the Hamiltonian $H = \vec{p}^2/(2m) = \Box/(2m)$ (if $\vec{p}$ and $\Box$ live in a Euclidean x space). Using $m = 1/2$ we get $H = \Box$ and therefore we can use quantum mechanics to write a path integral representation of the transition amplitude:

$$\langle y|e^{-\tau H}|x\rangle = \int_x^y \mathcal{D}x(t)e^{-\frac{1}{4}\int_0^\tau dt \dot{x}^2}. \qquad (7.11)$$

Since $H = \Box$, we use this representation to express the propagator of a massless relativistic particle in (7.10) as

$$\langle x|\Box^{-1}|y\rangle = \int_0^\infty d\tau \, \langle y|e^{-\tau\Box}|x\rangle = \int_0^\infty d\tau \int_x^y \mathcal{D}x(t)e^{-\frac{1}{4}S_p}, \qquad (7.12)$$

where $S_p = \int_0^\tau dt \dot{x}^2$ is the massless particle action. We have not met that yet, but we will see it in the next section.

So the particle action defines the propagator, and to complete the perturbative definition of the quantum field theory by Feynman diagrams we need to add the vertex rules specifying the interactions of the theory (for instance, in the $V = \lambda\phi^4$ example in Chapter 1 we had a vertex $-\lambda$), as well as the integration rules, which in the case of the particle are trivial.

We do the same for string theory: we define perturbative string theory by defining Feynman diagrams. We write a worldsheet action that will give the propagator, and then interaction rules and integration rules, which unlike the particle case, follow directly from the consistency of the theory.

7.2 First order particle action

As we saw above, in order to define perturbative quantum field theory we need to define a worldline action for the particle, but it is not the usual particle action, which is non-linear in Xs, but rather a quadratic action. To define the propagator, we need an $\int \dot{x}^2$ action. This action is obtained in a first order formalism for the worldline. As usual with first order formalisms, we need to introduce an extra (auxiliary) field. In this case, it is the (intrinsic) worldline metric field $\gamma_{\tau\tau}(\tau)$, obtained by considering the worldline as an intrinsic "surface," not defined by its embedding in spacetime. More precisely, we use the "vielbein" formalism, and introduce the "einbein" field (since we are in one dimension) $e(\tau) = \sqrt{-\gamma_{\tau\tau}(\tau)}$.

To write down the action in terms of the field X^μ, which from the point of view of the worldline is a scalar, we use the fact that in one dimension, $\sqrt{-\det\gamma} \times \gamma^{\tau\tau} = e^{-1}(\tau)$ and $\sqrt{-\det\gamma} = e(\tau)$ are the quantities used to integrate a 2-index tensor and a scalar. Then the action for the massive particle of mass m can only be

$$S_p = \frac{1}{2} \int d\tau \left(e^{-1}(\tau) \frac{dX^\mu}{d\tau} \frac{dX^\nu}{d\tau} \eta_{\mu\nu} - em^2 \right), \tag{7.13}$$

up to a relative number, since the first term is the quadratic action for a scalar in curved spacetime, and the second is a "cosmological constant" term. The action is spacetime Poincaré invariant and also worldine reparameterization invariant. The reparameterization transformation ("general coordinate invariance") is defined by $e'(\tau')d\tau' = e(\tau)d\tau$, which immediately leads to invariance of the second term in the action, and to $e'^{-1}(\tau')/d\tau' = e^{-1}(\tau)/d\tau$, which leads to invariance of the first term.

To check that the action is indeed the first order form of the action from the previous section, we write down the equation of motion for $e(\tau)$,

$$-\frac{1}{e^2}\dot{X}^2 - m^2 = 0 \Rightarrow e^2(\tau) = -\frac{\dot{X}^\mu \dot{X}_\mu}{m^2}. \tag{7.14}$$

Substituting in S_p we get

$$S_p = \frac{1}{2} \int d\tau \left[\frac{m}{\sqrt{-\dot{X}^2}} \dot{X}^2 - \frac{\sqrt{-\dot{X}^2}}{m} m^2 \right] = -m \int d\tau \sqrt{-\dot{X}^\mu \dot{X}_\mu} \equiv S_1, \tag{7.15}$$

therefore the action S_p in (7.13) is indeed a first order form of the action S_1 in the previous section. It follows that the actions S_p and S_1 are classically equivalent, though quantum mechanically they probably are not.

The action S_p is now much simpler, being only quadratic in the scalar fields X^μ. Also, now we can take the $m \to 0$ limit of S_p and obtain a nontrivial result,

$$S_{p,m=0} = \frac{1}{2} \int d\tau \left(e^{-1}(\tau) \frac{dX^\mu}{d\tau} \frac{dX^\nu}{d\tau} \eta_{\mu\nu} \right), \tag{7.16}$$

unlike the action S_1, when we get zero. The action S_p is also reparameterization invariant, which as we saw changes $e(\tau)$ to another function $e'(\tau')$, therefore we can fix a gauge for this symmetry and set $e(\tau)$ to anything, for instance $e(\tau) = 1$. In this gauge, the massless particle action becomes

$$S_{m=0,e=1} = \frac{1}{2} \int d\tau \frac{dX^\mu}{d\tau} \frac{dX^\nu}{d\tau} \eta_{\mu\nu}, \tag{7.17}$$

which is the result we used in the previous section, in the calculation of the massless particle propagator.

For this gauge-fixed action, the equation of motion for $X^\mu(\tau)$ is

$$\frac{d}{d\tau} \left(\frac{dX^\mu}{d\tau} \right) = 0, \tag{7.18}$$

as above. However, we now have to supplement it with a constraint, which is the equation of motion of $e(\tau)$ ($\delta S_p / \delta e(\tau) = 0$), since $e(\tau)$ was set to 1 in this gauge,

$$-\frac{ds^2}{d\tau^2} = \frac{dX^\mu}{d\tau} \frac{dX^\nu}{d\tau} \eta_{\mu\nu} \equiv T = 0. \tag{7.19}$$

This is just the statement that the particle is massless, and the constraint is the equivalent of the Gauss law constraint for electrodynamics in the gauge $A_0 = 0$.

7.3 A relativistic tensionful string: the Nambu–Goto action

We now go back to strings and mimic what we did for particles. A string is an object with a 1-dimensional spatial extension, so the moving string spans a 1+1-dimensional "worldsheet" as in Fig. 7.2, parameterized by intrinsic coordinates (σ, τ), with $\sigma =$ worldsheet length and $\tau =$ worldsheet time. The spacetime coordinates are now again scalars from the point of view of the worldsheet, called $X^\mu(\sigma, \tau)$.

Since in the particle case, the second order action was the length of the worldline, with the mass in front, now the second order action, due to Nambu and Goto, is the area of the worldsheet, with the tension (mass per unit length) in front, $T = 1/2\pi\alpha'$,

$$S = -T \int dA = -\frac{1}{2\pi\alpha'} \int dA. \tag{7.20}$$

Here α' has mass dimension -2, and the area of the string worldsheet is, as we saw in the general relativity chapter, $d^2\xi \sqrt{-\det(\gamma_{ab})}$, where $a, b = 1, 2$ and $\{\xi^a\} = (\sigma, \tau)$. Therefore we expect the action to be

$$S = -\frac{1}{2\pi\alpha'} \int d\sigma \, d\tau \sqrt{-\det(\gamma_{ab})} = -\frac{1}{2\pi\alpha'} \int_{\tau_i}^{\tau_f} d\tau \int_0^l d\sigma \sqrt{-\gamma_{11}\gamma_{22} + (\gamma_{12})^2}. \tag{7.21}$$

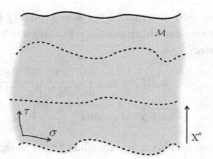

String moving in spacetime parameterized by X^μ spans a worldsheet $\mathcal{M}$ parameterized by σ (coordinate along the string) and τ (worldsheet time).

But we need to define the string through an embedding of the worldsheet surface into spacetime. We start by understanding the familiar case of a spatial 2-dimensional surface parameterized by ξ^i, $i = 1, 2$, embedded into Euclidean 3-dimensional space with metric $ds^2 = d\vec{X} \cdot d\vec{X}$. Consider $\vec{X}(\xi^i)$ as defining the embedding that describes the surface. Then

$$d\vec{X} = \frac{\partial \vec{X}}{\partial \xi^i} d\xi^i \Rightarrow$$

$$ds^2 = \left(\frac{\partial \vec{X}}{\partial \xi^i} \cdot \frac{\partial \vec{X}}{\partial \xi^j} \right) d\xi^i d\xi^j \equiv g_{ij}(\xi) d\xi^i d\xi^j, \tag{7.22}$$

therefore the *induced metric* (metric induced on the surface by the embedding in flat space) is

$$g_{ij}(\xi) = \frac{\partial \vec{X}}{\partial \xi^i} \cdot \frac{\partial \vec{X}}{\partial \xi^j}. \tag{7.23}$$

Now let us generalize to the case of embedding a 1+1-dimensional (Minkowskian) surface parameterized by ξ^a, into a curved spacetime of arbitrary dimension. The embedding $X^\mu(\xi^a)$ gives

$$ds^2 = g_{\mu\nu} dX^\mu dX^\nu = \eta_{\mu\nu} \frac{\partial X^\mu}{\partial \xi^a} \frac{\partial X^\nu}{\partial \xi^b} d\xi^a d\xi^b, \tag{7.24}$$

leading to the induced metric on the worldsheet, or the "pull-back" of the spacetime metric,

$$h_{ab}(\xi^a) = \partial_a X^\mu \partial_b X^\nu g_{\mu\nu}(X(\xi^a)). \tag{7.25}$$

Finally, the Nambu–Goto action is the area action written not in terms of the intrinsic metric $\gamma_{ab}(\xi^a)$, but the induced metric $h_{ab}(\xi^a)$,

$$S_{\text{NG}} = -\frac{1}{2\pi\alpha'} \int d\sigma\, d\tau \sqrt{-\det(h_{ab}(\xi))}. \tag{7.26}$$

The area written in terms of γ_{ab} was explicitly reparameterization (general coordinate) invariant, so we can check that the one written in terms of $h_{ab}(\xi)$ is also reparameterization invariant, since

$$h_{ab}(\xi) = g_{\mu\nu}\frac{\partial X^\mu}{\partial \xi'^c}\frac{\partial X^\nu}{\partial \xi'^d}\frac{\partial \xi'^c}{\partial \xi^a}\frac{\partial \xi'^d}{\partial \xi^b} = h_{cd}(\xi')\frac{\partial \xi'^c}{\partial \xi^a}\frac{\partial \xi'^d}{\partial \xi^b} \Rightarrow$$

$$\sqrt{\det(h_{ab})} = \sqrt{\det(h'_{ab})}\left|\frac{\partial \xi'}{\partial \xi}\right|. \tag{7.27}$$

This Nambu–Goto action then has $X^\mu(\xi)$ as the only variable, so it derives the form of the string worldsheet embedding into spacetime.

7.4 The Polyakov action

We can, however, also write a first order action, just as in the particle case. The action is called the Polyakov action (since it was discovered by Brink, DiVecchia, Howe, Deser, and Zumino), and in flat spacetime ($g_{\mu\nu} = \eta_{\mu\nu}$), it is

$$S_P[X, \gamma] = -\frac{1}{4\pi\alpha'}\int d\sigma d\tau \sqrt{-\gamma}\,\gamma^{ab}\partial_a X^\mu \partial_b X^\nu \eta_{\mu\nu}. \tag{7.28}$$

It depends on the intrinsic worldsheet metric γ_{ab} and the coordinates $X^\mu(\xi^a)$, in which it is quadratic, unlike the Nambu–Goto action.

The variation of the action with respect to γ^{ab} gives, using $\delta \det(\gamma_{ab})/\det(\gamma_{ab}) = -\gamma_{ab}\delta\gamma^{ab}$,

$$\delta S_P = -\frac{1}{4\pi\alpha'}\int d\sigma d\tau \sqrt{-\gamma}\delta\gamma^{ab}\left[\partial_a X^\mu \partial_b X^\nu \eta_{\mu\nu} - \frac{1}{2}\gamma_{ab}(\gamma^{cd}\partial_c X^\mu \partial_d X^\nu \eta_{\mu\nu})\right]. \tag{7.29}$$

Substituting $h_{ab} = \partial_a X^\mu \partial_b X^\nu \eta_{\mu\nu}$ into the above, the equation of motion of γ^{ab} is obtained to be

$$\gamma_{ab} = h_{ab}. \tag{7.30}$$

Then substituting into the Polyakov action, we obtain

$$\frac{h_{ab}}{\sqrt{-h}} = \frac{\gamma_{ab}}{\sqrt{-\gamma}} \Rightarrow S_P = -\frac{1}{2\pi\alpha'}\int d\tau d\sigma \sqrt{-\det(h_{ab})} = S_{NG}, \tag{7.31}$$

thus indeed, the Polyakov action is the first order form of the Nambu–Goto action.

The Polyakov action has the following invariances:

- Spacetime Poincaré invariance.
- Worldsheet diffeomorphism (general coordinate) invariance, defined by two transformations $(\sigma'(\sigma, \tau), \tau'(\sigma, \tau))$, that give $X'^\mu(\sigma', \tau') = X^\mu(\sigma, \tau)$.
- Worldsheet Weyl invariance: for an arbitrary $\omega(\sigma, \tau)$, we have the transformation

$$X'^\mu(\sigma, \tau) = X^\mu(\sigma, \tau); \quad \gamma'_{ab}(\sigma, \tau) = e^{2\omega(\sigma,\tau)}\gamma_{ab}(\sigma, \tau). \tag{7.32}$$

This gives $\sqrt{-\det \gamma_{ab}} \to e^{2\omega}\sqrt{-\det \gamma_{ab}}$, $h_{ab} \to h_{ab}$.

The Weyl invariance is very important in the following, and is not present in the Nambu–Goto action. Therefore the Polyakov form is more fundamental. Classically, the two actions are equivalent, as we saw. But quantum mechanically, they are not.

7.5 Equations of motion, constraints, and quantization in covariant gauge

Strings have spatial extension, but that means we also need boundary conditions for them. They can be *open*, in which case the endpoints of the string are at different points in spacetime, or *closed*. The correct boundary conditions can be found by varying the action, together with the equations of motion.

We first define the worldsheet energy-momentum tensor as usual in a general relativistic theory, with a conventional factor of 4π in front,

$$T^{ab}(\sigma, \tau) \equiv -4\pi \frac{1}{\sqrt{-\gamma}} \frac{\delta S_P}{\delta \gamma_{ab}} = +\frac{1}{\alpha'} \left(\partial^a X^\mu \partial^b X_\mu - \frac{1}{2} \gamma^{ab} \partial_c X^\mu \partial^c X_\mu \right). \quad (7.33)$$

It is conserved as usual, $\nabla_a T^{ab} = 0$. Then we see that the equation of motion for γ^{ab} is $T_{ab} = 0$. But moreover, the Weyl invariance of the action implies that $\gamma^{ab} \delta S / \delta \gamma^{ab} = 0$ *off-shell*, thus the energy-momentum tensor is traceless off-shell, $T^a{}_a = 0$.

The other equation of motion for S_P comes from varying with respect to X^μ, giving by a partial integration

$$\delta_X S_P = \frac{1}{2\pi\alpha'} \int d\tau \int_0^l d\sigma \sqrt{-\gamma} \delta X^\mu \nabla^2 X_\mu - \frac{1}{2\pi\alpha'} \int d\tau \sqrt{-\gamma} \delta X^\mu \partial_\sigma X_\mu \Big|_{\sigma=0}^{\sigma=l}. \quad (7.34)$$

The possible boundary conditions come from setting the boundary term to zero. For closed strings, the boundary condition is

$$X^\mu(\tau, l) = X^\mu(\tau, 0); \quad \gamma_{ab}(\tau, l) = \gamma_{ab}(\tau, 0), \quad (7.35)$$

but also more generally, $X^\mu(\tau, \sigma + l) = X^\mu(\tau, \sigma), \gamma_{ab}(\tau, \sigma + l) = \gamma_{ab}(\tau, \sigma)$ (periodicity). For open strings, the boundary condition is either Neumann,

$$\partial^\sigma X^\mu(\tau, 0) = \partial^\sigma X^\mu(\tau, l), \quad (7.36)$$

which as we see shortly means that the endpoints of the open string are free and move at the speed of light, or Dirichlet,

$$\delta X^\mu(\tau, 0) = \delta X^\mu(\tau, l) = 0, \quad (7.37)$$

which implies that the endpoints of the open string are constrained to be at a fixed point. We will show that the existence of the Dirichlet boundary condition is related to objects called D-branes, to be studied in Chapter 9.

The equation of motion for X^μ is found by setting the bulk variation of the action to zero, giving the wave equation in two dimensions,

$$\nabla^2 X^\mu = 0. \quad (7.38)$$

We saw that the Polyakov action has three local worldsheet invariances (defined by arbitrary functions of (σ, τ)): two diffeomorphisms ($\sigma'(\sigma, \tau)$ and $\tau'(\sigma, \tau)$) and one Weyl invariance ($\omega(\sigma, \tau)$). That means that we can choose the three independent elements of the symmetric matrix $h_{ab}(\sigma, \tau)$ (the worldsheet metric) to be anything we want. We should actually check that we can in fact reach a particular gauge we want, but we will not do it here. We choose the gauge

$$h_{ab} = \eta_{ab} = \begin{pmatrix} -1 & 0 \\ 0 & 1 \end{pmatrix}, \tag{7.39}$$

usually called the *conformal gauge*. Then the Polyakov action in flat spacetime becomes

$$S = -\frac{T}{2} \int d^2\sigma \, \eta^{ab} \partial_a X^\mu \partial_b X^\nu \eta_{\mu\nu}. \tag{7.40}$$

The X^μ equation of motion becomes the 2-dimensional flat space wave equation,

$$\Box X^\mu = \left(\frac{\partial^2}{\partial\sigma^2} - \frac{\partial^2}{\partial\tau^2} \right) X^\mu = -4\partial_+\partial_- X^\mu = 0. \tag{7.41}$$

Here we define

$$\sigma^\pm = \tau \pm \sigma; \quad \partial_\pm = \frac{1}{2}(\partial_\tau \pm \partial_\sigma). \tag{7.42}$$

Then the general solution of the 2-dimensional wave equation is

$$X^\mu(\sigma, \tau) = X_R^\mu(\sigma^-) + X_L^\mu(\sigma^+), \tag{7.43}$$

where we call $X_R(\sigma^-)$ the "right-moving mode" and $X_L(\sigma^+)$ the "left-moving mode."

The gauge-fixed Polyakov action in conformal gauge has a residual gauge invariance, a combination of reparameterization invariance and Weyl invariance called *conformal invariance*, studied in detail in the next chapter. This is a common feature of gauge theories: just because we have fixed a number of components using an equal number of arbitrary functions, does not mean that we cannot have an invariance with a restricted dependence on the (worldsheet, in this case) coordinates. A common such example is the gauge $A_0 = 0$ in electromagnetism, which still allows one to also fix $\vec{\nabla} \cdot \vec{A} = 0$, obtaining the radiation gauge. We can immediately check that both the equations of motion (7.41) and the gauge fixed action are invariant under the "conformal transformations"

$$\sigma^+ \to \tilde{\sigma}^+ = f(\sigma^+); \quad \sigma^- \to \tilde{\sigma}^- = g(\sigma^-), \tag{7.44}$$

where the functions f and g are arbitrary. Under this transformation, the flat metric on the worldsheet changes as

$$ds^2 = d\sigma^+ d\sigma^- = \frac{d\tilde{\sigma}^+}{f'} \frac{d\tilde{\sigma}^-}{g'} = (f'(\sigma^+)g'(\sigma^-))^{-1} d\tilde{\sigma}^+ d\tilde{\sigma}^-, \tag{7.45}$$

so we see that by combining this diffeomorphism (reparameterization) with a Weyl transformation $ds^2 \to (f'g')ds^2$, we obtain an invariance of the flat metric in conformal gauge.

Open string endpoints

The Neumann open string endpoints, as we said, move at the speed of light. To see that, consider the constraints in the conformal gauge. The energy-momentum tensor is

$$T_{ab} = \frac{1}{\alpha'} \left(\partial_a X^\mu \partial_b X_\mu - \frac{1}{2} \eta_{ab} \partial_c X^\mu \partial^c X_\mu \right). \tag{7.46}$$

In components, we have

$$\alpha' T_{01} = \alpha' T_{10} = \dot{X} \cdot X',$$
$$\alpha' T_{00} = \alpha' T_{11} = \frac{1}{2}(\dot{X}^2 + X'^2), \tag{7.47}$$

where dot means derivative with respect to τ and prime derivative with respect to σ. The constraints $T_{00} = T_{11} = 0$ become $\dot{X}^2 + X'^2 = 0$. But at endpoints, we have $X'^\mu(\sigma, \tau) = 0$, so we obtain $\dot{X}^2 = 0$. We can write it as $dX^\mu dX_\mu = 0$, i.e. motion of the endpoints at the speed of light.

Closed string modes

For closed strings with periodicity 2π, $X^\mu(\tau, 2\pi) = X^\mu(\tau, 0)$, so the general solution of the wave equation for X^μ (on a spatial circle) is a linear function in τ, plus Fourier modes on the circle. In terms of X_L and X_R, we have

$$X_R^\mu(\tau - \sigma) = \frac{1}{2}x^\mu + \frac{\alpha'}{2}p^\mu(\tau - \sigma) + \frac{i\sqrt{2\alpha'}}{2} \sum_{n \neq 0} \frac{1}{n} \alpha_n^\mu e^{-in(\tau - \sigma)},$$

$$X_L^\mu(\tau + \sigma) = \frac{1}{2}x^\mu + \frac{\alpha'}{2}p^\mu(\tau + \sigma) + \frac{i\sqrt{2\alpha'}}{2} \sum_{n \neq 0} \frac{1}{n} \tilde{\alpha}_n^\mu e^{-in(\tau + \sigma)}, \tag{7.48}$$

such that in total we have

$$X^\mu(\sigma, \tau) = x^\mu + \alpha' p^\mu \tau + i\frac{\sqrt{2\alpha'}}{2} \sum_{n \neq 0} \frac{1}{n} \left[\alpha_n^\mu e^{-in(\tau - \sigma)} + \tilde{\alpha}_n^\mu e^{-in(\tau + \sigma)} \right]. \tag{7.49}$$

From the condition that X_L and X_R must be real, we obtain that x^μ, p^μ are real, and

$$\alpha_{-n}^\mu = (\alpha_n^\mu)^\dagger; \quad \tilde{\alpha}_{-n}^\mu = (\tilde{\alpha}_n^\mu)^\dagger. \tag{7.50}$$

We can also define (in the closed string case)

$$\alpha_0^\mu = \sqrt{\frac{\alpha'}{2}} p^\mu = \tilde{\alpha}_0^\mu, \tag{7.51}$$

for later use. In the open string case we have $\alpha_0^\mu = \sqrt{2\alpha'} p^\mu$.

Free (Neumann) open string modes

For Neumann open strings, since $X'^{\mu}|_{\sigma=0,\pi} = 0$, the most general solution for X^{μ} is

$$X^{\mu}(\sigma,\tau) = x^{\mu} + 2\alpha'p^{\mu}\tau + i\sqrt{2\alpha'}\sum_{n\neq 0}\frac{1}{n}\alpha_n^{\mu}e^{-in\tau}\cos(n\sigma), \qquad (7.52)$$

or more succinctly put, we see that we identify $\alpha_n^{\mu} = \tilde{\alpha}_n^{\mu}$ in the closed string expansion.

Constraints and Hamiltonian

We saw that the constraints in the conformal gauge are $\alpha'T_{10} = \alpha'T_{01} = \dot{X}\cdot X' = 0$ and $\alpha'T_{00} = \alpha'T_{11} = 1/2(\dot{X}^2 + X'^2) = 0$. In terms of σ^+ and σ^-, they are

$$\alpha'T_{++} = \frac{\alpha'}{2}(T_{00} + T_{01}) = \partial_+X\cdot\partial_+X = \frac{1}{4}(\dot{X} + X')^2 = \dot{X}_L^2,$$

$$\alpha'T_{--} = \frac{\alpha'}{2}(T_{00} - T_{01}) = \partial_-X\cdot\partial_-X = \frac{1}{4}(\dot{X} - X')^2 = \dot{X}_R^2, \qquad (7.53)$$

where the last equality in both cases is valid only on-shell. These are called the *Virasoro constraints*.

To write down the worldsheet Hamiltonian, we first write the Polyakov action in conformal gauge as

$$S_P = \frac{1}{4\pi\alpha'}\int d\tau\int d\sigma(\dot{X}^2 - X'^2) = \int d\tau L. \qquad (7.54)$$

Then the worldsheet Hamiltonian is obtained in the usual way, defining first the worldsheet momentum[1]

$$P_{\tau}^{\mu} = \frac{\delta S}{\delta\dot{X}_{\mu}} = \frac{1}{2\pi\alpha'}\dot{X}^{\mu}, \qquad (7.55)$$

and then

$$H = \int_0^l d\sigma(\dot{X}_{\mu}P_{\tau}^{\mu} - L) = \frac{1}{4\pi\alpha'}\int_0^l d\sigma(\dot{X}^2 + X'^2) = \frac{1}{2\pi}\int_0^l d\sigma T_{00}. \qquad (7.56)$$

For example, for an on-shell open string, $l = \pi$ and using the orthonormality of cosines, $\int_0^{\pi}d\sigma\cos n\sigma\cos m\sigma = (\pi/2)\delta_{n+m}$, we get

$$H = \frac{1}{2}\sum_{n=-\infty}^{+\infty}\alpha_{-n}^{\mu}\alpha_n^{\mu}. \qquad (7.57)$$

For an on-shell closed string, $l = 2\pi$ and similarly, we get the sum of terms with α_n^{μ} and $\tilde{\alpha}_n^{\mu}$,

$$H = \frac{1}{2}\sum_{n=-\infty}^{+\infty}(\alpha_{-n}^{\mu}\alpha_n^{\mu} + \tilde{\alpha}_{-n}^{\mu}\tilde{\alpha}_n^{\mu}). \qquad (7.58)$$

In the sum we have included the $n = 0$ modes defined in (7.51).

We can now expand the constraints in Fourier modes.

---

[1] Note that indeed $\int_0^{\pi}d\sigma P_{\tau}^{\mu} = p^{\mu}$ for the open string, and $\int_0^{2\pi}P_{\tau}^{\mu} = p^{\mu}$ for the closed string.

For the *closed string*,

$$L_m = \frac{1}{2\pi} \int_0^{2\pi} d\sigma \, e^{-im\sigma} T_{--} = \frac{1}{2\pi\alpha'} \int_0^{2\pi} d\sigma \, e^{-im\sigma} \dot{X}_R^2 = \frac{1}{2} \sum_{n=-\infty}^{+\infty} \alpha_{m-n}^\mu \alpha_n^\mu,$$

$$\tilde{L}_m = \frac{1}{2\pi} \int_0^{2\pi} d\sigma \, e^{-im\sigma} T_{++} = \frac{1}{2\pi\alpha'} \int_0^{2\pi} d\sigma \, e^{-im\sigma} \dot{X}_L^2 = \frac{1}{2} \sum_{n=-\infty}^{+\infty} \tilde{\alpha}_{m-n}^\mu \tilde{\alpha}_n^\mu. \quad (7.59)$$

For the *open string*, we have

$$\begin{aligned} L_m &= \frac{1}{2\pi} \int_0^\pi d\sigma \left(e^{im\sigma} T_{++} + e^{-im\sigma} T_{--} \right) = \frac{1}{2\pi} \int_{-\pi}^\pi d\sigma \, e^{im\sigma} T_{++} \\ &= \frac{1}{8\pi\alpha'} \int_{-\pi}^{+\pi} d\sigma \, e^{im\sigma} (\dot{X} + X')^2 \\ &= \frac{1}{2} \sum_{n=-\infty}^{+\infty} \alpha_{m-n}^\mu \alpha_n^\mu. \end{aligned} \quad (7.60)$$

We then have

$$\begin{aligned} H &= L_0 & \text{open} \\ &= L_0 + \tilde{L}_0 & \text{closed.} \end{aligned} \quad (7.61)$$

The $H = L_0 = 0$ constraint for the open string translates into

$$M^2 \equiv -p_\mu p^\mu = -\frac{\alpha_0^2}{2\alpha'} = \frac{1}{\alpha'} \sum_{n \geq 1} \alpha_{-n}^\mu \alpha_n^\mu. \quad (7.62)$$

For the closed string, $H = L_0 + \tilde{L}_0 = 0$ translates into

$$M^2 \equiv -p_\mu p^\mu = -\frac{\alpha_0^2 + \tilde{\alpha}_0^2}{\alpha'} = \frac{2}{\alpha'} \sum_{n \geq 1} (\alpha_{-n}^\mu \alpha_n^\mu + \tilde{\alpha}_{-n}^\mu \tilde{\alpha}_n^\mu). \quad (7.63)$$

On the other hand, the constraint $0 = L_0 - \tilde{L}_0 \equiv P_\sigma$ (worldsheet momentum associated with translational invariance on the closed string, $\partial/\partial\sigma$: it should be trivial) translates into

$$\sum_{n \geq 1} \alpha_{-n}^\mu \alpha_n^\mu = \sum_{n \geq 1} \tilde{\alpha}_{-n}^\mu \tilde{\alpha}_n^\mu. \quad (7.64)$$

Quantization

For the Polyakov action in conformal gauge, the momentum is (7.55).

We then have the equal-time (classical) Poisson brackets for the Polyakov action,

$$\begin{aligned} [X^\mu(\sigma, \tau), X^\nu(\sigma', \tau)]_{\text{P.B.}} &= [P^\mu(\sigma, \tau), P^\nu(\sigma', \tau)]_{\text{P.B.}} = 0, \\ [P^\mu(\sigma, \tau), X^\nu(\sigma', \tau)]_{\text{P.B.}} &= -\delta(\sigma - \sigma')\eta^{\mu\nu}. \end{aligned} \quad (7.65)$$

From these brackets, substituting the expansion (7.48) for the closed string, we obtain the Poisson brackets for the coefficients,

$$[\alpha_m^\mu, \alpha_n^\nu]_{\text{P.B.}} = [\tilde{\alpha}_m^\mu, \tilde{\alpha}_n^\nu]_{\text{P.B.}} = -im\delta_{m+n,0}\eta^{\mu\nu},$$
$$[\alpha_m^\nu, \tilde{\alpha}_n^\nu]_{\text{P.B.}} = 0,$$
$$[p^\mu, x^\nu]_{\text{P.B.}} = \eta^{\mu\nu}. \tag{7.66}$$

To this, we must add the Virasoro constraints $(\dot{X} \pm X')^2 = 0$, with Fourier components $L_m = 0 = \tilde{L}_m$.

To quantize, as usual, we replace the Poisson brackets $[,]_{\text{P.B.}}$ by the commutator $-i[,]$. Therefore the basic commutators are

$$[X^\mu(\sigma,\tau), X^\nu(\sigma',\tau)] = [P^\mu(\sigma,\tau), P^\nu(\sigma',\tau)] = 0,$$
$$[P^\mu(\sigma,\tau), X^\nu(\sigma',\tau)] = -i\delta(\sigma-\sigma')\eta^{\mu\nu}. \tag{7.67}$$

From these we get the commutators for the Fourier coefficients

$$[\alpha_m^\mu, \alpha_n^\nu] = [\tilde{\alpha}_m^\mu, \tilde{\alpha}_n^\nu] = m\delta_{m+n,0}\eta^{\mu\nu},$$
$$[\alpha_m^\nu, \tilde{\alpha}_n^\nu] = 0,$$
$$[p^\mu, x^\nu] = -i\eta^{\mu\nu}. \tag{7.68}$$

We note that we can redefine the coefficients so they satisfy the creation/annihilation algebra,

$$\alpha_m^\mu = \sqrt{m}a_m^\mu; \quad \alpha_{-m}^\mu = \sqrt{m}a_m^{\dagger\mu}; \quad m > 0. \tag{7.69}$$

We must then impose the Fourier modes of the Virasoro constraints $L_m = \tilde{L}_m = 0$ on the physical states. But at the quantum level, we find ordering ambiguities for the constraints that affect $L_0, \tilde{L}_0$ (the only ones among (7.59) that contain products of non-commuting objects) by constants. Specifically, we get $(L_0 - a)|\psi\rangle = (\tilde{L}_0 - a)|\psi\rangle = 0$. The analysis of the spectrum will be done when we study quantization in light-cone gauge, in the next section.

7.6 Quantization in light-cone gauge; the bosonic string spectrum

We have seen that the Polyakov action in conformal gauge is invariant under conformal transformations (7.44). Then $\tilde{\tau} = 1/2(\tilde{\sigma}^+(\sigma^+) + \tilde{\sigma}^-(\sigma^-))$ satisfies

$$\left(\frac{\partial^2}{\partial\sigma^2} - \frac{\partial^2}{\partial\tau^2}\right)\tilde{\tau} = \left(\frac{\partial}{\partial\sigma} - \frac{\partial}{\partial\tau}\right)\left(\frac{\partial}{\partial\sigma} + \frac{\partial}{\partial\tau}\right)\tilde{\tau} = -4\frac{\partial}{\partial\sigma^+}\frac{\partial}{\partial\sigma^-}\tilde{\tau}(\sigma^+,\sigma^-) = 0, \tag{7.70}$$

i.e. $\tilde{\tau}$ satisfies the same wave equation as X^μ. Therefore, we can make $\tilde{\tau}$ equal or proportional to any of the X^μs on-shell. Defining the *light-cone coordinates* for spacetime,

$$X^\pm = \frac{X^0 \pm X^{D-1}}{\sqrt{2}}, \tag{7.71}$$

and $\mu = (+-i)$, we can choose $\tilde{\tau} = X^+/p^+ +$ constant, or in other words fix the *light-cone gauge* by

$$X^+(\sigma,\tau) = x^+ + p^+\tau. \tag{7.72}$$

This means that classically, the oscillator coefficients for X^+, $\alpha_n^+ = 0$ for $n \neq 0$. We also obtain that $\dot{X}^+ \pm X'^+ = p^+$ and then the Virasoro constraints $(\dot{X} \pm X')^2 = 0$ give

$$\dot{X}^- \pm X'^- = \frac{(\dot{X}^i \pm X'^i)^2}{2(\dot{X}^+ \pm X'^+)} = \frac{(\dot{X}^i \pm X'^i)^2}{2p^+}. \tag{7.73}$$

In terms of Fourier modes, for the open string we have

$$\alpha_n^- = \frac{\sqrt{2\alpha'}}{2p^+} \sum_{m \in \mathbb{Z}} \alpha_{n-m}^i \alpha_m^i, \tag{7.74}$$

and for the closed string we also have a similar relation with tildes. For $n = 0$, we obtain for the open string

$$M^2 \equiv 2p^+ p^- - p^i p^i = \frac{1}{\alpha'} \sum_{n \geq 1} \alpha_{-n}^i \alpha_n^i. \tag{7.75}$$

We see that $\alpha_n^+ = 0$ and α_n^- is given in terms of α_n^i, therefore the only independent oscillators are α_n^i.

Quantization

We first analyze the *open string* case.

At quantum level, as we mentioned in the previous section, we must care about normal ordering, since we have a_ns and $a_n^\dagger$s. As we saw, X^+ is gauge fixed, and X^- is fixed in terms of X^+ and X^i. Otherwise the quantization follows as in the covariant case for the X^is (thus for α_n^is). The constraint (7.74) becomes

$$\alpha_n^- = \frac{\sqrt{2\alpha'}}{p^+} \left[\frac{1}{2} \sum_{i=1}^{D-2} \sum_{m \in \mathbb{Z}} : \alpha_{n-m}^i \alpha_m^i : -a\delta_{n,0} \right], \tag{7.76}$$

where a is a constant. Now X^+ and X^- are eliminated from the theory (fixed) and there are no more constraints left, and we therefore quantize in a physical gauge, in terms of only independent physical oscillators α_n^i.

For $n = 0$, we obtain the *open string light-cone Hamiltonian*

$$H = p^- = \frac{p^i p^i}{2p^+} + \frac{1}{2\alpha' p^+}(N - a), \tag{7.77}$$

which in turn gives the mass squared

$$M^2 \equiv 2p^+ p^- - p^i p^i = \frac{1}{\alpha'}(N - a). \tag{7.78}$$

Here N is a kind of number operator,

$$N = \sum_{n \geq 1} \alpha_{-n}^i \alpha_n^i = \sum_{n \geq 1} n a_n^{\dagger i} a_n^i, \tag{7.79}$$

n is called the *level*, and as we can see it counts the contribution of a state with $a_n^{\dagger i}$ to M^2. Therefore the string contains an infinite number of different types of particles made up of combinations of the basic particles created by the creation operators $a_n^{\dagger i}$, and having mass squared equal to n/α'. As an analogy, we can think of phonons with different frequencies in a material, except that now the particles are really different, independent of the material or background they propagate in.

We can readily see that we need the normal ordering constant to be $a = 1$ in order to have a vector, with $D - 2$ physical degrees of freedom, be massless. But the number operator N appears as usual from the combination $(a_n^{\dagger i} a_n^i + a_n^i a_n^{\dagger i})/2$, therefore the normal ordering constant should actually be

$$\sum_{i=1}^{D-2} \sum_{n \geq 1} \frac{n}{2} = \frac{D-2}{2} \sum_{n \geq 1} n. \tag{7.80}$$

The infinite sum can be regularized using zeta function regularization. Namely, the zeta function $\zeta(s) = \sum_{n \geq 1} 1/n^s$ admits an analytical continuation in the complex plane to $s \to -1$, with $\zeta(-1) = -1/12$. Therefore the constant is actually

$$a = \frac{D-2}{24}, \tag{7.81}$$

and $a = 1$ only if $D = 26$. Hence the consistency of the bosonic string requires that it lives in 26 dimensions. $D = 26$ is known as the critical dimension of the bosonic string. There are other ways to see this, which will not be explained here, for instance requiring that there is no quantum anomaly for Weyl symmetry and spacetime Lorentz invariance, but the result is the same.

Bosonic open string spectrum

The vacuum of the open string will be a state with momentum, since the momentum operator is a zero mode of the string. Therefore the state is $|0; \vec{k}\rangle$, defined by

$$p^+ |0; \vec{k}\rangle = k^+ |0; \vec{k}\rangle; \quad p^i |0; \vec{k}\rangle = k^i |0; \vec{k}\rangle; \quad \alpha_m^i |0; \vec{k}\rangle = 0, \tag{7.82}$$

and is therefore tachyonic, since now $N|0; \vec{k}\rangle = 0$, so $M^2 = -1/\alpha'$.

A general state is obtained by acting on the vacuum with the creation operators, that is

$$|N; \vec{k}\rangle = \left[\Pi_{i=1}^{D-2} \Pi_{n \geq 1} \frac{(a_n^{\dagger i})^{N_{in}}}{\sqrt{N_{in}!}} \right] |0; \vec{k}\rangle = \left[\Pi_{i=1}^{D-2} \Pi_{n \geq 1} \frac{(\alpha_{-n}^{\dagger i})^{N_{in}}}{\sqrt{n^{N_{in}} N_{in}!}} \right] |0; \vec{k}\rangle. \tag{7.83}$$

Consider the next state, with $N = 1$, i.e. only one creation operator, at level 1. The state is then

$$a_1^{\dagger i} |0; \vec{k}\rangle = \alpha_{-1}^{\dagger i} |0; \vec{k}\rangle. \tag{7.84}$$

It is obviously a vector, but since i runs over only $D-2$ values, it has the number of on-shell degrees of freedom of a massless vector. As we indicated, this requires that $a = 1$.

All the other states are massive, so the only non-massive states of the bosonic open string are a scalar tachyonic vacuum and a massless vector.

Bosonic closed string spectrum

In the closed string case, the constraints are $L_0 - a = 0$ and $\tilde{L}_0 - \tilde{a} = 0$, and we find the mass squared operator

$$M^2 = \frac{4}{\alpha'} \sum_{n \geq 1} (\alpha^i_{-n} \alpha^i_n - 1) = \frac{4}{\alpha'} \sum_{n \geq 1} (\tilde{\alpha}^i_{-n} \tilde{\alpha}^i_n - 1). \tag{7.85}$$

The normal ordering constant is calculated in the same way as before, and in $D = 26$ dimensions we again obtain 1 inside the sum. We can now also define $\tilde{N} = \sum_{n \geq 1} \tilde{\alpha}^i_{-n} \tilde{\alpha}^i_n$, besides the same N, and the constraint $P_\sigma = L_0 - \tilde{L}_0 = 0$ becomes

$$N = \tilde{N} \Rightarrow \sum_{n \geq 1} \alpha^i_{-n} \alpha^i_n = \sum_{n \geq 1} \tilde{\alpha}^i_{-n} \tilde{\alpha}^i_n. \tag{7.86}$$

Here P_σ is the translation generator along σ, and the condition that it acts trivially on states is the condition of translational invariance along the string, $\sigma \to \sigma + a$.

We can then write the mass squared operator also as

$$M^2 = \frac{2}{\alpha'} \sum_{n \geq 1} (\alpha^i_{-n} \alpha^i_n + \tilde{\alpha}^i_{-n} \tilde{\alpha}^i_n - 2) = \frac{2}{\alpha'} (N + \tilde{N} - 2). \tag{7.87}$$

The vacuum of the closed string is now also a tachyon $|0, 0; \vec{k}\rangle$, defined by

$$p^+ |0, 0; \vec{k}\rangle = k^+ |0, 0; \vec{k}\rangle; \quad p^i |0, 0; \vec{k}\rangle = k^i |0, 0; \vec{k}\rangle; \quad \alpha^i_m |0, 0; \vec{k}\rangle = \tilde{\alpha}^i_m |0, 0; \vec{k}\rangle = 0, \tag{7.88}$$

but now with mass $M^2 = -4/\alpha'$. The first excited state has $N = \tilde{N} = 1$ (since we need $N = \tilde{N}$), therefore one level one excitation for both left and right movers. It is

$$a^{\dagger i}_1 \tilde{a}^{\dagger j}_1 |0, 0; \vec{k}\rangle = \alpha^i_{-1} \tilde{\alpha}^j_{-1} |0, 0; \vec{k}\rangle \equiv |ij\rangle, \tag{7.89}$$

and can be decomposed as $((ij))$, i.e. a symmetric traceless state, corresponding to the graviton g_{ij}, $[ij]$, i.e. an antisymmetric tensor, called the B_{ij} field, and ii, i.e. the trace, called the dilaton ϕ field.

7.7 Strings in background fields

We have written the Polyakov action in flat spacetime, $g_{\mu\nu} = \eta_{\mu\nu}$, and no other fields. The introduction of a curved spacetime is done by replacing $\eta_{\mu\nu}$ by $g_{\mu\nu}$ in the Polyakov action. But what is the more general interpretation of this procedure?

The theory of both open and closed bosonic strings has a tachyonic vacuum, of $M^2 = -1/\alpha'$ for open strings and $M^2 = -4/\alpha'$ for closed strings. The associated tachyonic field is a scalar, and $M^2 < 0$ means that we are perturbing a potential $V(\Phi)$, where Φ is the tachyon field, around a maximum, $V(\Phi) \simeq V_0 + M^2 (\delta\Phi)^2$; $M^2 < 0$ instead of a minimum. It then means that this vacuum will decay to the true vacuum, if such a true vacuum exists, or otherwise run to minus infinity if there is no vacuum. But this important fact is not known for sure, due to the non-perturbative nature of the calculations at large Φ. The bosonic string is thus not very well undestood.

Instead, we consider supersymmetric strings in the next section, for which the tachyons are absent, and the theory is defined in $D = 10$ spacetime dimensions instead of 26. But the massless states of the bosonic strings, while they also include other bosonic states, whose quantum theory is less understood, include the states of the massless closed bosonic string, the graviton $g_{\mu\nu}$, antisymmetric tensor $B_{\mu\nu}$, and dilaton ϕ. These states now form the vacuum of the supersymmetric string, above which we have modes of increasing mass, governed by the mass scale $1/\sqrt{\alpha'}$. Each string mode correponds to a spacetime field of a given mass. In the low energy limit, which can be thought of as the $\alpha' \to 0$ limit, these modes do not contribute, and we are left with a theory for the massless states, which acquire classical backgrounds with quantum corrections, i.e. VEVs. The way this happens is that these possible external states of the string theory can be created by "vertex operators" present in Feynman diagrams. But being massless, they can condense and form classical backgrounds for $g_{\mu\nu}, B_{\mu\nu}, \phi$.

One can show that the effect of this is to modify the Polyakov action to self-consistently include interactions with the massless modes of the closed string in flat background, $g_{\mu\nu}, B_{\mu\nu}, \phi$, and one can derive the form of the modified action. But we can avoid the derivation and basically guess the answer from the symmetries. We saw that the natural guess for introducing $g_{\mu\nu}$ is via the replacement $\eta_{\mu\nu} \to g_{\mu\nu}$, and it is indeed correct. The field $B_{\mu\nu}$ is antisymmetric, so instead of coupling with $\sqrt{-\gamma}\gamma^{ab}\partial_a X^\mu \partial_b X^\nu$, it should couple to the antisymmetric tensor density $\epsilon^{ab}\partial_a X^\mu \partial_b X^\nu$, obtaining a term in the action called a Wess–Zumino term. The scalar dilaton ϕ should couple to a worldsheet scalar that vanishes in a trivial case, so the only possibility is with $\sqrt{-\gamma}R^{(2)}(\gamma)$, the 2-dimensional Einstein action. Finally, the action is

$$S = -\frac{1}{4\pi\alpha'}\int d^2\sigma \left[\sqrt{-\gamma}\gamma^{ab}\partial_a X^\mu \partial_b X^\nu g_{\mu\nu}(X^\rho) + \alpha'\epsilon^{ab}\partial_a X^\mu \partial_b X^\nu B_{\mu\nu}(X^\rho) \right.$$
$$\left. -\alpha'\sqrt{-\gamma}\mathcal{R}^{(2)}\Phi(X^\rho)\right], \tag{7.90}$$

where $\mathcal{R}^{(2)}$ is the 2-dimensional Ricci scalar, and the quantity

$$\frac{1}{4\pi}\int d^2\sigma\sqrt{-\gamma}\mathcal{R}^{(2)} = \chi \tag{7.91}$$

is a topological invariant, i.e. a negative integer that counts the number of holes the topology of the 2-dimensional surface contains (times -2, specifically, $\chi = 2(1 - g)$). But e^{-S} then contains $e^{-\chi\Phi} = (e^\Phi)^{2(g-1)}$. Therefore, the addition of a hole to a worldsheet, which is interpreted as an extra loop in the quantum interaction of a string, as in Fig. 7.3a, gives a factor of $e^{2\Phi}$, prompting the identification of e^Φ with the string coupling constant, g_s.

This procedure, of putting the string in a background ("condensate") of its own ground state modes, needs a self-consistency condition: the procedure must preserve the original invariances of the action, specifically Weyl invariance (or conformal invariance, see next section) at the worldsheet quantum level. Imposing Weyl invariance of the action in fact turns out to give the equations of motion for $g_{\mu\nu}, B_{\mu\nu}, \phi$, the classical *spacetime* equations of motion, plus α' corrections from string worldsheet quantum loops.

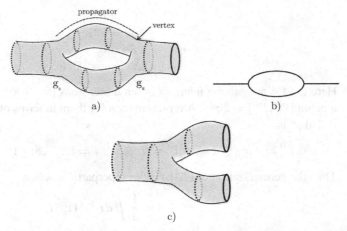

Figure 7.3 a) String loop diagram: the vertices are not pointlike, but are spread out, and have a coupling g_s; b) In comparison, a particle loop diagram; c) Basic string interaction: "pair of pants"= vertex for a string to split into two strings.

7.8 Supersymmetric strings

As we saw, the bosonic string has a tachyonic vacuum, and thus is unstable, so we need a new theory that will have a stable vacuum. Such a theory is found by introducing supersymmetry. A priori there are two types of supersymmetry that one could introduce, in spacetime and on the worldsheet of the string, but we will see that they are equivalent possibilities. The easiest to understand and argue for is the spacetime supersymmetry, since we want to eliminate the spacetime tachyon field from the theory. This formalism is called the Green–Schwarz formalism, or the superstring. An argument similar to the argument which led to $D = 26$ for the bosonic string, but more complicated, now gives $D = 10$ (critical dimension for the superstring) spacetime dimensions for the quantum consistency of the theory.

The superparticle

To understand it, we first look once again at the particle, and we generalize the first order action for the particle, (7.16). The generalization is done by introducing objects θ^A, $A = 1, 2, \ldots, N$ which are N spacetime spinors and worldsheet scalars, i.e. $\theta^{A\alpha}$, with α a spacetime spinor index in ten dimensions. The bosonic action is written in terms of the object $dX^\mu/d\tau$, so we write a supersymmetric generalization for it, namely

$$\Pi_\tau^\mu \equiv \frac{dX^\mu}{d\tau} + \bar{\theta}^A \Gamma^\mu \frac{d\theta^A}{d\tau}, \tag{7.92}$$

which is immediately seen to be invariant under the N spacetime supersymmetry transformations

$$\delta\theta^A = \epsilon^A,$$
$$\delta\bar{\theta}^A = \bar{\epsilon}^A,$$
$$\delta e = 0,$$
$$\delta X^\mu = -\bar{\epsilon}^A \Gamma^\mu \theta^A. \tag{7.93}$$

Here the Γ^μ are gamma matrices in ten dimensions (for $SO(9,1)$), satisfying the Clifford algebra $\{\Gamma^\mu, \Gamma^\nu\} = 2\eta^{\mu\nu}$. A representation for them in terms of gamma matrices of $SO(8)$ called γ^i is

$$\Gamma^0 = -i\sigma_2 \otimes \mathbb{1}_{16}; \quad \Gamma^i = \sigma_1 \otimes \gamma^i, \ i = 1, \ldots, 8; \quad \Gamma^9 = \sigma_3 \otimes \mathbb{1}_{16}. \tag{7.94}$$

Then the generalization of (7.16) is the superparticle action

$$S = \frac{1}{2} \int d\tau e^{-1} \Pi_\tau^\mu \Pi_{\tau\mu}. \tag{7.95}$$

But now this action has an additional fermionic symmetry, called kappa symmetry, with parameters $\kappa^{A\alpha}$ (N spacetime spinors), given by

$$\delta\theta^A = -\Gamma_\mu \Pi_\tau^\mu \kappa^A,$$
$$\delta X^\mu = -\bar{\theta}^A \Gamma^\mu \delta\theta^A,$$
$$\delta e = 4e\dot{\bar{\theta}}^A \kappa^A. \tag{7.96}$$

The invariance of the action is proven by noting that

$$\delta\Pi_\tau^\mu = -2\dot{\bar{\theta}}^A \Gamma^\mu \delta\theta^A \Rightarrow \delta\Pi_\tau^2 = -4\dot{\bar{\theta}}^A \Gamma_\mu \Pi_\tau^\mu \delta\theta^A = 4\Pi_\tau^2 \dot{\bar{\theta}}^A \kappa^A,$$
$$\delta e^{-1} = -4e^{-1} \dot{\bar{\theta}}^A \kappa^A. \tag{7.97}$$

The superstring

To write the action for the superstring, we want to generalize the Polyakov action (7.28), in terms of manifestly supersymmetric objects similar to Π_τ^μ above, but generalizing $\partial_a X^\mu$, not $dX^\mu/d\tau$. Thus we define

$$\Pi_a^\mu \equiv \partial_a X^\mu + \bar{\theta}^A \Gamma^\mu \partial_a \theta^A, \tag{7.98}$$

which are manifesly supersymmetric under the same N global supersymmetry transformations

$$\delta\theta^A = \epsilon^A; \quad \delta\bar{\theta}^A = \bar{\epsilon}^A,$$
$$\delta X^\mu = -\bar{\epsilon}^A \Gamma^\mu \theta^A. \tag{7.99}$$

So the naive guess would be

$$S_{\text{kin}} = -\frac{1}{4\pi\alpha'} \int d^2\sigma \sqrt{-\gamma} \gamma^{ab} \Pi_a^\mu \Pi_{b\mu}, \tag{7.100}$$

and is seen to be manifestly supersymmetric under the N global supersymmetries if $\delta\gamma_{ab} = 0$.

But now the action does not have kappa symmetry, and this is a very important property as we shall soon see. So we need to add another term to the action, and when we do so we will be able to obtain both supersymmetry and kappa symmetry in ten dimenions only for $N \leq 2$ and for Majorana–Weyl fermions. The action can be written in the general form of a super-Wess–Zumino term,

$$S_{\text{WZ}} = -\frac{1}{4\pi} \int d^2\sigma \, \epsilon^{ab} \Pi_a^M \Pi_b^N B_{MN}. \tag{7.101}$$

As we saw, the bosonic Wess–Zumino form appears when putting the bosonic string in a background $B_{\mu\nu}$. Here $M = (\mu, \alpha)$ is a superspace index. In flat spacetime $B_{\mu\nu} = 0$, but the fermionic components are nonzero. In components, the WZ term in flat spacetime is

$$S_{\text{WZ}} = \frac{1}{2\pi\alpha'} \int d^2\sigma \left[\epsilon^{ab} \partial_a X^\mu (\bar{\theta}^1 \Gamma_\mu \partial_b \theta^1 - \bar{\theta}^2 \Gamma_\mu \partial_b \theta^2) - \epsilon^{ab} (\bar{\theta}^1 \Gamma^\mu \partial_a \theta^1)(\bar{\theta}^2 \Gamma_\mu \partial_b \theta^2) \right]. \tag{7.102}$$

This WZ term is supersymmetry invariant only for $N = 2$, since we have only two θs, θ^1 and θ^2, but one can write an $N = 1$ invariant action as well.

Then the action $S_{\text{kin}} + S_{\text{WZ}}$, called the *Green–Schwarz action for the superstring* is also invariant under kappa symmetry with parameter $\kappa^{Aa\alpha}$, i.e. two ($A = 1, 2$) worldsheet vectors (a) and spacetime spinors (α), given by

$$\delta_\kappa \theta^A = -2\Gamma_\mu \Pi_a^\mu \kappa^{Aa},$$
$$\delta_\kappa X^\mu = -\bar{\theta}^A \Gamma^\mu \delta\theta^A,$$
$$\delta_\kappa(\sqrt{-\gamma}\gamma^{ab}) = -16\sqrt{-\gamma}(P_-^{ac} \bar{\kappa}^{1b} \partial_c \theta^1 + P_+^{ac} \bar{\kappa}^{2b} \partial_c \theta^2), \tag{7.103}$$

where we have defined the self-dual and anti-self-dual projection tensors

$$P_\pm^{ab} = \frac{1}{2}\left(\gamma^{ab} \pm \frac{\epsilon^{ab}}{\sqrt{-h}} \right). \tag{7.104}$$

We will not prove the supersymmetry and kappa symmetry under the above rules, it is left as an exercise for the reader (Exercises 5 and 6).

We can impose a gauge condition that fixes kappa symmetry. In terms of the light-cone gamma matrices

$$\Gamma^\pm = \frac{\Gamma^0 \pm \Gamma^9}{\sqrt{2}}, \tag{7.105}$$

the condition is given by the condition on the fermions

$$\Gamma^+ \theta^1 = \Gamma^+ \theta^2 = 0, \tag{7.106}$$

which halves the number of fermionic degrees of freedom.

One can show that by imposing the above light-cone gauge for the fermions, together with the usual light-cone gauge condition for the bosons, we can write the action in the equivalent form

$$S_{\text{lc}} = -\frac{1}{4\pi\alpha'} \int d^2\sigma \left[\partial_a X^i \partial^a X^i + 2\alpha' \bar{S}^m \gamma^a \partial_a S^m \right], \tag{7.107}$$

where now the gauge-fixed worldsheet scalars $\theta^{A\alpha}$ with α a spacetime spinor index of $SO(9, 1)$ have been regrouped as a 2-component Majorana worldsheet spinor S^m with m a spinor index of $SO(8)$ (the little group of $SO(9, 1)$), and the bar on the spinor in the action refers to 2-dimensional conjugation. Therefore the kappa-fixed action has now both spacetime and worldsheet supersymmetry, and the fermionic variable is a spinor of both the two worldsheet dimensions and the ten spacetime dimensions.

We can quantize the gauge-fixed GS action, but we cannot quantize covariantly the GS action for the superstring. For that, one needs to turn to a formalism called the *Berkovits formalism, or pure spinor formalism*, that will not be described here.

The gauge fixed Green–Schwarz action coincides with a gauge fixed version of an action with manifest worldsheet supersymmetry, to be studied next.

The spinning string

One can define another formalism, where the manifest supersymmetry is worldsheet one, called the Neveu–Schwarz–Ramond (NSR) action, and is sometimes called the spinning string, since it has internal (i.e. worldsheet) fermionic symmetry. It is written in terms of fermionic variables, which are now worldsheet spinors, and spacetime vectors, ψ^μ, and is given by

$$S = -\frac{1}{4\pi\alpha'} \int d^2\sigma \left[\partial_a X^\mu \partial^a X_\mu + \bar{\psi}^\mu \gamma^a \partial_a \psi_\mu \right], \tag{7.108}$$

where the Dirac matrices in 1+1 dimensions are

$$\gamma^0 = \begin{pmatrix} 0 & -1 \\ +1 & 0 \end{pmatrix} = -i\sigma_2; \quad \gamma^1 = \begin{pmatrix} 0 & 1 \\ 1 & 0 \end{pmatrix} = \sigma_1. \tag{7.109}$$

The action has worldsheet supersymmetry,

$$\delta X^\mu = \bar{\epsilon}\psi^\mu,$$
$$\delta\psi^\mu = \gamma^a \partial_a X^\mu \epsilon. \tag{7.110}$$

This is the same as the supersymmetry of the 2-dimensional Wess–Zumino model that we studied, for each μ value ($\delta\phi = \bar{\epsilon}\psi$, $\delta\psi = \partial\!\!\!/\phi\epsilon$).

When varying the action, besides the bulk term giving the equations of motion, and the usual bosonic boundary term, we also get a fermionic boundary term (in the open string case):

$$\psi_+ \delta\psi_+ - \psi_- \delta\psi_-|_0^\pi. \tag{7.111}$$

This means that we have to impose the boundary conditions $\psi_+ = \pm\psi_-$. We can put $\psi_+^\mu(0,\tau) = \psi_-^\mu(0,\tau)$ by redefining the fermions with a possible minus sign, but then we have two possibilities at the other endpoint:

$$\psi_+(\pi,\tau) = \pm\psi_-^\mu(\pi,\tau). \tag{7.112}$$

The condition with a + sign is called the Ramond (R) boundary condition, and leads to spacetime fermionic states, and the condition with a − sign is called Neveu–Schwarz (NS) boundary condition, and leads to spacetime bosonic states. In the case of the closed strings, we can independently put these boundary conditions for the left and right moving states, leading to NS–NS, R–R, NS–R and R–NS states, the first two being bosonic and the last two fermionic.

In a light-cone gauge, the equivalence of the NSR action with the GS action is due to the equivalence of the vector representation 8_V of $SO(8)$, to which ψ^i, the physical NSR fermions belong, and the spinor representation 8_S or $8'_S$ of $SO(8)$, to which S^m belongs. The equivalence of the three basic representations of $SO(8)$, $8_S, 8'_S$, and 8_V is called triality and is characteristic to $SO(8)$.

One thing we have not addressed until now is the chirality of the spacetime spinors θ^A in the Green–Schwarz formulation. For closed string theories, we have $N = 2$ supersymmetry, and we can choose the same chirality for both spinors θ^A, in which case we obtain the *type IIA* theory, or opposite chirality, in which case we obtain the *type IIB* theory. In the case where we also have open strings interacting with closed strings (we cannot have only open strings, since open strings can close and form a closed string), the theory has $N = 1$ supersymmetry, and open strings can have Yang–Mills indices associated with the endpoints, as we see in Chapter 9. This theory is called type I theory, and has gauge group $SO(32)$. We can also have a theory, called heterotic, that comes in two types corresponding to different gauge groups, $SO(32)$ and $E_8 \times E_8$.

Type IIA string theory at strong coupling, $g_s = e^{<\phi>} \to \infty$, is called *M-theory*, and in this limit *it becomes 11-dimensional*, with the radius R of the 11th dimension being $= g_s$ in string units, as we will see in the next subsection. About M-theory most of the known facts refer to its low energy, but it does not have a good (perturbative or non-perturbative) definition.

The NS–NS sector is common to all the string theories, and contains in the massless sector the $(g_{\mu\nu}, B_{\mu\nu}, \phi)$ fields already discussed. The NS–R and R–NS sectors contain fermions, and the RR massless sector contains antisymmetric $p + 1$-form fields $A_{\mu_1...\mu_{p+1}}$, different sets in different theories.

7.9 Supergravities in the $\alpha' \to 0$ limit and the duality web

As we noted, in the $\alpha' \to 0$ limit, we obtain a theory of the massless fields of string. In the case of supersymmetric string theories, these live in ten dimensions, have supersymmetry and include the metric $g_{\mu\nu}$, so they are described by 10-dimensional supergravities, specifically IIA and IIB supergravities. In the case of M-theory, the IIA string theory at strong coupling, the low energy is described by the unique 11-dimensional supergravity. The action for the NS–NS fields $g_{\mu\nu}, B_{\mu\nu}, \phi$ whose coupling to the string action is known, is found as we said by requiring invariance of the quantum action under the classical symmetries, in particular under worldsheet Weyl symmetry. The rest of the action can be found by requiring supersymmetry, and we find that the action matches the known supergravity action.

In the case of type IIA theory, the RR sector contains a gauge field A_μ, with field strength $F_{\mu\nu} = \partial_\mu A_\nu - \partial_\nu A_\mu$, also called F_2 in form language, and a 3-index antisymmetric tensor field $A_{\mu\nu\rho}$, with field strength $F_{\mu_1...\mu_4} = 4\partial_{[\mu_1} A_{\mu_2\mu_3\mu_4]}$, also called F_4 in form language. The field strength of the NS–NS field $B_{\mu\nu}$ is $H_{\mu\nu\rho} = 3\partial_{[\mu} B_{\nu\rho]}$, also called H_3 in form language. The bosonic part of the supergravity action *in string frame*, i.e. in terms of the metric naturally appearing in the Polyakov action for the string, is

$$S_{IIA} = \frac{1}{2\kappa_{10}^2} \int d^{10}x \left\{ \sqrt{-G} \left[e^{-2\phi} \left(R + 4\partial_\mu \phi \partial^\mu \phi - \frac{1}{2}|H_3|^2 \right) - \frac{1}{2}|F_2|^2 - \frac{1}{2}|\tilde{F}_4|^2 \right] \right.$$
$$\left. - \frac{1}{2} B_2 \wedge F_4 \wedge F_4 \right\}, \tag{7.113}$$

where $|F_n|^2 \equiv 1/n! \, F_{\mu_1 \ldots \mu_n} F^{\mu_1 \ldots \mu_n}$ and

$$\tilde{F}_4 = dA_3 - A_1 \wedge F_3. \tag{7.114}$$

For type IIB supergravity, the RR sector contains a scalar a also called A_0 (in "form" language), with "field strength" $F_\mu = \partial_\mu a$ also called F_1 in form language, a 2-index antisymmetric tensor field $A_{\mu\nu}$ with field strength $F_{\mu\nu\rho} = 3\partial_{[\mu} A_{\nu\rho]}$, also called F_3 in form language, and a 4-index antisymmetric tensor field $A^+_{\mu\nu\rho\sigma}$, with modified field strength $\tilde{F}^+_{\mu_1 \ldots \mu_5}$, also called $\tilde{F}^+_5$ in form language, which is self-dual,

$$\tilde{F}^+_{\mu_1 \ldots \mu_5} = \frac{1}{5!} \epsilon_{\mu_1 \ldots \mu_5}{}^{\mu_6 \ldots \mu_{10}} \tilde{F}^+_{\mu_6 \ldots \mu_{10}}. \tag{7.115}$$

Because of the existence of the self-dual field strength, there is no known fully covariant form for the type IIB action. But if one imposes the self-duality as a constraint after varying the action, we have the bosonic supergravity action

$$S_{IIB} = \frac{1}{2\kappa_{10}^2} \int d^{10}x \left\{ \sqrt{-G} \left[e^{-2\phi} \left(R + 4\partial_\mu \phi \partial^\mu \phi - \frac{1}{2}|H_3|^2 \right) \right. \right.$$
$$\left. \left. - \frac{1}{2}|F_1|^2 - \frac{1}{2}|\tilde{F}_3|^2 - \frac{1}{4}|\tilde{F}_5|^2 \right] - \frac{1}{2} A_4 \wedge H_3 \wedge F_3 \right\}, \tag{7.116}$$

where we have defined

$$\tilde{F}_3 = F_3 - A_0 \wedge H_3,$$
$$\tilde{F}_5 = F_5 - \frac{1}{2} A_2 \wedge H_3 + \frac{1}{2} B_2 \wedge F_3. \tag{7.117}$$

S-duality

Type IIB supergravity action is invariant under an $Sl(2, \mathbb{R})$ group. We consider the complex field $\tau = a + ie^{-\phi}$ and form the matrix

$$\mathcal{M}_{ij} \equiv \frac{1}{\mathrm{Im}(\tau)} \begin{pmatrix} |\tau|^2 & -\mathrm{Re}(\tau) \\ -\mathrm{Re}(\tau) & 1 \end{pmatrix}. \tag{7.118}$$

We also consider the NS–NS field strength H_3 and the RR field strength F_3 as forming a column vector

$$F_3^i = \begin{pmatrix} H_3 \\ F_3 \end{pmatrix}, \tag{7.119}$$

and change to the Einstein metric $G^{(E)}_{\mu\nu}$ by the usual conformal rescaling. Then the type IIB supergravity action becomes

$$S_{IIB} = \frac{1}{2\kappa^2_{10}} \int d^{10}x\sqrt{-G_E}\left[R_E - \frac{\partial_\mu\bar{\tau}\partial^\mu\tau}{2(\text{Im}\tau)^2} - \frac{M_{ij}}{2}F^i_3 \cdot F^j_3 - \frac{1}{4}|\tilde{F}_5|^2\right]$$
$$- \frac{\epsilon_{ij}}{8\kappa^2_{10}} \int d^{10}x A_4 \wedge F^i_3 \wedge F^j_4. \tag{7.120}$$

It is invariant under the $Sl(2, \mathbb{R})$ symmetry

$$\begin{pmatrix} a & b \\ c & d \end{pmatrix} \in Sl(2, \mathbb{R}); \quad ad - bc = 1, \tag{7.121}$$

that acts on the fields by

$$\tau' = \frac{a\tau + b}{c\tau + d}, \quad F'^i_3 = \Lambda^i_j F^j_3, \quad \Lambda^i_j = \begin{pmatrix} d & c \\ b & a \end{pmatrix}; \quad \tilde{F}'_5 = \tilde{F}_5 \quad G'_{\mu\nu}{}^{(E)} = G^{(E)}_{\mu\nu}, \tag{7.122}$$

and implies that the matrix $\mathcal{M}$ changes by $\mathcal{M} \to \mathcal{M}' = (\Lambda^{-1})^T \mathcal{M} \Lambda^{-1}$.

This $Sl(2, \mathbb{R})$ is an invariance of the classical type IIB supergravity, but an $Sl(2, \mathbb{Z})$ subgroup of it survives at the quantum level, as an invariance of full string theory. Included in it is the transformation

$$\begin{pmatrix} a & b \\ c & d \end{pmatrix} = \begin{pmatrix} 0 & -1 \\ 1 & 0 \end{pmatrix}, \tag{7.123}$$

that implies

$$\tau' = -1/\tau. \tag{7.124}$$

If $a = 0$, the transformation is

$$\phi' = -\phi \Rightarrow g'_s = \frac{1}{g_s}, \tag{7.125}$$

since $g_s = e^{<\phi>}$. But then this transformation is a nonperturbative transformation, or *duality*, called *S-duality*. This duality then also implies

$$H'_3 = F_3; \quad F'_3 = -H_3; \quad B'_2 = A_2; \quad A'_2 = -B_2; \quad A'_4 = A_4. \tag{7.126}$$

M-theory and 11-dimensional supergravity

As we mentioned, the type IIA string theory at strong coupling becomes an 11-dimensional theory called M-theory, whose low energy is the unique 11-dimensional supergravity, with the fields G_{MN}, $A_{MNP} \equiv A_{(3)}$ with field strength $F_{M_1...M_4} = 4\partial_{[M_1}A_{M_2M_3M_4]} \equiv F_{(4)}$. The bosonic action for 11-dimensional supegravity is

$$S_{11} = \frac{1}{2\kappa^2_{11}} \int d^{11}x\sqrt{-G}\left(R - \frac{1}{2}|F_4|^2\right) - \frac{1}{6\kappa^2_{11}} \int d^{11}x A_3 \wedge F_4 \wedge F_4. \tag{7.127}$$

To show the relation with type IIA supergravity in ten dimensions, we consider the KK dimensional reduction ansatz

$$ds^2 \equiv G^{(11)}_{MN} dX^M dX^N$$
$$= e^{-\frac{2}{3}\phi} G^{s,(10)}_{\mu\nu} dx^\mu dx^\nu + e^{\frac{4}{3}\phi}(dx^{10} + A_\mu dx^\mu)^2,$$
$$A_{MNP} : A_{\mu\nu 11} \equiv B^{(IIA)}_{\mu\nu}$$
$$A_{\mu\nu\rho} \equiv A^{(IIA)}_{\mu\nu\rho}. \tag{7.128}$$

One can check that it reduces the 11-dimensional supergravity action to the 10-dimensional type IIA supergravity action in string frame. We will not do it here, we will only check the parameters of the theory. Look at the coefficient of $(dx^{10})^2$ in the metric. It is $e^{\frac{4}{3}\phi}$, but it should equal $(R/l_P)^2$, the radius of the extra dimension squared in 11-dimensional Planck units. Here l_P is the 11-dimensional Planck length, defined by $2\kappa^2_{11} = (2\pi)^8 l_P^9$. Then

$$\left(\frac{R}{l_P}\right)^2 = e^{\frac{4}{3}<\phi>} = g_s^{4/3} \Rightarrow R = l_P g_s^{2/3}, \tag{7.129}$$

or $g_s = (R/l_P)^{3/2}$. Identifying the Einstein–Hilbert action term in 11 dimensions with the Einstein–Hilbert action term in ten dimensions in string frame under KK-dimensional reduction, we get

$$\frac{1}{2\kappa^2_{11}} \int d^{11}x R^{(11)} = \frac{2\pi R}{2\kappa^2_{11}} \int d^{10}x \sqrt{-g^{s,(10)}} e^{-2\phi} R^{(10)} + \dots$$
$$= \frac{2\pi R}{(2\pi)^8 l_P^9} \int d^{10}x \sqrt{-g^{s,(10)}} e^{-2\phi} R^{(10)} + \dots$$
$$= \frac{1}{g_s^2 (2\pi)^7 (\sqrt{\alpha'})^8} \int d^{10}x \sqrt{-g^{s,(10)}} e^{-2(\phi-<\phi>)} R^{(10)} + \dots$$
$$\equiv \frac{1}{2\kappa^2_{10}} \int d^{10}x \sqrt{-g^{s,(10)}} e^{-2(\phi-<\phi>)} R^{(10)} + \dots \tag{7.130}$$

Identifying the expressions gives $l_P^9 = R g_s^2 (\sqrt{\alpha'})^8$, and using $R = l_P g_s^{2/3}$, we obtain

$$l_P = g_s^{1/3} \sqrt{\alpha'} \Rightarrow R = g_s \sqrt{\alpha'}, \tag{7.131}$$

i.e. R is g_s in string units, as indicated.

The string duality web

There are five string theories, type IIA and type IIB closed string theories, type I open plus closed string theory, and heterotic $SO(32)$ and heterotic $E_8 \times E_8$, together with the 11-dimensional M-theory. But all of them are related by various dualities, so in reality there is only one string theory. Here we simply enumerate the duality relations, without much explanation.

We saw that M-theory on a circle gives type IIA supergravity, and that S-duality $g_s \rightarrow 1/g_s$ takes type IIB theory to itself. Besides reduction and S-duality, we have another duality, called T-duality, that is explained in more detail in Chapter 9. Here it suffices to say that it is a symmetry of string theory that relates a string theory compactified on a radius R with another theory compactified on the radius α'/R. It relates type IIA string theory with type IIB string theory, and the $SO(32)$ heterotic theory with the $E_8 \times E_8$ heterotic string theory. In turn, S-duality of $SO(32)$ heterotic string theory takes it to type I string theory. Finally, M-theory compactified on the interval $S^1/\mathbb{Z}_2$ (a circle identified under $x \sim -x$), with "M9-branes", i.e. 9+1-dimensional walls, at its endpoints gives the heterotic $E_8 \times E_8$ theory, completing the duality web relating the various theories.

The power of string theory is related to this fact, that there is really a unique string theory, and we always observe its various facets that may seem different, but are not.

7.10 Constructing S-matrices

The construction of S-matrices is complicated, and will not be explained here, but the physical ideas are simple, and follow the general pattern of the particle case, with some modifications.

To construct string theory perturbatively, as for the particle case, we construct S-matrices through Feynman diagrams, as in Fig. 7.3a. The basic interaction that gives the Feynman diagrams is the "pants diagram" in Fig. 7.3c relating three asymptotic closed string states. The Polyakov action defines the propagator as in the particle case. The external vertices of the theory (creating external states) are defined via so-called "vertex operators", and unlike the particle case, where we are free to introduce them as we wish, which gives rise to the various quantum field theories, in the string theory case the properties of the theory uniquely fix them. The internal vertices, describing the interaction of strings, are uniquely given by the topology of the diagram (the basic vertex looks like the pair of pants described above). This is as it should be, since as we saw, there is a unique string theory. Moreover, one can show that we reproduce supergravity vertices in the low energy limit $\alpha' \to 0$.

As we noted, one also needs to define integration carefully at each loop order, since the string interaction vertex is "smoothed out" by the extension of the string and cannot be localized at one point. This is at the root of the reason why string theory is interesting, since this smoothing out is responsible for the remarkable fact that, unlike quantum field theory, in string theory there are no UV infinities in any string Feynman diagrams (though they may be in the *sum* of the perturbative Feynman diagrams).

Important concepts to remember

- String theory is the theory of relativistic strings, with tension = energy/ length.
- For the Feynman diagram construction of quantum field theory, we need the particle action to define the propagator, the vertex factors to define the theory, and integration rules.
- The first order particle action $1/2 \int d\tau[e^{-1}(\tau)(\dot{X}^\mu)^2 - em^2]$ is more fundamental: it contains the massless case and is quadratic. We can fix a gauge $e = 1$ for reparameterization invariance. It contains interactions with the spacetime gravitational field.
- The Nambu–Goto string action is the area spanned by the moving string, and its minimization is due to its tension.

- The Polyakov string action is again more fundamental: it has more symmetries, specifically Weyl invariance, besides diffeomorphism and Poincaré invariance.

- By fixing a gauge, the closed string action reduces to free 2-dimensional bosons, which contain left and right moving wave modes. The residual gauge invariance is called conformal invariance.

- By quantizing these modes, we get the particle spectrum $M^2 = 1/\alpha'(N - a)$. The massless particles are the graviton, an antisymmetric tensor, and a scalar.

- The bosonic string must exist in 26 dimensions to have $a = 1$, corresponding to the existence of a massless vector.

- The bosonic string is unstable. The superstring is stable and exists in ten dimensions for the same reason the bosonic string exists in 26 dimensions, thus we need to use Kaluza–Klein dimensional reduction to get down to four dimensions.

- The superparticle action is $1/2 \int e^{-1}(\Pi_\tau^\mu)^2$, where Π_τ^μ is supersymmetric in spacetime. It is also kappa symmetric.

- The Green–Schwarz superstring has $-1/4\pi\alpha' \int (\Pi_a^\mu)^2$, where Π_a^μ is supersymmetric in spacetime. It also has a WZ term, needed for kappa symmetry.

- The NSR spinning string is worldsheet supersymmetric, and it matches the GS superstring in light-cone gauge, which has both spacetime and worldsheet supersymmetry.

- Self-consistent backgrounds for the string are given by the theory of the massless modes of the superstring, namely supergravity.

- The low energy limit ($\alpha' \to 0$) of string theory is supergravity.

- IIB supergravity has two minimal 10-dimensional fermions of the same chirality and has S-duality, and IIA has fermions of opposite chirality. IIA string theory at strong g_s becomes 11-dimensional M-theory, whose low energy limit is 11-dimensional supergravity.

- All string theories are related by dualities into a duality web, forming a unique theory.

- One knows how to construct string theory S-matrices from Feynman diagrams by defining the propagator, vertices, and integration rules.

References and further reading

Perhaps the best introduction to string theory (tailored to MIT undergraduates) is the book by Zwiebach [21]. A good advanced book, though lacking coverage of recent work (it doesn't cover any of the developments relating to the "second superstring revolution" dealing with D-branes and dualities) is the one by Green, Schwarz, and Witten [22]. It contains the most in-depth explanations of many classic string theory problems. A modern book that also contains both D-branes and dualities is the one by Polchinski [23], from one of the people who started the second superstring revolution. The most up-to-date book, containing many of the research problems being developed now (but as a result with less in-depth coverage of the basics of string theory), is [24]. Finally, a book focused more on D-branes is [25].

Exercises

1. Write down the worldline reparameterization invariance for the first order action for the particle, both the finite and infinitesimal versions.

2. Calculate the equation of motion of the free particle in a gravitational field (the geodesic equation), from the action (7.3), with $\eta_{\mu\nu} \rightarrow g_{\mu\nu}$, and specialize to the Newtonian limit to recover motion in Newtonian gravity.

3. Calculate $L_m, \tilde{L}_m$, and $L_0 + \tilde{L}_0$ for the bosonic string.

4. Derive the worldsheet momentum P_σ.

5. Write down the states of the first massive closed string level.

6. Prove the supersymmetry of the Wess–Zumino term of the superstring.

7. Prove the kappa symmetry of the action $S_{kin} + S_{WZ}$ for the superstring.

8. Show that the coupling to $B_{\mu\nu}$ is of the type of p-brane sources, thus a string is a 1-brane source for the field $B_{\mu\nu}$.

9. Find a representation for the 11-dimensional Γ matrices in terms of the Pauli σ^i matrix tensor products.

10. Check that the reduction ansatz (7.128) takes (7.127) to (7.113).

8

Elements of conformal field theory

We have already encountered conformal invariance in two dimensions, in the context of the string, as the residual gauge invariance obtained when we fix the conformal gauge for the Polyakov string, in order to fix general coordinate and Weyl invariance, (7.44). Then the conformal invariance is invariance under a specific combination of a general coordinate and a Weyl transformation that leaves the flat metric intact.

Conformal invariance is therefore invariance *of a theory in flat space* under a general coordinate transformation that, when acting on a flat space, changes the metric by a conformal factor, i.e. it is a (local) generalization of scale invariance,

$$x'^{\mu} = \alpha x^{\mu} \Rightarrow ds^2 = d\vec{x}'^2 = \alpha^2 d\vec{x}^2. \tag{8.1}$$

8.1 Conformal transformations and the conformal group

Scale invariance and the beta function

In quantum field theory, we can have theories that are classically scale invariant (under (8.1) or not, but generically at the quantum level the scale invariance is broken due to the dynamical appearance of the renormalization scale (dimensional transmutation).

The procedure of *renormalization* involves a cut-off ϵ and bare coupling λ_0 and mass m_0. For example, dimensional regularization of scalar field theory for $V(\phi) = m^2\phi^2/2 + \lambda\phi^4$ gives

$$\lambda_0 = \mu^{\epsilon}\left(\lambda + \sum_{k=1}^{\infty}\frac{a_k(\lambda)}{\epsilon^k}\right); \quad m_0^2 = m^2\left(1 + \sum_{k=1}^{\infty}\frac{b_k(\lambda)}{\epsilon^k}\right), \tag{8.2}$$

where μ is the renormalization scale, out of which we extract the renormalized coupling $\lambda = \lambda(\mu, \epsilon; \lambda_0, m_0)$, which in general depends on scale.

This *running of the coupling constant* with the scale is characterized by the β function,

$$\beta(\lambda, \epsilon) = \mu\frac{d\lambda}{d\mu}\bigg|_{m_0, \lambda_0, \epsilon}. \tag{8.3}$$

A quantum mechanically scale invariant theory (i.e. a theory independent of α in (8.1) at the quantum level) must be μ-independent, and thus must have a zero β function. There are two ways in which this can happen:

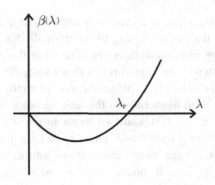

Figure 8.1 $\beta(\lambda)$ for the case of an IR stable point.

- $\beta = 0$ everywhere, which means a cancellation of Feynman diagrams that implies there are no infinities. OR

- A nontrivial interacting theory: the β function is nontrivial, but has a zero (fixed point) away from $\lambda = 0$, at which a nontrivial (nonperturbative) theory emerges: a conformal field theory. For the case in Fig. 8.1, λ_F is called an IR stable point. Indeed, if $\lambda > \lambda_F$, $\beta(\lambda) > 0$, thus λ decreases if μ decreases (thus in the IR). And if $\lambda < \lambda_F$, $\beta(\lambda) < 0$, thus λ increases if μ again decreases (in the IR). That means that if we go to the IR, wherever we start, we are driven to $\lambda = \lambda_F$, that has $\beta(\lambda_F) = 0$.

In condensed matter physics, a quantum scale invariant theory would be obtained near a phase transition (critical point).

Classical scale invariance is a global symmetry, and in global symmetries other than Poincaré (which is needed for causality) we can have quantum anomalies, i.e. breaking at the quantum level by certain Feynman diagrams (in four dimensions, one-loop triangle diagrams). But conformal invariance and Weyl invariance are local symmetries (similar to gauge symmetries), giving a local generalization of scale invariance, and as such cannot be broken at the quantum level, as we argued in Chapter 1. So one must require the absence of quantum anomalies with respect to Weyl and conformal invariance, which gives consistency conditions that constrain the theory.

On the other hand, most theories that are *quantum mechanically* scale invariant, and thus have $\beta = 0$, are also invariant under the larger invariance called conformal invariance. In fact it is an ongoing problem, one that has generated a lot of research in the last few years, whether there exist theories that are quantum mechanically scale invariant, but not conformal invariant.

In flat d dimensions, i.e. on $R^{1,d-1}$, conformal transformations are defined by $x_\mu \to x'_\mu(x)$ such that

$$ds^2 = dx'_\mu dx'^\mu = [\Omega(x)]^{-2} dx_\mu dx^\mu, \tag{8.4}$$

thus indeed a local generalization of (8.1).

Note once again that conformal invariance is NOT the same as general coordinate invariance (though conformal *transformations* obviously are a *subclass* of general coordinate transformations), since it is an invariance of *flat space* field theories, but the metric is modified, from flat $ds^2 = dx'_\mu dx'_\mu$ to "conformally flat" $ds^2 = [\Omega(x)]^{-2} dx_\mu dx_\mu$. The point is then that we can absorb this overall factor in the action, such that the transformed action can be written in the same form, with the same *flat* metric.

To give an example, consider the case of conformal transformations in two dimensions (to be explained more fully in the next section), which are holomorphic transformations, $z \to f(z)$, $\bar{z} \to \bar{f}(\bar{z})$, and two terms in the action, the kinetic term for a real scalar, $\int d^2z(\partial_\mu\phi)^2 = \int dz d\bar{z}\, \partial\phi\bar{\partial}\phi$, and a mass term $\int dz d\bar{z}\, m^2\phi^2$. Under the conformal transformation, the integration measure transforms as $\int dz d\bar{z} \to \int dz' d\bar{z}'(f'\bar{f}')^{-1}$ (which equals in fact $\int d^2z\sqrt{g}$, as it should), but the kinetic term transforms as $\partial\phi\bar{\partial}\phi \to (f'\bar{f}')\partial'\phi\bar{\partial}'\phi$, whereas the mass term does not transform $\phi^2 \to \phi^2$. Therefore the kinetic term is conformally invariant, but the mass term is not. Note that from the point of view of a general relativistically invariant theory, the mass term is also written in a general relativistically invariant way, *but not with the flat metric, like the kinetic term.* The difference is that when writing a general relativistically invariant version of the flat space theory, the kinetic term is also Weyl invariant, whereas the mass term is not. Thus indeed, conformal invariance arises because of the existence of a generalization of the theory that is both general relativistically invariant and Weyl invariant.

In conclusion, conformal transformations are generalizations of scale transformations (8.1) of flat space that change the distance between points by a local factor.

The infinitesimal version of the conformal transformation is

$$x'_\mu = x_\mu + v_\mu(x); \quad \Omega(x) = 1 - \sigma_v(x)$$

$$\Rightarrow \partial_\mu v_\nu + \partial_\nu v_\mu = 2\sigma_v\delta_{\mu\nu} \Rightarrow \sigma_v = \frac{1}{d}\partial \cdot v. \tag{8.5}$$

The solution $d = 2$ is special, and is analyzed separately. But except for $d = 2$, the most general solution to this equation is

$$v_\mu(x) = a_\mu + \omega_{\mu\nu}x_\nu + \lambda x_\mu + b_\mu x^2 - 2x_\mu b \cdot x, \tag{8.6}$$

with $\omega_{\mu\nu} = -\omega_{\nu\mu}$ (antisymmetric) and $\sigma_v(x) = \lambda - 2b \cdot x$, a statement that is left as an exercise to check (Exercise 1).

Thus the parameters of conformal transformations are $\lambda, a_\mu, b_\mu, \omega_{\mu\nu}$, corresponding respectively to scale transformations, translations, a new type of transformations, and rotations. The new type of transformations parameterized by b_μ are called "special conformal transformations." Together there are $1 + d + d + d(d-1)/2 = (d+1)(d+2)/2$ components for the parameters of conformal transformations.

These transformations together form a symmetry group. Its generators are: P_μ for a_μ and $J_{\mu\nu}$ for $\omega_{\mu\nu}$, forming together the Poincaré group, as expected. For them, we have the particular case of $\Omega(x) = 1$, i.e. no local scale transformation. The new generators are K_μ for the special conformal transformations, b_μ, and dilatation generator D for λ. Counting shows that we can assemble these generators in a group defined by an antisymmetric $(d + 2) \times (d + 2)$ matrix,

$$\bar{J}_{MN} = \begin{pmatrix} J_{\mu\nu} & \bar{J}_{\mu,d+1} & \bar{J}_{\mu,d+2} \\ -\bar{J}_{\nu,d+1} & 0 & D \\ -\bar{J}_{\nu,d+2} & -D & 0 \end{pmatrix}, \qquad (8.7)$$

where

$$\bar{J}_{\mu,d+1} = \frac{K_\mu - P_\mu}{2}; \quad \bar{J}_{\mu,d+2} = \frac{K_\mu + P_\mu}{2}; \quad \bar{J}_{d+1,d+2} = D. \qquad (8.8)$$

By looking at the Lie algebra of $\bar{J}_{MN}$ we find that the metric in the $d+2$ direction is negative, thus the symmetry group is $SO(2,d)$. So conformal invariance in flat $(1, d-1)$ dimensions $(d > 2)$ corresponds to the symmetry group $SO(2,d)$, the same as the symmetry group of $d+1$-dimensional Anti-de Sitter space, AdS_{d+1}.

This is in fact the first hint of a relation between d-dimensional conformal field theory, i.e. a field theory on d-dimensional Minkowski space that is invariant under the conformal group, and a gravity theory in $d+1$-dimensional Anti-de Sitter space. The precise relation between the two will be AdS/CFT, defined in Chapter 10.

A comment is in order here. Strictly speaking, $SO(2,d)$ is a group that only contains elements continuously connected to the identity; however, the conformal group is an extension that also contains the *inversion*, defined by

$$I : x'_\mu = \frac{x_\mu}{x^2} \Rightarrow \Omega(x) = x^2. \qquad (8.9)$$

In fact, all conformal transformations can be generated by combining the inversion with the rotations and translations. The finite version of the special conformal transformation is

$$x^\mu \rightarrow \frac{x^\mu + b^\mu x^2}{1 + 2x^\nu b_\nu + b^2 x^2}, \qquad (8.10)$$

and the finite version of the scale transformation is $x^\mu \rightarrow \lambda x^\mu$.

Since we are defining AdS/CFT in Euclidean space, we should note that the conformal group on $\mathbb{R}^d$ (Euclidean space) is $SO(1, d+1)$, as expected.

8.2 Conformal fields in two Euclidean dimensions

As we mentioned, $d = 2$ is special. In $d = 2$, the conformal group is much larger: in fact, it has an infinite set of generators.

To describe conformal fields in Euclidean $d = 2$, we use complex coordinates $(z, \bar{z})$,

$$ds^2 = dz d\bar{z}. \qquad (8.11)$$

It is easy then to see that the most general solution of the conformal transformation condition (8.5) is a general *holomorphic* transformation, i.e. $z' = f(z)$ (but not a function of $\bar{z}$), as we claimed in the previous section. Then the transformation on the metric is

$$ds'^2 = dz' d\bar{z}' = \frac{\partial z'}{\partial z} \frac{\partial \bar{z}'}{\partial \bar{z}} dz d\bar{z} = \Omega^{-2}(z, \bar{z}) dz d\bar{z}. \qquad (8.12)$$

The conformal transformation, as we see, is a general coordinate transformation of a special type: holomorphic. Therefore it is useful if we first define general relativity tensors, and then think of what this means for theories in flat space. A covariant tensor is an object $T_{i_1...i_n}$ that transforms (in Euclidean Cartesian coordinates $z_1, z_2, i_1, ..., i_n = 1, 2$) as

$$T_{i_1...i_n}(z_1, z_2) = T'_{j_1...j_n}(z'_1, z'_2)\frac{\partial z'^{j_1}}{\partial z^{i_1}} ... \frac{\partial z'^{j_n}}{\partial z^{i_n}}. \tag{8.13}$$

Now we consider the complex coordinates $z = z_1 + iz_2, \bar{z} = z_1 - iz_2$, and the holomorphic transformations $z' = z'(z)$ that are conformal. We generalize the notion of tensor to objects called *primary fields or tensor operators*, where the "number of covariant indices" is not necessarily an integer, i.e. a primary field of dimensions $(h, \tilde{h})$ is an object that transforms as

$$T_{z...z\bar{z}...\bar{z}} = T'_{z...z\bar{z}...\bar{z}}\left(\frac{dz'}{dz}\right)^h \left(\frac{d\bar{z}'}{d\bar{z}}\right)^{\tilde{h}}, \tag{8.14}$$

but not necessarily with integer $(h, \tilde{h})$. Indeed, under a complex global scale transformation $z' = \zeta z$ (i.e., a combination of scale transformation and rotation), there is a basis of local operators that diagonalize the dilatation operator, so scale as

$$\mathcal{O}(z, \bar{z}) = \zeta^h \bar{\zeta}^{\tilde{h}} \mathcal{O}'(z', \bar{z}'), \tag{8.15}$$

and for a general operator at the quantum level we have noninteger $h, \tilde{h}$, with $h + \tilde{h} \equiv \Delta$ being the dimension of $\mathcal{O}$ (giving the behavior under usual – real – scaling) and $h - \tilde{h}$ being the spin of $\mathcal{O}$ (giving the behavior under rotations).

The primary fields are denoted by $\phi^{(h,\tilde{h})}(z, \bar{z})$. The infinitesimal transformation $z' = z + \epsilon(z)$ implies for the primary field

$$\delta_\epsilon \phi^{(h,\tilde{h})}(z, \bar{z}) = \epsilon(z)\partial\phi^{(h,\tilde{h})}(z, \bar{z}) + h\partial\epsilon(z)\phi^{(h,\tilde{h})}(z, \bar{z}) + h.c. \tag{8.16}$$

In two dimensions, a conformally invariant tensor is traceless $T^i_i = 0$, as we have already seen in the previous chapter, since in flat Euclidean space $T_{ij} = \delta S/\delta g^{ij}$, and in a conformally invariant theory $g^{ij}\delta S/\delta g^{ij} = 0$. In complex coordinates, that translates into $T_{z\bar{z}}(= (T_{00} + T_{11})/4) = 0$. Then the conservation equation $\partial^a T_{ab} = 0$ becomes $\bar{\partial}T_{zz} = \partial T_{\bar{z}\bar{z}} = 0$, so that we have $T(z) \equiv T_{zz}(z)$ and $\tilde{T}(\bar{z}) = T_{\bar{z}\bar{z}}(\bar{z})$.

In a general quantum field theory, and in particular in two dimensions, we can define the *operator product expansion* (OPE). The product of two operators at different points can be approximated to any desired accuracy by a sum of coefficients times all possible operators, i.e. (calling arbitrary operators by $\mathcal{O}_k$),

$$\mathcal{O}_i(x_i)\mathcal{O}_j(x_j) = \sum_k c^k_{ij}(x_i - x_j)\mathcal{O}_k(x_j). \tag{8.17}$$

This is an operatorial statement, i.e. it holds on all the correlators of the theory. In particular, normalizing the operators of the conformal field theories so that their 2-point function, fixed by the behavior under scaling up to a constant, is

$$\langle\mathcal{O}_i(x_i)\mathcal{O}_i(x_j)\rangle = \frac{1}{|x_i - x_j|^{2\Delta_i}}, \tag{8.18}$$

where Δ_i is the scaling dimension, in higher correlators we have

$$\langle \mathcal{O}_i(x_i)\mathcal{O}_j(x_j)\ldots\rangle = \sum_k \frac{c^k{}_{ij}}{|x_i - x_j|^{\Delta_i + \Delta_j - \Delta_k}} \left\langle \mathcal{O}_k\left(\frac{x_i + x_j}{2}\right)\ldots\right\rangle, \tag{8.19}$$

so only numerical coefficients $c^k{}_{ij}$ are undetermined for the OPE. In particular, if $\mathcal{O}_j = \mathcal{O}_k$, we have a spacetime dependent factor $1/|x_i - x_j|^{2\Delta_i}$.

That means that two terms in the OPE of the energy-momentum tensor $T(z)$ with a general operator are determined up to constants which can also be found, as

$$T(z)\mathcal{O}(0,0) = \cdots + \frac{h}{z^2}\mathcal{O}(0,0) + \frac{1}{z}\partial\mathcal{O}(0,0) + \cdots, \tag{8.20}$$

since the scaling dimension of $T(z)$ in two dimensions is $\Delta_T = 2$.

For a primary field, we in fact have

$$T(z)\phi_i^{(h_i,\tilde{h}_i)} = \frac{h_i}{z^2}\phi^{(h_i,\tilde{h}_i)} + \frac{1}{z}\partial\phi_i^{(h_i,\tilde{h}_i)} + \text{nonsingular}. \tag{8.21}$$

Example: free massless scalar fields

The most relevant example of conformal field theory is the case of several free massless scalar fields, with (Euclidean) action,

$$S_E = \frac{1}{4\pi\alpha'}\int d^2\sigma[\partial_1 X^\mu \partial_1 X_\mu + \partial_2 X^\mu \partial_2 X_\mu]. \tag{8.22}$$

As we can see, this is nothing but the Polyakov string action in conformal gauge. In fact, we have seen that by choosing the conformal gauge we obtain a residual gauge invariance, a combination of Weyl and diffeomorphism invariance which is the conformal invariance.

Using complex coordinates

$$z = \sigma^1 + i\sigma^2; \quad \bar{z} = \sigma^1 - i\sigma^2; \quad \partial \equiv \partial_z = \frac{\partial_1 - i\partial_2}{2}; \quad \bar{\partial} \equiv \partial_{\bar{z}} = \frac{\partial_1 + i\partial_2}{2}, \tag{8.23}$$

we get the action

$$S = \frac{1}{2\pi\alpha'}\int d^2 z \partial X^\mu \bar{\partial} X_\mu, \tag{8.24}$$

which has the equation of motion

$$\partial\bar{\partial}X^\mu(z,\bar{z}) = 0, \tag{8.25}$$

with the general solution

$$X^\mu(z,\bar{z}) = X^\mu(z) + X^\mu(\bar{z}), \tag{8.26}$$

where we can expand in a Laurent series

$$X^\mu(z) = \frac{x^\mu}{2} - i\alpha'\frac{p^\mu}{2}\ln z + i\sqrt{\frac{\alpha'}{2}}\sum_{m\in\mathbb{Z},m\neq 0}\frac{\alpha_m^\mu}{mz^m}. \tag{8.27}$$

The continuation to Minkowski space is done by $\sigma^2 = i\sigma^0 = i\tau$, and under it a holomorphic function (function of w only) becomes a function of $-(\tau - \sigma)$, i.e. right-moving, and an anti-holomorphic function (function of $\bar{w}$ only) becomes a function of $\bar{w} = \tau + \sigma$, i.e. left-moving. We thus recover the Minkowski space treatment of the string from the previous chapter after the transformation $z = e^{-iw}$.

Then $T(z)$ corresponds to T_{--}, and $\tilde{T}(\bar{z})$ to T_{++} from the previous chapter. Their moments were defined by

$$L_m = \frac{1}{2\pi} \int_0^{2\pi} e^{-im\sigma} T_{--} d\sigma,$$

$$\tilde{L}_m = \frac{1}{2\pi} \int_0^{2\pi} e^{im\sigma} T_{++} d\sigma, \tag{8.28}$$

so now we can define them equivalently as Laurent coefficients of T_{zz} and $\tilde{T}_{\bar{z}\bar{z}}$, namely

$$T_{zz}(z) = \sum_{m \in \mathbb{Z}} \frac{L_m}{z^{m+2}}; \quad \tilde{T}_{\bar{z}\bar{z}}(\bar{z}) = \sum_{m \in \mathbb{Z}} \frac{\tilde{L}_m}{\bar{z}^{m+2}}. \tag{8.29}$$

By commuting the operators L_m, one finds the *Virasoro algebra*,

$$[L_m, L_n] = (m - n)L_{m+n} + \frac{c}{12}(m^3 - m)\delta_{m,-n}, \tag{8.30}$$

and similarly for the $\tilde{L}_m$s. Here as usual $(L_n)^\dagger = L_{-n}$.

The algebra at $c = 0$ is the classical part, and the term with c is a quantum correction. Here c is called a *central charge* and in general is a parameter of the theory. In string theory it has a given value for each component, such that the total central charge of the theory is zero.

The Virasoro algebra is the equivalent of the conformal group in two dimensions, which means that the L_ms are conserved charges, corresponding to symmetry operators. But it is not really a usual Lie algebra, since it has an infinite number of generators and, more importantly, the algebra contains a constant term (proportional to c), therefore the algebra does not close in the usual sense. However, L_0, L_1, and L_{-1} form a closed algebra without central charge:

$$[L_1, L_{-1}] = 2L_0; \quad [L_0, L_1] = -L_1; \quad [L_0, L_{-1}] = L_{-1}, \tag{8.31}$$

which is the algebra of the group $Sl(2, \mathbb{C})$, whose finite transformations act on z as

$$z \to \frac{az + b}{cz + d}. \tag{8.32}$$

This is a subalgebra of the Virasoro algebra that is sometimes called, by an abuse of notation, the conformal algebra in two dimensions.

The representations of the Virasoro algebra are given in terms of a "highest weight state" $|h\rangle$ that can be thought of as equal to $\lim_{z \to 0} \phi^h(z)|0\rangle$, i.e. obtained by acting with the primary field at $z = 0$ on a vacuum. Then as usual in the method of induced representations, we have:

$$L_0|h\rangle = h|h\rangle; \quad L_n|h\rangle = 0, n > 0; \tag{8.33}$$

so h is the eigenvalue of L_0 ("energy"), therefore the rest of the fields in the representation, called *descendants*, are obtained by acting with L_{-n}, which increases the value of the energy h by n, i.e.

$$L_0(L_{-n}|h\rangle) = (h + n)(L_{-n}|h\rangle). \tag{8.34}$$

We see then that the term "highest weight", used because of historical reasons, is a misnomer, and it should really be "lowest weight." Then the representation is $|h\rangle, L_{-1}|h\rangle$, $(L_{-1})^2|h\rangle, L_{-2}|h\rangle, \ldots$ and is called a *Verma module*.

One observation we should make is that we have talked here about states, but before we talked about operators. In a conformal field theory there is an important relation called the operator-state correspondence, which can be roughly understood as above, namely a state corresponds to an operator at 0 acting on a vacuum, $|h\rangle = \lim_{z \to 0} \phi^h(z)|0\rangle$. Therefore from now on we use interchangeably the words operator and state when talking about conformal field theories.

We can also calculate the two-point correlators of the free scalars. Defining the partition function in the case of one scalar (putting $2\pi\alpha' = 1$),

$$Z[J] = \int \mathcal{D}X e^{-\int d^2 z \partial X \bar{\partial} X + \int d^2 z X J}, \tag{8.35}$$

we can calculate the 2-point function

$$\langle X(z)X(w) \rangle = \frac{\delta}{\delta J(z)} \frac{\delta}{\delta J(w)} Z[J] \bigg|_{J=0} = -\frac{1}{4\pi} \log(|z - w|^2). \tag{8.36}$$

Then it follows that we also have

$$\langle \partial X(z) \partial X(w) \rangle = -\frac{1}{4\pi (z - w)^2}. \tag{8.37}$$

8.3 Conformal fields and correlators in $d > 2$

As we saw, there is a big difference between two dimensions and higher, since in two dimensions we have an infinite dimensional algebra, that imposes very strong constraints on the theory, whereas in higher dimensions we just have the $SO(d, 2)$ Lie algebra. However, many features are common.

We can define primary operators in $d > 2$ also, but in a slightly different manner.

Representations of the conformal group are defined in terms of eigenfunctions of the scaling operator D with eigenvalue $-i\Delta$, where Δ is the scaling dimension, i.e. under $x \to \lambda x$, the field transforms as

$$\phi(x) \to \phi'(x) = \lambda^\Delta \phi(\lambda x). \tag{8.38}$$

Therefore D now plays the role of L_0 in $d = 2$, giving the "energy" of the state. On the other hand, instead of L_n and L_{-n} we now have K_μ and P_μ.

Then Δ is increased by P_μ, since the $SO(d, 2)$ conformal algebra described before acts as

$$[D, P_\mu] = -iP_\mu \Rightarrow D(P_\mu \phi) = P_\mu(D\phi) - iP_\mu \phi = -i(\Delta + 1)(P_\mu \phi), \tag{8.39}$$

and is decreased by K_μ, since

$$[D, K_\mu] = iK_\mu, \tag{8.40}$$

thus we can think of K_μ as an annihilation operator a and P_μ as a creation operator $a^\dagger$.

Since P_μ and K_μ are symmetry operators, by their successive action we get other states in the theory. The representation then is built as if using creation/annihilation operators P_μ / K_μ, as in the $d = 2$ case.

There will be an operator of lowest dimension, Φ_0, in the representation of the conformal group. Then, it follows that $K_\mu \Phi_0 = 0$, and Φ_0 is called the primary operator. The representation is obtained from Φ_0 and operators obtained by acting successively with P_μ ($\sim a^\dagger$) on Φ_0 ($\sim |0>$).

We have seen that the conformal transformations in $d > 2$ are the $SO(d, 2)$ transformations and the inversion. For a finite conformal transformation $x'^\mu(x^\nu)$, we can define an orthogonal matrix (in $O(d)$):

$$R_{\mu\nu}(x) = \Omega(x) \frac{\partial x'^\mu}{\partial x^\nu}; \quad \Rightarrow (RR^T)_{\mu\nu} = R_{\mu\rho} R_{\nu\rho} = \delta_{\mu\nu}. \tag{8.41}$$

For an inversion, we have $x'^\mu = x^\mu / x^2$ and $\Omega(x) = x^2$, giving the orthogonal matrix representing inversion,

$$R_{\mu\nu}(x) \equiv I_{\mu\nu}(x) = \delta_{\mu\nu} - 2 \frac{x^\mu x^\nu}{x^2}. \tag{8.42}$$

Then $I_{\mu\nu}(x - y)$ can be thought of as performing parallel transport, since it transforms under a conformal transformation as

$$I_{\mu\nu}(x' - y') = R_{\mu\rho} R_{\nu\sigma} I_{\rho\sigma}(x - y); \quad (x' - y')^2 = \frac{(x - y)^2}{\Omega(x)\Omega(y)}. \tag{8.43}$$

Two-point correlators

For scalar operators, we have already seen the general form allowed by conformal invariance,

$$\langle \mathcal{O}_i(x) \mathcal{O}_j(y) \rangle = \frac{C \delta_{ij}}{|x - y|^{2\Delta_i}}, \tag{8.44}$$

and we can normalize the operators such that $C = 1$.

We are also interested in conserved currents J_μ^a, that transform under the inversion as corresponding to $\Delta = d - 1$, namely $J_\mu^a(x) \rightarrow (x'^2)^{d-1} I_{\mu\nu}(x') J_\nu^a(x')$. Their 2-point function needs to be covariant under the inversion and Poincaré transformations, for conformal invariance of the theory. That restricts the function to be

$$\langle J_\mu^a(x) J_\nu^b(y) \rangle = C \frac{\delta^{ab} I_{\mu\nu}(x - y)}{|x - y|^{2(d-1)}}. \tag{8.45}$$

Three-point correlators

For scalar operators, we have also already given part of the form allowed by conformal invariance when describing the OPE, (8.19). The rest can be found by permutations, giving

$$\langle \mathcal{O}_i(x)\mathcal{O}_j(y)\mathcal{O}_k(z) \rangle = \frac{C_{ijk}}{|x-y|^{\Delta_i+\Delta_j-\Delta_k}|y-z|^{\Delta_j+\Delta_k-\Delta_i}|z-x|^{\Delta_k+\Delta_i-\Delta_j}}. \quad (8.46)$$

We will leave a description of the 3-point functions of conserved currents for when we compute them using AdS/CFT.

8.4 $\mathcal{N} = 4$ Super Yang–Mills as a conformal field theory

In $d = 4$, $\mathcal{N} = 4$ Super Yang–Mills theory is a representation of the conformal group. $\mathcal{N} = 4$ Super Yang–Mills theory with $SU(N)$ gauge group has the fields $\{A_\mu^a, \psi_\alpha^{ai}, \phi_{[ij]}^a\}$, as we saw in Chapter 3. Here we have used $SU(4)$ notation ($i \in SU(4)$) and $a \in SU(N)$. Indeed, one can calculate the β function of the theory and find that it is zero, thus the theory has quantum scale invariance. In fact, it is quantum mechanically invariant under the full conformal group.

On one hand, $\mathcal{N} = 4$ SYM is a gauge theory, with a well-defined perturbation theory, so objects like amplitudes and S-matrices would be natural to define. Indeed, we see in Part III of the book that we can calculate gluon amplitudes at strong coupling via AdS/CFT using a prescription due to Alday and Maldacena, and in perturbation theory one can use Feynman diagrams to define them, but in that case the amplitudes are calculated using an (IR) regularization like dimensional regularization or mass regularization. But since we are in a conformal field theory that a priori has no notion of scale, and therefore no notion of asymptotic separation of particle wavefunctions, amplitudes and S-matrices are tricky to define.

Natural objects to define and study in a conformal field theory are instead correlators and conformal dimensions of operators. These are best understood using AdS/CFT, and have been extensively studied. We can also break conformal invariance using finite temperature (since the temperature is an energy scale), so we can calculate many finite temperature quantities as well.

Note on dimensions

We should note that the quantum conformal dimension (scaling dimension) Δ need not be the same as the free (at coupling $g = 0$) scaling dimension for an operator in $\mathcal{N} = 4$ Super Yang–Mills, since $\beta = 0$ just means that there are no infinities, but there still can be finite renormalizations giving nontrivial quantum effects, so that $\Delta = \Delta_0 + \mathcal{O}(g)$.

Classically, the fundamental fields have dimensions $[A_\mu^a] = 1$, $[\psi_\alpha^{ai}] = 3/2$, $[\phi_{[ij]}^a] = 1$, and we form operators out of them, for instance Tr $F_{\mu\nu}^2$ (which will have classical dimension four). For some of these, the classical dimension will be exact, for others it will have quantum corrections.

Important concepts to remember

- Conformal transformations are coordinate transformations that act on flat space and give a space-dependent scale factor $[\Omega(x)]^{-2}$, thus conformal invariance is an invariance of flat space.
- A quantum mechanical scale invariant theory (with zero beta function) is generally conformal invariant. The absence of anomalies requires consistency conditions on the theory.
- In $d > 2$ Minkowski dimensions, the conformal group is $SO(d, 2)$, the same as the invariance group of AdS_{d+1}. The inversion is also part of the group, and together with the Poincaré group generates all transformations.
- In two dimensions, conformal invariance is an infinite algebra, the Virasoro algebra, of a more general type (with a constant term). A normal subgroup is $Sl(2, \mathbb{C})$.
- Primary fields of dimensions $(h, \bar{h})$ in two dimensions scale under $z \to \lambda z, \bar{z} \to \lambda \bar{z}$ as $\phi \to (\lambda)^{-(h+\bar{h})}\phi$, and in four dimensions primary fields of dimension Δ scale as $\phi \to \phi(\lambda)^{-\Delta}$.
- The operator product expansion (OPE) defines the correlators up to numerical constants $C^k{}_{ij}$.
- In $d = 4$, a representation of the conformal algebra is obtained by acting with P_μ on the primary field.
- The representative $I_{\mu\nu}(x - y)$ of inversion helps us define the 2-point function of conserved currents.

References and further reading

For an introduction to renormalization, see any quantum field theory book, in particular [1] and [2]. For an introduction to conformal field theory in the context of string theory, see Polchinski [23]. For conformal field theory in four dimensions, in the context of AdS/CFT, see the AdS/CFT review [26].

Exercises

1. Check that

$$v_\mu = a_\mu + \omega_{\mu\nu}x_\nu + \lambda x_\mu + b_\mu x^2 - 2x_\mu b \cdot x,$$

$$\partial_\mu v_\nu + \partial_\nu v_\mu = 2\sigma_v \delta_{\mu\nu}; \quad \sigma_v = \frac{1}{d}\partial \cdot v, \tag{8.47}$$

and that if $x'_\mu = x_\mu + v_\mu$, then the conformal factor is $\Omega(x) = 1 - \sigma_v(x)$.

2. Derive the conformal algebra in terms of $P_\mu, J_{\mu\nu}, K_\mu, D$ from the $SO(d, 2)$ algebra, given that $J_{\mu,d+1} = (K_\mu - P_\mu)/2$, $J_{\mu,d+2} = (K_\mu + P_\mu)/2$, $J_{d+1,d+2} = D$.

3. Prove that the special conformal transformation

$$x^\mu \rightarrow \frac{x^\mu + b^\mu x^2}{1 + 2x^\nu b_\nu + b^2 x^2} \tag{8.48}$$

can be obtained by an inversion, followed by a translation, and another inversion.

4. Prove that a circle $(x_\mu - c_\mu)^2 = R^2$ remains a circle after a general finite conformal transformation.

5. Using the definitions (8.29) or equivalently (8.28) of L_m and $\tilde{L}_m$ and the quantization conditions (7.68), derive the classical part of the Virasoro algebra (8.30) (for $c = 0$).

6. Check the transformation law of $I_{\mu\nu}(x - y)$, (8.43).

7. Check that

$$Z_\mu \equiv \frac{1}{2} \frac{\partial}{\partial z^\mu} \ln \frac{(z - y)^2}{(z - x)^2} = \frac{(x - z)_\mu}{(x - z)^2} - \frac{(y - z)_\mu}{(y - z)^2} \tag{8.49}$$

satisfies, under a conformal transformation,

$$Z'_\mu = \Omega(x) R_{\mu\nu}(z) Z_\nu. \tag{8.50}$$

We can also define X_μ and Y_μ by permutations.

8. Check the composition law of $I_{\mu\nu}$,

$$I_{\mu\nu}(x - y) I_{\nu\rho}(z - y) = I_{\mu\rho}(x - y) + 2(x - y)^2 X_\mu Y_\rho. \tag{8.51}$$

D-branes

9.1 Dirichlet boundary conditions and D-branes

We saw in Chapter 7 that open strings can have two types of boundary conditions, Neumann, $\partial^\sigma X^\mu(\tau, 0) = \partial^\sigma X^\mu(\tau, l) = 0$, and Dirichlet, $\delta X^\mu(\tau, 0) = \delta X^\mu(\tau, l) = 0$. The Dirichlet boundary conditions mean that the endpoints of open strings are fixed, and the Neumann boundary conditions mean they are free, and move at the speed of light. But we can consider these conditions independently for each coordinate, in particular we can choose $p + 1$ Neumann boundary conditions for p spatial dimensions and time, and $d - p - 1$ Dirichlet boundary conditions. This means that the endpoints of the string are constrained to exist on a $p + 1$-dimensional wall in spacetime, a "D-p-brane." As we saw, the word p-brane comes from a generalization of the word mem-brane. But different string endpoints could be on different walls (D-p-branes), as in Fig. 9.1a. We will call the directions on which both endpoints of the open strings are Neumann (NN directions) X^+, X^-, X^a (sometimes all of them by X^a) and the directions on which they are Dirichlet (DD directions) X^i.

Dai, Leigh and Polchinski [27] proved that in fact this wall where open strings end is dynamical, i.e. it can fluctuate and respond to external interactions, and that it has degrees of freedom existing on it. The wall was then called a D-brane, from Dirichlet-brane (as in Dirichlet boundary conditions). For $p = 2$, we would have a Dirichlet mem-brane. By extension, we have a Dirichlet p-brane, or D-p-brane.

A dynamical p-brane should have an action minimizing its "worldvolume," the same way as the particle has an action minimizing its worldline proper time, and a string has an action minimizing the area of its worldsheet, and the coefficient should be a p-brane tension, i.e. energy (mass) per unit volume, namely

$$S_p = -T_p \int d^{p+1}\xi \sqrt{-\det(h_{ab})}, \tag{9.1}$$

where h_{ab} is the induced metric on the worldvolume, or pull-back of the spacetime metric, namely

$$h_{ab} = \frac{\partial X^\mu}{\partial \xi^a} \frac{\partial X^\nu}{\partial \xi^b} g_{\mu\nu}(X). \tag{9.2}$$

This is an obvious generalization of the case of the Nambu–Goto string action, which is given in terms of the induced metric on the string worldsheet.

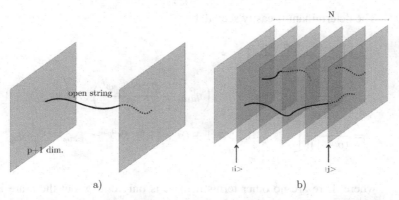

Figure 9.1 a) Open string between two D-p-branes ($p+1$-dimensional "walls"); b) The endpoints of the open string are labelled by the D-brane they end on (out of N D-branes), here $|i\rangle$ and $|j\rangle$.

9.2 D-brane fluctuations: fields, action, and tension

We have seen the generic action for a p-brane coupled to a spacetime metric. But moreover, the combination that appears in string theory is actually $g_{\mu\nu}(X) + \alpha'B_{\mu\nu}$, as we saw for instance in the string action in background fields, (7.90). We can also understand this from the decomposition of the physical massless states $\alpha^i_{-1}\tilde{\alpha}^j_{-1}|0,k\rangle$ into the symmetric traceless $g_{\mu\nu}$ and the antisymmetric $B_{\mu\nu}$ (and the dilaton trace).

Then the action for coupling to $g_{\mu\nu}$ and $B_{\mu\nu}$ should be

$$S = -T_p \int d^{p+1}\xi \sqrt{-\det\left(\frac{\partial X^\mu}{\partial\xi^a}\frac{\partial X^\nu}{\partial\xi^b}(g_{\mu\nu} + \alpha'B_{\mu\nu})\right)}. \tag{9.3}$$

The action has reparameterization (general coordinate) invariance on the worldvolume, which can be fixed by choosing a *static gauge*, where $X^a = \xi^a$ for $a = 0, 1, \ldots, p$. Then

$$X^i(\xi^a) \equiv \frac{\phi^i(\xi^a)}{\sqrt{T_p}} \tag{9.4}$$

are fields on the worldvolume. Writing $g_{\mu\nu}(X) = \eta_{\mu\nu} + 2\kappa_N h_{\mu\nu}(X)$, we get the action for ϕ^i at $h_{\mu\nu} = 0, B_{\mu\nu} = 0$ (in flat spacetime) as

$$S_p = -T_p \int d^{p+1}\xi \sqrt{-\det\left(\eta_{ab} + \frac{\partial_a\phi^i\partial_b\phi^i}{T_p}\right)}. \tag{9.5}$$

This is called the *scalar Dirac–Born–Infeld (DBI) action*. Expanding the determinant, we get

$$S_p = -T_p \int d^{p+1}\xi$$

$$\sqrt{\frac{1}{(p+1)!}\epsilon^{a_1\ldots a_{p+1}}\epsilon^{b_1\ldots b_{p+1}}(\eta_{a_1b_1} + \partial_{a_1}\phi^i\partial_{b_1}\phi^i)\ldots(\eta_{a_{p+1}b_{p+1}} + \partial_{a_{p+1}}\phi^i\partial_{b_{p+1}}\phi^i)}. \tag{9.6}$$

The determinant is easily seen to be

$$
\det\left(\eta_{ab} + \frac{\partial_a\phi^i\partial_b\phi^i}{T_p}\right)
$$

$$
= \frac{1}{(p+1)!}\epsilon^{a_1\ldots a_{p+1}}\epsilon^{b_1\ldots b_{p+1}}\left(\eta_{a_1b_1} + \frac{\partial_{a_1}\phi^i\partial_{b_1}\phi^i}{T_p}\right)\ldots\left(\eta_{a_{p+1}b_{p+1}} + \frac{\partial_{a_{p+1}}\phi^i\partial_{b_{p+1}}\phi^i}{T_p}\right)
$$

$$
= \frac{1}{(p+1)!}\left[\epsilon_{a_1\ldots a_{p+1}}\epsilon^{a_1\ldots a_{p+1}} + (p+1)\epsilon^{a_1\ldots a_p a_{p+1}}\epsilon_{a_1\ldots a_p}{}^{b_{p+1}}\frac{1}{T_p}\partial_{a_{p+1}}\phi^i\partial_{b_{p+1}}\phi^i + \ldots\right],
$$

$$(9.7)$$

where there are no other terms if there is only one ϕ, but there are in the case of more ϕ's. Since $\epsilon^{a_1\ldots a_{p+1}}\epsilon_{a_1\ldots a_{p+1}} = (p+1)!$ and $\epsilon^{a_1\ldots a_p a_{p+1}}\epsilon_{a_1\ldots a_p}{}^{b_{p+1}} = p!\,\eta^{a_{p+1}b_{p+1}}$, we get the action

$$
S_p = -T_p\int d^{p+1}\xi\sqrt{1 + \frac{\partial^a\vec\phi\partial_a\vec\phi}{T_p} + \ldots} \simeq -\int d^{p+1}\xi\left[T_p + \frac{1}{2}\partial^a\vec\phi\partial_a\vec\phi + \ldots\right]. \quad (9.8)
$$

Therefore, we see that to first order we get a canonical scalar kinetic term.

We can also find the first order coupling of the scalar to gravity by expanding the action in $h_{\mu\nu}$ and ϕ^i,

$$
\mathcal{L}_p \simeq -T_p\sqrt{-\det\left(\eta_{ab} + 2\kappa_N h_{ab} + 4\kappa_N h_{ai}\partial_b\frac{\phi^i}{\sqrt{T_p}}\right)}
$$

$$
= -T_p\sqrt{\frac{1}{(p+1)!}\left[\ldots + 4(p+1)\kappa_N h_{ai}\frac{\partial_b\phi^i}{\sqrt{T_p}}p!\,\delta_a^b + \ldots\right]}
$$

$$
\simeq -T_p\sqrt{\left(1 + \ldots + 4\kappa_N h_{ai}\frac{\partial^a\phi}{\sqrt{T_p}} + \ldots\right)}. \quad (9.9)
$$

The scalar–graviton action is then

$$
S_p = -\int d^{p+1}\xi\left[T_p + \frac{1}{2}\partial^a\vec\phi\partial_a\vec\phi + 2\kappa_N\sqrt{T_p}h_{ai}\partial^a\phi^i + \ldots\right]. \quad (9.10)
$$

Therefore, we find the coupling of the scalar ϕ^j with momentum k^μ, with the graviton h_i^a, is $2\kappa_N\sqrt{T_p}\,ik^a\delta_i^j$.

This is a coupling of a closed string mode, $g_{\mu\nu}(X)$, that exists everywhere, i.e. in the bulk of spacetime, with $\phi^i(\xi)$, or $X^i(\xi)$, which is a mode that lives on the worldvolume of the D-brane only. The X^i mode is an open string mode, since open strings can have both ends on the D-brane, hence their modes belong to the worldvolume of the D-brane. Therefore, the interpretation of this coupling is that a closed string collides with a D-brane, and excites an open string mode X^i corresponding to the position of the D-brane, by breaking at a point, and makes the D-brane vibrate, as in Fig. 9.2a.

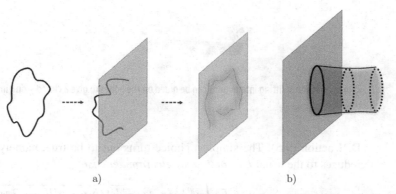

Figure 9.2   a) Closed string colliding with a D-brane, exciting an open string mode and making it vibrate; b) String worldsheet corresponding to a), with a closed string tube coming from infinity and ending on the D-brane as an open string boundary.

There is a corresponding string theory calculation for this process, where a string world-sheet connects an open string boundary fixed on a D-brane with a closed string tube coming from infinity, as in Fig. 9.2b. This calculation (which will not be done here) is matched with the above calculation of the scalar–graviton coupling on the D-brane, which fixes the p-brane tension T_p in terms of string theory parameters as

$$T_p = \frac{1}{(2\pi\alpha')^2 g_{p+1}^2},\tag{9.11}$$

where g_{p+1} is the coupling of the $p + 1$-dimensional theory,

$$g_{p+1}^2 = (2\pi)^{p-2} g_s \alpha'^{\frac{p-3}{2}}.\tag{9.12}$$

The last closed string mode in the NS–NS sector is the dilaton ϕ, which as we saw comes from the trace of the massless physical mode $a_1^{\dagger i} a_1^{\dagger i}|0, k\rangle$. The dependence of S_p on ϕ is found as follows. The closed string action has $1/g_s^2$ in front, where g_s is the (closed) string coupling, $g_s = e^{\langle\phi\rangle}$, but a (closed string) is an (open string)^2. More precisely, as the diagram in Fig. 9.3 shows, the closed string coupling g_s is given in terms of the open string coupling g_o as $g_s = (g_o)^2$. Therefore the D-brane action, which is an open string action, should have in front $1/g_o^2 = 1/g_s = e^{-\langle\phi\rangle}$. Then the action also including ϕ is

$$S_p = T_p \int d^{p+1}\xi\, e^{-\phi} \sqrt{-\det\left(\frac{\partial X^\mu}{\partial \xi^a}\frac{\partial X^\nu}{\partial \xi^b}(g_{\mu\nu} + \alpha' B_{\mu\nu})\right)}.\tag{9.13}$$

But there are other fields existing on the worldvolume of the D-brane. In particular, we have seen that the open strings with Neumann boundary conditions have a massless vector $a^{\dagger i}|0, k\rangle$. This means that the open string with $p + 1$ Neumann boundary conditions in the X^a directions has a massless vector A_a corresponding to $a^{\dagger a}|0, k\rangle$. For the action, we know that in the weak field limit we expect the usual massless vector action, the Maxwell action $-1/4 F_{ab}F^{ab}$. We also expect that it should fit somehow inside the square root of the scalar

Two open string splitting interactions can be glued on the edges to give a closed string interaction ("pair of pants"), therefore $g_{YM}^2 = g_s$.

DBI action (9.5). The simplest choice turns out to be true, namely the combination that reduces to the usual *DBI action for electromagnetism*,

$$S_{p,\text{DBI}} = T_p \int d^{p+1}\xi e^{-\phi} \sqrt{-\det(h_{ab} + \alpha' B_{ab} + 2\pi\alpha' F_{ab})}. \qquad (9.14)$$

This is generically called the *DBI action*. The DBI action for electromagnetism (at $B_{\mu\nu} = h_{\mu\nu} = 0$, $X^i = 0$) is a remarkable action, with many unique properties. The reason it was introduced by Born and Infeld in 1934 (it was much studied later by Dirac up to the 1950s – when it was considered a crazy thing to do by most physicists, who perhaps thought Dirac was past his glory days – hence his name is usually attached to it) is that it has a maximum value for the electric field and the energy density, and thus avoids the usual infinities of Maxwell theory. It was proven by Plebanski that it is the unique nonlinear completion of Maxwell electrodynamics (without higher derivatives, i.e. only in terms of $F_{\mu\nu}$) that is both causal and has only one characteristic surface ("wave front"), since generically (other than Maxwell and DBI) there are two such surfaces.

We have not yet given the coupling of the D-p-brane to the RR antisymmetric n-form fields. It is given by a "Wess–Zumino" term. Part of it can be guessed, since as we saw in Chapter 6, the source term for a p-brane in D dimensions is

$$\int d^D x j^{\mu_1...\mu_{p+1}} A_{\mu_1...\mu_{p+1}} = \mu_p \int d^{p+1}\xi A_{01...p} \equiv \mu_p \int_{p+1} A_{(p+1)}, \qquad (9.15)$$

where μ_p is a p-brane charge. Thus we use form integration to integrate the $p + 1$-form $A_{(p+1)}$ on a $p + 1$-dimensional worldvolume. But given that on the worldvolume of a D-p-brane we also have the fields F_{ab} (intrinsic) and B_{ab} (induced from spacetime), and the fact that as we saw, the RR fields appear in ps jumping by two units (in IIA we have $A_{(1)}$ and $A_{(3)}$ and in IIB we have $A_{(0)}$, $A_{(2)}$ and $A_{(4)}^+$) we can in fact form other $p + 1$-forms by wedge product with the combination $\alpha' B_{(2)} + 2\pi\alpha' F_{ab}$ any number of times. Formally then, the full Wess–Zumino term is

$$S_{p,\text{WZ}} = \mu_p \int_{p+1} \left[\exp(\alpha' B_{ab} + 2\pi\alpha' F_{ab}) \wedge \sum_n A_{(n)} \right], \qquad (9.16)$$

where for each given $A_{(n)}$ we keep only the term in the expansion of the exponential that completes to a $p + 1$-form, and of course the exponential is understood formally in terms of wedge product.

Finally, in a supersymmetric situation, e.g. for a flat D-brane in flat spacetime, we expect no cosmological constant on the worldvolume of the D-brane, since it would gravitate, and break supersymmetry. But since the action (9.10) has a constant term $-\int T_p$, to cancel it with a constant WZ term we need to have $\mu_p = T_p$. This is indeed the result in flat spacetime, where $\phi = \langle \phi \rangle$, and in general we have $\mu_p = T_p e^{\phi - \langle \phi \rangle}$.

The final bosonic D-p-brane action is $S_{p-\text{DBI}} + S_{p,\text{WZ}}$. Of course, the action involves fermions also, and can be found by imposing supersymmetry. It has a quadratic first order form analogous to the Polyakov action, which can be made both supersymmetric and kappa symmetric, but its form is too complicated, and will not be given here.

9.3 Chan–Patton factors and quantization of open strings on D-branes

On the endpoints of open strings we can add labels $|i\rangle$, with i going from 1 to N, called *Chan–Patton factors*. Then the open strings, with one end in the N representation of $U(N)$ and the other in the $\bar{N}$ representation of $U(N)$, can be considered to be in the adjoint representation. Considering the matrices λ_{ij}^A in the adjoint of $U(N)$, the open string wavefunctions

$$|k; A\rangle = \sum_{i,j=1}^{N} |k, ij\rangle \lambda_{ij}^A, \tag{9.17}$$

when describing massless vector states, give $U(N)$ gauge fields A_a^A. We have seen that there are $U(1)$ gauge fields on a single D-brane, but now the $|i\rangle$ index is easily identified as labelling D-branes on which each string endpoint can end, as in Fig. 9.1b. Therefore, when there are N D-branes we expect a $U(N)$ gauge theory. Of course, for the gauge fields existing on the D-branes, that is a natural situation, but for the scalars X^i or ϕ^i corresponding to the positions of the D-branes that seems strange, since we expect to have only N positions for them, not N^2. Naturally, the diagonal components of the $U(N)$ matrices X^i are identified with the D-brane positions, but the off-diagonal components are purely quantum variables, that describe a sort of quantum geometry viewed by D-branes.

At quadratic level, we expect the *nonabelian* action for ϕ^i and A_a^A to be

$$S_p = \int d^{p+1}\xi (-2) \text{Tr} \left[-\frac{1}{2} D_a \vec{\phi} \cdot D^a \vec{\phi} - \frac{1}{4} F_{ab} F^{ab} + \dots \right], \tag{9.18}$$

based on the fact that the abelian version is correct.

Quantization of bosonic open strings on D-p-branes

Quantization of open strings stretched between two D-branes at X_1^i and X_2^i proceeds as expected.

In the purely Neumann case, the Virasoro constraints for X^- give us (7.73), which can now be written as

$$\dot{X}^- \pm X'^- = \frac{1}{2p^+}(\dot{X}^I \pm X'^I)^2 = \frac{1}{2p^+}\left[(\dot{X}^i \pm X'^i)^2 + (\dot{X}^a \pm X'^a)^2\right]. \qquad (9.19)$$

On the other hand, for the NN (Neumann) coordinates X^a, we still have the relation (7.52), which gives

$$\dot{X}^a \pm X'^a = \sqrt{2\alpha'}\sum_{n\in\mathbb{Z}}\alpha_n^a e^{-in(\tau\pm\sigma)}. \qquad (9.20)$$

But as we saw, all X^μ satisfy the equation of motion

$$\partial_+\partial_- X^\mu = 0, \qquad (9.21)$$

which is solved by $X^\mu(\tau,\sigma) = f(\tau+\sigma) + g(\tau-\sigma)$. But for the Dirichlet directions of the open string, we have the boundary conditions

$$X^i(\tau,\sigma)\big|_{\sigma=0} = x_1; \quad X^i(\tau,\sigma)\big|_{\sigma=\pi} = x_2. \qquad (9.22)$$

The boundary condition at $\sigma = 0$ means that $f = -g$ if we add the constant x_1, so we can write (rescaling the functions by 1/2)

$$X^i(\tau,\sigma) = x^i + \frac{1}{2}\left(f^i(\tau+\sigma) - f^i(\tau-\sigma)\right). \qquad (9.23)$$

Then the boundary condition at $\sigma = \pi$ gives

$$f^i(\tau+\pi) - f^i(\tau-\pi) = 2(x_2^i - x_1^i) \Rightarrow f^i(x+2\pi) = f^i(x) + 2(x_2^i - x_1^i). \qquad (9.24)$$

We can thus expand the functions f^i in Fourier modes as

$$f^i(x) = \frac{x}{\pi}(x_2^i - x_1^i) + \sum_{n\geq 1}(f_n^i \sin(nx) + \tilde{f}_n^i \cos(nx)), \qquad (9.25)$$

leading to the expansion for X^i,

$$X^i(\tau,\sigma) = x_1^i + \frac{(x_2^i - x_1^i)}{\pi}\sigma + \sum_{n\geq 1}(f_n^i \cos(n\tau) - \tilde{f}_n^i \sin(n\tau))\sin(n\sigma). \qquad (9.26)$$

We can rewrite this as a sum over positive and negative modes:

$$X^i(\tau,\sigma) = x_1^i + \frac{(x_2^i - x_1^i)}{\pi}\sigma + \sqrt{2\alpha'}\sum_{n\neq 0}\frac{1}{n}\alpha_n^i e^{-in\tau}\sin(n\sigma). \qquad (9.27)$$

We can include the σ term as an $n = 0$ term (or rather, $n \to 0$) by defining

$$\alpha_0^i = \frac{1}{\pi\sqrt{2\alpha'}}(x_2^i - x_1^i). \qquad (9.28)$$

Then we obtain again, as in the Neumann case (except for an overall $\pm$ sign),

$$X'^i \pm \dot{X}^i = \sqrt{2\alpha'}\sum_{n\in\mathbb{Z}}\alpha_n^i e^{-in(\tau\pm\sigma)}. \qquad (9.29)$$

Quantization is also done in the same way, leading to the same commutation relations for the coefficients,

$$[\alpha_m^i, \alpha_n^j] = m\delta^{ij}\delta_{m+n,0}, \tag{9.30}$$

for $m, n \neq 0$. The Virasoro constraints again lead to

$$2p^+p^- = \frac{1}{\alpha'}\left[\alpha'p^ap^a + \frac{1}{2}\alpha_0^i\alpha_0^i + \sum_{n\geq 1}(\alpha_{-n}^a\alpha_n^a + \alpha_{-n}^i\alpha_n^i) - 1\right], \tag{9.31}$$

which is left as an exercise (Exercise 4) to prove, which in turn gives the spectrum of the string, since

$$M^2 \equiv 2p^+p^- - p^ap^a = \frac{1}{2\alpha'}\alpha_0^i\alpha_0^i + \frac{1}{\alpha'}\left[\sum_{n\geq 1}(\alpha_{-n}^a\alpha_n^a + \alpha_{-n}^i\alpha_{-n}^i) - 1\right]$$

$$= \left(\frac{x_2^i - x_1^i}{2\pi\alpha'}\right)^2 + \frac{1}{\alpha'}(N^\perp - 1), \tag{9.32}$$

where we have defined the (transverse) number operator

$$N^\perp \equiv \sum_{n\geq 1}\sum_i na_n^{a\dagger}a_n^a + \sum_{m\geq 1}\sum_a ma_m^{i\dagger}a_m^i. \tag{9.33}$$

The ground state is

$$|p^+, p^a; [ij]\rangle, \tag{9.34}$$

where $i, j = 1, 2$ corresponds to the two D-branes in between which stretches the string, but we can generalize to the case where they take N values, $i, j = 1, \ldots, N$ if there are N D-branes. The ground state has a mass squared

$$M^2 = -\frac{1}{\alpha'} + \left(\frac{x_2^i - x_1^i}{2\pi\alpha'}\right)^2. \tag{9.35}$$

This means that we can have a massless scalar ground state if $|x_2^i - x_1^i| = 2\pi\sqrt{\alpha'}$. Then a general state is obtained by acting with $a_n^{\dagger i}$ and $a_n^{\dagger a}$ on the ground state, thus having

$$|\psi\rangle = \left[\prod_i \prod_{n\geq 1}\frac{(a_n^{\dagger i})^{N_{in}}}{\sqrt{N_{in}!}}\right]\left[\prod_a \prod_{m\geq 1}\frac{(a_n^{\dagger a})^{N_{an}}}{\sqrt{N_{an}!}}\right]|p^+, p^a; [ij]\rangle. \tag{9.36}$$

Therefore the first excited state has

$$M^2 = \left(\frac{x_2^i - x_1^i}{2\pi\alpha'}\right)^2, \tag{9.37}$$

and is composed of

$$a_1^{\dagger i}|p^+, p^a; [ij]\rangle; \quad a_1^{\dagger a}|p^+, p^a; [ij]\rangle . \tag{9.38}$$

The first set are scalars ϕ^i from the point of view of the D-brane worldvolume, and the second set fills a vector state A_a. Together they form the bosonic states we have described before as existing on the D-brane worldvolume.

In the supersymmetric case, as before, we just eliminate the tachyonic ground state, but otherwise we keep the same states at the first excited level. For coincident branes, these states are massless, so we get the gauge fields A_a and scalars ϕ^i.

9.4 The action of multiple D3-branes and the $\mathcal{N} = 4$ SYM limit

We saw in the previous section that the massless fields for coincident D-branes are the scalars ϕ^i and gauge fields A_a, and the quadratic action for them is (9.18).

Consider D3-branes in type IIB superstring theory (which are sourced by $A^+_{\mu\nu\rho\sigma}$), meaning $p = 3$ and $D = 10$. Then there are $10 - 4 = 6$ scalars ϕ^i and the gauge fields on the worldvolume, together with the fermions that fill the supersymmetric multiplet.

We have six scalars that have six on-shell degrees of freedom, and one 4-dimensional gauge field with two on-shell degrees of freedom, for a total of eight bosonic on-shell degrees of freedom. On the other hand, a minimal 4-dimensional fermion (e.g. Majorana) has two on-shell degrees of freedom. That means that for a supersymmetric theory we need four fermions $\psi^I, I = 1, \ldots, 4$. All the fields are, as we saw, in the adjoint representation A of the $U(N)$ gauge group, so the total field content is $A_a^A, \phi^{iA}, \psi^{IA}$, which matches the one of $\mathcal{N} = 4$ SYM theory. We can also find explicitly that the D3-brane action is invariant under four supersymmetries in four dimensions, i.e. 16 supercharges, which corresponds to half of the total supercharges of type IIB superstring theory: For D-branes we have the condition $\Gamma_0 \Gamma_1 \ldots \Gamma_p \epsilon = \epsilon$ on the supersymmetry parameters ϵ, which halves their number of components from the 32 of the type IIB theory in ten dimensions.

Then we must have that the quadratic action on N D3-branes is in fact $\mathcal{N} = 4$ SYM. The full action on multiple coincident D-branes is in fact not known exactly, so we will not attempt to describe anything about it here. The full action on a single D3-brane is the DBI + WZ action. On a flat background with a constant RR 4-form field, giving a WZ term $\int T_3$, we have the bosonic DBI action

$$S_{\text{bosonic}} = -T_3 \int d^{3+1}x \left(\sqrt{-\det(\eta_{ab} + 2\pi\alpha' F_{ab})} - 1 \right). \tag{9.39}$$

Putting $2\pi\alpha' = 1$ temporarily, the matrix whose determinant we have above is, in terms of $\vec{E}$ and $\vec{B}$ ($F_{0i} = -E_i$ and $F_{ij} = \epsilon_{ijk}B_k$),

$$M_{ab} = \eta_{ab} + F_{ab} = \begin{pmatrix} -1 & -E_1 & -E_2 & -E_3 \\ E_1 & +1 & B_3 & -B_2 \\ E_2 & -B_3 & +1 & B_1 \\ E_3 & B_2 & -B_1 & +1 \end{pmatrix}. \tag{9.40}$$

Its determinant is

$$- \det M_{ab} = 1 - (\vec{E}^2 - \vec{B}^2) - (\vec{E} \cdot \vec{B})^2. \qquad (9.41)$$

But since

$$\frac{\vec{E}^2 - \vec{B}^2}{2} = -\frac{1}{4} F_{\mu\nu} F^{\mu\nu}; \quad \vec{E} \cdot \vec{B} = -\frac{1}{8} \epsilon^{\mu\nu\rho\sigma} F_{\mu\nu} F_{\rho\sigma} \equiv -\frac{1}{4} \tilde{F}^{\mu\nu} F_{\mu\nu}, \qquad (9.42)$$

the bosonic action of a single D3-brane in flat space is written as (reintroducing $2\pi\alpha'$)

$$S = -T_3 \int d^{3+1}x \left[\sqrt{1 + (2\pi\alpha')^2 \frac{F_{\mu\nu} F^{\mu\nu}}{2} - (2\pi\alpha')^4 \left(\frac{1}{4} \tilde{F}^{\mu\nu} F_{\mu\nu} \right)^2} - 1 \right]. \qquad (9.43)$$

For consistency with the quadratic action, we see that we need $T_3 = 1/[(2\pi\alpha')^2 g_{3+1}^2]$, which can also be found from the general formula (9.11). So the single D3-brane bosonic action in flat space becomes

$$S = -\frac{1}{(2\pi\alpha')^2 g_{3+1}^2} \int d^{3+1}x \left[\sqrt{1 - (2\pi\alpha')^2 (\vec{E}^2 - \vec{B}^2) - (2\pi\alpha')^4 (\vec{E} \cdot \vec{B})^2} - 1 \right]. \qquad (9.44)$$

At $\vec{B} = 0$, from the condition that the factor inside the square root be positive (in order to have a real action), we obtain the bound

$$|\vec{E}| \leq \frac{1}{2\pi\alpha'} \equiv E_{\text{crit.}}, \qquad (9.45)$$

which means that as we stated, in Born–Infeld theory there is a maximal electric field, which also leads to a maximal energy density, and no infinities.

Important concepts to remember

- D-branes are $(p+1)$-dimensional endpoints of strings, that act as dynamical walls.
- The generic action of a p-brane would be $-T_p \int d^{p+1}x \sqrt{-\det(h_{ab})}$, with h_{ab} the induced metric on the brane worldvolume.
- The DBI term in the D-brane action is obtained by the replacement $h_{ab} \rightarrow h_{ab} + \alpha' B_{ab} + 2\pi\alpha' F_{ab}$, and with an $e^{-\phi}$ in front, since the D-brane action has to have a $1/g_s = 1/(g_o)^2$ factor.
- The D-brane tension is found by matching the bulk gravity (closed string)–worldvolume scalar (open string) vertex with the corresponding string theory calculation.
- The WZ term is $\mu_p \int_{p+1} A_{(p+1)}$, plus other corrections, can be written succinctly as $\mu_p \int_{p+1} \exp[\alpha' B_{ab} + 2\pi\alpha' F_{ab}] \wedge \sum_n A_{(n)}$.
- In the presence of Dirichlet boundary conditions, the M^2 acquires a term $(T\Delta X)^2$, corresponding to a classical string stretching between the two D-branes, and the rest is as usual.
- The bosonic massless modes on coincident D-branes are transverse scalars ϕ^i and gauge fields A_a.
- N coincident D-branes give a $U(N)$ gauge group, and the theory on the D3-branes (in four dimensions) is $\mathcal{N} = 4$ SYM.

- The full bosonic action on a single D3-brane is the DBI action, and at zero scalars it reduces to the Born–Infeld action for electromagnetism, which has a critical electric field at $\vec{B} = 0$, namely $E_{\text{crit}} = 1/(2\pi\alpha')$.

References and further reading

D-branes were found to be dynamical objects (not fixed walls in spacetime) in [27], but their importance was not understood until the seminal paper of Polchinski [28]. For an introductory treatment of D-branes, see [23]. The book by Clifford Johnson [25] contains the most extensive information on D-branes.

Exercises

1. Prove that the action (9.3) is reparameterization invariant, and that the static gauge fixes this reparameterization.
2. Prove that in the case of several transverse scalars ϕ^i, all depending on a single combination of worldvolume coordinates, for instance $s = x^2 - t^2$, the form of the scalar DBI action on the field configurations reduces to

$$S_p = -T_p \int d^{p+1}\xi \sqrt{1 + \frac{\partial^a \vec{\phi} \cdot \partial_a \vec{\phi}}{T_p}}, \tag{9.46}$$

without any other terms $(\ldots)$ inside the square root.
3. Find the vertex coupling for two worldvolume ϕ^i fields and a metric perturbation $h_{\mu\nu}$, i.e. $\phi^i \phi^j h_{\mu\nu}$ coming from the DBI action, following the procedure for one ϕ and a $h_{\mu\nu}$ in (9.9).
4. Prove that we get the quantization relation for the coefficients α_n^i in the expansion of the Dirichlet directions X^i, (9.30), and the Virasoro constraints (9.31) in terms of the α_ns, by following the procedure in Chapter 7 for the Neumann strings.
5. Consider the 4-dimensional *Born–Infeld Lagrangean* (9.44) at zero magnetic field $\vec{B} = 0$ and define the "electric induction"

$$\vec{D} \equiv \frac{\partial \mathcal{L}}{\partial \vec{E}}. \tag{9.47}$$

Write down the "Maxwell equations" for $\vec{E}$ and $\vec{D}$. Solve them to find the basic object, the "electron" (point charge = source) profile for $\vec{D}$ and $\vec{E}$. Define the "charge densities," "external" and "in the material",

$$\rho_{\text{ext}} \equiv \frac{1}{4\pi}\vec{\nabla} \cdot \vec{D}; \qquad \rho_{\text{mat}} \equiv \frac{1}{4\pi}\vec{\nabla} \cdot \vec{E}, \tag{9.48}$$

and show that then

$$Q_{\text{tot}} \equiv \int dV \rho_{\text{ext}} = \int dV \rho_{\text{mat}}. \tag{9.49}$$

6. Consider the equivalent ansatz to the one in the previous exercise, now for the scalar DBI action (9.5) with only one scalar ϕ, and no other fields, and for static configurations. Define the "electric field" $\vec{F} \equiv \vec{\nabla}\phi$ and the "electric induction"

$$\vec{C} \equiv \frac{\partial \mathcal{L}}{\partial \vec{F}}. \tag{9.50}$$

Write down the "Maxwell equations" for $\vec{F}$ and $\vec{C}$ and solve them for the basic object, the "catenoid" solution, with source = point scalar charge. Find the profile for $\vec{C}$ and $\vec{F}$.

We can join two of these solutions (with different asymptotic regions) on their apparent singularities. Think of an interpretation for this patched solution.

PART II

BASICS OF AdS/CFT FOR $\mathcal{N} = 4$ SYM VS. $AdS_5 \times S^5$

10

The AdS/CFT correspondence: motivation, definition, and spectra

We have finally arrived at the point where we can define the AdS/CFT correspondence. We have already seen several hints for its existence. For the particular case that is the subject of the second part of the book, the duality between $\mathcal{N} = 4$ SYM in four dimensions and string theory in $AdS_5 \times S^5$ background, the symmetries are an important argument: the isometry (invariance) group of AdS_5 is $SO(4,2)$, as seen in Chapter 2, which also matches the symmetry group of 4-dimensional conformal field theories, as seen in Chapter 8, whereas the isometry group of S^5 is $SO(6) \simeq SU(4)$, which is the same as the R-symmetry group of $\mathcal{N} = 4$ SYM, as seen in Chapter 3.

But more generally, we can say that the isometry group of AdS_{d+1} is $SO(d,2)$, the same as the conformal group in $(d-1,1)$ dimensions. Moreover, the boundary of Euclidean AdS_{d+1} (Lobachevsky) space in Poincaré coordinates is $\mathbb{R}^d$, and in global coordinates is $\mathbb{R}_t \times S^{d-1}$, which is conformal to (therefore equivalent for conformal field theories) $\mathbb{R}^d$. The Wick rotation to Minkowski signature was a bit subtle, but in Euclidean signature things are clear. It then seems natural to assume that gravitational theories in AdS space are holographic, and equivalent to field theories on the boundary of AdS space.

The deeper reason for this was also explained in Chapter 2. A light signal starting at a generic point in AdS space takes a finite time to reach the boundary ($\int dt = \int^\infty e^{-y} dy < \infty$), and then reflects back and returns in finite time. On the other hand, it takes an infinite time to reach the center of AdS space in global coordinates, whose Penrose diagram is an infinite (in the time direction) cylinder, or to reach the false part of the boundary in Poincaré coordinates. So if we were to excise the middle of the cylinder (a thin cylinder in the middle of the Penrose diagram), we would get a space for which light travels a finite time between boundaries, which is morally like a quantum mechanical box.

The fact that light can reach the boundary in finite time means there is a good chance for the theory to be holographic, since its boundary is in causal contact with the interior. Moreover, for nonholographic theories we define S-matrices by considering asymptotic states separated at infinity, and scattering them to get S-matrices. Because of the fact that the boundary is a finite time away, the notion of S-matrix is not well defined in AdS space, and instead the well-defined observables are correlators of fields with sources on the boundary. In fact we study these observables in the next chapter.

As for the notion that light travelling a finite time to the boundary makes space like a quantum mechanical box, it can be made more precise as follows. Consider a metric of generic type with Lorentz invariance in $t, \vec{x}$, the metric independent of $t, \vec{x}$ and with a radial coordinate ρ,

$$ds^2 = g_{tt}dt^2 + g_{\vec{x}\vec{x}}d\vec{x}_{d-1}^2 + g_{\rho\rho}d\rho^2 + g_{ab}dx^a dx^b$$
$$= |g_{tt}(\rho)|(-dt^2 + d\vec{x}_{d-1}^2) + g_{\rho\rho}(\rho)d\rho^2 + g_{ab}(\rho, x^a)dx^a dx^b, \qquad (10.1)$$

and the equation of motion for a massless scalar, the Klein–Gordon equation in this space, given by the Box operator:

$$
\begin{aligned}
\Box &= \frac{1}{\sqrt{-g}}\partial_\mu g^{\mu\nu}\sqrt{-g}\partial_\nu = (-\partial_t^2 + \partial_{\vec{x}}^2)|g^{tt}| + \frac{1}{\sqrt{-g}}\frac{\partial}{\partial\rho}g^{\rho\rho}\sqrt{-g}\frac{\partial}{\partial\rho} + \frac{1}{\sqrt{\tilde{g}}}\partial_a g^{ab}\sqrt{\tilde{g}}\partial_b \\
&= |g^{tt}|\left[-\partial_t^2 + \partial_{\vec{x}}^2 + \frac{1}{\sqrt{\tilde{g}}(\sqrt{|g_{tt}|})^{d-1}}\sqrt{\frac{|g_{tt}|}{g_{\rho\rho}}}\frac{\partial}{\partial\rho}(\sqrt{\tilde{g}}(\sqrt{|g_{tt}|})^{d-1})\sqrt{\frac{|g_{tt}|}{g_{\rho\rho}}}\frac{\partial}{\partial\rho}\right] \\
&\quad + \frac{1}{\sqrt{\tilde{g}}}\partial_a g^{ab}\sqrt{\tilde{g}}\partial_b \\
&\rightarrow |g^{tt}|\left[-k^2 + \sqrt{\frac{|g_{tt}|}{g_{\rho\rho}}}\frac{\partial}{\partial\rho}\sqrt{\frac{|g_{tt}|}{g_{\rho\rho}}}\frac{\partial}{\partial\rho}\right. \\
&\quad \left. + \left[\frac{1}{\sqrt{\tilde{g}}(\sqrt{|g_{tt}|})^{d-1}}\sqrt{\frac{|g_{tt}|}{g_{\rho\rho}}}\frac{\partial}{\partial\rho}(\sqrt{\tilde{g}}(\sqrt{|g_{tt}|})^{d-1})\right]\sqrt{\frac{|g_{tt}|}{g_{\rho\rho}}}\frac{\partial}{\partial\rho} + \cdots\right].
\end{aligned} \qquad (10.2)
$$

Here $\sqrt{-g} = \sqrt{\tilde{g}}\sqrt{g_{\rho\rho}}\sqrt{|g_{tt}|}(\sqrt{g_{\vec{x}\vec{x}}})^{d-1}$ and $\tilde{g}$ is the determinant of the metric g_{ab}, and we have used the fact that the metric is $t, \vec{x}$ independent, and $g_{tt} = 1/g^{tt}$, $g_{\rho\rho} = 1/g^{\rho\rho}$. In the last line we went to momentum space in the $t, \vec{x}$ directions. But as we see,

$$\int d\rho\sqrt{\frac{g_{\rho\rho}}{|g_{tt}|}} \equiv \int dx \qquad (10.3)$$

is the *time of flight* for a light ray (moving on ds^2). In terms of this variable, the massless KG equation $\Box\Psi = 0$ becomes, after a redefinition of the wavefunction that removes the term with a single derivative (see Eq. 21.38), of the type

$$\left[-\frac{d^2}{dx^2} + (V(x) - E)\right]\tilde{\Psi}(x) = 0, \qquad (10.4)$$

i.e. in the form of a 1-dimensional quantum mechanical problem in a potential. The time of flight to infinity in AdS space being finite means that the support of $V(x)$ is finite on the corresponding side (though it is infinite on the side corresponding to the interior of AdS, unless we cut off the divergence). A finite support for $V(x)$ leads to a discrete spectrum for $E = m^2 \equiv -k^2$. So a finite time of flight to infinity truly acts from the physics point of view like putting the system in a box, as stated, and therefore the argument for holography is stronger.

But these were general arguments, which we revisit in Part III of the book (in Chapter 21), when we look for more phenomenological versions of AdS/CFT. However, in the best understood cases, we have a heuristic derivation, which we turn to next.

10.1 D-branes = extremal p-branes

A first step towards deriving the AdS/CFT correspondence is the observation that in string theory D-branes are the same as (extremal) p-branes. We recall that D-branes were defined abstractly as dynamical objects on which open strings can end, whereas p-branes were solutions of supergravity (the low energy limit of string theory) that gravitate, thus curving space. The low energy limit on D-branes was found to be $\mathcal{N} = 4$ SYM. In fact, we saw that the bosonic fields of $\mathcal{N} = 4$ SYM exist on the D-branes, but we have not yet described in detail why there are 16 supercharges. We can understand that the boundary condition on open strings reduces the supersymmetry by half, from the 32 supercharges of type IIB theory to 16, hence the same happens on D-branes.

On the other hand, if D-branes are the same as extremal p-branes, we can calculate how many supersymmetries remain from the condition that the variation of the gravitini is zero, $\delta\psi_\mu^i = 0$. The fact that this is a condition for supersymmetry follows from the statement that in a classical background all of the fermions are zero, hence all of their susy variations must be zero also, which in turn imposes a condition on the bosons. We saw in Chapter 6 that for p-branes we have $A_{01\ldots p} = H(r)$, where $H(r)$ is a harmonic function depending only on the radial coordinate away from the brane, and in Chapter 4 that generically in supergravities we have $\delta\psi_\mu^i = D_\mu\epsilon^i + c\Gamma^{\mu_1\ldots\mu_{p+1}}F_{\mu\mu_1\ldots\mu_{p+1}} + \ldots$ Then in order to solve $\delta\psi_\mu^i = 0$ on the p-brane solution we need to impose $\epsilon = \epsilon(r) = f(r)\epsilon_0$, obtaining

$$D_r\epsilon(r) + c\Gamma^{01\ldots p}\partial_r H(r)\epsilon(r) = 0 \Rightarrow \Gamma^{01\ldots p}\epsilon_0 = \pm\epsilon_0, \tag{10.5}$$

(together with an equation for $f(r)$). Here $\Gamma^{01\ldots p} = \Gamma^0\ldots\Gamma^p$ and the condition $\Gamma^{01\ldots p}\epsilon_0 = \pm\epsilon_0$ relates half the components of ϵ_0 to the other half, hence reduces the supersymmetry from 32 supercharges to 16.

These *extremal p-branes* saturate the BPS bound, which, as we saw in Chapter 3, can be derived from the supersymmetry algebra, and for the extremal solutions (saturating the bound) we have $Q|\psi\rangle = 0$ for half of the Qs, as was found above. But the BPS bound for the extremal p-brane solutions was also derived from the condition to have no "naked singularities," namely for the singularity to be hidden behind a horizon (for $M > Q$ we have a horizon outside the singularity, and for $M = Q$ the horizon coincides with the singularity). The extremal p-branes are a kind of soliton, therefore nonperturbative objects, with the mass proportional to $1/g_s$, unlike the usual solitons which have masses proportional to $1/g^2$, or fundamental objects with masses independent of the coupling. The $1/g_s$ behavior matches what is expected from D-branes, since we saw that the D-brane action has a $1/g_s$ in front.

In 1995, Polchinski showed that D-branes and p-branes are the same, by calculating the tensions and charges of the D-branes from string theory and matching with the p-brane (solutions of supergravity) results. This gave rise to the "second superstring revolution," since after that one could do nonperturbative calculations involving D-branes. We will not reproduce the tension and charge calculation here.

Thus D-branes curve space and N D3-branes (with $p = 3$) correspond to the p-brane supergravity solution (here $F_5 = F_{\mu_1\ldots\mu_5}dx^{\mu_1} \wedge \ldots \wedge dx^{\mu_5}$):

$$ds^2 = H^{-1/2}(r)d\vec{x}_{\parallel}^2 + H^{1/2}(r)(dr^2 + r^2 d\Omega_5^2),$$

$$F_5 = (1 + *)dt \wedge dx_1 \wedge dx_2 \wedge dx_3 \wedge (dH^{-1}),$$

$$H(r) = 1 + \frac{R^4}{r^4}; \quad R^4 = 4\pi g_s N\alpha'^2; \quad Q = g_s N. \tag{10.6}$$

But if we go a little away from the extremal limit $Q = M$ by adding a small mass δM, this solution develops an event horizon at a small $r_0 > 0$, and like the Schwarzschild black hole, it emits "Hawking radiation," i.e. thermal radiation produced by the event horizon.

10.2 Motivation: near-horizon limit, Hawking radiation, and the two points of view

If the extremal p-brane supergravity solution is also a D-brane, one can derive the Hawking radiation of the *near-extremal p*-brane in the D-brane picture from a unitary quantum process: two open strings existing on a D-brane collide to form a closed string, which as a result is not bound to the D-brane anymore and can peel off the D-brane and move away as Hawking radiation, as in Fig. 10.1. The Hawking radiation then corresponds to all the closed string fields that can move in the bulk of spacetime.

Therefore there should be a relation between the theory of open strings living on N D-branes, i.e. $\mathcal{N} = 4$ SYM with $SU(N)$ gauge group, and the gravitational theory corresponding to fields of the Hawking radiation, existing in the background curved by the D-branes (10.6).

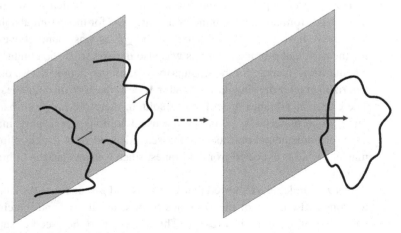

Figure 10.1 Two open strings existing on a D-brane collide and form a closed string, that can then peel off and go away from the brane.

The heuristic derivation motivating the AdS/CFT correspondence follows from describing string theory in the presence of the D-branes from the two points of view (endpoints of open strings and gravitational solution), taking into account the Hawking radiation process described above.

Point of view no.1

Consider the D-branes viewed as endpoints of open strings. Then string theory with D3-branes has three ingredients:

- The open strings existing on the D3-branes, giving a theory that reduces to $\mathcal{N} = 4$ Super Yang–Mills in the low energy limit.
- The closed strings existing in the bulk (the whole) of spacetime, giving a theory that is supergravity coupled to the massive modes of the string. In the low energy limit, only supergravity remains.
- The interactions between the two, giving for instance Hawking radiation through the process just described above.

Thus the action of these strings will be something like

$$S = S_{\text{bulk}} + S_{\text{brane}} + S_{\text{interactions}}. \tag{10.7}$$

In the low energy limit $\alpha' \to 0$, the massive string modes drop out, and $S_{\text{bulk}} \to S_{\text{supergravity}}$, also $S_{\text{brane}} \to S_{\mathcal{N}=4\text{SYM}}$. Moreover,

$$S_{\text{int}} \propto \kappa_N \sim g_s \alpha'^2, \tag{10.8}$$

(the g_s dependence is because in front of the Einstein action we must have $1/\kappa_N^2$, but also $1/g_s^2$, and α'^2 for dimensional reasons) where $\alpha' \to 0$, whereas g_s is the string coupling and stays fixed. Then we see that $S_{\text{int}} \to 0$ and moreover, since the Newton constant $8\pi G_N = k_N^2 \to 0$, gravity – and hence supergravity also – becomes free. Thus in this limit we get two decoupled (non-interacting) systems:

- Free gravity in the bulk of spacetime;
- 4-dimensional $\mathcal{N} = 4$ gauge theory on the D3-branes.

Point of view no.2.

We now replace the D3 brane by the supergravity solution (p-brane).

Then the energy E_p measured at a point $r \to 0$ and the energy E measured at infinity are related by

$$E_p \to i\frac{d}{d\tau} = \frac{i}{\sqrt{-g_{00}}}\frac{d}{dt} \to \frac{1}{\sqrt{-g_{00}}}E$$
$$\Rightarrow E = H^{-1/4}E_p \propto rE_p. \tag{10.9}$$

Therefore, for fixed E_p, as $r \to 0$, the energy observed at infinity, E, goes to zero, i.e. we are in the low energy regime at infinity.

Thus from this point of view, we also have two decoupled low energy systems of excitations:

- At large distances ($\delta r \to \infty$), or equivalently at low energies (since energy $\sim 1/\text{length}$), gravity becomes free: the gravitational coupling has dimensions, therefore the effective dimensionless coupling is $G_N E^2$ and this $\to 0$ as $E \to 0$. Thus again we have free gravity at large distances, i.e. away from the p-brane, in the bulk of spacetime.
- At small distances $r \to 0$, we also have low energy excitations, as we saw.

The fact that these two systems are decoupled can be seen in a couple of ways. One can calculate that waves of large r have a vanishing absorption cross-section on D-branes. One can also show that conversely, the waves at $r = 0$ cannot climb out of the gravitational potential and escape at infinity. We will not show this here.

Thus in the second point of view we again have two decoupled low energy systems, one of which is again free gravity at large distances (in the bulk of spacetime).

Therefore, we can identify the other low energy system in the two points of view and obtain that:

The 4-dimensional gauge theory on the D3-branes, i.e.
$\mathcal{N} = 4$ *Super Yang–Mills with gauge group $SU(N)$, at large N is*
= *gravity theory at $r \to 0$ in the D-brane background, if we take $\alpha' \to 0$.*

This is called the AdS/CFT correspondence, but at this moment it is just a vague statement, that needs to be made more precise.

10.3 Definition: limit, validity, operator map

We now turn to defining better the correspondence hinted at in the previous section.

We first define the gravitational background. If we take $r \to 0$, then the harmonic function $H \simeq R^4/r^4$, and we obtain the supergravity background solution

$$ds^2 \simeq \frac{r^2}{R^2}(-dt^2 + d\vec{x}_3^2) + \frac{R^2}{r^2}dr^2 + R^2 d\Omega_5^2. \qquad (10.10)$$

By changing the coordinates $r/R \equiv R/x_0$, we get

$$ds^2 = R^2 \frac{-dt^2 + d\vec{x}_3^2 + dx_0^2}{x_0^2} + R^2 d\Omega_5^2, \qquad (10.11)$$

which is the metric of $AdS_5 \times S_5$, i.e. 5-dimensional Anti-de Sitter space times a 5-sphere of the same radius R, where AdS_5 is in Poincaré coordinates.

From the point of view of the supergravity background solution, the gauge theory exists in the original metric, before taking the $r \to 0$ limit, which after taking the limit corresponds to $r \to \infty$. Therefore in the new $AdS_5 \times S_5$ limit space we can say that the gauge theory lives at $r \to \infty$, or $x_0 \to 0$, which as we have proven when analyzing AdS space, is part of the real boundary of global AdS space. In Poincaré coordinates, $x_0 \to 0$ is a Minkowski space.

Then the gravity theory exists in $AdS_5 \times S_5$, whereas the Super Yang–Mills theory exists on the 4-dimensional Minkowski boundary of AdS_5 parameterized by t and $\vec{x}_3$.

We still need to understand the $\alpha' \to 0$ limit. We want to keep arbitrary excited string states at position r as we take $r \to 0$ (to obtain the low energy limit). Therefore the energy at point p in string units, $E_p \sqrt{\alpha'}$, needs to be fixed. Since $H \simeq R^4/r^4 \propto \alpha'^2/r^4$, the energy measured at infinity is

$$E = E_p H^{-1/4} \propto E_p r/\sqrt{\alpha'}. \qquad (10.12)$$

But at infinity we have the gauge theory, therefore the energy measured at infinity (in the gauge theory) must also stay fixed. Since $E_p \sqrt{\alpha'} \sim E \alpha'/r$ must be fixed, it follows that

$$U \equiv \frac{r}{\alpha'} \qquad (10.13)$$

is fixed as $\alpha' \to 0$ and $r \to 0$, and can be thought of as an energy scale in the gauge theory (since we said that E/U was fixed). The metric is then ($R^4 = \alpha'^2 4\pi g_s N$),

$$ds^2 = \alpha' \left[\frac{U^2}{\sqrt{4\pi g_s N}} (-dt^2 + d\vec{x}_3^2) + \sqrt{4\pi g_s N} \left(\frac{dU^2}{U^2} + d\Omega_5^2 \right) \right], \qquad (10.14)$$

where $\alpha' \to 0$, but everything inside the brackets is finite.

We should also explain what happens to the field strength F_5. In the near horizon limit, we obtain

$$F_5 = (1+*)4\frac{r^3}{R^4}dx^0 \wedge \ldots \wedge dx^3 \wedge dr = 4R^4(1+*)\epsilon_{(5)} = 16\pi g_s \alpha'^2 N(1+*)\epsilon_{(5)}. \qquad (10.15)$$

Here R is the radius of AdS_5 and S^5 and

$$\epsilon_{(5)} = \frac{\sqrt{-g}}{R^5}dx^0 \wedge \ldots \wedge dx^3 \wedge dr \qquad (10.16)$$

is the volume form, that integrates to 1 on AdS_5, and g is the determinant of the AdS_5 metric. Therefore we see that the flux is quantized in units of N.

Here in the gravity theory N is the number of D3-branes and g_s is the string coupling. In the Super Yang–Mills gauge theory, N is the rank of the $SU(N)$ gauge group, for the low energy theory on the N D3-branes. And g_s is related to the Yang–Mills coupling by

$$4\pi g_s = g_{YM}^2, \qquad (10.17)$$

since g_{YM} is the coupling of the gauge field A_μ^a, which we argued is the massless mode of the open string existing on the D3-branes, so $g_{YM} = g_o$, and we have already argued that $g_s \sim (g_o)^2$. Indeed, out of two open strings we can make a closed string, therefore out of two open string splitting interactions, governed by the $g_{YM} = g_o$ open string coupling, we can make one closed string splitting interaction, governed by the g_s coupling, as in Fig. 9.3.

The last observation that one needs to make is that in the limit $\alpha' \to 0$, string theory becomes its low energy limit, supergravity.

Therefore the AdS/CFT correspondence relates string theory, in its supergravity limit, in the background (10.14), with $\mathcal{N} = 4$ SYM with gauge group $SU(N)$ existing in $d = 4$, at the boundary of AdS_5.

Limits of validity

To be able to use the supergravity approximation of string theory, we must have:

- The curvature of the background (10.14) must be large compared to the string length, i.e. $R = \sqrt{\alpha'}(4\pi g_s N)^{1/4} \gg \sqrt{\alpha'} = l_s$, to avoid α', or string worldsheet quantum corrections. That means that we are in the limit $g_s N \gg 1$, or $g_{YM}^2 N \gg 1$.
- Quantum string corrections, governed by g_s, are small, thus $g_s \to 0$.

Therefore, for supergravity to be valid, we need to have $g_s \to 0, N \to \infty$, but $\lambda = g_s N = g_{YM}^2 N$ must be fixed and large ($\gg 1$).

Another way to understand the requirement that quantum string corrections must be small is by imposing that $R \gg l_P$, where the 10-dimensional (reduced) Planck length $l_{P,10} = M_{P,10}^{-1}$ is given from the coefficient $1/(2l_{P,D}^{D-2})$ of the Einstein–Hilbert action, as $l_{P,10}^8 = \frac{1}{2}(2\pi)^7 g_s^2 \alpha'^4$, leading to

$$\frac{R}{l_{P,10}} = 2^{-1/4}\pi^{-5/8}N^{1/4}. \tag{10.18}$$

The requirement of no quantum string corrections is then $N \to \infty$, which, given that $g_s N$ is kept fixed and large, implies $g_s \to 0$.

But in the large N limit, 't Hooft showed that gauge theories with adjoint fields have as expansion parameters the *effective, or 't Hooft coupling* $\lambda = g_{YM}^2 N$ and $1/N$. The dependence of the amplitudes on the expansion parameters in a gauge theory with only adjoint fields is[1]

$$\sim (g^2 N)^L N^{1-2h}, \tag{10.19}$$

where g is the YM coupling (the 3-gluon vertex has a factor of g, the 4-gluon vertex a factor of g^2), L is the number of loops of the Feynman diagram, and $\chi = 2 - 2h - l$ ($l = 1$ is the number of external or "quark" index loops, 1 in the case of external adjoint fields) is the Euler characteristic of the surface on which we can draw the diagram, with h the number of handles of this surface. For *planar diagrams*, we can draw them on a

---

[1] The proof in the case of diagrams with no external lines goes as follows. The factor is $g^{V_3+2V_4}N^I$, where V_3 is the number of 3-vertices, V_4 the number of 4-vertices, and I the number of (fundamental) index traces. Define also $V = \sum_n V_n = V_3 + V_4$ and call F the number of faces of the Feynman diagram viewed as a figure drawn on a surface with h handles, with P the number of propagators, and l the number of external or "quark" index loops, which equals 1 for a diagram of only adjoint fields, as we can verify. Then each propagator ends on two vertices, and each n-vertex connects n propagators, giving $2P = \sum_n nV_n = 3V_3 + 4V_4$. Then also $2P - 2V = V_3 + 2V_4$. Moreover, we see that $F = I + l$, so the couplings factor becomes $g^{2P-2V}N^{F-l}$. But a theorem by Euler for an object with F faces, P edges, and V vertices gives that $F - P + V = 2 - 2h$, so the factor is $(g^2 N)^{P-V}N^{2-2h-l}$. For a diagram with no external lines, the number of loops (or independent momentum integrations) is $L = P - V$, since there are P momentum integrations (one for each propagator) constrained by V delta functions (one at each vertex). For a diagram with external lines, the formula would be $L = P_I - V + 1$, where P_I is the number of internal lines, since the number of loops equals the number of integrations P_I, constrained by the V delta functions, minus the one for giving the conservation of external momentum. Then $P - V = V_3/2 + V_4 = L$, and with $l = 1$, we get a factor of $(g^2 N)^L N^{1-2h}$. In the case of E external lines, writing P_I for the number of internal propagators, the relation $2P = \sum_n nV_n$ becomes $E + 2P_I = \sum_n nV_n$, since external lines connect to only one vertex. Then the factor is $g^{E+2P_I-2V}N^{F-l}$, which can be rewritten as $g^{E-2}(g^2 N)^{P_I-V+1}N^{2-2h-l} = g^{E-2}(g^2 N)^L N^{2-2h-l}$.

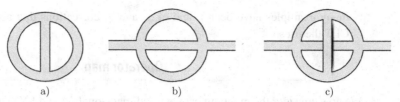

a)   b)   c)

Figure 10.2   a) Planar 1-loop diagram with two 3-point vertices; b) Planar 2-loop diagram with two 4-point vertices; c) Nonplanar 3-loop diagram.

sphere, with $h = 0$. In Fig. 10.2 are examples of Feynman diagrams, so we can understand the above formula. Drawing the diagrams, we use 't Hooft's double line notation, where adjoint fields are denoted with a double line, to acknowledge the fact that they have two (fundamental) matrix indices, A_μ^{ij}, which are contracted throughout the diagram, with the external lines. A propagator from a state (ij) to a state (kl) will have $\delta^{ik}\delta_{jl}$, etc. Then an *index loop* (closed line in the double line notation) corresponds to $\delta_i^i = N$, so for every index loop we have a factor of N. We see for instance that a planar 1-loop diagram with two 3-point vertices as in Fig. 10.2a has a factor of g^2N, whereas a 2-loop diagram with two 4-point vertices as in Fig.10.2b has a factor of $(g^2)^2N^2 = (g^2N)^2$. The nonplanar 3-loop diagram with two external 4-point vertices and two internal 3-point vertices in Fig. 10.2c has only one index loop and thus a factor $(g^2)^2g^2N = (g^2N)^3N^{-2}$, as needed for a surface with one handle, as we can check that the Feynman diagram shows.

Therefore in the $N \to \infty$ limit with g^2N held fixed, known as the 't Hooft limit, planar diagrams dominate (the nonplanar diagrams are subleading by factors of $1/N^2$), and then the effective coupling is λ. 't Hooft considered the perturbative expansion in λ, i.e. $\lambda \ll 1$ and fixed, but in the case of AdS/CFT we see that we need $\lambda \gg 1$ and fixed instead.

Then we find that AdS/CFT is a *duality*, since the two descriptions, gauge theory perturbation theory and supergravity in $AdS_5 \times S_5$ are valid in opposite regimes, $\lambda \ll 1$ and $\lambda \gg 1$, respectively. This means that such a duality will be hard to check, since in one regime we can use a description to calculate, but not the other.

At this point, the definition of AdS/CFT is of a duality between supergravity on $AdS_5 \times S_5$ as in (10.14) and 4-dimensional $\mathcal{N} = 4$ SYM with $SU(N)$ gauge group, existing at the AdS_5 boundary, with $g_s \to 0, N \to \infty$ and g_sN fixed and large.

But in fact the issue of validity of AdS/CFT is more complicated. We can have several possible versions for the duality:

- The weakest version is the one just described: AdS/CFT is valid only at large $4\pi g_sN = \lambda$, when we have just the supergravity approximation of string theory in the background (10.14). If we go to the full string theory, i.e. away from large g_sN, we may find disagreements.
- A stronger version would be that the AdS/CFT duality is valid at any finite g_sN, but only if $N \to \infty$ and $g_s \to 0$, which means that string worldsheet α' corrections, given by the ratio $\alpha'/R^2 = 1/\sqrt{4\pi g_sN} = 1/\sqrt{\lambda}$ agree, but $4\pi g_s = g_{YM}^2 = \lambda/N^2$ corrections may not.
- The strongest version would be that the duality is valid at any g_s and N, even if we can only make calculations within certain limits. This is what is believed to be true, since

many examples have been found of α' and g_s corrections that agree between AdS and CFT theories.

Operator map

We now consider the mapping between 4-dimensional $\mathcal{N} = 4$ SYM and AdS_5 supergravity, and we start with the relation between *gauge invariant* operators in SYM and fields in supergravity.

A gauge invariant operator $\mathcal{O}$ in the $\mathcal{N} = 4$ SYM CFT will be characterized by certain conformal dimensions Δ and a representation index I_n for the $SO(6) = SU(4)$ R-symmetry. Note that the operator must be gauge invariant, since there is no gauge group on the supergravity side. As already mentioned, the $SO(6)$ R-symmetry corresponds to the symmetry of the S^5 sphere in $AdS_5 \times S^5$, whereas the conformal symmetry $SO(4,2)$ becomes the isometry group of AdS_5.

In the string theory in $AdS_5 \times S_5$, $\mathcal{O}$ corresponds to a field, but we restrict this discussion to the supergravity limit. Then we have supergravity on $AdS_5 \times S_5$, where S_5 is a compact space, and thus we can apply the Kaluza–Klein procedure of compactification: we expand the supergravity fields in spherical harmonics (Fourier-like modes) on the sphere.

We recall that a scalar field would be expanded as

$$\phi(x, y) = \sum_n \sum_{I_n} \phi_{(n)}^{I_n}(x) Y_{(n)}^{I_n}(y), \tag{10.20}$$

where n is the level, the analog of the n in $e^{inx/R}$ for a Fourier mode around a circle of radius R. I_n is an index in a representation of the symmetry group, x is a coordinate on AdS_5 and y a coordinate on S_5, and the spherical harmonic $Y_{(n)}^{I_n}(y)$ is the analog of $e^{inx/R}$ for a Fourier mode.

Then the field $\phi_{(n)}^{I_n}$ existing in AdS_5, of mass m, corresponds to an operator $\mathcal{O}_{(n)}^{I_n}$ (of the same symmetry) in 4-dimensional $\mathcal{N} = 4$ SYM, of dimension Δ. The relation between m and Δ in AdS_{d+1} is (we derive this at the beginning of the next chapter)

$$\Delta = \frac{d}{2} + \sqrt{\frac{d^2}{4} + m^2 R^2}. \tag{10.21}$$

Thus a tower of KK modes of increasing mass labelled by n corresponds to a tower of operators of increasing conformal dimension labelled by the same.

For completeness, we note that the general formula for a p-form field in AdS_{d+1} comes from $(\Delta - p)(\Delta + p - d) = m^2 R^2$, giving

$$\Delta = \frac{d}{2} + \sqrt{\frac{d^2}{4} + m^2 R^2 - p(p - d)}. \tag{10.22}$$

The dimensional reduction on S_5, i.e. keeping only the lowest mode in the Fourier-like expansion, should give a supergravity theory in AdS_5. But as we mentioned in Chapter 4, supergravity theories that admit AdS backgrounds (with a cosmological constant) are actually gauged supergravity theories, so the dimensional reduction gives maximal 5-dimensional gauged supergravity.

We should note that m^2 in an AdS space need not be ≥ 0 as in flat spacetime. The condition comes basically from normalizability at an infinity of localized excitations, i.e. from stability: in flat space, $e^{ik \cdot x}$ has an e^{imt} factor that can behave like $e^{+t\sqrt{-m^2}}$. In AdS space, there is an analogous condition on m^2 known as the *Breitenlohner–Freedman bound*, for normalizability of perturbations. For scalars, it amounts to reality of Δ in (10.21), i.e. positivity of the argument of the square root,

$$m^2 R^2 \geq -\frac{d^2}{4}. \tag{10.23}$$

Scalars satisfying this condition are stable.

10.4 Spectra and "experimental" evidence

For our first test of the AdS/CFT correspondence, we test the tower of KK modes of the supergravity fields against towers of operators in SYM. On the supergravity side, we have fields $\phi_{(n)}$ obtained by the spherical harmonic expansion of 10-dimensional supergravity around the background solution $AdS_5 \times S_5$, and on the CFT side we have a set of operators that belong to definite representations of the symmetry groups. We match the representations and then the masses against dimensions.

However, this is not as simple as it sounds, since we mentioned that even though $\mathcal{N} = 4$ SYM has zero beta function, there are still quantum corrections to the conformal dimensions Δ of operators. Since we are working in the deeply nonperturbative gauge theory regime, of effective coupling $\lambda \gg 1$, it would seem that we have no control over the result for the quantum value of Δ of a given operator.

But we are saved by the large amount of symmetry available. Supersymmetry together with the conformal group $SO(2,4)$ gives the superconformal group $SU(2,2|4)$.

Representations of the conformal group are given as we said by a primary operator $\mathcal{O}$ and its "descendants," obtained by acting on them with P_μ like a creation operator on the vacuum, $(P_{\mu_1} \ldots P_{\mu_n} \mathcal{O})$.

Representations of the superconformal group are correspondingly larger (there are more symmetries, which must relate to more fields, or put another way, there are more creation operators in the method of induced representations), so they will include many primary operators of the conformal group (there was one primary field, or "vacuum" of the creation operators per representation of the conformal group). In particular, there are 2^{16} primary operators for a generic representation of $\mathcal{N} = 4$ supersymmetry in $d = 4$, since there are 16 supercharges, each giving a creation operator $a^\dagger$, and in a state it can be present or not, as we saw in Chapter 3.

However, again as we saw in Chapter 3 when discussing representations of supersymmetry, there are special, *short* representations of the superconformal group, that are generated by so-called *chiral primary operators*. These are primary operators that are annihilated by some combination of Qs (thus they preserve some supersymmetry *by themselves*), i.e. $[Q \text{ comb.}]\mathcal{O}_{\text{ch.pr}} = 0$.

The conformal dimension Δ of chiral primary operators is uniquely determined by the R-symmetry charge. This fact comes out of the superconformal algebra, and is the equivalent of the saturation of the BPS bound ("extremality"). The relation $Q\mathcal{O} = 0$ means that the conformal dimension, equivalent to mass, is given in terms of R-symmetry charge, in the same way as $M = Z$ appeared for the usual BPS bound. We then cannot have quantum corrections to Δ since it would imply there would need to be quantum corrections to the R-charge as well, which is impossible, or otherwise we would break the BPS bound and change the number of states of the representation, again impossible. Then the $\lambda \gg 1$ value of Δ is the same as the $\lambda = 0$ value, and we can check it using AdS-CFT!

The representations I_n of the symmetry groups for supergravity fields are small. Then Kaluza–Klein supergravity modes in AdS_5 correspond to chiral primary fields in SYM, with dimensions protected against quantum corrections. On the other hand, non-supergravity string fields will in general belong to large representations.

KK scalar fields in AdS_5 belong to six families, and correspondingly we find six families of chiral primary scalar representations (the $\mathcal{N} = 4$ SYM fields are denoted as $\{\phi^I, \lambda_{\alpha A}, \bar{\lambda}_{\dot{\alpha}}^A, A_\mu\}$):

- $\mathcal{O}_n \equiv \mathrm{Tr}\,(\phi^{(I_1}\ldots\phi^{I_n)})$ (in the symmetric representation), which therefore have dimensions $\Delta = n$ (there are n fields of dimension 1), and by the above relation we expect them to correspond to KK fields of masses $m^2 R^2 = n(n-4), n \geq 2$.
- $Q^2\mathcal{O}_{n+2}(\equiv \epsilon^{\alpha\beta}\{Q_\alpha, [Q_\beta, \mathcal{O}_n]\}) = \mathrm{Tr}\,(\epsilon^{\alpha\beta}\lambda_{\alpha A}\lambda_{\beta B}\phi^{I_1}\ldots\phi^{I_n})$, of dimensions $\Delta = n+3$ (λ has dimension 3/2), therefore corresponding to KK fields of masses $m^2 R^2 = (n+3)(n-1), n \geq 0$.
- $Q^4\mathcal{O}_{n+2} = \mathrm{Tr}\,(F_{\mu\nu}F^{\mu\nu}\phi^{I_1}\ldots\phi^{I_n})$, of dimensions $\Delta = n+4$ (A_μ has dimension 1), corresponding to $m^2 R^2 = n(n+4), n \geq 0$.
- $Q^2\bar{Q}^2\mathcal{O}_{n+4} = \mathrm{Tr}\,(\epsilon^{\alpha\beta}\epsilon^{\dot{\alpha}\dot{\beta}}\lambda_{\alpha A_1}\lambda_{\beta A_2}\bar{\lambda}_{\dot{\alpha}}^{B_1}\lambda_{\dot{\beta}}^{B_2}\phi^{I_1}\ldots\phi^{I_n})$ of dimensions $\Delta = n+6$, corresponding to KK fields of masses $m^2 R^2 = (n+6)(n+2), n \geq 0$.
- $Q^4\bar{Q}^2\mathcal{O}_{n+4} = \mathrm{Tr}\,(\epsilon^{\alpha\beta}\lambda_{\alpha A}\lambda_{\beta B}F_{\mu\nu}^2\phi^{I_1}\ldots\phi^{I_n})$, of dimensions $\Delta = n+7$, corresponding to KK fields of masses $m^2 R^2 = (n+3)(n+7), n \geq 0$.
- $Q^4\bar{Q}^4\mathcal{O}_{n+4} = \mathrm{Tr}\,(F_{\mu\nu}^4\phi^{I_1}\ldots\phi^{I_n})$, of dimensions $\Delta = n+8$, corresponding to KK fields of masses $m^2 R^2 = (n+4)(n+8), n \geq 0$.

We find that indeed the KK families have such masses, therefore we have "experimental evidence" for AdS/CFT.

10.5 Global AdS/CFT; dimensional reduction on S^3

Up to now we have obtained AdS/CFT in the Poincaré patch of AdS space. But AdS space is larger, and it is natural to assume that there is a relation to the full global AdS space. In fact, we have already seen in Chapter 2 that the boundary of global AdS is related to the boundary of Poincaré AdS by a conformal transformation. Let us review this a little. We consider the Euclidean version, since we have seen that while the Wick rotation in the bulk of global AdS is related simply to the Wick rotation in the bulk of Poincaré AdS, on the boundary of global AdS we have a sort of radial time Wick rotation.

The metric of global $AdS_5 \times S^5$ is

$$ds^2 = \frac{R^2}{\cos^2\theta}(d\tau^2 + d\theta^2 + \sin^2\theta \, d\Omega_3^2) + R^2 d\Omega_5^2. \tag{10.24}$$

If we are a distance ϵ from the boundary, at $\theta = \pi/2 - \epsilon$, then $\cos^2\theta \simeq \epsilon^2$, so the boundary metric is approximately (since the S^5 is infinitely small with respect to the AdS_5 we can drop it):

$$ds^2 = \frac{R^2}{\epsilon^2}(d\tau^2 + d\Omega_3^2). \tag{10.25}$$

On the other hand, in Poincaré coordinates, we have

$$ds^2 = R^2 \frac{d\vec{x}^2 + dx_0^2 + x_0^2 d\Omega_5^2}{x_0^2}. \tag{10.26}$$

At a distance ϵ from the boundary, at $x_0 = \epsilon$, we obtain

$$ds^2 = \frac{R^2}{\epsilon^2} d\vec{x}^2, \tag{10.27}$$

i.e. flat space. But the $\mathbb{R}_t \times S^3$ and the flat space $\mathbb{R}^4$ are related by a conformal transformation, irrelevant for a CFT, since

$$ds^2 = d\vec{x}^2 = dx^2 + x^2 d\Omega_3^2 = x^2((d\ln x)^2 + d\Omega_3^2) = x^2(d\tau^2 + d\Omega_3^2). \tag{10.28}$$

This suggests that string theory in global $AdS_5 \times S^5$ is dual to $\mathcal{N} = 4$ SYM on the *cylinder* $\mathbb{R}_t \times S^3$ on its boundary, which is in fact true.

Operator-state correspondence in conformal field theory

However, the rules are a bit different in global coordinates, so it is worth describing them. The reason is that when applying a conformal transformation on flat space to go to the cylinder, we relate operators $\mathcal{O}$ to states $|\mathcal{O}\rangle$. This *operator–state correspondence* is an important concept in conformal field theory, though it is more familiar in two dimensions, so we start there first. Then we apply it to four dimensions for the case of global AdS/CFT.

Consider the complex plane parameterized by z and the transformation $z = e^{-iw}$. In w consider the semi-infinite cylinder $\text{Im}(w) \leq 0$, $0 \leq \text{Re}(w) \leq 2\pi$, with the identification $w \sim w + 2\pi$. It is mapped to the unit disk in z, $|z| \leq 1$. This can be thought of as a representation of the string worldsheet, a semi-infinite closed string cylinder, mapped to a unit disk. The center of the disk corresponds to the point $w = -i\infty$, i.e. the asymptotic past region. So operators at the origin of the disk will correspond to something acting in the asymptotic past region of the cylinder, which can be thought of as initial states.

On the complex plane, as we saw in (8.27), the closed string scalar operator X^μ admits a Laurent series expansion. The unit operator can be thought of as mapping to the vacuum state in the asymptotic region of the cylinder. But the general string asymptotic states are created by acting with α^μ_{-m} and $\tilde{\alpha}^\mu_{-m}$, and from (8.27) we have

$$\alpha^\mu_{-m} = \sqrt{\frac{2}{\alpha'}}\frac{i}{(m-1)!}\partial^m X^\mu(0). \qquad (10.29)$$

Therefore we can think of the asymptotic state $\alpha^\mu_{-m}|0,0\rangle$ on the cylinder in w as corresponding to the operator

$$\sqrt{\frac{2}{\alpha'}}\frac{i}{(m-1)!}\partial^m X^\mu(0) \qquad (10.30)$$

on the plane in z. On the other hand, on the cylinder in w, the expansion (7.48) corresponding to (8.27) is a Fourier expansion, or in general terms, a KK expansion on S^1, and it corresponds to the Taylor expansion terms (10.30). Then the operators for the Taylor expansion of X^μ correspond to states of the string.

On the string states of the type $\alpha^\mu_{-m}|0,0\rangle$, we can act with the Hamiltonian calculated in (7.58). In this way, we have effectively reduced the case of free (no string interactions) propagating strings on the cylinder, via KK expansion on S^1, to a 0+1-dimensional (quantum mechanical) system of states, acted upon by a Hamiltonian.

Note that the Hamiltonian on $\mathbb{R}_\tau$ is the same as the "dilatation operator" in the original plane, the generator of the scaling symmetry $\vec{r} \to \lambda\vec{r}$, because of the 2-dimensional version of (10.28), since

$$H_{\mathbb{R}_\tau} = i\partial_\tau = ir\partial_r = D_{\mathbb{R}^2}, \qquad (10.31)$$

where $d\vec{r}^2 = dr^2 + r^2 d\theta^2$.

Operator–state correspondence in higher dimensions

The one important difference from the higher dimensional case is that the 2-dimensional cylinder, obtained after the conformal transformation from the plane, is flat (has no curvature), whereas in higher dimensions it is curved. But on a curved space, in order to preserve conformal invariance of scalars, we need to add a conformal coupling $R\phi^2$ to the action. Thus whereas in two dimensions this does not change the action since $R = 0$, in higher dimensions it does. In particular, on $\mathbb{R}_t \times S^3$, this gives a unit mass term, $-\int \phi^2/2$ in the action.

Otherwise the analysis proceeds as above. For scalars Z on the plane $\mathbb{R}^4$, the Taylor expansion terms

$$Z^{(m)}_{\alpha_1...\alpha_m} \sim (\partial_{\alpha_1} \dots \partial_{\alpha_m})Z \qquad (10.32)$$

correspond to states for the KK expansion on $\mathbb{R} \times S^3$. Due to the mass term, the constant term (0th order in the expansion) has an energy of 1, i.e. it is described by a harmonic oscillator of unit frequency. For more general operators, we can again map them to some kind of states.

On these states we can again define a Hamiltonian, integrating over the S^3, thus writing it in terms of the KK expansion. In general, it is quite complicated, but in a special limit (the "pp wave limit") described in Chapter 17, it becomes simple.

As in the 2-dimensional case, the Hamiltonian H on $\mathbb{R}_\tau$ is the same as the dilatation operator D on the original plane, generating the symmetry $\vec{r} \to \lambda\vec{r}$, since

$$H_{\mathbb{R}_\tau} = i\partial_\tau = ir\partial_r = D_{\text{plane}}. \tag{10.33}$$

We will see that this identification allows us to write a Hamiltonian acting on operators through Feynman diagrams, by defining its action as the action of the dilatation operator on the CFT on the plane.

Important concepts to remember

- D-branes are the same as (extremal) p-branes, and we have $\mathcal{N} = 4$ Super Yang–Mills with gauge group $SU(N)$ on the worldvolume of N D3-branes.
- AdS/CFT states that the $\mathcal{N} = 4$ SYM with gauge group $SU(N)$ at large N equals string theory in the $\alpha' \to$ limit, in the $r \to 0$ of the D3-brane metric, which is $AdS_5 \times S_5$.
- The most conservative statement of AdS/CFT relates supergravity in $AdS_5 \times S_5$ with $\mathcal{N} = 4$ SYM with gauge group $SU(N)$ and $g_{\text{YM}}^2 = 4\pi g_s$ at $g_s \to 0$, $N \to \infty$ and $\lambda = g_{\text{YM}}^2 N$ fixed and large ($\gg 1$).
- The strongest version of AdS/CFT is believed to hold: string theory in $AdS_5 \times S_5$ is related to $\mathcal{N} = 4$ SYM with gauge group $SU(N)$ at any $g_{\text{YM}}^2 = 4\pi g_s$ and N, but away from the above limit it is hard to calculate anything.
- AdS/CFT is a duality, since weak coupling calculations in string theory $\alpha' \to 0$, $g_s \to 0$ are strong coupling (large $\lambda = g_{\text{YM}}^2 N$) in $\mathcal{N} = 4$ SYM, and vice versa.
- Supergravity fields in $AdS_5 \times S_5$, Kaluza–Klein dimensionally reduced on S_5, correspond to operators in $\mathcal{N} = 4$ SYM, and the conformal dimension of operators is related to the mass of supergravity fields.
- Chiral primary operators are primary operators that preserve some supersymmetry, and belong to special (short) representations of the superconformal group. The dimension of chiral primary operators matches with what is expected from the mass of the corresponding AdS_5 fields.
- AdS/CFT is actually defined in global AdS space, which has an $S^3 \times \mathbb{R}_t$ boundary. The $\mathcal{N} = 4$ SYM theory exists at this boundary, which is conformally related to $\mathbb{R}^4$.

References and further reading

The most complete review of AdS/CFT is [26], though it appeared in 2000. It also assumes a lot of information, much more than is assumed here, so it works mainly as a reference tool. Another useful review is [29]. The AdS/CFT correspondence was started by Maldacena in [30], but the paper is a demanding read. The correspondence was then made more concrete first in [31] and then in the paper by Witten [32]. In particular, the state map and the "experimental evidence" are found in [32]. A comparison with the spectrum of 10d IIB supergravity on $AdS_5 \times S^5$ is found in [33]. This dimensional reduction is only at the linear level. The full nonlinear reduction on S^5 is not yet done. For the other two cases of interest (discussed only in Part III of this book) of AdS/CFT, $AdS_4 \times S^7$ and $AdS_7 \times S^4$, a nonlinear reduction was done in [34] (though it is not totally complete) for $AdS_4 \times S^7$, and in [35, 36] (completely) for the $AdS_7 \times S^4$ case.

Exercises

1. The metric for an "M2 brane" solution of $d = 11$ supergravity is given by

$$ds^2 = H^{-2/3}(d\vec{x}_3)^2 + H^{+1/3}(dr^2 + r^2 d\Omega_7^2); \quad H = 1 + \frac{2^5 \pi^2 l_P^6}{r^6}. \tag{10.34}$$

   Check that the same limit taken for D3 branes gives M theory on $AdS_4 \times S_7$ if $l_P \to 0$, $U \equiv r^2/l_P^3$ fixed.

2. Check that the $r \to 0$ limit of the D-p-brane metric gives $AdS_{p+2} \times S_{8-p}$ only for $p = 3$.

3. String corrections to the gravity action come about as g_s corrections to terms already present and α' corrections appear generally as $(\alpha' \mathcal{R})^n$, with $\mathcal{R}$ the Ricci scalar, or some particular contraction of Riemann tensors. What then do α' and g_s string corrections correspond to in SYM via AdS/CFT (in the $N \to \infty$, $\lambda = g_{YM}^2 N$ fixed and large limit)?

4. Show that the time it takes a light ray to travel from a finite point in AdS to the real boundary of space and back is finite, but the times it takes to reach the center of AdS ($x_0 = \infty$, or $r = 0$, or $\rho = 0$) is infinite. Try this in both Poincaré and global coordinates.

5. Consider a metric that interpolates in the radial coordinate r between AdS_4 with radius R and $AdS_2 \times S^2$ with radius $R/2$. Is a scalar that is marginally stable in AdS_4 (saturates the BF bound) also stable in $AdS_2 \times S^2$? How about if the $AdS_2 \times S^2$ radius is $R/3$?

6. Write down towers of chiral primary operators corresponding to massive vectors in AdS_5, based on $\mathcal{O}_n$ (by acting with Qs and $\bar{Q}$s), and predict the vector masses $m_k^2 R^2$.

11

Witten prescription and 3-point correlator calculations

In Chapter 10, we saw that there should be a correspondence between string theory, at least in the supergravity limit, on $AdS_5 \times S^5$, and 4-dimensional $\mathcal{N} = 4$ SYM on its boundary, but we only defined the relation between fields in AdS_5 and operators in the CFT, and we have not defined yet the relation between observables. We also saw that the natural observables to define in AdS space are not S-matrices, but correlators with sources on the boundary. In this chapter we present a prescription for the calculation of correlators using AdS/CFT, and then use it to calculate 2-point and 3-point correlators. The calculations are done in Euclidean signature, relating Euclidean AdS_5 (5-dimensional Lobachevski space) with the CFT on the Euclidean $\mathbb{R}^4$ boundary. The Wick rotation to Minkowskian signature is tricky, and it involves new features; some of its aspects are studied in the next chapter.

11.1 Witten prescription for correlation functions

The prescription to calculate correlation functions in x space in AdS/CFT that we explain here was first described by Witten, following work in momentum space by Gubser, Klebanov, and Polyakov [31].

We have seen that operators $\mathcal{O}$ of conformal dimension Δ in the CFT are related to fields ϕ in AdS_5 with mass m related to $\mathcal{O}$ through (10.21). For a massless field, $m = 0$ gives $\Delta = d$ for the operator corresponding to it in the CFT.

In this chapter, we only treat the massless field case, since for massive scalars, the correct method involves the so-called *holographic renormalization*, discussed in Chapter 22, and so we will postpone the treatment of the massive 2-point function until then. One can treat the massive case correctly using the Witten method by taking great care about regularization of infinities near the boundary of AdS space, as was done by Freedman, Mathur, Matusis, *et al.* [38], but the systematic treatment of Chapter 22 makes it much clearer how to deal correctly with infinities in general.

For a massless field ϕ, we can solve the Klein–Gordon (KG) equation near the boundary of AdS_{d+1} space in Poincaré coordinates,

$$0 = \frac{1}{\sqrt{g}} \partial_\mu \sqrt{g} g^{\mu\nu} \partial_\nu \phi = \frac{1}{R^2} \left[x_0^{d+1} \partial_0 x_0^{1-d} \partial_0 \phi + x_0^2 \partial_i^2 \phi \right] = 0, \qquad (11.1)$$

and find that there is a non-normalizable solution where $\phi \to \phi_0$, i.e. it goes to a constant (x_0-independent) value on the boundary. More precisely, the two solutions are

$\phi \to x_0^{d-\Delta}\phi_0$, with $\Delta = d$ and 0. If we also add a mass term, i.e. $\Box\phi - m^2\phi = 0$, we get $\phi \to x_0^{d-\Delta}\phi_0$ and $x_0^{\Delta}\phi_0$, with Δ the conformal dimension of the dual operator, (10.21). Bringing the two solutions together, we can write

$$\phi \sim x_0^{\Delta_\pm}\phi_0, \quad \Delta_\pm = \frac{d}{2} \pm \sqrt{\frac{d^2}{4} + m^2 R^2}, \tag{11.2}$$

since the KG equation reduces to

$$\Delta_\pm(\Delta_\pm - d) - m^2 R^2 = 0. \tag{11.3}$$

The natural interpretation of ϕ_0 is as a source for the operator $\mathcal{O}$, that also exists on the boundary and has the same representations for the symmetry groups. Also, since there is no gauge group in gravity, the field ϕ has no gauge indices, i.e. is gauge invariant, which means that the operator $\mathcal{O}$ must also be gauge invariant. In turn, this also means that it is composite, since the basic fields are in the adjoint representation of the gauge group.

We are then led to consider the partition function with sources for the composite operator $\mathcal{O}$, $Z_\mathcal{O}[\phi_0]$, which is a generating functional of correlation functions of $\mathcal{O}$, as we discussed in Chapter 1.

In Euclidean space, the partition function with sources for the composite operators $\mathcal{O}$ is written as

$$Z_\mathcal{O}[\phi_0] = \int \mathcal{D}[\text{SYM fields}] \exp\left(-S_{\mathcal{N}=4 \text{ SYM}} + \int d^4x \mathcal{O}(x)\phi_0(x)\right), \tag{11.4}$$

and is the generating functional of correlators for the operators $\mathcal{O}$, since we obtain

$$\langle\mathcal{O}(x_1)\ldots\mathcal{O}(x_n)\rangle = \frac{\delta^n}{\delta\phi_0(x_1)\ldots\delta\phi_0(x_n)}Z_\mathcal{O}[\phi_0]\Big|_{\phi_0=0}. \tag{11.5}$$

The natural prescription to calculate the partition function is to make it equal to the partition function for ϕ in string theory, with boundary value ϕ_0, i.e.

$$Z_\mathcal{O}[\phi_0]_{\text{CFT}} = Z_\phi[\phi_0]_{\text{string}}. \tag{11.6}$$

We can understand Witten's prescription as this natural assumption.

However, we can simplify things since we are in the limit $g_s \to 0, \alpha' \to 0, R^4/\alpha'^2 = 4\pi g_s N \gg 1$, i.e. we have no string worldsheet corrections or quantum string corrections, therefore the classical supergravity is a good approximation. Then the partition function $Z[\phi_0]$ of the field ϕ with source ϕ_0 on the boundary, in the classical supergravity (sugra) limit, i.e. for $\phi \to \phi_0$ on the boundary, becomes

$$Z[\phi_0] = \exp[-S_{\text{sugra}}[\phi[\phi_0]]], \tag{11.7}$$

since quantum fluctuations are damped, and the path integral is dominated by the minimum action, i.e. the classical on-shell supergravity action. To calculate it, one finds the classical solution as a function of the boundary source, $\phi[\phi_0]$, and replaces it in S_{sugra}.

Then the AdS/CFT prescription for the generating functional of correlation functions of operators corresponding to (massless) scalars in AdS_5 is

$$Z_\mathcal{O}[\phi_0]_{\text{CFT}} = \int \mathcal{D}[\text{fields}]e^{-S+\int d^4x\mathcal{O}(x)\phi_0(x)} = Z_{\text{class}}[\phi_0]_{AdS} = e^{-S_{\text{sugra}}[\phi[\phi_0]]}. \tag{11.8}$$

11.2 Set-up: the 2-point function of scalars in x space

To calculate n-point functions in the CFT using AdS/CFT, we need to calculate $S_{\text{sugra}}[\phi[\phi_0]]$. For that, we can use a diagrammatic technique, using a specialization of Feynman diagrams, called Witten diagrams. We explain the method here, but we will also present an alternative, simpler, method for the two-point function.

The first step is to understand how to write the classical solution $\phi[\phi_0]$. For that, one can define a classical AdS_5 Green's function, here in the Euclidean Poincaré patch (in the next chapter we consider other cases). Since we want to calculate the field with sources on the boundary, we define the *bulk to boundary propagator*

$$"\Box_{\vec{x},x_0}"K_B(\vec{x}, x_0; \vec{x}') = \delta^4(\vec{x} - \vec{x}'), \qquad (11.9)$$

where "$\Box_{\vec{x},x_0}$" is the kinetic operator and the delta function is a source on the flat 4-dimensional boundary of AdS_5. In the case of a scalar field, the kinetic operator is really $\Box_{\vec{x},x_0}$, but for other fields it is different.

Then the field ϕ is written as

$$\phi(\vec{x}, x_0) = \int d^4\vec{x}' K_B(\vec{x}, x_0; \vec{x}')\phi_0(\vec{x}'), \qquad (11.10)$$

and one replaces it in $S_{\text{sugra}}[\phi]$, after which the partition function is $Z = \exp\left[-S_{\text{sugra}}[\phi[\phi_0]]\right]$.

For a scalar field, the bulk to boundary propagator in AdS_{d+1}, solution to (11.9) (as one can check), is[1]

$$K_{B,\Delta}(\vec{x}, x_0; \vec{x}') = \frac{\Gamma(\Delta)}{\pi^{\frac{d}{2}}\Gamma\left(\Delta - \frac{d}{2}\right)}\left[\frac{x_0}{x_0^2 + (\vec{x} - \vec{x}')^2}\right]^\Delta \equiv C_d\left[\frac{x_0}{x_0^2 + (\vec{x} - \vec{x}')^2}\right]^\Delta. \quad (11.11)$$

On the boundary, $x_0 \to 0$, this satisfies

$$K_{B,\Delta}(\vec{x}, x_0; \vec{x}') \to x_0^{d-\Delta}\delta(\vec{x} - \vec{x}'), \qquad (11.12)$$

as it should because of (11.10). We can check this since at $x_0 \to 0$, $K_B = 0$ for $\vec{x} - \vec{x}' \neq 0$ and for $\vec{x} = \vec{x}'$, $K_B \to \infty$.

We now turn to calculating the simplest possible example, a 2-point function of an operator corresponding to a massless scalar field. From Witten's prescription, we obtain

$$\langle \mathcal{O}(x_1)\mathcal{O}(x_2)\rangle = \frac{\delta^2}{\delta\phi_0[x_1]\delta\phi_0[x_2]}e^{-S_{\text{sugra}}[\phi[\phi_0]]}\bigg|_{\phi_0=0}. \qquad (11.13)$$

---

[1] The solution in the massless case is found as follows. We will shortly see that inversion invariance in AdS plays the same role as inversion invariance in a conformal field theory. We then first look for a solution for K_B that depends only on x_0. Then the KG equation (11.1) has solution $K_B = C_d x_0^d$. Applying the inversion $x^\mu \to x^\mu/x^2$, we get $K_B = C_d x_0^d/(x_0^2 + \vec{x}^2)^d$. In the presence of a mass term, the power d is replaced by Δ.

But since

$$S_{\text{sugra}}[\phi[\phi_0]] = \frac{1}{2} \int (d^5 x \sqrt{g}) \int d^4 \vec{x}'$$
$$\times \int d^4 \vec{y}' \, \partial_\mu K_B(\vec{x}, x_0; \vec{x}') \phi_0(\vec{x}') \partial^\mu K_B(\vec{x}, x_0; \vec{y}') \phi_0(\vec{y}') + O(\phi_0^3), \quad (11.14)$$

where $\frac{1}{\sqrt{g}} \partial_\mu \sqrt{g} \partial^\mu = \Box$ is the kinetic operator, we obtain

$$S_{\text{sugra}}[\phi[\phi_0]]|_{\phi_0=0} = 0; \quad \left. \frac{\delta S_{\text{sugra}}[\phi[\phi_0]]}{\delta \phi_0} \right|_{\phi_0=0} = 0, \quad (11.15)$$

and only second derivatives and higher give a nonzero result. Then

$$\langle \mathcal{O}(x_1) \mathcal{O}(x_2) \rangle = \left. \frac{\delta}{\delta \phi_0[x_1]} \left(-\frac{\delta S_{\text{sugra}}}{\delta \phi_0[x_2]} e^{-S_{\text{sugra}}} \right) \right|_{\phi_0=0} = \left. -\frac{\delta^2 S_{\text{sugra}}[\phi[\phi_0]]}{\delta \phi_0(x_1) \delta \phi_0(x_2)} \right|_{\phi_0=0}$$
$$= -\frac{\delta^2}{\delta \phi_0[x_1] \delta \phi_0[x_2]} \frac{1}{2} \int d^5 x \sqrt{g} \int d^4 \vec{x}' \int d^4 \vec{y}' \, \partial_{\mu \vec{x}, x_0} K_B(\vec{x}, x_0; \vec{x}') \phi_0(\vec{x}')$$
$$\times \partial^\mu_{\vec{x}, x_0} K_B(\vec{x}, x_0; \vec{y}') \phi_0(\vec{y}') = -\int d^5 x \sqrt{g} \partial_{\mu \vec{x}, x_0} K_B(\vec{x}, x_0; \vec{x}_1) \partial^\mu_{\vec{x}, x_0} K_B(\vec{x}, x_0; \vec{x}_2). \quad (11.16)$$

This is the general approach one can use for any n-point function, but in the particular case of the 2-point function the problem simplifies, and the integral that needs to be done is simpler, so we will not continue to calculate it this way.

Instead, we rewrite the on-shell action. Note first that, as we just saw, we only need to consider the quadratic part of the action, since we take two ϕ_0 derivatives and put $\phi_0 = 0$ after that. Therefore we consider a free scalar field, satisfying $\Box \phi = 0$, and its kinetic (quadratic part of the) action is on-shell:

$$S = \frac{1}{2} \int d^5 x \sqrt{g} (\partial_\mu \phi) \partial^\mu \phi = -\frac{1}{2} \int d^5 x \sqrt{g} \phi \Box \phi + \frac{1}{2} \int d^5 x \sqrt{g} \partial_\mu (\phi \partial^\mu \phi)$$
$$= \frac{1}{2} \int_{\text{boundary}} d^4 x \sqrt{h} (\phi \vec{n} \cdot \vec{\nabla} \phi), \quad (11.17)$$

where h is the metric on the boundary. Here $\sqrt{h} = x_0^{-d}$; $\vec{n} \cdot \vec{\nabla} = x_0 \partial / \partial x_0$, and

$$x_0 \frac{\partial}{\partial x_0} \phi(\vec{x}, x_0) = x_0 \frac{\partial}{\partial x_0} \int d^d \vec{x}' K_B(\vec{x}, x_0; \vec{x}') \phi_0(\vec{x}') \to C_d d x_0^d \int d^d \vec{x} \frac{\phi_0(\vec{x}')}{|\vec{x} - \vec{x}'|^{2d}} \quad (11.18)$$

as $x_0 \to 0$. We then obtain for the on-shell kinetic action:

$$S_{\text{kinetic,sugra}}[\phi] = \lim_{x_0 \to 0} \int d^d x x_0^{-d} \phi(\vec{x}, x_0) x_0 \frac{\partial}{\partial x_0} \phi(\vec{x}, x_0)$$
$$= \frac{C_d d}{2} \int d^d \vec{x} \int d^d \vec{x}' \frac{\phi_0(\vec{x}) \phi_0(\vec{x}')}{|\vec{x} - \vec{x}'|^{2d}}, \quad (11.19)$$

which leads to the 2-point function

$$\langle \mathcal{O}(x_1) \mathcal{O}(x_2) \rangle = -\frac{C_d d}{|\vec{x} - \vec{x}'|^{2d}}, \quad (11.20)$$

which is the correct behavior for a field of conformal dimension $\Delta = d$. As we said, the massless scalar field should indeed correspond to an operator of protected dimension $\Delta = d$, so we have our first check of AdS/CFT!

We also note that were it not for the trick we used to turn the kinetic integral into a boundary term, the calculation would have been difficult. But one does not necessarily need to continue the calculation until the end, since one can take advantage of the symmetries of the problem.

We have seen in Chapter 8 that an arbitrary conformal transformation can be obtained by composing the Poincaré group (Lorentz rotations and translations) with the inversion. Since we are interested in Poincaré invariant theories, for 4-dimensional conformal invariance of a result we only need to check invariance under the 4-dimensional inversion. But the 5-dimensional inversion, $x'^\mu = x^\mu/x^2$, where $x^\mu = (x_0, \vec{x}) \equiv (x_0, x_i)$, is an invariance of the Euclidean AdS_5 (Lobachevski) metric in Poincaré coordinates,

$$ds^2 = \frac{R^2}{x_0^2}(dx_0^2 + d\vec{x}^2), \tag{11.21}$$

as we can directly check. It reduces to the 4-dimensional inversion on the boundary $x_0 \to 0$, so in a sense we already expect the AdS results to be 4-dimensional conformal invariant. But it could be that when writing an on-shell result, we lose the invariance. So we need to check the 4-dimensional inversion invariance of $\phi[\phi_0]$, and we also need to check the dimension of the source ϕ_0. In the 4-dimensional action, we add a term $\int d^d x \mathcal{O} \phi_0$, and since we claimed that $\mathcal{O}$ has conformal dimension Δ in (10.21), ϕ_0 should have dimension $d - \Delta$, so under the 4-dimensional inversion we should have $\phi_0'(\vec{x}) = |\vec{x}'|^{2(d-\Delta)}\phi_0(\vec{x}')$. We also obtain under the full 5-dimensional inversion at $x_0 \to 0$,

$$d^d x = \frac{d^d x'}{|\vec{x}|^{2d}},$$

$$\left[\frac{x_0}{x_0^2 + (\vec{x} - \vec{y})^2}\right]^\Delta = \left[\frac{x_0'}{(x_0')^2 + (\vec{x}' - \vec{y}')^2}\right]^\Delta |\vec{x}'|^{2\Delta}. \tag{11.22}$$

Then the AdS field ϕ is AdS inversion invariant, $\phi(y') = \int d^d x K_B(y_0, \vec{y}; \vec{x})\phi'(\vec{x})$.

In general then, we can use inversion invariance to bring AdS integrals to more manageable forms, so we can calculate the on-shell supergravity action, and from it the n-point functions.

11.3 Set-up: 2-point function of gauge fields in *x* space and momentum space

Gauge fields in AdS_{d+1} space correspond to conserved currents on the boundary, i.e. we have a coupling $\int d^d x J_i(\vec{x}) a_i(\vec{x})$, where the source $a_i(\vec{x})$ is the boundary value of the bulk gauge field $A_\mu(x_0, \vec{x})$.

The action for YM fields in five dimensions is a kinetic term $F \wedge *F$ and a 5-dimensional Chern–Simons term $\sim dA \wedge dA \wedge A + A^4$ terms $+ A^5$ terms. The quadratic and cubic terms can be written as

$$
S[A] = \frac{1}{2g^2} \int_{B_5} \left[\delta_{ab} dA^a \wedge *dA^b + f_{abc} dA^a \wedge *\{A^b \wedge A^c\} \right] + \frac{ik}{72\pi^2} \int_{B_5} d_{abc} A^a \wedge dA^b \wedge dA^c,
$$
(11.23)

where k is an integer. For our $AdS_5 \times S^5$ background, one obtains $k = N^2$. Here the antihermitian generators T_a in the fundamental representation are normalized as $\mathrm{Tr}\,[T_a T_b] = -1/2\delta_{ab}$, and their product is

$$
T_a T_b = \frac{1}{2}(f_{ab}{}^c - id_{ab}{}^c)T_c,
$$
(11.24)

where f_{abc} are the antisymmetric structure constants and the d_{abc} are symmetric. The linearized equation of motion coming from the quadratic term in the action is the same as the equation of motion for a Maxwell field in Euclidean AdS space,

$$
\frac{1}{\sqrt{g}} \partial_\mu (\sqrt{g} F^{\mu\nu}) = 0.
$$
(11.25)

For a solution that depends only on x_0 (like the case of the scalar in the previous subsection), $A = f(x_0)dx^i$, we get

$$
\frac{d}{dx_0} \left[(x_0)^{3-d} \frac{df}{dx_0} \right] = 0,
$$
(11.26)

with the solution $f = C(x_0)^{d-2}$. After an AdS inversion $x^\mu \to x^\mu/x^2$, the solution is

$$
A = C \left(\frac{x_0}{x_0^2 + \vec{x}^2} \right)^{d-2} d \left(\frac{x^i}{x_0^2 + \vec{x}^2} \right).
$$
(11.27)

More precisely, this $A \equiv A_\mu dx^\mu$, depending also on the index i, is the *gauge field bulk to boundary propagator* $G_{\mu i}$. In general, we write

$$
A_\mu^a(z) = \int d^4\vec{x}\, G_{\mu i}(z, \vec{x}) a_i^a(\vec{x}),
$$
(11.28)

by analogy with the scalar case. One important difference is that A_μ is now a gauge field, subject to a gauge transformation $\delta A_\mu^a = \partial_\mu \Lambda^a$ at the linear level (otherwise we have a covariant derivative), meaning that the propagator is also subject to gauge transformations, i.e.

$$
G_{\mu i}(z, \vec{x}) \to G_{\mu i}(z, \vec{x}) + \frac{\partial}{\partial z_\mu} \Lambda_i(z, \vec{x}).
$$
(11.29)

The propagator above is in a gauge where it is conformally covariant on the boundary, such that we can take advantage of the transformation properties under conformal transformations. It can be written explicitly as

$$
G_{\mu i}(z, \vec{x}) = C^d \left(\frac{z_0}{z_0^2 + (\vec{z} - \vec{x})^2} \right)^{d-2} \partial_\mu \left(\frac{(\vec{z} - \vec{x})_i}{z_0^2 + (\vec{z} - \vec{x})^2} \right) = C^d \frac{z_0^{d-2}}{[z_0^2 + (\vec{z} - \vec{x})^2]^{d-1}} I_{\mu i}(z - \vec{x}),
$$
(11.30)

where C^d is a constant, normalized such that $A_i \to a_i$ on the boundary, giving

$$C^d = \frac{\Gamma(d)}{2\pi^{\frac{d}{2}}\Gamma\left(\frac{d}{2}\right)}. \tag{11.31}$$

From now on we will call the quantity $z_0^2 + (\vec{z} - \vec{x})^2 \equiv R$ to simplify notation. Changing momentarily the normalization by writing $C^d \equiv \tilde{C}^d(d-1)/(d-2)$, we have for the propagator written in form language

$$G = \tilde{C}^d \frac{d-1}{d-2}\left(\frac{z_0}{R}\right)^{d-2} d\left(\frac{(\vec{z}-\vec{x})^i}{R}\right). \tag{11.32}$$

We can find the propagator in another gauge by adding the gauge transformation

$$d\Lambda = -\tilde{C}^d \frac{1}{d-2} d\left(\frac{(\vec{z}-\vec{x})_i z_0^{d-2}}{R^{d-1}}\right), \tag{11.33}$$

after which the propagator becomes

$$G' = \tilde{C}^d \left[\frac{z_0^{d-2}}{R^{d-1}} dz_i - \frac{z_0^{d-3}(\vec{z}-\vec{x})_i}{R^{d-1}} dz_0\right]. \tag{11.34}$$

Writing it in terms of a derivative (note that we have $\partial_{z^i} = -\partial_{x^i}$ on R), we have

$$G'_{\mu i}(z_0, \vec{z}; \vec{x}) dz^\mu = \tilde{C}^d \left[\frac{z_0^{d-2}}{R^{d-1}} dz_i - \frac{1}{2(d-2)}\partial_{x^i}\left(\frac{z_0^{d-3}}{R^{d-2}}\right) dz_0\right]. \tag{11.35}$$

The kinetic (quadratic part of the) action is written again, as in the scalar case, as a total derivative, i.e. a boundary term (below $F \equiv dA$),

$$S[A] = \frac{1}{2}\int_{AdS_{d+1}} F \wedge *F = \frac{1}{2}\int_{AdS_{d+1}} d(A \wedge *F) = \frac{1}{2}\lim_{\epsilon \to 0}\int_{\partial AdS_{d+1}, x_0 = \epsilon} A \wedge *F. \tag{11.36}$$

This can be written as

$$S[A] = \lim_{\epsilon \to 0}\frac{1}{d-1}\int_{x_0 = \epsilon} d^d x\, x_0^{3-d}\delta^{ij} A_i(x_0, \vec{x}) F_{0j}(x_0, \vec{x}), \tag{11.37}$$

and considering that $A_i \to a_i$ as $x_0 \to 0$ (the propagator was constructed to satisfy this), we obtain, calculating from (11.34)

$$
\begin{aligned}
F &= dA = d \wedge \int d^d x\, G'\, a_i(\vec{x}) \\
&= (d-1) z_0^{d-3}\tilde{C}^d \left[\int d^d x \frac{dz_0 \wedge dz_i}{R^{d-1}}\left(\delta_{ij} - \frac{(\vec{z}-\vec{x})_i(\vec{z}-\vec{x})_j}{R}\right) a_j(\vec{x}) + \mathcal{O}(z_0^2)\right],
\end{aligned}
\tag{11.38}
$$

which implies

$$S[A] = \int d^d x\, d^d x'\, a_i(\vec{x}) a_j(\vec{x}')\frac{\tilde{C}^d}{(\vec{x}-\vec{x}')^{2d-2}} I_{ij}(\vec{x}-\vec{x}'). \tag{11.39}$$

Therefore the 2-point function is obtained by taking the double derivative with respect to a_i and a_j at $a_k = 0$, and is given by

$$\langle J_i(\vec{x})J_j(\vec{x}')\rangle = \frac{\tilde{C}^d}{(\vec{x}-\vec{x}')^{2d-2}}I_{ij}(\vec{x}-\vec{x}'), \tag{11.40}$$

which as we saw in Chapter 8 is the form required by conformal invariance.

But we can calculate this on-shell action also in momentum space, in a way which generalizes to n-point functions, for which there is no way to write the n-point AdS terms as boundary terms. However, the calculation in momentum space is more involved, its advantage being only that we can readily relate this to easier and more familiar momentum space calculations in SYM theory. We only show the procedure on the AdS side, since it is the case of interest (it is a new calculation).

Indeed, using the propagator G' in (11.35), we find for dA

$$dA^a(z_0, \vec{z}) = \tilde{C}^d dz_i \wedge dz_j \int d^d x a_i^a(\vec{x})\partial_{x^j}\frac{z_0^{d-2}}{R^{d-1}}$$
$$-\tilde{C}^d dz_0 \wedge dz_i \int d^d x a_j^a(\vec{x})\frac{1}{2(d-2)}[\delta_{jk}\delta_{li} - \delta_{lk}\delta_{ji}]\partial_{x^k}\partial_{x^l}\frac{z_0^{d-3}}{R^{d-2}}. \tag{11.41}$$

Then the quadratic part of the action becomes

$$S[A] = \frac{1}{2g^2}\int d^d x_1 \int d^d x_2 a_i^a(\vec{x}_1)a_j^b(\vec{x}_2)\delta_{ab}$$
$$\times \left[(\partial_1^i \partial_1^j - \delta^{ij}\partial_1^2)I_{d-2,d-2}^{d/2-1} - \frac{1}{8(d-2)^2}(\partial_1^i \partial_1^j - \delta^{ij}\partial_1^2)\partial_1^2 I_{d-3,d-3}^{d/2-2}\right]. \tag{11.42}$$

Here we have defined the integrals

$$I_{mn}^f(\vec{x}_1, \vec{x}_2) = \int dz_0 d^d z \frac{z_0^{2f+1}}{R_1^{m+1}R_2^{n+1}}, \tag{11.43}$$

which are convergent only for $m + n > f + d/2 - 1$, which is satisfied in the case of both of the integrals appearing above in the kinetic action. Note that by translational invariance, I_{mn}^f is a function of only $\vec{x}_1 - \vec{x}_2$, so when acting on it, $\partial_1 = -\partial_2$.

Taking the double functional derivative with respect to a_i and a_j to find the current 2-point function, and then going to momentum space, we find after some algebra

$$\langle J_a^i(\vec{q}_1)J_b^j(\vec{q}_2)\rangle = \int d^4\vec{x}_1 d^4\vec{x}_2 e^{i\vec{q}_1\vec{x}_1+i\vec{q}_2\vec{x}_2}\langle J_a^i(\vec{x}_1)J_b^j(\vec{x}_2)\rangle$$
$$= \frac{\delta_{ab}}{g^2}(q_1^i q_1^j - \delta^{ij}\vec{q}_1^2)\left[I_{d-2,d-2}^{d/2-1} + \frac{\vec{q}_1^2}{8(d-2)^2}I_{d-3,d-3}^{d/2-2}\right]. \tag{11.44}$$

The integrals I_{mn}^f are now in momentum space. By shifting the integrations over $\vec{x}_1, \vec{x}_2$ by the original $\vec{z}$, we obtain

$$I_{mn}^f(\vec{q}_1, \vec{q}_2) = \int dz_0 \int d^d x_1 d^d x_2 e^{i\vec{q}_1\vec{x}_1+i\vec{q}_2\vec{x}_2}\frac{z_0^{2f+1}(2\pi)^d\delta^d(\vec{q}_1 + \vec{q}_2)}{(z_0^2 + \vec{x}_1^2)^{m+1}(z_0^2 + \vec{x}_2^2)^{n+1}}. \tag{11.45}$$

We could evaluate these integrals completely, but we want to show the method that can be extended to 3-point functions and higher, so instead we write them in an integral form that matches with the SYM momentum space calculations. Specifically, we use multiple Schwinger parameters for the denominators, using the basic identity

$$
I_m(z_0, \vec{q}_1) = \int d^d x_1 \frac{e^{i \vec{q}_1 \vec{x}_1}}{(z_0^2 + \vec{x}_1^2)^{m+1}} = \frac{1}{\Gamma(m+1)} \int_0^\infty d\tau\, \tau^m \int d^d \vec{x}_1 e^{i \vec{x}_1 \vec{q}_1 - \tau(z_0^2 + \vec{x}_1^2)}
$$

$$
= \frac{\Omega(d)\Gamma(d/2)}{2\Gamma(m+1)} \int_0^\infty d\tau\, \tau^{m-d/2} e^{-\tau z_0^2 - \vec{q}_1^2/4\tau}, \quad (11.46)
$$

where $\Omega(d) = 2\pi^{d/2}/\Gamma(d/2)$ is the solid angle in d dimensions. Then the integral I_{mn}^f becomes

$$
I_{mn}^f = \frac{\pi^d}{\Gamma(m+1)\Gamma(n+1)} \int dx_0 d\tau_1 d\tau_2 x_0^{2f+1} \tau_1^{m-d/2} \tau_2^{n-d/2} e^{-x_0 \bar{\tau} - \sum q_r^2/4\tau_r}
$$

$$
= \frac{\pi^d \Gamma(f+1))}{2\Gamma(m+1)\Gamma(n+1)} \int d\tau_1 d\tau_2 \frac{\tau_1^{m-d/2} \tau_2^{n-d/2}}{\bar{\tau}^{f+1}} e^{-\sum q_r^2/4\tau_r}, \quad (11.47)
$$

where we have defined $\bar{\tau} = \sum_r \tau_r$. Substituting in the 2-point function and rescaling $\tau_i \to \tau_i/4$ we get

$$
\langle J_a^i(\vec{q}_1) J_b^j(\vec{q}_2) \rangle = \frac{8\pi^{2d} \Gamma(d/2)}{g^2 \Gamma(d-1)^2} \delta_{ab} \delta^d(q_1 + q_2)
$$

$$
\times \int d\tau_1 d\tau_2 \frac{(\tau_1 \tau_2)^{d/2-2}}{\bar{\tau}^{d/2}} e^{-\sum q_r^2/4\tau_r} \left[q_1^i q_1^j - \delta^{ij} q_1^2 \right] \left(1 + q_1^2 \frac{\bar{\tau}}{(d-2)\tau_1 \tau_2} \right).
$$

$$(11.48)$$

Writing q_1^2 as a derivative on the exponent, integrating by parts and writing $\tau_1 = \alpha\tau$ and $\tau_2 = (1-\alpha)\tau$, we can perform the α integration and obtain

$$
\langle J_a^i(\vec{q}_1) J_b^j(\vec{q}_2) \rangle = \frac{4\pi^{2d}}{g^2} \frac{\Gamma(d/2)}{\Gamma(d-1)^2} \delta_{ab} \left[q_1^i q_1^j - \delta^{ij} q_1^2 \right] \int d\bar{\tau}\, \bar{\tau}^{d/2-3} e^{-q^2/\bar{\tau}}. \quad (11.49)
$$

This integral result can be reproduced from the SYM side in a similar fashion: write the denominators with Schwinger parameters, then do the momentum integrations, after which redefine the Schwinger parameters and do all of the integrations except the last one. We will not do it here, but it can be found in [39].

11.4 3-point functions; example: R-current anomaly

Having described how to calculate 2-point functions, we now turn to 3-point functions. Whereas the 2-point functions are fixed by conformal invariance given their conformal dimensions, and the numerical constant is simply a normalization, the 3-point functions are the first ones that can provide a test of the dynamics. Their functional form is still fixed by conformal invariance, but the overall coefficient is not. However,

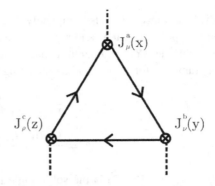

Figure 11.1   Triangle diagram contributing to the $\langle J_i^a(x)J_j^b(y)J_k^c(z)\rangle$ correlator. Chiral fermions run in the loop.

as for the conformal dimension, the numerical coefficient of the 3-point function can in principle receive quantum corrections, so we need to find quantities that are not renormalized.

One obvious candidate is the R-current anomaly. As we reviewed in Chapter 1, anomalies in general are one-loop exact, so the first nonzero calculation, the one-loop triangle, gives the complete result. The R-currents J_i^a are conserved currents in $\mathcal{N} = 4$ SYM corresponding to the $SU(4) = SO(6)$ global symmetry; they are composite and gauge invariant operators coupling with the gauge fields A_μ^a, with boundary values a_i^a. The A_μ^a arise from the KK reduction on S^5 and the symmetry of the 5-sphere, $SO(6)$, becomes the R-symmetry. The R-currents for the symmetry,

$$\delta\psi^a = \epsilon^a(T_a^\psi)\frac{1+\gamma_5}{2}\psi; \quad \delta\phi = \epsilon^a(T_a^\phi)\phi, \tag{11.50}$$

are

$$J_a^i(x) = \frac{1}{2}\phi(x)T_a^\phi(\overleftrightarrow{\partial}^i + 2gA^i(x))\phi(x) - \frac{1}{2}\bar{\psi}(x)T_a^\psi\gamma^i\frac{1+\gamma_5}{2}\psi(x). \tag{11.51}$$

Their anomaly in four dimensions is given by the triangle diagram in Fig. 11.1, where the loop (triangle) is formed by chiral fermions. The anomaly means that we have

$$\frac{\partial}{\partial x_i}\langle J_i^a(x)J_j^b(y)J_k^c(z)\rangle \neq 0. \tag{11.52}$$

In order to calculate general n-point functions in AdS space, we must calculate the on-shell (classical) supergravity action with boundary values and differentiate with respect to these boundary values. The resulting AdS n-point functions are given by a sum of tree level Feynman diagrams (since we use classical supergravity) in x space, with external points on the boundary. These are known as Witten diagrams. To exemplify them, consider the simplest case, of a 3-point function of scalars coming from an interaction term in the Lagrangean $\mathcal{L}_{int} = \lambda\phi_1\phi_2\phi_3$. Using that in AdS space in Poincaré coordinates $\sqrt{-g} = x_0^{-d-1}$ and that *to leading order* we have $\phi_i(z_0, \vec{z}) = \int d^dx K_{B,\Delta_i}(z_0, \vec{z}; \vec{x})\phi_{0,i}(\vec{x})$ (in general we have an extra term $-\lambda \int d^dy dy_0 G(z_0, \vec{z}; y_0, \vec{y})\phi^2(y_0, \vec{y})$, where G is the bulk to bulk propagator, which can be self-consistently solved in perturbation theory by

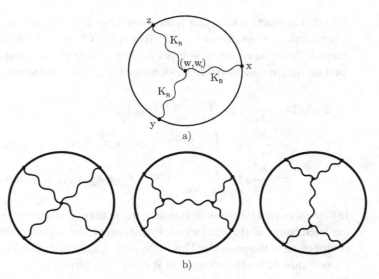

Figure 11.2 a) Tree level "Witten diagram" for the 3-point function in AdS space; b)Tree level Witten diagrams for the 4-point function in AdS space.

replacing the leading order result for ϕ, etc. on the right-hand side), we find the 3-point function

$$\langle \mathcal{O}(x_1)\mathcal{O}(x_2)\mathcal{O}(x_3)\rangle = \frac{\delta^3}{\delta\phi_{0,1}(x_1)\delta\phi_{0,2}(x_2)\delta\phi_{0,3}(x_3)}e^{-S_{\mathrm{sugra}}[\phi_{0,i}]}\bigg|_{\phi_{0i}=0}$$

$$= -\frac{\delta^3}{\delta\phi_{0,1}(x_1)\delta\phi_{0,2}(x_2)\delta\phi_{0,3}(x_3)}S_{\mathrm{int}}[\phi_{0i}]$$

$$= -\lambda \int \frac{d^dz dz_0}{z_0^{d+1}}K_{B,\Delta_1}(z_0,\vec{z};\vec{x}_1)K_{B,\Delta_2}(z_0,\vec{z};\vec{x}_2)K_{B,\Delta_3}(z_0,\vec{z};\vec{x}_3).$$

$$(11.53)$$

The corresponding "Witten diagram" (Feynman diagram in x space with points on the boundary) is given in Fig. 11.2a: three bulk to boundary propagators connecting the points on the boundary with a point in the bulk, are integrated over. For the 4-point function, in the case of a theory with 4-point interaction terms in the Lagrangean as well as the 3-point interaction terms, we have the three Witten diagrams (tree diagrams), given in Fig. 11.2b.

Note that since we are interested in *connected supergravity tree diagrams* only, we actually have in general

$$\langle \mathcal{O}(x_1)\ldots\mathcal{O}(x_n)\rangle = (-1)^n\frac{\delta^n}{\delta\phi_0(x_1)\ldots\delta\phi_0(x_n)}S_{\mathrm{sugra,on\text{-}shell}}[\phi_0]. \qquad (11.54)$$

Coming back to the R-current anomaly mentioned above, it is in general proportional to $d_{abc} = -2i\mathrm{Tr}\ (T_a\{T_b,T_c\})$, which is totally symmetric under the interchange of a,b,c; it is also antisymmetric in the indices i,j,k. To obtain the anomaly from AdS space we need a 3-point interaction term that is antisymmetric in the spacetime indices μ, ν, ρ and gives a contribution proportional to d_{abc}. This is the Chern–Simons term in the 5-dimensional gauged (AdS) supergravity. It gives the last term in (11.23), and it is easiest to write as a total derivative term in six dimensions, giving a 5-dimensional boundary term,

$$
\begin{aligned}
S_{\mathrm{CS}}(A) &= \frac{N^2}{144\pi^2}\mathrm{Tr}\int_{M_6}\epsilon^{\mu\nu\rho\sigma\tau\epsilon}F_{\mu\nu}F_{\rho\sigma}F_{\tau\epsilon}\\
&= \frac{N^2}{18\pi^2}\mathrm{Tr}\int_{M_6}\epsilon^{\epsilon\mu\nu\rho\sigma\tau}\partial_\epsilon(A_\mu(\partial_\nu A_\rho)\partial_\sigma A_\tau + A^4\text{ terms} + A^5\text{ terms})\\
&= \frac{N^2}{18\pi^2}\mathrm{Tr}\int_{B_5=\partial M_6}\epsilon^{\mu\nu\rho\sigma\tau}(A_\mu(\partial_\nu A_\rho)\partial_\sigma A_\tau + A^4\text{ terms} + A^5\text{ terms}).\quad (11.55)
\end{aligned}
$$

In the first form it is obvious it is symmetric in the a, b, c indices, since it is symmetric under the interchange of the Fs. Indeed, by peforming the trace we obtain d_{abc}, so we obtain a contribution to the anomaly. This is the only d_{abc} contribution in the 3-point interaction of gauge fields, hence the anomaly of R-currents is given by

$$
\langle J^{ia}(x_1)J^{jb}(x_2)J^{kc}(x_3)\rangle_{\mathrm{CFT},\,d_{abc}\text{ part}} = -\left.\frac{\delta^3 S^{3-\text{pnt vertex}}_{\mathrm{CS,sugra}}[A^a_\mu[a^d_i]]}{\delta a^a_i(x_1)\delta a^b_j(x_2)\delta a^c_k(x_3)}\right|_{a=0}.\quad (11.56)
$$

We could continue by substituting $A^a_\mu[a^d_i]$ and doing the integrals and differentiations, but there is a simpler way in the case of the anomaly.

The gauge variation in AdS space,

$$
\delta A^a_\mu = (D_\mu\Lambda)^a = \partial_\mu\lambda^a + gf^a{}_{bc}A^b_\mu\lambda^c,\quad (11.57)
$$

of the Chern–Simons term gives

$$
\begin{aligned}
\delta_\Lambda S_{\mathrm{CS}} &= \frac{N^2}{24\pi^2}\mathrm{Tr}\int_{B_5}d^5x\,\epsilon^{\mu\nu\rho\sigma\tau}(\delta A_\mu F_{\nu\rho}F_{\sigma\tau})\\
&= \frac{iN^2}{96\pi^2}d_{abc}\int_{B_5}d^5x\,\epsilon^{\mu\nu\rho\sigma\tau}(D_\mu\Lambda)^a F^b_{\nu\rho}F^c_{\sigma\tau}\\
&= -\frac{iN^2}{96\pi^2}d_{abc}\int_{B_5}d^5x\,\epsilon^{\mu\nu\rho\sigma\tau}\partial_\tau\left[\Lambda^a\partial_\mu\left(A^b_\nu\partial_\rho A^c_\sigma + \frac{1}{4}f^c{}_{de}A^b_\nu A^d_\rho A^e_\sigma\right)\right]\\
&= -\frac{iN^2}{96\pi^2}d_{abc}\int_{\text{boundary}}d^4x\,\epsilon^{ijkl}\Lambda^a\partial_i\left(A^b_j\partial_k A^c_l + \frac{1}{4}f^c{}_{de}A^b_j A^d_k A^e_l\right),\quad (11.58)
\end{aligned}
$$

where in the third line we have used partial integration and $D_{[\mu}F_{\nu\rho]} = 0$, and in the last expression we can substitute A^a_is with their boundary values a^a_i.

But the AdS/CFT prescription implies that

$$
\begin{aligned}
\delta_\Lambda S_{\mathrm{class}}[a^a_i] &= \delta_\Lambda(-\ln Z[a^a_i]) = \int d^4x\,\delta a^{ai}(x)J^a_i(x) = \int d^4x(D^i\Lambda)^a J^a_i(x)\\
&= -\int d^4x\,\Lambda^a[D^i J_i]^a.\quad (11.59)
\end{aligned}
$$

Substituting $\delta_\Lambda S_{CS}$ on the left-hand side we get (at leading order in N)

$$
(D^i J_i)^a(x) \equiv \frac{\partial}{\partial x^i} J_i^a + f^a{}_{bc} a^{ib} J_i^c
$$
$$
= \frac{iN^2}{96\pi^2} d_{abc} \epsilon^{ijkl} \partial_i \left(a_j^b \partial_k a_l^c + \frac{1}{4} f^c{}_{de} a_j^b a_k^d a_l^e \right), \tag{11.60}
$$

which is exactly the operator equation for the R-current anomaly in the CFT (coming from the 1-loop CFT computation). On the other hand, at $a = 0$, the 1-loop result for the anomaly of the 3-point function in the CFT is

$$
\frac{\partial}{\partial z^k} \langle J_i^a(x) J_j^b(y) J_k^c(z) \rangle_{\text{CFT},d_{abc}} = -\frac{(N^2 - 1) i d_{abc}}{48\pi^2} \epsilon^{ijkl} \frac{\partial}{\partial x_k} \frac{\partial}{\partial y_l} \delta(x - y) \delta(y - z), \tag{11.61}
$$

which indeed matches with the above holographic calculation at leading order in N (and a careful analysis matches also at subleading order).

11.5 Calculation of full 3-point function of R-currents

We want now to calculate the full 3-point function, not only the anomalous part. Since the d_{abc} part is anomalous, the other group invariant that appears in the 3-point vertex (11.23) is f_{abc}, which will thus give the non-anomalous part of the 3-point function. As we mentioned in the case of the 2-point function, this calculation could in principle be done in x space and in p space. The p space calculation is more familiar in field theory, but in gravity it is somewhat more involved, so we will describe the x-space calculation.

In x-space we can use conformal invariance to simplify the calculations. It dictates that the 3-point function of currents should have the general form

$$
\langle J_i^a(x) J_j^b(y) J_k^c(z) \rangle_{f_{abc}} = f_{abc} (k_1 D_{ijk}^{\text{sym}}(x, y, z) + k_2 C_{ijk}^{\text{sym}}(x, y, z)), \tag{11.62}
$$

where k_1, k_2 are arbitrary coefficients and C_{ijk}^{sym} and D_{ijk}^{sym} stand for the symmetrized version (adding cyclic permutations) of the objects

$$
D_{ijk}(x, y, z) = \frac{1}{(x-y)^2 (z-y)^2 (x-z)^2} \frac{\partial}{\partial x^i} \frac{\partial}{\partial y^j} \log(x-y)^2 \frac{\partial}{\partial z^k} \log\left(\frac{(x-z)^2}{(y-z)^2}\right),
$$
$$
C_{ijk}(x, y, z) = \frac{1}{(x-y)^4} \frac{\partial}{\partial x^i} \frac{\partial}{\partial z^l} \log(x-z)^2 \frac{\partial}{\partial y^j} \frac{\partial}{\partial z_l} \log(y-z)^2 \frac{\partial}{\partial z^k} \log\left(\frac{(x-z)^2}{(y-z)^2}\right). \tag{11.63}
$$

By conformal invariance we can fix one point, e.g. to $z = 0$, and another, e.g. to $y \to \infty$. Then the form of the two structures becomes

$$
D_{ijk}(x, y, 0) \overset{y \to \infty}{\to} \frac{-4}{y^6 x^4} I_{jl}(y) \left\{ \delta_{ik} x_l - \delta_{il} x_k - \delta_{kl} x_i - 2 \frac{x_i x_k x_l}{x^2} \right\},
$$
$$
C_{ijk}(x, y, 0) \overset{y \to \infty}{\to} \frac{8}{y^6 x^4} I_{jl}(y) \left\{ \delta_{ik} x_l - \delta_{il} x_k - \delta_{kl} x_i + 4 \frac{x_i x_k x_l}{x^2} \right\}. \tag{11.64}
$$

On the other hand, in AdS space the 3-point correlator comes from the 3-point vertex proportional to f_{abc}, which is

$$\frac{1}{2g^2} \int \frac{d^d w dw_0}{w_0^{d+1}} i f_{abc} \partial_{[\mu} A^a_{\nu]}(w) w_0^4 A^b_\mu(w) A^c_\nu(w). \tag{11.65}$$

Using the conformal bulk to boundary propagator (11.30) for consistency, since we used conformal invariance on the boundary, we obtain

$$\langle J^a_i(x) J^b_j(y) J^c_k(z)\rangle_{f_{abc}} = -\frac{i f_{abc}}{2g^2} 2 F^{\text{sym}}_{ijk}(\vec{x}, \vec{y}, \vec{z}),$$

$$F_{ijk}(\vec{x}, \vec{y}, \vec{z}) \equiv \int \frac{d^d w dw_0}{w_0^{d+1}} \partial_{[\mu} G_{\nu]i}(w, \vec{x}) w_0^4 G_{\mu j}(w, \vec{y}) G_{\nu k}(w, \vec{z}). \tag{11.66}$$

After some algebra, one finds

$$F_{ijk}(\vec{x}, \vec{y}, \vec{z}) = -\tilde{K}^d \frac{I_{jl}(\vec{y} - \vec{x})}{|\vec{y} - \vec{x}|^{2(d-1)}} \frac{I_{km}(\vec{z} - \vec{x})}{|\vec{z} - \vec{x}|^{2(d-1)}} \frac{1}{|\vec{t}|^d}$$

$$\times \left(\delta_{lm} t_i + (d-1)\delta_{il} t_m + (d-1)\delta_{im} t_l - d \frac{t_i t_m t_l}{|\vec{t}|^2} \right), \tag{11.67}$$

where $\tilde{K}^d$ is a given constant, and

$$\vec{t} \equiv (\vec{y} - \vec{x})' - (\vec{z} - \vec{x})' \quad \text{and} \quad (\vec{w})' \equiv \frac{\vec{w}}{w^2}. \tag{11.68}$$

We can now put $\vec{z} = 0$ and $|\vec{y}| \to \infty$ in this result and compare with the CFT result (11.62) and (11.64). We obtain

$$F^{\text{sym}}_{ijk}(\vec{x}, \vec{y}, \vec{z}) = \frac{1}{\pi^4} \left(D^{\text{sym}}_{ijk}(\vec{x}, \vec{y}, \vec{z}) - \frac{C^{\text{sym}}_{ijk}(\vec{x}, \vec{y}, \vec{z})}{8} \right). \tag{11.69}$$

One can in fact check that this matches the *1-loop* result of CFT, even though we are at strong coupling ($\lambda \equiv g^2 N \gg 1$). That implies that there should exist some non-renormalization theorem at work, similar to the one for the quantum anomaly. In fact, such a theorem was proved for 3-point functions, using superconformal symmetry. Thus in fact, in $\mathcal{N} = 4$ SYM the 3-point functions of currents are 1-loop exact and match with the AdS space calculation!

Important concepts to remember

- The Witten prescription states that the exponential of (minus) the supergravity action for fields ϕ with boundary values ϕ_0 is the partition function for operators $\mathcal{O}$ corresponding to ϕ, and with sources ϕ_0.
- The bulk to boundary propagator, together with the AdS supergravity (gauged supergravity) vertices, define "Witten diagrams" from which we calculate the boundary (2-, 3-, 4-,...point) correlators.
- The 2-point functions match, but they are kinematic. Dynamics is encoded in 3-point functions and higher.

- Gauge field propagators are subject to gauge transformations. For the bulk to boundary propagator, $G_{\mu i} \to G_{\mu i} + \partial_\mu \Lambda_i$.
- Calculations in momentum space can be made to match between AdS and SYM by writing the resulting integrals with Schwinger parameters.
- To compare both sides of the duality, we need correlators that do not get renormalized. The R-current anomaly is such an object.
- The R-current anomaly in field theory is given by a one-loop triangle Feynman diagram contribution to the 3-point function of R-currents, and comes from the AdS (gauged) supergravity Chern–Simons term. The coefficient matches.
- Even the full 3-point function of R-currents matches with the AdS space calculation of gauge field 3-point function. It has later become understood to come from non-renormalization theorems.

References and further reading

The prescription for calculating CFT correlators was developed by Witten in [32], as well as the calculation of scalar 2-point functions and the anomaly in the R-current 3-point function. The 3-point functions were calculated in [37–39]. Three-point functions of scalars were calculated in [37]; in [38] 3-point functions of scalars and R-currents were calculated using x-space and conformal invariance (the method described in the text), and in [39] a momentum space method for the calculation of R-current 3-point function was used, matching the usual p-space quantum field theory calculation.

Exercises

1. Knowing that parts of the gauge terms Tr $F_{\mu\nu}^2$ and S_{CS} used for the AdS/CFT calculation of the 3-point function of R-currents come from the 10-dimensional Einstein term $\sim \frac{1}{g_s^2} \int d^{10}x \sqrt{G^{(10)}} \, \mathcal{R}$ (here $\mathcal{R}$ is the 10-dimensional Ricci scalar), prove that the overall factor in $S_{\text{sugra}}[A_\mu(a_\rho)]$, and thus in the 3-point function of R-currents, is N^2 (no g_{YM} factors). Use that $R_{AdS_5} = R_{S^5} = \sqrt{\alpha'}(g_s N)^{1/4}$.

2. Consider the equation $(\Box - m^2)\phi = 0$ in the Poincaré patch of AdS_{d+1}. Check that near the boundary $x_0 = 0$, the two independent solutions go like $x_0^{2h_\pm}$, with

$$2h_\pm = \frac{d}{2} \pm \sqrt{\frac{d^2}{4} + m^2 R^2}, \tag{11.70}$$

(so that $2h_+ = \Delta$, the conformal dimension of the operator dual to ϕ).

3. Check that near $x_0 = 0$, the massless scalar field $\phi = \int K_B \phi_0$, with

$$K_B(\vec{x}, x_0; \vec{x}') = c \left(\frac{x_0}{x_0^2 + |\vec{x} - \vec{x}'|^2} \right)^d, \tag{11.71}$$

goes to a constant, ϕ_0. Then check that for the massive scalar case, replacing the power d by $2h_+$ in K_B, we have $\phi \to x_0^{2h_-}\phi_0$ near the boundary.

4. Check that the (1-loop) anomaly of R-currents is proportional to N^2 at leading order, by doing the trace over indices in the diagram.

5. Write down the classical equations of motion for the 5-dimensional Chern–Simons action for A_μ^a.

6. Consider a scalar field ϕ in AdS_5 supergravity, with action

$$S = \int d^5x \sqrt{-g} \left[\frac{1}{2}(\partial_\mu\phi)^2 + \frac{1}{2}m^2\phi^2 + \lambda\frac{\phi^3}{3} \right]. \tag{11.72}$$

Is the 4-point function of operators $\mathcal{O}$ sourced by ϕ, $\langle \mathcal{O}(x_1) \ldots \mathcal{O}(x_4) \rangle$, zero or nonzero, and why?

We have seen that the AdS/CFT correspondence for correlators is easy to define in Euclidean AdS space (Lobachevsky space), and we have described calculations in Poincaré coordinates. However, new concepts appear in Lorentzian signature, and in the case of global coordinates calculations are also somewhat different, so we treat them separately in this chapter. For that, we first take another look at the Euclidean AdS case.

We first solve for the *bulk-to-bulk propagator*. Indeed, we saw that in the Witten diagrams (tree level Feynman diagrams with external points on the boundary), we have in general both bulk-to-boundary propagators and bulk-to-bulk propagators. The bulk-to-bulk propagators appear the first time in the 4-point correlator, through the last two diagrams in Fig.11.2, involving only 3-point vertices. The bulk-to-bulk propagator for a massive scalar in Euclidean Poincaré AdS_{d+1} satisfies the equation

$$(\Box_x - m^2)G(x,y) = \frac{1}{R^2}x_0^{d+1}\partial_{x^\mu}\left[x_0^{-d+1}\partial_{x^\mu}G(x,y)\right] - m^2 G(x,y)$$

$$= -\frac{1}{\sqrt{g_y}}\delta^{d+1}(x-y) = -\delta^d(\vec{x}-\vec{y})\delta(x_0-y_0)\frac{y_0^{d+1}}{R^{d+1}}. \quad (12.1)$$

Defining $G(x,y) = x_0^{d/2}\tilde{G}(x,y)$, the operator $(\Delta_\nu + \partial_i^2)$ that acts on $\tilde{G}$ ($\Delta_\nu = \partial_0^2 + \frac{1}{x_0}\partial_0 - \frac{\nu^2}{x_0^2}$) has eigenfunctions $J_\nu(wx_0)e^{i\vec{k}\cdot\vec{x}}$ with eigenvalues $-(w^2 + \vec{k}^2)$, where $\nu^2 = m^2R^2 + d^2/4$. Here J_ν are Bessel functions. Using also the decomposition $\delta(x_0 - y_0) = y_0\int_0^\infty dw\, wJ_\nu(wx_0)J_\nu(wy_0)$, we find

$$G(x,y) = (x_0y_0)^{d/2}\int\frac{d^dk}{(2\pi)^d}\int_0^\infty dw\, w\frac{1}{w^2 + \vec{k}^2}e^{i\vec{k}\cdot(\vec{x}-\vec{y})}J_\nu(wx_0)J_\nu(wy_0)$$

$$= (x_0y_0)^{d/2}\int\frac{d^dk}{(2\pi)^d}e^{i\vec{k}\cdot(\vec{x}-\vec{y})}I_\nu(kx_0^<)K_\nu(kx_0^>), \quad (12.2)$$

where $x_0^<$ ($x_0^>$) is the smallest (largest) among x_0 and y_0.

This formula is of a usual kind when solving for Green's functions of operators, being constructed out of the two possible solutions to the homogeneous equation $(\Box - m^2)\Phi = 0$. The first solution is

$$\Phi \propto e^{i\vec{k}\cdot\vec{x}}x_0^{d/2}K_\nu(kx_0)\phi_0(\vec{k}), \quad (12.3)$$

which is regular everywhere, behaving near the boundary as the non-normalizable mode $x_0^{\Delta_-}$, or more precisely a combination of the non-normalizable mode $x_0^{\Delta_-}$ and the normalizable mode $x_0^{\Delta_+}$. The second solution is with the Bessel function K_ν replaced by I_ν, and it behaves near the boundary as the normalizable mode $x_0^{\Delta_+}$, but it blows up exponentially in the center of AdS (at $x_0 = \infty$). The propagator is then $\int d^d k \Phi_{1,\vec{k}}(x_0^>) \Phi_{2,\vec{k}}(x_0^<)$.

Since $I_\nu(x) \sim x^\nu$, $K_\nu(x) \sim x^{-\nu}$ when $x \to 0$, the bulk to bulk propagator behaves when one point goes near the boundary as

$$G(x,y) \sim (x_0)^{\frac{d}{2}+\nu} = (x_0)^{\Delta_+}. \tag{12.4}$$

The bulk-to-boundary propagator can then be defined by taking out this x_0 behavior as

$$K_B(\vec{x}, x_0; \vec{y}) = \lim_{y_0 \to 0} (y_0)^{-\Delta_+} G(x,y), \tag{12.5}$$

and we can check that we obtain the same result as we found before in (11.11). As we saw before, near the boundary, the on-shell field obtained using K_B behaves as

$$\phi(\vec{x}, x_0) = \int d^d y K_B(\vec{x}, x_0; \vec{y}) \phi_0(\vec{y}) \sim x_0^{d-\Delta} \phi_0(\vec{x}) = x_0^{\Delta_-} \phi_0(\vec{x}). \tag{12.6}$$

Using the on-shell kinetic term for ϕ calculated in (11.19), we can calculate the one-point function for a free massless AdS scalar in the presence of nonzero ϕ_0 from

$$\langle e^{\int_{\partial M} \phi_0 \mathcal{O}} \rangle = e^{-S_{\text{sugra}}[\phi(\phi_0)]} \Rightarrow$$

$$\langle \mathcal{O}(\vec{x}) \rangle_{\phi_0} = -\frac{\delta S_{\text{sugra}}[\phi(\phi_0)]}{\delta \phi_0(\vec{x})}$$

$$= -\frac{C_d d}{2} \int d^d y \frac{\phi_0(\vec{y})}{|\vec{x} - \vec{y}|^{2d}}. \tag{12.7}$$

In the case of a massive scalar, the exponent $2d$ is replaced by $2\Delta_+$. This is an operator VEV in the presence of a source ϕ_0, so is not an independent quantity. Note that this result matches what one obtains from an expansion in the CFT:

$$\langle \mathcal{O}(\vec{x}) \rangle_{\phi_0} = \langle \mathcal{O}(\vec{x}) e^{\int \phi_0 \mathcal{O}} \rangle \approx \int d^d x' \phi_0(\vec{x}') \langle \mathcal{O}(\vec{x}) \mathcal{O}(\vec{x}') \rangle = -\frac{C_d d}{2} \int d^d \vec{x}' \frac{\phi_0(\vec{x}')}{|\vec{x} - \vec{x}'|^{2\Delta_-}}, \tag{12.8}$$

where in the last relation we have substituted the 2-point function, fixed from conformal invariance.

12.1 Mode and propagator calculations in Lorentzian signature for Poincaré coordinates

In Lorentzian signature, the situation is more complicated, since we have *independent* normalizable modes appearing as well. In Lorentzian signature, for boundary momenta $k^2 > 0$ (which can now be split into time and space components, by an abuse of notation calling the spatial momentum again $\vec{k}$, $k^2 = -\omega^2 + \vec{k}^2$), the solutions are the same as in the Euclidean case, (12.3) and the one with I_ν instead of K_ν, and no regular normalizable solutions.

However, in the case $k^2 = -\omega^2 + \vec{k}^2 < 0$, which is the physical one for positive m^2, since we want $k^2 = -m^2$ on-shell, we see that we need to analytically continue the Euclidean

AdS solutions to imaginary $|k|$, so the I_ν and K_ν Bessel functions are replaced by $J_{\pm\nu}$, giving the solutions

$$\Phi^\pm \propto e^{ik\cdot x}(x_0)^{d/2}J_{\pm\nu}(|k|x_0), \tag{12.9}$$

when $\nu = \sqrt{d^2 + 4m^2R^2}/2$ is not an integer. When ν is an integer, J_ν and $J_{-\nu}$ are equivalent, and the two independent solutions are Φ^+, with J_ν, and Φ^- with Y_ν instead of $J_{-\nu}$.

Since $J_\nu(x) \sim x^\nu$ near $x = 0$, Φ^- behaves like $(x_0)^{\Delta_-}$ and is non-normalizable, the same as in the Euclidean AdS solution. But now we also have Φ^+, which behaves like $(x_0)^{\Delta_+}$ and is normalizable, and also well defined in the interior, unlike the Euclidean case.

In conclusion, the new feature in the case of Lorentzian signature is the appearance of the normalizable and regular solutions.

As for the propagators (bulk-to-bulk and bulk-to-boundary), they are the obvious analytical continuation for $x_d \to it$ of the Euclidean Poincaré propagators.

12.2 Prescription for holography in Lorentzian signature for Poincaré coordinates

In the Euclidean case we had only non-normalizable modes in AdS, which meant that their meaning was clear, giving sources on the boundary for the operator dual to the bulk field. Now we have also *independent* normalizable modes, which have the natural interpretation as states in the boundary theory, but we need to understand the details of the mapping.

First of all, Wick rotating the partition function leads to the Lorentzian version of the partition function map,

$$Z_{\text{sugra}}[\phi_0] = e^{iS_{\text{sugra}}[\phi(\phi_0)]} = Z_{\text{CFT}}[\phi_0] = \langle s|e^{i\int_{\partial M}\phi_0\mathcal{O}}|s\rangle, \tag{12.10}$$

where unlike the Euclidean case we have considered the possibility that we have a nontrivial state $|s\rangle$, not just the vacuum, since as we saw we can obtain nontrivial states from normalizable modes. Then by differentiation, we obtain

$$\frac{\delta}{\delta\phi_0(\vec{x})}S_{\text{sugra}}[\phi(\phi_0)] = \langle\mathcal{O}(\vec{x})\rangle_{\phi_0}. \tag{12.11}$$

Therefore the one-point function is now in a nontrivial state, and is again obtained by differentiation with respect to the boundary source of the supergravity action. For the example of a free massless scalar, the on-shell solution is now a combination of the non-normalizable mode obtained by integrating a source with the bulk to boundary propagator, and an independent normalizable mode, i.e.

$$\phi(x_0, \vec{x}) = \phi_n(x_0, \vec{x}) + \int d^dy\, K_B(x_0, \vec{x}; \vec{y})\phi_0(\vec{y})$$

$$= \phi_n(x_0, \vec{x}) + c\int d^dy\, \frac{x_0^d}{(x_0^2 + (\vec{x} - \vec{y})^2)^d}\phi_0(\vec{y}). \tag{12.12}$$

For a massive scalar, we would change the power $d \to \Delta_+$. Since for the normalizable mode near the boundary $\phi_n(x_0, \vec{x}) \to (x_0)^d\tilde{\phi}_n(\vec{x})$, we can repeat the calculation of the on-shell action (11.19) in a slightly different way

$$\delta S(\phi) = \int_{\partial M} d\Sigma^0 \frac{\partial \phi}{\partial x_0} \delta \phi$$

$$\Rightarrow \frac{\partial \phi}{\partial x_0}\Big|_{\partial M} = d(x_0)^{d-1}\tilde{\phi}_n(\vec{x}) + cd(x_0)^{d-1}\int d^d\vec{x}' \frac{\phi_0(\vec{x}')}{|\vec{x} - \vec{x}'|^{2d}} \Rightarrow$$

$$\langle \tilde{\phi}_n|\mathcal{O}(\vec{x})|\tilde{\phi}_n\rangle_{\phi_0} = d\tilde{\phi}_n(\vec{x}) + cd\int d^dx' \frac{\phi_0(\vec{x}')}{|\vec{x} - \vec{x}'|^{2d}}. \tag{12.13}$$

Therefore, now, unlike the Euclidean case, the operator gets a VEV from two independent sources: the dependent VEV due to the explicit source ϕ_0, as in the Euclidean case, but now also from the excited state $|\tilde{\phi}_n\rangle$, which is a coherent state on the boundary in which states have nontrivial expectation values.

We can explain (12.13) also as an operator statement. Think first of the normalizable bulk mode as a quantum mode expansion in on-shell modes $\phi_{n,k}$,

$$\hat{\phi}_n = \sum_k [a_k \phi_{n,k} + a^+ \phi^*_{n,k}]. \tag{12.14}$$

The relation (12.13) says that this is mapped to the operator expansion (acting as a field)

$$\hat{\mathcal{O}} = \sum_k [b_k \tilde{\phi}_{n,k} + b_k^+ \tilde{\phi}^*_{n,k}], \tag{12.15}$$

where $\tilde{\phi}_{n,k}$ are arbitrary boundary terms for the normalizable modes, $\phi_n(x_0, \vec{x}) \rightarrow (x_0)^d \tilde{\phi}_n(\vec{x})$. Then we can identify the creation operators in the expansion $a_k = b_k$.

The short form of the above statement is that *non-normalizable modes are mapped to sources and normalizable modes to VEVs (or states)*, so that

$$\phi \sim \alpha_i(x_0)^{d-\Delta} + \beta_i(x_0)^\Delta, \tag{12.16}$$

where $\Delta = d$ for massless fields, and β_i is the coefficient of the normalizable mode, whereas α_i is the coefficient of the non-normalizable mode. This implies

$$H = H_{CFT} + \alpha_i \mathcal{O}_i, \tag{12.17}$$

and

$$\langle \beta_i|\mathcal{O}|\beta_i\rangle = \beta_i + (\alpha_i \text{ piece}). \tag{12.18}$$

The very important consequence is that if we set the non-normalizable mode to zero (meaning that we have no sources) and look at a bulk configuration or probe which corresponds to a combination of normalizable modes (maybe with non-normalizable components as well), it will get mapped to a VEV of the dual operator.

12.3 Mode and propagator calculations in global Lorentzian coordinates

We now consider physics in the global Lorentzian coordinates,

$$ds^2 = \frac{R^2}{\cos^2 \rho}(-dt^2 + d\rho^2 + \sin^2 \rho\, d\Omega^2_{d-1}). \tag{12.19}$$

As we saw, the boundary is at $\rho = \pi/2$ and the center of AdS is at $\rho = 0$. We want first to solve the KG equation, $(\Box - m^2)\Phi = 0$. We first parameterize the solutions by expanding

$$\Phi = e^{-i\omega t} Y_{l,\{m\}}(\Omega)\chi(\rho), \tag{12.20}$$

where $Y_{l,\{m\}}$ are spherical harmonics on S^{d-1}, satisfying

$$\nabla^2_{S^{d-1}} Y_{l,\{m\}} = -l(l+d-2)Y_{l,\{m\}}. \tag{12.21}$$

Then we find (after some algebra) that the equation satisfied by $\chi(\rho)$ is

$$\frac{1}{(\tan \rho)^{d-1}} \partial_\rho \left[(\tan \rho)^{d-1} \partial_\rho \right] + \left[\omega^2 - l(l+d-2)\csc^2 \rho - m^2 \sec^2 \rho \right] \chi = 0. \tag{12.22}$$

The solutions relevant for AdS/CFT must be regular at the origin, such that the on-shell boundary term $\int d\Omega dt \sqrt{-g} g^{\rho\rho} \Phi \partial_\rho \Phi$ gives no contribution at $\rho = 0$. One finds that of the two possible solutions, only one satisfies this condition,

$$\Psi_1 = e^{-i\omega t} Y_{l,\{m\}}(\Omega)(\cos \rho)^{\Delta_+}(\sin \rho)^l {}_2F_1 \left(\frac{\Delta_+ + l + \omega}{2}, \frac{\Delta_+ + l - \omega}{2}; l + \frac{d}{2}; \sin^2 \rho \right). \tag{12.23}$$

At the boundary, we find two possible solutions, written as a function of $\cos^2 \rho$ instead of $\sin^2 \rho$, namely

$$\Phi^+ = e^{-i\omega t} Y_{l,\{m\}}(\Omega)(\cos \rho)^{\Delta_+}(\sin \rho)^l$$
$$\times {}_2F_1 \left(\frac{\Delta_+ + l + \omega}{2}, \frac{\Delta_+ + l - \omega}{2}; \Delta_+ + 1 - \frac{d}{2}; \cos^2 \rho \right),$$

$$\Phi^- = e^{-i\omega t} Y_{l,\{m\}}(\Omega)(\cos \rho)^{\Delta_-}(\sin \rho)^l$$
$$\times {}_2F_1 \left(\frac{\Delta_- + l + \omega}{2}, \frac{\Delta_- + l - \omega}{2}; \Delta_- + 1 - \frac{d}{2}; \cos^2 \rho \right). \tag{12.24}$$

We see that $\Phi^\pm \sim (\cos \rho)^{\Delta_\pm}$ at the boundary, so Φ^+ is a normalizable mode, whereas Φ^- is a non-normalizable mode. The solution that is well behaved at $\rho = 0$, Ψ_1, is a combination of the two solutions defined at the boundary,

$$\Psi_1 = C^+ \Phi^+ + C^- \Phi^-, \tag{12.25}$$

where

$$C^+ = \frac{\Gamma\left(l + \frac{d}{2}\right)\Gamma(-\nu)}{\Gamma\left(\frac{\Delta_- + l + \omega}{2}\right)\Gamma\left(\frac{\Delta_- + l - \omega}{2}\right)},$$

$$C^- = \frac{\Gamma\left(l + \frac{d}{2}\right)\Gamma(\nu)}{\Gamma\left(\frac{\Delta_+ + l + \omega}{2}\right)\Gamma\left(\frac{\Delta_+ + l - \omega}{2}\right)}. \tag{12.26}$$

We see that in general the regular solution is non-normalizable, since it has a Φ^- component. But in the particular case that $C^- = 0$ we also have normalizable modes, since then

Ψ_1 has no non-normalizable component. The condition requires that one of the gamma functions in the denominator has a negative integer argument, i.e.

$$\omega_{nl} = \pm(\Delta_+ + l + 2n), \tag{12.27}$$

where n is a natural number (positive integer or zero). At these particular frequencies, we have normalizable modes, behaving at the boundary as $(\cos \rho)^{\Delta_+}$. For these particular ω_{nl}s, the solutions $\Phi^\pm$ can be written in terms of Jacobi polynomials:

$$\Phi^\pm_{nl\{m\}} = e^{-i\omega t} Y_{l,\{m\}}(\Omega)(\cos \rho)^{\Delta_\pm}(\sin \rho)^l P_n^{l+d/2-1,\nu}(\cos 2\rho). \tag{12.28}$$

The normalizable mode is called $\phi_{nl\{m\}}$.

Therefore, whereas in Poincaré coordinates the Lorentzian wave equation has continuous normalizable as well as non-normalizable solutions, in global coordinates the general solution is non-normalizable, but particular cases are normalizable.

The bulk-to-bulk propagator (Green's functions) can be written as an infinite sum over the normalizable modes using the general Green's function formula

$$iG(x,y) = \int \frac{d\omega}{2\pi} \sum_{nl\vec{m}} e^{i\omega(t-t')} \frac{\phi^*_{nl\vec{m}}(\vec{x})\phi_{nl\vec{m}}(\vec{y})}{\omega_{nl}^2 - \omega^2 - i\epsilon}. \tag{12.29}$$

The sum can be explicitly performed, with the result

$$iG(x,y) = \frac{C_B}{\left(\cosh^2 \frac{s}{R}\right)^{\frac{\Delta_+}{2}}} {}_2F_1\left(\frac{\Delta_+}{2}, \frac{\Delta_+ + 1}{2}; \nu + 1; \frac{1}{\cosh^2 \frac{s}{R}} - i\epsilon\right). \tag{12.30}$$

Here C_B is a constant and s is the *geodesic distance* (minimal distance between two points in curved space) in AdS, satisfying

$$\cosh\left(\frac{s}{R}\right) = \frac{\cos(t - t') - \sin \rho \sin \rho' \Omega \cdot \Omega'}{\cos \rho \cos \rho'}. \tag{12.31}$$

We can obtain the bulk-to-boundary propagator as a limit of the bulk-to-bulk propagator. Indeed, we know that a bulk field is obtained as $\phi(x) = \int_{\partial M} db K_B(b,x)\phi_0(b)$ (where b is a boundary point (t, Ω) and x is a bulk point) and that it should satisfy $\phi \to (\cos \rho)^{\Delta_-}\phi_0(b)$ for $\rho \to \pi/2$, i.e. behave like the non-normalizable mode. On the other hand, we have also $\phi(x) = \int_M dy \sqrt{-g}\, iG(x,y)$, and since at the boundary $\sqrt{-g} \propto \cos \rho^{-d}$, we must have $iG(x,y) \cos \rho^{-d} \propto K_B(b,x)$. Then we can define the bulk-to-boundary propagator by taking out this scaling factor, as

$$K_B(b,x) = 2\nu R^{d-1} \lim_{\rho \to \pi/2} (\cos \rho)^{-\Delta_-} iG(x,y). \tag{12.32}$$

The numerical prefactor is just due to the change in normalization. One obtains from (12.30) after the limit

$$K_B(b,x) = C_B \left[\frac{\cos \rho'}{\cos(t - t') - \sin \rho' \Omega \cdot \Omega' + i\epsilon}\right]^{\Delta_+}. \tag{12.33}$$

One can also take the limit on (12.29) and obtain

$$K_B(b,x) = 2\nu R^{d-1} \int \frac{d\omega}{2\pi} \sum_{nl\vec{m}} e^{i\omega(t-t')} \frac{k_{nl} Y^*_{l\vec{m}}(\Omega)\phi_{nl\vec{m}}(\vec{x})}{\omega_{nl}^2 - \omega^2 - i\epsilon}. \tag{12.34}$$

We can also take the second point on the boundary, and find the boundary-to-boundary propagator,

$$G_\partial(b, b') \propto \lim_{\rho' \to \pi/2} \cos \rho'^{-\Delta_+} K_B(b, x').$$ (12.35)

This gives for the sum formula

$$G_\partial(b, b') \propto \int_{\partial M} \frac{d\omega}{2\pi} \sum_{nl\vec{m}} e^{i\omega(t-t')} \frac{k_{nl}^2 Y_{l\vec{m}}^*(\Omega) Y_{l\vec{m}}(\Omega')}{\omega_{nl}^2 - \omega^2 - i\epsilon},$$ (12.36)

and for the final formula

$$G_\partial(b, b') \propto \frac{1}{\left[(\cos(t - t') - \Omega \cdot \Omega')^2 + i\epsilon\right]^{\frac{\Delta_+}{2}}}.$$ (12.37)

Note that the interval $x_{12}^2 = |x_1 - x_2|^2$ in Poincaré coordinates becomes, when transforming to global coordinates,

$$x_{12}^2 = \frac{2(\cos(t_1 - t_2) - \Omega \cdot \Omega')}{(\cos \tau_1 - \Omega_1^d)(\cos \tau_2 - \Omega_2^d)},$$ (12.38)

so we can define a natural interval $s_{12}^2 \equiv \cos(t_1 - t_2) - \Omega \cdot \Omega'$. The boundary-to-boundary propagator is written in terms of it, but note that it is not quite the Poincaré propagator!

The moral we must take from this calculation is that between coordinate systems, and in particular between the Poincaré patch and the global coordinates, a simple change of coordinates does not take us from one propagator to another, we must define all objects from the start in every coordinate set.

12.4 Holography in global Lorentzian coordinates: interpretation

As in the global Euclidean case, the Lorentzian global AdS has as a boundary the cylinder $S_3 \times \mathbb{R}_t$, and the conformal field theory in $\mathbb{R}^4$ is mapped to the cylinder, and dimensionally reduced on S^3 to a quantum mechanical Hamiltonian (for just the time direction).

Non-normalizable AdS modes, obtained for general frequencies ω correspond, as before, to sources for the conformal field theory operators, whereas at special AdS frequencies, we have normalizable modes that correspond to the states of the conformal field theory on the cylinder, with discrete energies

$$\omega_{nl} = \Delta + 2n + l.$$ (12.39)

Important concepts to remember

- Solutions to the KG equation in Euclidean Poincaré AdS space, are: a non-normalizable solution near the boundary, written in terms of the Bessel function K_ν, and a solution in terms of I_ν which is normalizable at the boundary but blows up in the bulk, hence is excluded.

- The bulk-to-bulk propagator is found by integrating a product of the two solutions, and the bulk-to-boundary propagator is a limit of the bulk-to-bulk propagator.
- In Euclidean Poincaré AdS there is a one-point function, but it is a dependent one, induced by a source ϕ_0 for the operator $\mathcal{O}$.
- In the Lorentzian Poincaré case, the solutions to the KG equation are written in terms of $J_{\pm\nu}$, and one solution is non-normalizable, but now there is also a normalizable regular solution.
- The Lorentzian propagators are analytical continuations of the Euclidean propagators.
- Besides the field in the bulk induced by a source on the boundary, now there is a normalizable mode component, corresponding to a state on the boundary.
- Therefore, non-normalizable modes correspond to sources, and normalizable modes to states or operator VEVs.
- In global Lorentzian coordinates, for general AdS frequencies, we only have a regular non-normalizable mode (more precisely, a combination of the normalizable and non-normalizable modes). For special frequencies, $\pm(\Delta_+ + l + 2n)$, we have regular normalizable modes.
- The bulk-to-bulk propagator can be written as a sum over normalizable modes.
- The propagators in global and Poincaré coordinates are not related simply by a change of coordinates, but involve extra factors.

References and further reading

The Lorentzian version of AdS/CFT was proposed in [66] and developed in [65]. In [66] we can also find mode calculations in Lorentzian AdS, whereas in [67] we can find more details on the mode and propagator calculations in the same.

Exercises

1. Complete the missing steps in the derivation of the bulk-to-bulk propagator (12.2).
2. Write down explicit formulas for the bulk-to-bulk and bulk-to-boundary propagators in Lorentzian Poincaré coordinates and write them in terms of the solutions to the homogeneous KG equation.
3. Check that the equation for $\chi(\rho)$ for solutions to the KG equation in global coordinates is (12.22).
4. Check that (12.23) satisfies (12.22) and the condition that the boundary term $\int d\Omega dt \sqrt{-g} g^{\rho\rho} \Phi \partial_\rho \Phi$ at $\rho = 0$ is zero.
5. Prove that (12.30) is a bulk-to-bulk propagator for AdS space, i.e. that it satisfies the Poisson equation in Lorentzian AdS space in global coordinates.
6. Prove that the distance s between two points in AdS defined by (12.31) is indeed a geodesic distance, i.e. a minimum distance between two points.

Until now we have described the propagation and correlations of fundamental objects, but one of the areas where AdS/CFT can be tested is in obtaining the mapping of solitonic objects on the two sides of the correspondence. In gauge theories we have objects like instantons and baryons which are solitonic in nature, so must correspond to some other solitonic objects in the AdS space, and we will see that natural candidates for these solitonic objects are branes.

In this chapter we analyze three types of solitonic objects: instantons, which are Euclidean solutions of pure Yang–Mills, baryons, which can be thought of as solitons in theories with quarks, and fuzzy spheres, which are ground state (soliton) solutions in $SU(N)$ gauge theories with $N \to \infty$. They will all be described as branes in the gravity dual $AdS_5 \times S^5$ space.

13.1 Instantons vs. D-instantons

Instantons in gauge theories

The most common soliton in gauge theories is the instanton, defined in $SU(2)$ theories. In theories with bigger gauge groups ($G \supset SU(2)$), the instantons are still the $SU(2)$ BPST instantons (after Belavin, Polyakov, Schwartz, and Tyupkin; its quantum properties were described by 't Hooft) described here, embedded in the higher gauge group.

The integral of Tr $[F_{\mu\nu} * F^{\mu\nu}] = $ Tr $[\epsilon^{\mu\nu\rho\sigma} F_{\mu\nu} F_{\rho\sigma}]$ is a topological invariant. We can see this since the integrand is a total derivative. In form language, Tr $[F_{\mu\nu} * F^{\mu\nu}] = 4$Tr $[F \wedge F]$ and then

$$\text{Tr } [F \wedge F] = d\mathcal{L}_{\text{CS}} = d\left[AdA + \frac{2}{3}A \wedge A \wedge A\right], \tag{13.1}$$

which means that it is a topological invariant. We mentioned in Chapter 1 that the above Chern–Simons Lagrangean, when multiplied by $g^2/4\pi^2$, gives a topological number called a winding number. Therefore the object

$$n = \frac{g^2}{16\pi^2} \int d^4x \text{Tr } \left[F_{\mu\nu} * F^{\mu\nu}\right] \tag{13.2}$$

is a topologically invariant number, called the instanton number or the *Pontryagin index*. Moreover, considering a 4-volume bounded by $x_4 = -\infty$ and $x_4 = +\infty$, the instanton number is the difference between the winding numbers at $-\infty$ and $+\infty$.

On the other hand, we can always write (in both Euclidean and Minkowski signature)

$$\frac{1}{4g^2} \int d^4x (F^a_{\mu\nu})^2 = \int d^4x \left[\frac{1}{4g^2} F^a_{\mu\nu} * F^{a\mu\nu} + \frac{1}{8g^2}(F^a_{\mu\nu} - *F^a_{\mu\nu})^2 \right], \qquad (13.3)$$

but only in Euclidean space can we have solutions to the self-duality constraint, $F^a_{\mu\nu} = *F^a_{\mu\nu}$. In that case, the above rewriting means that we have a BPS bound, which is minimized by the instanton solutions to the self-duality constraint.

The instanton solution is

$$A^a_\mu = \frac{2}{g} \frac{\eta^a_{\mu\nu}(x - x_i)_\nu}{(x - x_i)^2 + \rho^2}, \qquad (13.4)$$

where x_i is the instanton position and $\eta^a_{\mu\nu}$ is the 't Hooft symbol,

$$\eta^a_{ij} = \epsilon^{aij}; \quad \eta^a_{i4} = \delta^a_i; \quad \eta^a_{4i} = -\delta^a_i. \qquad (13.5)$$

Here ρ is the width of the instanton profile. A short calculation gives for the YM Lagrangean on the solution (Tr $(T^a T^b) = -\delta^{ab}/2$):

$$-\frac{1}{2}\text{Tr}\left[F_{\mu\nu} F^{\mu\nu} \right] = \frac{48}{g^2} \frac{\rho^4}{[(x - x_i)^2 + \rho^2]^4}. \qquad (13.6)$$

Integrating this, we find that the action is independent of ρ^2, giving the topological invariant

$$S_{\text{inst}} = \frac{8\pi^2}{g^2}. \qquad (13.7)$$

Thus the instanton is a solution to the Euclidean action, and as such does not have the interpretation of a real particle, rather it gives transition probabilities. Transition probabilities between static configurations are given by Euclidean path integrals bounded by the static configurations, which are given in the first approximation by the configuration of minimum action in a given topological sector (for a given Pontryagin index), i.e. the instanton. Thus the path integral giving the probability for tunneling transition between different winding numbers at $x_4 = -\infty$ and $x_4 = +\infty$ is approximated by $e^{-S_{\text{inst}}}$.

D-instantons in string theory

D-p-branes are objects with extension in p spatial dimensions and time, on which open strings can end, thus with Neumann boundary conditions in $p + 1$ directions and Dirichlet boundary conditions in the others. But one possibility that still exists is to consider Dirichlet boundary conditions in all dimensions, *including time*, thus obtaining a D(-1)-brane, or D-instanton.

As we saw, a D-p-brane is a source for a $p + 1$-form A_{p+1}, therefore a D-instanton is a source for the type IIB axion scalar a. Since in general in flat background $e^{-2\phi} = H_p^{(p-3)/2}$

and $A_{01...p} = (H_p^{-1} - 1)$, for a D-instanton we have $e^\phi = H_{-1}$ and $a - a_\infty = H_{-1}^{-1} - 1 \propto e^{-\phi} - 1/g_s$.

Now consider D-instantons in $AdS_5 \times S^5$, located at a point in AdS_5. The solution near the boundary of AdS_5 ($x_0 = 0$) was found to be

$$
e^\phi = g_s + \frac{24\pi}{N^2} \frac{x_0^4 \tilde{x}_0^4}{[\tilde{x}_0^2 + |\vec{x} - \vec{x}_a|^2]^4} + \ldots
$$
$$
a = a_\infty + e^{-\phi} - \frac{1}{g_s}. \tag{13.8}
$$

We will not reproduce the calculation here. Here $\tilde{x}_0$ is the position of the D-instanton in the fifth coordinate.

D-instantons as instantons

We now show that we can derive the instanton profile from the D-brane profile using the AdS/CFT prescription from the previous chapters. The 5-dimensional dilaton action contains a canonical kinetic term

$$
S = -\frac{1}{4\kappa_5^2} \int d^5x \, \sqrt{g} g^{\mu\nu} \partial_\mu \phi \partial_\nu \phi \,, \tag{13.9}
$$

where κ_5 is the 5-dimensional Newton constant κ_N. Varying this on-shell action, we get

$$
\delta S = -\frac{1}{2\kappa_5^2} \int d^4x \, \frac{R^3}{z^3} \delta\phi \, \partial_z \phi|_{z=0} \,. \tag{13.10}
$$

Using $R^3 = \kappa_5^2 N^2 / (4\pi^2)$ and the dilaton profile of the D-instanton (13.8), we obtain

$$
\frac{\delta S}{\delta \phi_0(\vec{x})} = -\frac{48}{4\pi g_s} \frac{\tilde{z}^4}{[\tilde{z}^2 + |\vec{x} - \vec{x}_a|^2]^4}. \tag{13.11}
$$

On the other hand, we saw in Chapter 10 that the dilaton ϕ couples through AdS/CFT to the Lagrangean itself, i.e. the operator $-\mathrm{Tr} \, [F_{\mu\nu}^2]/(2g_{YM}^2)$. That means that

$$
\frac{\delta S}{\delta \phi_0(\vec{x})} = \frac{1}{2g_{YM}^2} \langle \mathrm{Tr} \, [F_{\mu\nu}^2(\vec{x})] \rangle. \tag{13.12}
$$

Since we have $4\pi g_s = g_{YM}^2$, equating (13.11) with (13.12) gives

$$
-\frac{1}{2g_{YM}^2} \langle \mathrm{Tr} \, [F_{\mu\nu}^2(\vec{x})] \rangle = \frac{48}{g_{YM}^2} \frac{\tilde{z}^4}{[\tilde{z}^2 + |\vec{x} - \vec{x}_a|^2]^4} \,, \tag{13.13}
$$

which is exactly the instanton background, if we identify the instanton scale ρ with the D-instanton position in the fifth dimension, $\tilde{z}$. That is an example of the equivalence of the fifth dimension (radius) with the energy scale in the field theory.

13.2 Baryons in gauge theories and via AdS/CFT

Baryons in gauge theories, QCD and $\mathcal{N} = 4$ SYM

Baryons are gauge invariant objects made up of more than two quarks in gauge theories. In QCD and in general $SU(N_c)$ gauge theories, we can always define the mesons $M^{IJ} = \bar{q}_i^I q^{Ji}$ (here I, J are flavor indices), since the $SU(N)$ groups admit the invariant tensor $\delta_i^{\bar{j}}$ connecting the fundamental N representation and the antifundamental (conjugate) $\bar{N}$ representation.

For real world QCD, the $SU(3)_c$ group admits also the invariant ϵ_{ijk}, with i, j, k in the fundamental representation, so one can define the baryons

$$B^{IJK} = \epsilon_{ijk} q^{Ii} q^{Jj} q^{Kk} , \tag{13.14}$$

where I, J, K are flavor indices ($SU(N_f)$) and i, j, k are $SU(3)_c$ indices.

For a general $SU(N_c)$ theory, we can define baryons by contracting N quarks with the invariant tensor $\epsilon_{i_1...i_N}$,

$$B^{I_1...I_N} = \epsilon_{i_1...i_N} q^{I_1 i_1} \ldots q^{I_N i_N}. \tag{13.15}$$

In a $SU(N_c)$ theory without quarks, like $\mathcal{N} = 4$ SYM, we can still define a *baryon vertex*, connecting N external (very heavy, non-dynamical) quarks, introduced by hand in the theory, formally $\epsilon_{i_1...i_N}$ from the above. The baryon vertex has an energy, understood as a solitonic energy, even in the presence of external quarks only.

We mention in passing also the case of the $SO(2k)$ gauge group, for instance $\mathcal{N} = 4$ SYM theory with such a gauge group, where there is a true solitonic object, made up of gluons only (no external quarks), the Pfaffian particle $\text{Pf}(\Phi) = 1/k! \, \epsilon^{a_1...a_{2k}} \Phi_{a_1 a_2} \ldots \Phi_{a_{2k-1} a_{2k}}$, since in this case it is clearer that this is a solitonic object. We will not be describing its gravity dual, though it is also a wrapped brane in a certain geometry.

Baryons as solitons: Skyrme model

To understand the baryons as solitonic objects, we can look at the Skyrme model. The Skyrme model is a low energy model for QCD, described in terms of the pions $\vec{\pi}$, the lowest energy excitations. Together with the σ, one constructs a group element in $SO(4) \simeq SU(2)_L \times SU(2)_R$, with generators $(1, \tau_i)$ (τ_i are the Pauli matrices),

$$U = \exp\left[\frac{i}{f_\pi}(\sigma + \vec{\pi} \cdot \vec{\tau})\right] , \tag{13.16}$$

transforming from the left in $SU(2)_L$ and from the right in $SU(2)_R$. Defining also $L_\mu = U^{-1}\partial_\mu U$, the low energy QCD action is written in terms of it, with kinetic term

$$\mathcal{L}_{\text{kin}} = \frac{f_\pi^2}{4} \text{Tr} \, [L_\mu L^\mu]. \tag{13.17}$$

Then one can add various higher order interaction terms to it that lead to the same fact, namely a topological soliton solution. The term originally considered by Skyrme was

$$\mathcal{L}_{\text{int}} = \frac{\epsilon^2}{4} \text{Tr} \left([L_\mu, L_\nu]^2 \right),$$ (13.18)

but a variety of other higher order completions work. There is a topological current

$$B^\mu = \frac{1}{24\pi^2} \epsilon^{\mu\nu\rho\sigma} \text{Tr} \left[L_\nu L_\rho L_\sigma \right],$$ (13.19)

with the *integer* topological charge (one can prove that the object below only admits integer values due to the topology of the embedding, though we will only suggest why it is so below)

$$B = \frac{1}{24\pi^2} \int d^3x \, \epsilon^{ijk} \text{Tr} \left[L_i L_j L_k \right].$$ (13.20)

The current B^μ is conserved, since from $L_\mu = U^{-1} \partial_\mu U$ we can prove that $\partial_{[\mu} L_{\nu]} = -L_{[\mu} L_{\nu]}$ directly, which in turn means that $\partial_\mu B^\mu$ is a sum of terms of the type $\epsilon^{\mu\nu\rho\sigma} \text{Tr} \left[L_\mu L_\nu L_\rho L_\sigma \right]$, which are zero by cyclicity of the trace and antisymmetry of $\epsilon^{\mu\nu\rho\sigma}$ under cyclical permutations. A "hedgehog" configuration

$$U = \exp \left[iF(r) \vec{n} \cdot \vec{\tau} \right], \quad n \equiv \frac{\vec{r}}{r},$$ (13.21)

can be found in the case with suitable nonlinear terms like the example above, that has a nonzero topological charge. This topological charge can be identified with the baryonic charge, hence baryons can be thought of as solitons of the low energy QCD fields, the pions. For small fields $\vec{\pi}$, we obtain for B (using $\text{Tr} \left[\sigma_a \sigma_b \sigma_c \right] = 2i\epsilon_{abc}$)

$$B \simeq \frac{1}{12\pi^2 f_\pi^3} \epsilon^{ijk} \epsilon_{abc} \partial_i \pi^a \partial_j \pi^b \partial_k \pi^c + \ldots,$$ (13.22)

and we see that it is cubic in the fields. Moreover, we see that it is a map between the $SU(2) = SO(3)$ of the gauge group with index $a = 1, 2, 3$ and the rotational $SO(3)$ at spatial infinity with index $i = 1, 2, 3$, which is characterized by an integer number for the times the gauge $SO(3)$ wraps the spatial $SO(3)$. This topological number is B.

In conclusion, in the Skyrme model the baryons appear as solitons of the low energy fields, the pions, and to first approximation are cubic in the pions.

Baryons as wrapped branes

To consider a baryonic vertex in $\mathcal{N} = 4$ SYM, we need to have external quarks. We saw that strings ending on D-branes have states with Chan–Patton indices, $|\bar{i}j\rangle$, that correspond to adjoint fields, since for $U(N)$ the adjoint is an $N \times N$ matrix, obtained as the representation $N \otimes \bar{N}$. Then it can be inferred that a field in the fundamental representation of the gauge group, i.e. a quark, would come from a string with only one end on a D-brane, and the other free. But if we want *external* quarks, we need a very massive, thus very long, string. In the context of AdS/CFT, a string stretching from the boundary at infinity to a point in the interior would qualify.

The baryon vertex must then be a place in the interior of AdS where N fundamental strings can terminate, which suggests it has to be a D-brane. These strings have (electric) charge and extend from points $x_1, \ldots, x_N$ at infinity to the interior, so the D-brane should be located in the middle.

The natural candidate then is a D5-brane (present in type IIB string theory), whose spatial dimensions wrap the S^5 in $AdS_5 \times S^5$ and situated at a point in AdS_5, more precisely $S^5 \times \mathbb{R}$, where $\mathbb{R}$ corresponds to a timelike curve in AdS_5. But in order to show this is a good baryonic vertex, we must show that this brane *needs N strings to end on it, and only N strings*, which would make it the sought-after solitonic object.

To see this, consider the fact that the self-dual 5-form field strength of type IIB theory, $F^+_{\mu_1 \ldots \mu_5}$, has N units of flux on S^5, since the S^5 was the sphere at infinity around the N D3-branes that were used to obtain $AdS_5 \times S^5$. That means that

$$\int_{S^5} \frac{F^+_5}{2\pi} = \int_{S^5} d^5 x \epsilon^{\mu_1 \ldots \mu_5} \frac{F^+_{\mu_1 \ldots \mu_5}}{2\pi} = N. \tag{13.23}$$

But we saw in Chapter 9 that on D-branes we have couplings $\int_{M_p} e^{\wedge F/2\pi} \wedge \sum_n A_n$, with A_n all the n-forms. In particular, for the D5-brane and the 4-form field, we have a term $\int_{M_6} F \wedge A^+_4 = \int_{M_6} A \wedge F^+_5$, written as

$$\frac{1}{2\pi} \int_{S^5 \times \mathbb{R}} d^6 x \epsilon^{\mu_1 \ldots \mu_6} A_{\mu_1} F^+_{\mu_2 \ldots \mu_6}. \tag{13.24}$$

Then because of the N units of F^+_5 flux, we have N units of electric (A) charge on S^5, but since the total electric charge must be zero on a closed space, there must be a source of $-N$ flux coming from somewhere. Therefore we see that we do need N strings (and only N strings), corresponding to N quarks, to come from infinity and end on the D5-brane, which completes the identification of the wrapped D5-brane as a baryonic vertex.

Finally, notice the scaling with N of the energy of the baryonic vertex. The tension of the D5-brane, as for all D-branes, behaves like $\propto 1/g_s$. But as $N \to \infty$, $g_s N$ is fixed and large. Then the tension goes as $1/g_s \sim N$. On the other hand, the volume of the S^5 on which the D5-brane is wrapped is $\propto R^5 \propto (g_s N)^{5/4}$, constant as $N \to \infty$. All in all, the energy of the baryonic vertex = wrapped D5-brane is tension $\times$ volume $\propto N$, which is the correct scaling expected for a baryonic vertex as $N \to \infty$.

13.3 The D3–D5 system and the fuzzy 2-sphere via AdS/CFT

Finally, the last kind of solitonic object we describe via AdS/CFT is one that appears only in large N theories, the fuzzy 2-sphere.

The fuzzy S^2

A fuzzy sphere is a kind of matrix approximation to a classical sphere. On a classical sphere, we can expand any field in spherical harmonics $Y_{lm}(\theta, \phi)$ of arbitrarily high l and

m, i.e. an infinite set, and equivalently describe the field on the classical sphere as the infinite set of coefficients of the Y_{lm}s. On a fuzzy sphere, imagine keeping only the lms up to a maximum number, such that the total number of Y_{lm}s is in one to one correspondence with an $N \times N$ matrix, i.e. N^2. Then the fuzzy sphere comes from a theory of $N \times N$ matrices, and in the $N \to \infty$ limit, is supposed to exactly describe the classical sphere.

The precise construction is as follows. We consider the $SU(2)$ algebra, defined by

$$[X^i, X^j] = i\epsilon^{ijk}X^k. \tag{13.25}$$

As we know, the X^is can be thought of as spins J_i, so the irreducible representations of $SU(2)$ are spin j representations, with

$$X^i X^i \equiv J^i J^i = j(j+1) \equiv \frac{R^2}{r^2} = \text{constant}, \tag{13.26}$$

and of dimension $N = 2j+1$. That means that we have representations in terms of $N \times N = (2j+1) \times (2j+1)$ matrices, satisfying $X^i X^i = R^2/r^2 = $ constant, and naturally invariant under $SU(2) = SO(3)$, which is the rotational symmetry invariance of the S^2, acting on X^i, which are thought of as embedding Euclidean coordinates for the S^2. Therefore we have found a way to describe the fuzzy sphere using the X^is. To take the classical limit, consider taking $N \to \infty$ as the physical radius r is kept constant. Then define $x_i \equiv 2X^i/N$, such that $x^i x^i = 1$ in the limit. This implies also

$$[x^i, x^j] = \frac{2}{N}i\epsilon^{ijk}x^k \to 0, \tag{13.27}$$

so we recover classical commuting coordinates on the sphere in this limit.

Now consider spherical harmonics made up of the Euclidean coordinates $X^i \equiv J^i$ exactly in the same way as we would write them for the classical S^2, namely

$$Y_{lm}(J^i) = f_{lm}^{((i_1 \ldots i_l))}(J^{(i_1} \ldots J^{i_l)} - \text{traces}). \tag{13.28}$$

Here $(())$ means symmetric traceless. Since $x^i = 2J^i/N$ becomes the classical coordinate, $Y_{lm}(J^i) \to (2/N)^l Y_{lm}^{\text{classical}}(x_i)$ in the limit. But now, since J^i are $N \times N = (2j+1) \times (2j+1)$ matrices, not all of the $Y_{lm}(J_i)$ are independent, only an N^2 set of them, given by $0 \leq l \leq 2j = N - 1$. Indeed, since for each l there are $2l + 1$ values for m giving independent Y_{lm}s, the total number of independent Y_{lm}s is $\sum_{l=0}^{N-1}(2l + 1) = N^2$, as it should be.

A general matrix in the adjoint of $U(N)$, like for instance fluctuations of some field around the fuzzy sphere solution (13.25), can be expanded in these spherical harmonics as

$$A = \sum_{l=0}^{N-1} \sum_{m=-l}^{l} a^{lm} Y_{lm}(J_i). \tag{13.29}$$

The last ingredient is to verify that the field theory of these $N \times N$ fluctuations reduces to a field theory of fluctuations on the classical S^2 in the limit, which is usually nontrivial, but depends on each theory.

The fuzzy S^2 funnel as a solution in $\mathcal{N} = 4$ SYM

In order to find the fuzzy S^2 as a solution to some field theory, we only need to have an $SU(N)$ gauge theory with $N \to \infty$, and a matrix field satisfying (13.25). This is obtained in $\mathcal{N} = 4$ SYM as follows. In the action for the scalars X^I, $i = 1, \ldots, 6$, have a potential $g^2/2 \sum_{I<J}[X^I, X^J]^2$, and a kinetic term $-1/2\partial_\mu X^I \partial^\mu X^I$, thus if all the other fields are zero we have an equation of motion

$$\partial_\mu \partial^\mu X^I = g^2 \left[[X^I, X^J], X_J\right]. \tag{13.30}$$

This can be solved by keeping only X^i, $i = 1, 2, 3$ nonzero and satisfying

$$X^i = R(z)J^i , \tag{13.31}$$

where J^i are $N \times N$ (where $N = 2j + 1$, and j is the spin, which can be half integer) dimensional representations of the $SU(2)$ algebra,

$$[J^i, J^j] = i\epsilon^{ijk}J^k , \tag{13.32}$$

and $R(z)$ is a function of only one of the spatial coordinates of the $\mathcal{N} = 4$ SYM gauge theory. Substituting this ansatz in the equation of motion, we get

$$R''(z)J^i = g^2 R(z)^3 J^i , \tag{13.33}$$

solved by

$$R(z) = \frac{\sqrt{2}}{g(z - z_0)} \Rightarrow X^i = \frac{\sqrt{2}}{g(z - z_0)}J^i. \tag{13.34}$$

This is a *fuzzy funnel* solution, since at fixed z we have a fuzzy S^2 defined by J^i, but whose radius depends on the spatial coordinate z as $R \sim 1/(z - z_0)$, being 0 at $z = \infty$ and infinite at $z = z_0$.

Fuzzy S^2 funnel from brane in gravity dual

We want to understand this fuzzy funnel from the point of view of the gravity theory in AdS space.

First of all, what is the interpretation in the asymptotically flat spacetime background on which the D3-branes giving $\mathcal{N} = 4$ live? At $z = z_0$ (set to zero in the following) we have two extra dimensions appearing, first as a sphere, that eventually gets an infinite radius. Together with the original 3+1 dimensions on the N D3-branes, whose near-horizon limit gives AdS/CFT, we have 5+1 dimensions, for the worldvolume of a D5-brane.

Therefore the interpretation is of D3-branes ending on D5-branes, with configuration (the number refers to whether the worldvolume spans that particular coordinate):

$$\begin{array}{cccccccc} D3 & 0 & 1 & 2 & 3 & & & \\ D5 & 0 & 1 & 2 & & 4 & 5 & 6 \end{array} \tag{13.35}$$

Here the coordinate "3" is z, since at a given z_0 we have the new worldvolume appearing. The coordinates 4, 5, 6 are written in spherical coordinates as r, θ, ϕ. The unit 2-sphere J_i is parameterized by θ, ϕ, and $r = R$ is the radius of the sphere.

Thus from the point of view of the D3-brane, we have $r = R(z)$, together with θ, ϕ parameterizing the emerging D5-brane worldvolume. But now we can switch viewpoints and consider physics from the point of view of the D5-brane. The D5-brane wraps an S^2, but since it must be stabilized by some flux on its worldvolume, we need to assume that the $U(1)$ gauge field A living on the worldvolume of the D5-brane has a (magnetic type of) flux with n units, i.e.

$$F = (2\pi\alpha')n \sin\theta d\theta \wedge d\phi. \tag{13.36}$$

The only other variable in the action is the profile $z = z(r)$.

The DBI action for the D5-brane in this background is then

$$S = T_5 \int dr\, d\theta\, d\phi\, d^3x \sqrt{-\det\left(g_{\mu\nu}\partial_a X^\mu \partial_b X^\nu + F_{ab}\right)}. \tag{13.37}$$

The determinant splits into three factors, one for the coordinates X^0, X^1, X^2, with determinant -1, one for the coordinate r, with determinant $1 + (z'(r))^2$, and one for the θ, ϕ directions, with determinant

$$\det\begin{pmatrix} r^2 & (2\pi\alpha'n)\sin\theta \\ -(2\pi\alpha')n\sin\theta & r^2\sin\theta \end{pmatrix} = \sin^2\theta(r^4 + (2\pi\alpha'n)^2). \tag{13.38}$$

Moreover, integrating T_5 (the 5-brane tension) over the S^2 volume gives us T_3, the 3-brane tension, so $T_5 \int \sin\theta d\theta d\phi = T_3$. Then we obtain

$$S = T_3 \int dr\, d^3x \sqrt{\left(1 + (z'(r))^2\right)\left(r^4 + (2\pi\alpha'n)^2\right)}. \tag{13.39}$$

The action is independent of z (depends only on z'), so by the usual argument we get the conservation of the "momentum" $\delta S/\delta z'$, i.e.

$$\frac{\delta S}{\delta z'(r)} \equiv T_3(2\pi\alpha')c \tag{13.40}$$

is constant in the "time" r (conserved), and we have parameterized the constant in a convenient way. Then we can set the constant $c = -n$ to find the particular solution

$$z'(r) = -2\pi\alpha' \frac{n}{r^2} \Rightarrow z(r) = \frac{2\pi\alpha'n}{r}. \tag{13.41}$$

From the point of view of the D5-brane, at $r = 0$ we have a spike $z \to \infty$, corresponding to the D3-brane, as expected from the general picture of a D3-brane ending on a D5-brane.

This analysis was in a flat background, but the same analysis follows in the AdS background. In fact, one can find it even in the more general D3-brane background, before taking the near-horizon limit,

$$ds^2 = H^{-1/2}(r)(-dt^2 + d\vec{x}_3^2) + H^{1/2}(r)(dr^2 + r^2 d\Omega_2^2 + d\vec{y}_3^2),$$
$$A_4^+ = (H^{-1} - 1)dt \wedge dx^1 \wedge dx^2 \wedge dz. \tag{13.42}$$

Here $\vec{x}_3 = (x_1, x_2, z)$, $\Omega_2 = \Omega_2(\theta, \phi)$, and $\vec{y}_3 = (x_7, x_8, x_9)$. In this background, we want to consider a D5-brane at $\vec{y} = 0$ and with the spike located at $r = 0$. The calculation proceeds in the same way as before, just with the harmonic function $H(r)$, whose explicit form we do not need, as we shall see. Also, there is now a nonzero WZ term to be added to the DBI action, coming from $\int F \wedge A_4^+$. The 4-form pulled back on the worldvolume gives $A_4^+ = (H^{-1} - 1)dt \wedge dx^1 \wedge dx^2 \wedge z'(r)dr$.

Then the action is

$$S = T_3 \int dr \, d^3x \left[\sqrt{\left(1 + H^{-1}(r)(z'(r))^2\right)\left(r^4 + H^{-1}(r)(2\pi\alpha'n)^2\right)} - 2\pi\alpha'n(H^{-1} - 1)z'(r) \right].$$

(13.43)

Again it is independent of z, leading to (13.40), with the same (13.41) solution. The harmonic function $H(r)$ did not appear in the solution, therefore we can take the near-horizon limit and find the same solution is valid in $AdS_5 \times S^5$.

In terms of the variable $U = r/\alpha'$ natural for AdS/CFT, we have

$$z(U) = \frac{2\pi n}{U} \Rightarrow U(z) = \frac{2\pi n}{z} ,$$

(13.44)

which up to a constant rescaling by numerical factors was the same solution obtained from the point of view of the D3-brane in (13.34).

In conclusion, we see that the fuzzy funnel solution of $\mathcal{N} = 4$ SYM is interpreted in the gravity dual as a D5-brane solution of a spike type, wrapping a transverse S^2.

Important concepts to remember

- An instanton is a topological soliton in four Euclidean dimensions, solution of the self-duality constraint and carrying an instanton number (Pontryagin index).
- The instanton minimizes the action in a sector with a given Pontryagin index, giving a BPS bound, and the instanton action via $e^{-S_{\text{inst}}}$ gives the transition probability between spatial (static) configurations with different winding number.
- A D-instanton is a D(-1)-brane, i.e. a point in spacetime, when all the spacetime dimensions have Dirichlet boundary conditions.
- A D-instanton in AdS_5 implies on the boundary an instanton VEV for the Lagrangean.
- Whereas mesons are quark–antiquark bilinear $M = \bar{q}_i q^i$, baryons in $SU(N_c)$ theories are antisymmetric products of N quarks, $B^{I_1...I_N} = \epsilon_{i_1...i_N} q^{I_1 i_1} \ldots q^{I_N i_N}$.
- In $SU(N_c)$ gauge theories with external quarks only, we can define a baryonic vertex as the object connecting N external quarks, and it has an intrinsic energy, scaling as N in the large N limit.
- The Skyrme model is a low energy model for QCD in terms of the pion field π^i (and a σ), described by $U = \exp\left[i(\sigma + \vec{\pi} \cdot \vec{\tau})/f_\pi\right]$, a kinetic term $f_\pi^2 q/4\text{Tr}\,[L_\mu L^\mu]$, with $L_\mu = U^{-1}\partial_\mu U$, and various possible interaction terms.
- There is a topological charge $U = \exp\left[iF(r)(\vec{r}/r) \cdot \vec{\tau}\right]$, carried by "hedgehog" configurations and which counts wrappings of the $SU(2) = SO(3)$ gauge group over the rotational $SO(3)$ at spatial infinity.

- The topological charge is identified with the baryon charge, thus baryons appear as solitons of the low energy fields of QCD, the pions.
- In AdS/CFT, the baryonic vertex is a D5-brane wrapped on S^5, that needs N strings to end on it and cancel its charge. It has energy scaling as N for $N \to \infty$.
- A fuzzy S^2 is defined by the $SU(2)$ algebra $[X^i, X^j] = i\epsilon^{ijk}X^k$, whose spin j representations satisfy $X^iX^i = $ constant and whose rescaled $x_i = 2X^i/N$ ($x^ix^i = 1$) commute in the $N \to \infty$ limit.
- Spherical harmonics $Y_{lm}(x_i)$ are defined by analogy with their classical counterparts, but now only N^2 of them are independent, so we have a fuzzy approximation for the sphere.
- $\mathcal{N} = 4$ SYM has a fuzzy funnel solution, with radius $R(z) \propto 1/z$.
- In the gravity dual, the fuzzy S^2 is described by a D5-brane spike wrapping a transverse S^2.

References and further reading

The matching of instantons with D-instantons was understood in [65], and the description of baryons as coming from wrapped branes was understood in [68].

Exercises

1. Prove (13.6) and integrate it to obtain the instanton action.
2. Prove $R^3 = \kappa_5^2 N^2/(4\pi^2)$ and then using (13.8), prove (13.11).
3. Expand the action (13.17) + (13.18) up to 4th order in π^is and σ.
4. Substitute (13.21) into (13.20) to find the topological charge B in terms of $F(r)$. What are the needed conditions on $F(r)$?
5. Consider an $AdS_6 \times S^4$ background in type IIA theory, sourced by a 4-form field strength $F_{(4)}$. Repeat the argument in the text to find a baryonic N-vertex in the corresponding $SU(N)$ field theory.
6. Consider the spin $j = 1$ representation of $SU(2)$. Construct the nine independent $Y_{lm}(J_i)$s and check that one can decompose a general 3×3 matrix in terms of them.
7. Consider an $SU(N)$ matrix theory with the potential

$$V = \text{Tr}\left[\frac{1}{2}\sum_{i=1,2,3}\left(\frac{\mu X^i}{3}\right)^2 + \frac{\mu i}{3}\sum_{j,k,l=1}^{3}X^jX^kX^l - \frac{1}{4}\sum_{j,k=1,2}[X_j, X_k]^2\right]. \quad (13.45)$$

Does it have a fuzzy sphere ground state?

Quarks and the Wilson loop

In this chapter we study how to introduce external quarks in $\mathcal{N} = 4$ SYM, and how to calculate the Wilson loop, an important quantity in gauge theories, whose VEV decides, among other things, if a theory is confining or not, being a kind of order parameter for the confinement phase transition.

14.1 External quarks in QCD: Wilson loops

External quarks

In real QCD we have dynamical quarks, which means they are light degrees of freedom appearing in the action. But we can also consider external quarks, i.e. (infinitely) heavy quarks, not in the action, that are introduced in the theory as external probes.

QCD is confining, which means light quarks are not free in the vacuum, they appear in pairs, each together with an antiquark. Thus even if we consider external quarks (that are not in the quantum theory), we do not expect to be able to put a single quark in the vacuum, we need at least two: a quark and an antiquark.

Since the external quarks are very heavy, they will stay fixed, i.e. the distance between q and $\bar{q}$ will stay fixed in time, as in Fig. 14.1a. The question then is how do we measure the interaction potential between two such quarks, $V_{q\bar{q}}(L)$? We need to define physical observables that can measure it. One such physical, gauge invariant object is called the Wilson loop.

Wilson lines and loops

We first define the path ordered exponential from x to y in a general gauge theory

$$\Phi(y, x; P) = P \, \exp \left\{ i \int_x^y A_\mu(\xi) d\xi^\mu \right\} \equiv \lim_{n \to \infty} \prod_n e^{iA_\mu(\xi_n^\mu - \xi_{n-1}^\mu)}, \qquad (14.1)$$

where $A_\mu \equiv A_\mu^a T_a$.

We start the analysis with a $U(1)$ **(abelian) gauge field** A_μ. Under a gauge transformation $\delta A_\mu = \partial_\mu \chi$,

$$e^{iA_\mu d\xi^\mu} \to e^{iA_\mu d\xi^\mu + i\partial_\mu \chi d\xi^\mu} = e^{iA_\mu d\xi^\mu} e^{i\chi(x+dx) - i\chi(x)}, \qquad (14.2)$$

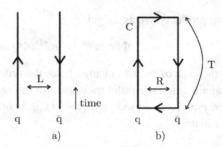

Figure 14.1   a) Heavy quark and antiquark staying at a fixed distance L; b) Wilson loop contour C for the calculation of the quark–antiquark potential.

the path ordered exponential transforms as

$$\Phi(y, x; P) = \Pi e^{iA_\mu d\xi^\mu} \rightarrow \Pi(e^{iA_\mu d\xi^\mu} e^{i\chi(x+dx) - i\chi(x)})$$
$$= e^{i\chi(y)} \left(\Pi e^{iA_\mu d\xi^\mu} \right) e^{-i\chi(x)} = e^{i\chi(y)} \Phi(y, x; P) e^{-i\chi(x)}. \tag{14.3}$$

If we have a complex field ϕ charged under this $U(1)$, i.e. transforming as

$$\phi(x) \rightarrow e^{i\chi(x)} \phi(x), \tag{14.4}$$

then the multiplication by $\Phi(y, x; P)$ gives

$$\Phi(y, x; P)\phi(x) \rightarrow e^{i\chi(y)} \Phi(y, x; P) e^{-i\chi(x)} e^{i\chi(x)} \phi(x) = e^{i\chi(y)} \left(\Phi(y, x; P)\phi(x) \right), \tag{14.5}$$

thus it defines parallel transport, i.e. the field $\phi(x)$ was parallel transported to the point y.

On the other hand, for a closed curve, i.e. for $y = x$, we have

$$\Phi(x, x; P) \rightarrow e^{i\chi(x)} \Phi(x, x; P) e^{-i\chi(x)} = \Phi(x, x; P), \tag{14.6}$$

i.e. the path ordered exponential is a gauge invariant object.

We next consider a **nonabelian gauge field**, for which the gauge transformation is

$$A_\mu \rightarrow \Omega(x) A_\mu \Omega^{-1}(x) - i(\partial_\mu \Omega)\Omega^{-1}. \tag{14.7}$$

An infinitesimal transformation $\Omega(x) = e^{i\chi(x)}$ for small $\chi(x) = \chi^a T_a$ gives

$$\delta A_\mu = D_\mu \chi = \partial_\mu \chi - i[A_\mu, \chi], \tag{14.8}$$

which implies

$$e^{iA_\mu d\xi^\mu} \simeq (1 + iA_\mu d\xi^\mu) \rightarrow 1 + \Omega(iA_\mu d\xi^\mu)\Omega^{-1} + d\xi^\mu(\partial_\mu \Omega)\Omega^{-1}$$
$$= [e^{i\chi(x)}(1 + iA_\mu d\xi^\mu) + d\xi^\mu \partial_\mu e^{i\chi(x)}]e^{-i\chi(x)}$$
$$\simeq e^{i\chi(x+dx)}(1 + iA_\mu d\xi^\mu)e^{-i\chi(x)} \simeq e^{i\chi(x+dx)} e^{iA_\mu d\xi^\mu} e^{-i\chi(x)}, \tag{14.9}$$

where we have neglected terms of order $O(dx^2)$.

By taking products, we again get

$$\Phi(y, x; P) \rightarrow e^{i\chi(y)} \Phi(y, x; P) e^{-i\chi(x)}, \tag{14.10}$$

but unlike the case of the $U(1)$ gauge field, the order of the terms matters now. Therefore, $\Phi(y, x; P)$ again defines parallel transport, for the same reason.

However, now for a closed path ($y = x$), Φ is not gauge invariant any more, but rather gauge covariant:

$$\Phi(y, x; P) \rightarrow e^{i\chi(x)} \Phi(x, x; P) e^{-i\chi(x)} \neq \Phi(x, x; P). \tag{14.11}$$

But now the trace of this object is gauge invariant, since the trace is cyclic. Thus we define the *Wilson loop*

$$W(C) = \mathrm{Tr}\ \Phi(x, x; C), \tag{14.12}$$

which is gauge invariant and independent of the particular point x on the closed curve C, since

$$\mathrm{Tr}\ \left[e^{i\chi(x)} \Phi e^{-i\chi(x)} \right] = \mathrm{Tr}\ [\Phi]. \tag{14.13}$$

14.2 The Wilson loop and the interquark potential

In the abelian case, for $x = y$ (closed path) we can use the Stokes theorem to put Φ in an explicitly gauge invariant form,

$$\Phi_C = e^{i \int_{C = \partial \Sigma} A_\mu d\xi^\mu} = e^{i \int_\Sigma F_{\mu\nu} d\sigma^{\mu\nu}}. \tag{14.14}$$

In the nonabelian case, we can do something similar, but we have corrections. If we take a small square of side a in the plane defined by directions μ and ν, we get

$$\Phi_{\square_{\mu\nu}} = e^{ia^2 F_{\mu\nu}} + O(a^4). \tag{14.15}$$

Since $F_{\mu\nu}$ transforms covariantly:

$$F_{\mu\nu} \rightarrow \Omega(x) F_{\mu\nu} \Omega^{-1}(x), \tag{14.16}$$

then the Wilson loop, defined for convenience with a $1/N$ since there are N terms in the trace for a $SU(N)$ gauge field, becomes

$$W_{\square_{\mu\nu}} = \frac{1}{N} \mathrm{Tr}\ \{\Phi_{\square_{\mu\nu}}\} = 1 - \frac{a^4}{2N} \mathrm{Tr}\ \{F_{\mu\nu} F_{\mu\nu}\} + O(a^6), \tag{14.17}$$

where we do not have a sum over the indices μ, ν. Here $\mathrm{Tr}\ \{F_{\mu\nu} F_{\mu\nu}\}$ is a gauge invariant operator (even if it is not summed over μ, ν), thus to first nontrivial order this is explicitly gauge invariant, and moreover the term we obtain is the kinetic term in the action for A_μ. This is the first example of the fact that we can describe the dynamics of gauge theories from the Wilson loops.

The object of interest is therefore

$$W[C] = \text{Tr } P \exp\left[\int iA_\mu d\xi^\mu\right],$$ (14.18)

and for the calculation of the static quark–antiquark potential we are interested in a loop as in Fig. 14.1b, a rectangle with length T in the time direction and R in the spatial direction, with $T \gg R$.

One can prove that the VEV of the Wilson loop in Fig. 14.2 behaves (as $T \to \infty$) as

$$\langle W(C) \rangle_0 \propto e^{-V_{q\bar{q}}(R)T},$$ (14.19)

if $T \to \infty$. A simple way to understand this is as follows. Adding infinitely heavy quarks amounts to adding to the action a term $\int j^\mu A_\mu = \int dt[eA_0(x(q)) - eA_0(x(\bar{q}))]$, and $eA_0(x(q))$ is like the potential at the position of q. This gives a term in e^{iS} of $e^{iT(V_q - V_{\bar{q}})}$, where $T = $ time. When rotating to Euclidean space, we get $e^{-TV_{q\bar{q}}}$, where T is Euclidean time.

We mentioned that the Wilson loop provides a good criterion for confinement. The reason is related to the above asymptotic behavior of the Wilson loop VEV. Indeed, the statement of confinement is that there is a constant force that resists when pulling the quark and the antiquark away, therefore that

$$V_{q\bar{q}}(R) \sim \sigma R,$$ (14.20)

i.e. a linear potential, with σ called the (QCD) string tension. The "QCD string" is really a confined flux tube (approximated by a string) for the QCD color electric flux, as in Fig. 14.2. It is not a fundamental object, but an effective description due to the confinement, which forces the flux lines to stay (to be confined) in a tube.

On the other hand, for QED with infinitely massive (external) quarks, we have the Coulomb static potential

$$V_{q\bar{q}}(R) \sim \frac{\alpha}{R},$$ (14.21)

and this model is scale invariant and in fact conformal. This is the result expected in a conformal theory, since then the Wilson loop VEV for the rectangular contour can only depend on the scale invariant quantity T/R. Since the VEV is proportional to $\exp[-V_{q\bar{q}}T]$, we can only have $V_{q\bar{q}} \propto 1/R$. This is the kind of potential we therefore expect in $\mathcal{N} = 4$ SYM, which is a conformal theory.

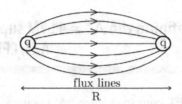

Figure 14.2   Between a quark and an antiquark in QCD, flux lines are confined: they exist in a flux tube.

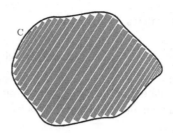

Figure 14.3 Approximation of a curve C by infinitely thin rectangles.

Then in a confining theory like QCD we get

$$\langle W(C)\rangle_0 \propto e^{-\sigma T \cdot R} = e^{-\sigma A}, \tag{14.22}$$

where $A = $ area, thus this behavior is known as the area law. In fact, since

$$W(C_1 \cup C_2) = W(C_1)W(C_2), \tag{14.23}$$

we can extend the area law to any smooth curve C, not just to the infinitely thin rectangle analyzed here, since we can approximate any area by such infinitely thin rectangles, as in Fig. 14.3.

Therefore, confinement means that for any smooth curve C,

$$\langle W(C)\rangle_0 \propto e^{-\sigma A(C)}. \tag{14.24}$$

On the other hand, in conformally invariant cases like QED with external quarks or our $\mathcal{N} = 4$ SYM with external quarks, we find, as we said, the scale invariant result for the infinitely thin curve

$$\langle W(C)\rangle_0 \propto e^{-\alpha \frac{T}{R}}, \tag{14.25}$$

and for more complicated curves we do not have an answer, but we just know that the answer must be scale invariant (independent of the overall size of the curve).

We can then think of $\langle W(C)\rangle_{0,T\to\infty}$ as an order parameter for the confinement/deconfinement transition, since $\langle W(C)\rangle_{0,T\to\infty} \to 0$ in the confined phase, whereas $\langle W(C)\rangle_{0,T\to\infty} \neq 0$ in a deconfined (Coulomb) phase.

Finally, although here we have only shown how to extract the quark–antiquark potential and the YM kinetic term from the Wilson loop and its VEVs, the Wilson loop is actually a very important object. We can in principle extract all the dynamics of the theory if we know the (complete operator) Wilson loop.

14.3 Defining the $\mathcal{N} = 4$ SYM supersymmetric Wilson loop via AdS/CFT

In the AdS/CFT correspondence, we obtain a $U(N)$ gauge group from a large number ($N \to \infty$) of D-branes situated at the same point. In this case, strings with two ends on different branes are massless, since there is no physical separation between the D-branes, and correspond to gauge fields, $A_\mu^a = (\lambda^a)_{ij}|i\rangle \otimes |j\rangle \otimes |\mu\rangle$.

If we consider $N+1$ D-branes, giving a $U(N+1)$ gauge group, and take one of the D-branes and separate it from the rest, as in Fig. 14.4, it means that we are breaking the gauge group, via a Higgs-like mechanism, to $U(N) \times U(1)$ (where $U(N)$ corresponds to the N D-branes that are still at the same point).

The strings that have one end on one of the N D-branes and one end on the extra D-brane will be massive, with *mass = string tension × D-brane separation*. These strings have a state

$$|i_0\rangle \otimes |i\rangle = |N+1\rangle \otimes |i\rangle, \qquad (14.26)$$

which is therefore in the fundamental representation of the unbroken $U(N)$ gauge group, since i is a fundamental index. The mass of the state is

$$M = \frac{1}{2\pi\alpha'}r = \frac{U}{2\pi}. \qquad (14.27)$$

This string behaves as a "W boson" from the point of view of the original $U(N+1)$ gauge group. In the Standard Model, the W boson is a vector (gauge) field made massive by the Higgs mechanism (breaking $SU(2)_L \times U(1)_Y \rightarrow U(1)_{em}$) by "eating" a degree of freedom of the scalar Higgs (a 4-dimensional massless vector has two degrees of freedom, whereas a massive vector has three). In our case the Higgs mechanism breaks $U(N+1) \rightarrow U(N) \times U(1)$, and the string mode (vector) eats scalar modes to become massive. On the other hand, the string state, or rather its endpoint with Chan–Patton factor $|i\rangle$, acts from the point of view of the $U(N)$ gauge theory as a source for the $U(N)$ gauge fields, or as a quark, therefore, in the fundamental representation of $U(N)$.

From (14.27), to get an infinite mass we need to take $U \rightarrow \infty$. Therefore, the introduction of an infinitely massive external quark is obtained by having a string stretched in AdS space, in the metric (10.14), between infinity in U and a finite point.

Since infinity in (10.14) is also where the $\mathcal{N}=4$ SYM gauge theory exists, we put the Wilson loop contour C at infinity, as a boundary condition for the string. So the string

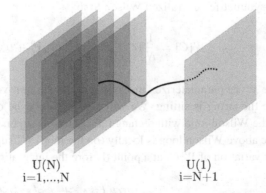

$$U(N) \qquad\qquad\qquad U(1)$$
$$i=1,...,N \qquad\qquad\quad i=N+1$$

Figure 14.4   One D-brane separated from the rest of the (N) D-branes acts as a probe on which the Wilson loop is located.

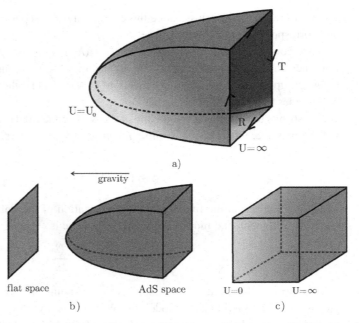

Figure 14.5 a) The Wilson loop contour C is located at $U = \infty$ and the string worldsheet ends on it and stretches down to $U = U_0$; b) In flat space, the string worldsheet would form a flat surface ending on C, but in AdS space 5-dimensional gravity pulls the string inside AdS; c) The free "W bosons" are strings that would stretch in all of the AdS space, from $U = \infty$ to $U = 0$, straight down, forming an area proportional to the perimeter of the contour C.

worldsheet stretches between the contour C at infinity down to a finite point in AdS, forming a smooth surface, as in Fig. 14.5a.

But there is a subtlety. Strings must also be situated on the S_5, parameterized by coordinates θ^I, since the dual of $\mathcal{N} = 4$ SYM is $AdS_5 \times S_5$, not just AdS_5. And θ^I correspond to the scalars X^I of $\mathcal{N} = 4$ SYM, which transform under the $SO(6)$ symmetry group (R symmetry of $\mathcal{N} = 4$ SYM and invariance symmetry of S_5). Because of that, we expect that the string worldsheet described above is not a source for the usual Wilson loop, but for the supersymmetric generalized Wilson loop:

$$W[C] = \frac{1}{N} \text{Tr} \; P \exp\left[\oint \left(iA_\mu \dot{x}^\mu + \theta^I X^I(x^\mu) \sqrt{\dot{x}^2} \right) d\tau \right], \qquad (14.28)$$

where $x^\mu(\tau)$ parameterizes the loop and θ^I is a unit vector that gives the position on S_5 where the string is sitting. We will consider only the case of $\theta^I = $ constant, since in this case the Wilson loop with rectangular contour is supersymmetric.

The above Wilson loop is locally (on the loop) 1/2 supersymmetric, since the supersymmetry variation of $W[C]$ at a point (before the integration) is given by

$$\sim W[C] \left(i\delta A_\mu \dot{x}^\mu + \theta^I \delta X^I(x^\mu) \sqrt{\dot{x}^2} \right), \qquad (14.29)$$

and putting it to zero turns into a condition on ϵ independent of the fermions ψ, as we can check from the SYM variations. The details are left as an exercise (Exercise 6). Then the variations at each point must commute and/or give the same result for the loop to be globally supersymmetric. This happens in the case of a straight line in $x^\mu(\tau)$ (thus for the infinitely thin rectangular Wilson loop it happens also) and constant θ^I. There are other contours that are supersymmetric, 1/2 or less so, for instance the circular Wilson loop, but we will not study them here.

From the above discussion, it follows that the prescription for calculating $\langle W[C] \rangle$ is as a partition function for the string with boundary on C. In the supergravity limit ($g_s \to 0$, $g_s N$ fixed and large) we obtain

$$\langle W[C] \rangle = Z_{\text{string}}[C] = e^{-S_{\text{string}}[C]}, \tag{14.30}$$

where S_{string} is the string worldsheet action, i.e. $1/(2\pi\alpha')\times$ the area of the worldsheet (area in $AdS_5 \times S_5$, not area of the 4-dimensional projection!). That, however, does not necessarily give the 4-dimensional area law for C, since the worldsheet has an area bigger than the 4-dimensional area enclosed by C. In fact, we expect that in this conformal case we *do not* obtain the area law, rather we obtain the conformal invariant law $\sim T/R$.

The string has tension, and it wants to have a minimum area. In flat space, that would mean that it would span just the flat surface enclosed by C, giving the area law (see Fig. 14.5b). However, in AdS space, we have a gravitational field

$$ds^2 = \alpha' \frac{U^2}{R^2/a'}(-dt^2 + d\vec{x}^2) + \dots \tag{14.31}$$

To understand the physics, we compare with the Newtonian approximation, though it is not a good approximation now, but we do get the correct qualitative picture,

$$ds^2 = (1 + 2V)(-dt^2 + \dots), \tag{14.32}$$

where V is the Newton potential. Newtonian gravity means that the string would go to the minimum V. In our case, that would mean the minimum position U. Therefore, the string worldsheet with boundary at $U = \infty$ "drops" down to $U = U_0$ as in Fig. 14.5b and is eventually stopped (held back) by its tension.

But the prescription is not complete yet, since the area of the worldsheet stretching from $U = \infty$ to $U = U_0$ is divergent, so naively we would get $\langle W[C] \rangle = 0$. In fact, we must remember that we said the string stretched between the $|i\rangle$ and $|N + 1\rangle$ D-branes, and therefore also between $U = \infty$ and $U = U_0$ in AdS_5, represents an infinitely massive "W boson," whose mass ϕ we must now subtract. The "free W boson" would stretch along all of AdS_5, thus from $U = \infty$ to $U = 0$, in a straight line, parallel with C, as in Fig. 14.5c. Thus the action that we must subtract is ϕl, where l is the length (perimeter) of the loop C and ϕ is the free W boson (free string) mass, $U/(2\pi)$. Then we have the AdS/CFT Wilson loop prescription

$$\langle W[C] \rangle = e^{-(S_\phi - l\phi)}. \tag{14.33}$$

14.4 Calculating the interquark potential

We now calculate explicitly the Wilson loop VEV for the contour C given by the infinitely thin rectangle, with $T \to \infty$, and a quark q at $x = -L/2$ and an antiquark $\bar{q}$ at $x = +L/2$. The metric of Euclidean AdS is

$$ds^2 = \alpha' \left[\frac{U^2}{R^2/\alpha'}(dt^2 + d\vec{x}^2) + \frac{R^2}{\alpha'}\frac{dU^2}{U^2} + \frac{R^2}{\alpha'}d\Omega_5^2 \right]; \quad R^2 = \alpha'\sqrt{4\pi g_s N}. \tag{14.34}$$

To calculate the Wilson loop VEV from AdS/CFT, we need to calculate the string action in this background. As we saw in Chapter 7, the *Euclidean space* Nambu–Goto action for the string is

$$S_{\text{string}} = \frac{1}{2\pi\alpha'} \int d\tau d\sigma \sqrt{\det(G_{MN}\partial_a X^M \partial_b X^N)}. \tag{14.35}$$

We choose a gauge where the worldsheet coordinates equal two spacetime coordinates, specifically $\tau = t$ and $\sigma = x$. This choice is known as a *static gauge*, and it is consistent to take it since we are looking for a static solution. Then we approximate the worldsheet to be translationally invariant in the time direction, which is only a good approximation if $T/L \to \infty$ (otherwise the curvature of the worldsheet near the corners becomes important). Since we are also looking at a static configuration, we have a single variable for the worldsheet, $U = U(\sigma)$, which becomes $U = U(x)$.

We calculate $h_{ab} = G_{MN}\partial_a X^M \partial_b X^N$ and obtain ($\tilde{R}^2 \equiv R^2/\alpha'$; $1 \equiv \tau, 2 \equiv \sigma$):

$$h_{11} = \alpha'\frac{U^2}{\tilde{R}^2}\left(\frac{dt}{d\tau}\right)^2 = \alpha'\frac{U^2}{\tilde{R}^2}; \quad h_{12} = 0,$$

$$h_{22} = \alpha'\frac{U^2}{\tilde{R}^2}\left(\frac{dx}{d\sigma}\right)^2 + \alpha'\frac{\tilde{R}^2}{U^2}\left(\frac{dU}{d\sigma}\right)^2 = \alpha'\left(\frac{U^2}{\tilde{R}^2} + \frac{\tilde{R}^2}{U^2}U'^2\right), \tag{14.36}$$

therefore

$$S_{\text{string}} = \frac{1}{2\pi}T \int dx \sqrt{(\partial_x U)^2 + \frac{U^4}{\tilde{R}^4}}, \tag{14.37}$$

and we have reduced the problem to a 1-dimensional mechanics problem.

We define U_0 as the minimum of $U(x)$ and $y = U/U_0$. Then we can check that the solution is defined by

$$x = \frac{\tilde{R}^2}{U_0} \int_1^{U/U_0} \frac{dy}{y^2\sqrt{y^4 - 1}}, \tag{14.38}$$

which gives $x(U, U_0)$ and inverted gives $U(x, U_0)$. To find U_0 we note that at $U = \infty$ we have $x = L/2$, therefore

$$\frac{L}{2} = \frac{\tilde{R}^2}{U_0} \int_1^{\infty} \frac{dy}{y^2\sqrt{y^4 - 1}} = \frac{\tilde{R}^2}{U_0}\frac{\sqrt{2}\pi^{3/2}}{\Gamma(1/4)^2}. \tag{14.39}$$

Then from the Wilson loop prescription (14.33) and the general way to extract the interquark potential from the Wilson loop (14.19), we have

$$S_\phi - l\phi = TV_{q\bar{q}}(L), \tag{14.40}$$

and we regularize this formula (which subtracts two infinities) by integrating only up to U_{max}. We have $l \simeq 2T$ and the mass of the free string (i.e., string stretching from $U = 0$ to U_{max}) is written as

$$\phi = \frac{U_{max} - U_0}{2\pi} + \frac{U_0}{2\pi} = \frac{U_0}{2\pi} \int_1^{y_{max}} dy + \frac{U_0}{2\pi}. \tag{14.41}$$

Then the object in the exponential in (14.33) is (there is a factor of 2 since we integrate from U_{max} to U_0 and then from U_0 to U_{max})

$$TV_{q\bar{q}}(L) = T\frac{2U_0}{2\pi} \left[\int_1^\infty dy \left(\frac{y^2}{\sqrt{y^4 - 1}} - 1 \right) - 1 \right]. \tag{14.42}$$

Finally, by substituting U_0 and R^2, we get

$$V_{q\bar{q}}(L) = -\frac{4\pi^2}{\Gamma(1/4)^4} \frac{\sqrt{2g_{YM}^2 N}}{L}. \tag{14.43}$$

So we do get $V_{q\bar{q}}(L) \propto 1/L$ as expected for a conformally invariant theory. However, we also find that $V_{q\bar{q}}(L) \propto \sqrt{g_{YM}^2 N}$ which is a nonpolynomial, therefore nonperturbative result. That means that this cannot be obtained by a finite loop order calculation. For example, the 1-loop result would be proportional to $g_{YM}^2 N$. The result at all couplings, leading to this at strong coupling, was obtained, however, by matrix model calculations, but we will not describe them here.

So we have obtained the conformal invariant result for the interquark potential, the Coulomb potential. But under what conditions can we obtain the area law? The basic reason for obtaining the Coulomb potential was that the area of the string worldsheet is not proportional to the area bounded by the contour C on which it ends. But that could happen asymptotically in the following situation. Consider cutting off AdS space at a small U_m, such that for sufficiently large separation L between the quarks at infinity, the string would fall down to U_m. Then there is a portion of the string worldsheet (in the middle) that will lay straight on the cut-off at $U = U_m = $ constant. That in turn means that a percentage of $\ln\langle W[C]\rangle$, that gets greater with L until it dominates, obeys the area law. This suggests that confinement is associated with having a cut-off at a minimum U in AdS space. We discuss this further in Part III of the book.

14.5 Nonsupersymmetric Wilson loop

We have presented up to now how to calculate the 1/2 supersymmetric Wilson loop using AdS/CFT, but there are other kinds of Wilson loops that can be calculated. In particular, we can also calculate the gravity dual of the regular (bosonic) Wilson loop in (14.18). The formula was proposed by Alday and Maldacena in 2007, and it passed several tests. The gravity dual of the regular Wilson loop is still given by (14.33), but the string worldsheet is different. In fact, the string worldsheet now also ends on the contour C at the boundary of AdS space, but unlike the 1/2 supersymmetric one, it has Neumann boundary conditions on the S^5 (as opposed to Dirichlet, i.e. fixed θ^I).

Important concepts to remember

- Introducing external quarks in the theory, we can measure the quark–antiquark potential between heavy sources.
- The Wilson loop, $W[C] = \text{Tr } P \exp\left[\int iA_\mu dx^\mu\right]$ is gauge invariant.
- By choosing the contour C as a rectangle with two sides in the time direction, of length T, and two sides in a space direction, of length $R \ll T$, we have a contour from which we can extract $V_{q\bar{q}}(R)$ by $\langle W(C)\rangle_0 = \exp[-V_{q\bar{q}}(R)]$.
- In a confining theory like QCD, $V_{q\bar{q}}(R) \sim \sigma R$, thus we have the area law: $\langle W(C)\rangle_0 \propto \exp(-\sigma A(C))$ for any smooth curve C, and conversely, if we find the area law the theory is confining. $\langle W(C)\rangle_{0,T\to\infty}$ acts as an order parameter for the confinement/deconfinement transition.
- In a conformally invariant theory like QED with external quarks, $V_{q\bar{q}}(R) = \alpha/R$ and the Wilson loop is conformally invariant. For the above C, $\langle W(C)\rangle_0 \propto \exp(-\alpha T/R)$.
- In AdS/CFT, the 1/2 supersymmetric Wilson loop one finds also has coupling to scalars (and fermions), and is defined by $\langle W(C)\rangle_0 = \exp(-S_{\text{string}}(C))$, where the string worldsheet ends at $U = \infty$ on the curve C, is at a fixed point on the S^5, and drops inside AdS space. One needs to subtract the mass of the free strings extending straight down over the whole space.
- The result or the calculation is nonperturbative (proportional to $\sqrt{\lambda}$), but has the expected conformal (Coulomb) behavior.
- If we have a cut-off at small U in AdS space, the theory will be confining, since asymptotically at large interquark separation L we have the area law for the Wilson loop.
- The gravity dual of the regular Wilson loop is the same as for the 1/2 supersymmetric Wilson loop, except with Neumann boundary conditions on S^5 for the string worldsheet.

References and further reading

For more on Wilson loops, see any QCD textbook, e.g. [40]. The Wilson loop in AdS/CFT was defined and calculated in [41, 42]. I have followed Maldacena's [41] derivation here.

Exercises

1. Check that in the nonabelian case, for a closed square contour of side a, in a plane defined by $\mu\nu$, we have

$$\Phi_{\square_{\mu\nu}} = e^{ia^2 F_{\mu\nu}} + \mathcal{O}(a^4). \tag{14.44}$$

2. Check that if a free relativistic string in four flat dimensions is stretched between q and $\bar{q}$ and we use the AdS/CFT prescription for the Wilson loop, $W[C] = e^{-S_{\text{string}}[C]}$, we get the area law.

3. Consider a circular Wilson loop C, of radius R. Give an argument to show that $W[C]$ in $\mathcal{N} = 4$ SYM, obtained from AdS/CFT as in the rectangular case, is also conformally invariant, i.e. independent of R.

4. Check that if AdS_5 terminates at a fixed $U = U_m$ and strings are allowed to reach U_m and get stuck there, then we get the area law for $\langle W[C]\rangle$ at large interquark separation L (this is similar to what happens in the case of finite temperature AdS/CFT), by using the argument at the end of Section 14.4 and calculating the scaling of the string areas at $U = U_m$ and at $U > U_M$ for $L \to \infty$.

5. Finish the steps left out in the calculation of the quark–antiquark potential to get the final result for $V_{q\bar{q}}(L)$.

6. Verify that the Wilson loop (14.29) is 1/2 supersymmetric, substituting the $\mathcal{N} = 4$ SYM susy variations.

In this chapter we describe how to introduce finite temperature into AdS/CFT, with the goal of getting close to real world applications. Indeed, we will see that there is a sort of universality for finite temperature phenomena, which allows us to use some calculations in $\mathcal{N} = 4$ SYM at finite temperature to describe real world QCD at finite temperature, and to compare with heavy ion collisions, where such a system is relevant.

15.1 Finite temperature in field theory: periodic time

First we want to understand how to put a field theory at finite temperature. This discussion appears in most textbooks on quantum field theory, but we will nevertheless review it. Finite temperature field theory is best understood starting from the Feynman–Kac formula for quantum mechanical transition amplitudes.

In quantum mechanics, we write down a transition amplitude between points q, t and q', t' (between Heisenberg states) as

$$
\begin{aligned}
\langle q', t'|q, t\rangle &= \langle q'|e^{-i\hat{H}(t'-t)}|q\rangle \\
&= \sum_{nm} \langle q'|n\rangle \langle n|e^{-i\hat{H}(t'-t)}|m\rangle \langle m|q\rangle \\
&= \sum_n \psi_n(q')\psi_n^*(q)e^{-iE_n(t'-t)}.
\end{aligned} \tag{15.1}
$$

On the other hand, it can also be written as a path integral,

$$
\langle q', t'|q, t\rangle = \int \mathcal{D}q(t)e^{iS[q(t)]}. \tag{15.2}
$$

If we perform a Wick rotation to Euclidean space by $t \to -it_E$, $t'-t \to -i\beta$, $iS \to -S_E$, such that

$$
iS[q] = i\int_0^{t_E=\beta} (-idt_E)\left[\frac{1}{2}\left(\frac{dq}{d(-it_E)}\right)^2 - V(q)\right] \equiv -S_E[q], \tag{15.3}
$$

we obtain

$$
\langle q', \beta|q, 0\rangle = \sum_n \psi_n(q')\psi_n^*(q)e^{-\beta E_n}. \tag{15.4}
$$

Then we note that if we specialize to the case $q' = q$ and integrate over this value of q, we obtain the statistical mechanics partition function of a system at a temperature T, with $kT \equiv 1/\beta$,

$$\int dq \langle q, \beta | q, 0 \rangle = \int dq \sum_n |\psi_n(q)|^2 e^{-\beta E_n} = \text{Tr}\{e^{-\beta \hat{H}}\} = Z[\beta]. \qquad (15.5)$$

This corresponds in the path integral to taking closed paths of Euclidean time length $\beta = 1/(kT)$, since $q' \equiv q(t_E = \beta) = q(t_E = 0) \equiv q$. Then we obtain the *Feynman–Kac formula*,

$$Z(\beta) = \text{Tr}\{e^{-\beta \hat{H}}\} = \int \mathcal{D}q e^{-S_E[q]}|_{q(t_E+\beta)=q(t_E)}, \qquad (15.6)$$

where the path integral is then taken over *all closed paths of Euclidean time length β*.

Similarly, in field theory we obtain for the Euclidean partition function

$$Z_E[\beta] = \int_{\phi(\vec{x},t_E+\beta)=\phi(\vec{x},t_E)} \mathcal{D}\phi e^{-S_E[\phi]} = \text{Tr} \ (e^{-\beta \hat{H}}). \qquad (15.7)$$

Therefore the partition function at finite temperature T is expressed again as a Euclidean path integral over paths of periodic Euclidean time. One can then extend this formula by adding sources and calculating propagators and correlators, exactly as for zero temperature field theory.

Thus the finite temperature field theory, for static quantities only (time-independent!), is obtained by considering periodic imaginary time, with period $\beta = 1/T$.

15.2 Quick derivation of Hawking radiation

We can use the formulation developed in the last subsection for field theory at finite temperature to deduce that black holes radiate thermally at a given temperature T, a process known as Hawking radiation.

We want to describe quantum field theory in the black hole background. As always, it is best described by performing a Wick rotation to Euclidean time. The Wick-rotated Schwarzschild black hole in four spacetime dimensions is

$$ds^2 = +\left(1 - \frac{2MG_N}{r}\right) d\tau^2 + \frac{dr^2}{1 - \frac{2MG_N}{r}} + r^2 d\Omega_2^2. \qquad (15.8)$$

Having now a Euclidean signature, it does not make sense to go inside the horizon, at $r < 2MG_N$, since then the signature will not be Euclidean anymore (unlike a Lorentz signature, when the only thing that happens is that the time t and radial space r change roles), but will be $(-- ++)$ instead.

Therefore, if the Wick-rotated Schwarzschild solution represents a Schwarzschild black hole, the horizon must not be singular, yet there must not be a continuation inside it, i.e. it must be smoothed out somehow. This is possible since in Euclidean signature one can have a conical singularity if

$$ds^2 = d\rho^2 + \rho^2 d\theta^2, \qquad (15.9)$$

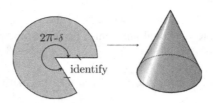

Figure 15.1   A flat cone is obtained by cutting out an angle from flat space, so that $\theta \in [0, 2\pi - \Delta]$ and identifying the cut.

but $\theta \in [0, 2\pi - \Delta]$. If $\Delta \neq 0$, then $\rho = 0$ is a singular point, and the metric describes a cone, as in Fig.15.1, therefore $\rho = 0$ is known as a conical singularity. However, if $\Delta = 0$ we do not have a cone, thus no singularity, and we have a (smooth) Euclidean space.

A similar situation applies to the Wick-rotated Schwarzschild black hole. Near $r = 2MG_N$, we have

$$ds^2 \simeq \frac{\tilde{r}}{2MG_N}d\tau^2 + 2MG_N\frac{d\tilde{r}^2}{\tilde{r}} + (2MG_N)^2d\Omega_2^2, \qquad (15.10)$$

where $r - 2MG_N \equiv \tilde{r}$. By defining $\rho \equiv \sqrt{\tilde{r}}$ we get

$$ds^2 \simeq 8MG_N\left(d\rho^2 + \frac{\rho^2 d\tau^2}{(4MG_N)^2}\right) + (2MG_N)^2 d\Omega_2^2, \qquad (15.11)$$

so near the horizon the metric looks like a cone. If τ has no restrictions, the metric does not make much sense. It must be periodic for it to make sense, but for a general period we get a cone, with $\rho = 0(r = 2MG_N)$ a singularity. Only if $\tau/(4MG_N)$ has period 2π we avoid the conical singularity and we have a smooth space, that cannot be continued inside $r = 2MG_N$.

Therefore we have periodic Euclidean time, with period $\beta_\tau = 8\pi MG_N$. By the analysis of the previous section, this corresponds to finite temperature quantum field theory at temperature,

$$T_{\text{BH}} = \frac{1}{\beta_\tau} = \frac{1}{8\pi MG_N}. \qquad (15.12)$$

We can then say that quantum field theory in the presence of a black hole has a temperature T_{BH} or that black holes radiate thermally at temperature T_{BH}.

Does that mean that we can put a quantum field theory at finite temperature by adding a black hole? Not quite, since the specific heat of the black hole is

$$C = \frac{\partial M}{\partial T} = -\frac{1}{8\pi T^2 G_N} < 0, \qquad (15.13)$$

therefore the black hole is thermodynamically unstable, and it does not represent an equilibrium situation.

But we will see that in Anti-de Sitter space we have a different situation. Adding a black hole does provide a thermodynamically stable system, which therefore does represent an equilibrium situation.

15.3 Black holes and supersymmetry breaking

Spin structure in black hole background

We now want to understand what happens to fermions in the presence of a black hole background.

At $r \to \infty$, the solution is $ds^2 \simeq d\tau^2 + d\vec{x}^2$, i.e. $\mathbb{R}^3 \times S^1$, since τ is periodic, which is a Kaluza–Klein vacuum. That is, it is a background solution around which we can expand the fields in Fourier modes (in general, we have spherical harmonics, but for compactification on a circle we have actual Fourier modes) and perform a dimensional reduction by keeping only the lowest modes.

But fermions do not necessarily need to be periodic, they can acquire a phase $e^{i\alpha}$ when going around the S^1_τ circle at infinity:

$$\psi \to e^{i\alpha}\psi. \tag{15.14}$$

These are known as "spin structures," and $\alpha = 0$ and $\alpha = \pi$ are always OK, since the Lagrangean always has terms with an even number of fermions, thus such a phase would still leave it invariant. If $\mathcal{L}$ has additional symmetries, there could be other values of α allowed.

At the horizon $r = 2MG_N$, the metric is (15.11), which is $\mathbb{R}^2 \times S^2$, where Ω_2 is the S^2 metric and the $\mathbb{R}^2$ is from the would-be cone, with $\theta \equiv \rho/(4MG_N)$. But $\mathbb{R}^2 \times S^2$ is simply connected, which means that there are no nontrivial cycles, or that any loop on $\mathbb{R}^2 \times S^2$ can be smoothly shrunk to zero. That means that there cannot be nontrivial fermion phases as you go in around any loop on $\mathbb{R}^2 \times S^2$, or that there is a *unique spin structure*.

We must therefore find to what does this unique spin structure correspond at infinity? The relevant loop at infinity is $\tau \to \tau + \beta_\tau$, which near the horizon is $\theta \to \theta + 2\pi$, i.e. a rotation in the 2-dimensional plane $\mathbb{R}^2$. Under such a rotation a fermion picks up a minus sign.

Indeed, a fermion can be defined as an object that gives a minus sign under a complete spatial rotation, i.e. an object that is periodic under 4π rotations instead of 2π. In four dimensions, the way to see that is that the spatial rotation $\psi \to S\psi$ around the axis defined by $\vec{v}$ is given by

$$S(\vec{v}, 0) = \cos\frac{\theta}{2}I + i\vec{v}\cdot\vec{\Sigma}\sin\frac{\theta}{2}; \quad \vec{\Sigma} = \begin{pmatrix} \vec{\sigma} & 0 \\ 0 & \vec{\sigma} \end{pmatrix}, \tag{15.15}$$

where $\vec{\sigma}$ are Pauli matrices. We can see that under a 2π rotation, $S = -1$.

Therefore the unique spin structure in the Euclidean Schwarzschild black hole background is one that makes the fermions antiperiodic at infinity, around the Euclidean time direction.

KK masses and supersymmetry breaking

That can only happen if they have some Euclidean time dependence, $\psi = \psi(\theta)$. That in turns means that the fermions at infinity get a nontrivial mass under dimensional reductions, since the 4-dimensional free flat space equation (valid at $r \to \infty$) gives

$$\not{\partial}\psi = 0 \Rightarrow \not{\partial}^2\psi = \square_{4d}\psi = 0. \tag{15.16}$$

Under dimensional reduction, $\square_{4d} = \square_{3d} + \partial^2/\partial\theta^2$, so we obtain

$$0 = \square_{4d}\psi = \left(\square_{3d} + \frac{\partial^2}{\partial\theta^2}\right)\psi = (\square_{3d} - m^2)\psi, \tag{15.17}$$

where $m^2 \neq 0$ is a 3-dimensional spinor mass squared. Therefore, in the presence of the black hole, fermions become massive, from the point of view of the reduced 3-dimensional theory.

Bosons on the other hand have no such restrictions on them, and we can have bosons that are periodic at infinity under $\theta \to \theta + 2\pi$, thus also the simplest case of bosons that are independent of θ. Then at infinity for a boson, e.g. a scalar ϕ,

$$0 = \square_{4d}\phi = \left(\square_{3d} + \frac{\partial^2}{\partial\theta^2}\right)\phi = \square_{3d}\phi, \tag{15.18}$$

and therefore scalars can be massless in three dimensions.

But supersymmetry in flat 3-dimensional Euclidean space requires that $m_{\text{scalar}} = m_{\text{fermion}}$. That is not the case in the presence of the black hole, since we can have $m_{\text{fermion}} \neq 0$, but $m_{\text{boson}} = 0$, therefore the presence of the black hole breaks supersymmetry.

In fact, one can prove that finite temperature always breaks supersymmetry, in any field theory.

It follows that one of the ways to break the unwelcome $\mathcal{N} = 4$ supersymmetry in AdS/CFT and get to more realistic field theories is by having finite temperature, specifically by putting a black hole in AdS space. We discuss this prescription in the following. Of course, in the way shown above we obtain a non-supersymmetric 3-dimensional field theory, but there are ways to obtain a 4-dimensional nonsupersymmetric theory in a similar manner.

15.4 The AdS black hole and Witten's finite temperature prescription

Now we finally have the tools to understand the way to put AdS/CFT at finite temperature. The prescription was proposed by Witten, and corresponds to introducing a black hole in AdS_5.

As we have seen, the metric of global Anti-de Sitter space can be written as

$$ds^2 = -\left(\frac{r^2}{R^2} + 1\right)dt^2 + \frac{dr^2}{\frac{r^2}{R^2} + 1} + r^2 d\Omega^2. \tag{15.19}$$

Then the black hole in $(n + 1)$-dimensional Anti-de Sitter space is

$$ds^2 = -\left(\frac{r^2}{R^2} + 1 - \frac{w_n M}{r^{n-2}}\right)dt^2 + \frac{dr^2}{\frac{r^2}{R^2} + 1 - \frac{w_n M}{r^{n-2}}} + r^2 d\Omega_{n-1}^2, \tag{15.20}$$

where from (6.52), we get

$$w_n = \frac{8\pi G_N^{(n+1)}}{(n-2)\Omega_{n-1}}, \tag{15.21}$$

and Ω_{n-1} is the volume of the unit sphere in $n-1$ dimensions. For $n=3$ (AdS_4), $\Omega_2 = 4\pi$ and $w_3 = 2G_N$. For AdS_5, $n=4$, $\Omega_3 = 2\pi^2$ and $w_4 = 2G_N^{(5)}/\pi$.

We now repeat the usual Hawking temperature analysis, for the Euclidean version of the AdS_5 black hole. We define the (outer) horizon r_+ as the largest solution of the equation

$$\frac{r^2}{R^2} + 1 - \frac{w_n M}{r^{n-2}} = 0. \tag{15.22}$$

The Euclidean metric near this outer horizon is (here $\delta r = r - r_+$)

$$ds^2 \simeq \left(\frac{2r_+}{R^2} + \frac{(n-2)w_n M}{r_+^{n-1}}\right)\delta r \, dt^2 + \frac{(d\delta r)^2}{\delta r\left(\frac{2r_+}{R^2} + \frac{(n-2)w_n M}{r_+^{n-1}}\right)} + r_+^2 d\Omega_2^2. \tag{15.23}$$

Then as in the flat 4-dimensional case, we obtain that the metric is singularity free only if the time t is periodic, with period

$$\beta = \frac{4\pi}{\frac{2r_+}{R^2} + \frac{(n-2)w_n M}{r_+^{n-1}}} = \frac{4\pi}{\frac{nr_+}{R^2} + \frac{(n-2)}{r_+}}, \tag{15.24}$$

where in the last equality we have used the definition of r_+ in (15.22). Then finally the temperature of the AdS black hole is

$$T = \frac{nr_+^2 + (n-2)R^2}{4\pi R^2 r_+}. \tag{15.25}$$

Then $T(M)$ (considering $T = T(r_+)$ and $r_+ = r_+(M)$) looks like in Fig.15.2. To find the position of the minimum, we must put to zero $dT/dM = (dT/dr_+)(dr_+/dM)$. Differentiating (15.22) with respect to M, we get

$$\frac{dr_+}{dM}\left[\frac{nr_+^{n-1}}{R^2} + (n-2)r_+^{n-3}\right] = w_n, \tag{15.26}$$

which implies $dr_+/dM > 0$. Then the minimum of T is found from $dT/dr_+ = 0$, which gives the horizon size r_+ at the minimum

$$r_+ = R\sqrt{\frac{n-2}{n}}. \tag{15.27}$$

In turn, this implies the temperature at the minimum is

$$T_{\min} = \frac{nr_+}{2\pi R} = \frac{\sqrt{n(n-2)}}{2\pi R}. \tag{15.28}$$

The low M branch has $C = \partial M \partial T < 0$, therefore is thermodynamically unstable, like the Schwarzschild black hole in flat space. Then this solution is in fact a small perturbation of that solution, since the black hole is small compared to the radius of curvature of AdS space.

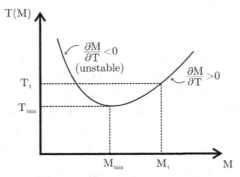

Figure 15.2 $T(M)$ for the AdS black hole. The lower M branch is unstable, having $\partial M/\partial T < 0$. The higher M branch has $\partial M/\partial T > 0$, and above T_1 it is stable.

The high M branch, however, has $C = \partial M/\partial T > 0$, thus is thermodynamically stable. We also need to check that the free energy of the black hole solution, F_{BH}, is smaller than the free energy of pure AdS space, F_{AdS}.

The free energy F is defined by

$$Z = e^{-\beta F}, \tag{15.29}$$

where $\beta = 1/T$ (if $k = 1$). But in a gravitational theory, the free energy is given by the gravitational action,

$$Z_{\text{grav}} = e^{-S}, \tag{15.30}$$

where S is the Euclidean action for gravity. We have seen this for example when defining correlators in AdS/CFT. Then we have that

$$S(\text{Euclidean action}) = \frac{F}{T}, \tag{15.31}$$

and therefore we need to compare zero against

$$F_{\text{BH}} - F_{\text{AdS}} = T(S_{\text{BH}} - S_{\text{AdS}}), \tag{15.32}$$

and an explicit calculation (that will not be done here) shows that it is < 0 only if

$$T > T_1 \equiv \frac{n-1}{2\pi R} > T_{\text{min}}. \tag{15.33}$$

There is one more problem. At $r \to \infty$, the metric is

$$ds^2 \simeq \left(\frac{r}{R}dt\right)^2 + \left(\frac{R}{r}dr\right)^2 + r^2 d\Omega_{n-1}^2, \tag{15.34}$$

therefore the Euclidean time direction is a circle of radius $(r/R) \times (1/T)$, and the transverse $n-1$-dimensional sphere has radius r. Thus both are proportional to $r \to \infty$; however, the $\mathcal{N} = 4$ SYM gauge theory that exists at $r \to \infty$ has conformal invariance, therefore, only relative scales are relevant for it, meaning that we can drop the overall r. It follows that the topology at infinity, where $\mathcal{N} = 4$ SYM exists, is $S^{n-1} \times S^1$, but we want to have a theory defined on $R^{n-1} \times S^1$ instead, namely n-dimensional flat space at finite temperature (with periodic Euclidean time).

That means that we need to scale the ratio of sizes to infinity

$$\frac{r}{\frac{r}{R}\frac{1}{T}} = R \cdot T \to \infty, \tag{15.35}$$

which implies that we must take $T \to \infty$, only possible if $M \to \infty$, and we must rescale the coordinates to get finite quantities. We find that $M \propto r^n$ and $r^2 dt^2$ must be finite. Including also finite factors of w_n and R, the needed rescaling is

$$r = \left(\frac{w_n M}{R^{n-2}}\right)^{1/n} \rho; \quad t = \left(\frac{w_n M}{R^{n-2}}\right)^{-1/n} \tau, \tag{15.36}$$

and $M \to \infty$. Under this rescaling, the metric becomes ($dx_i = (w_n M / R^{n-2})^2 d\Omega_i$),

$$ds^2 = \left(\frac{\rho^2}{R^2} - \frac{R^{n-2}}{\rho^{n-2}}\right) d\tau^2 + \frac{d\rho^2}{\frac{\rho^2}{R^2} - \frac{R^{n-2}}{\rho^{n-2}}} + \rho^2 \sum_{i=1}^{n-1} dx_i^2, \tag{15.37}$$

and the period of τ is

$$\beta_1 = \frac{4\pi R}{n}. \tag{15.38}$$

Since for $\rho \to \infty$ we get

$$ds^2_{\rho \to \infty} \simeq \rho^2 \left(\frac{d\tau^2}{R^2} + d\vec{x}^2\right), \tag{15.39}$$

considering string theory in the metric (15.37) puts $\mathcal{N} = 4$ SYM at the constant finite temperature

$$T = \frac{R}{\beta_1} = \frac{n}{4\pi}. \tag{15.40}$$

In four gauge theory dimensions, we obtain $T = 1/\pi$.

The metric (15.37) for the relevant case of $n = 4$ dimensions can be rewritten as

$$ds^2 = \frac{\rho^2}{R^2}\left[\left(1 - \frac{R^4}{\rho^4}\right) d\tau^2 + R^2 d\vec{x}^2\right] + R^2 \frac{d\rho^2}{\rho^2\left(1 - \frac{R^4}{\rho^4}\right)}. \tag{15.41}$$

Making the redefinitions

$$\frac{\rho}{R} = \frac{U}{U_0}; \quad \tau = t\frac{U_0}{R}; \quad \vec{x} = \vec{y}\frac{U_0}{R^2}, \tag{15.42}$$

adding the dropped factor $R^2 d\Omega_5^2$ to the gravity dual, and Wick rotating to Lorentzian signature, we get

$$ds^2 = \frac{U^2}{R^2}\left[-f(U)dt^2 + d\vec{y}^2\right] + R^2 \frac{dU^2}{U^2 f(U)} + R^2 d\Omega_5^2,$$
$$f(U) \equiv 1 - \frac{U_0^4}{U^4}. \tag{15.43}$$

We note that this metric is the nonextremal version of the $AdS_5 \times S^5$ metric (10.14). But (10.14) was obtained as the near-horizon limit of the D3-brane metric, and one can make a D-brane metric nonextremal by adding an $f(r)$ in front of dt^2 and $1/f(r)$ in front of dr^2, as we saw at the end of Chapter 6. It follows immediately that the same (15.43) above can also be obtained directly from the near-horizon limit of the near-extremal D3-brane metric, as we can easily check.

Finally, also redefining as usual $U/R = R/z$ and $U_0/R = R/z_0$, the metric becomes

$$ds^2 = \frac{R^2}{z^2}\left[-f(z)dt^2 + d\vec{y}^2 + \frac{dz^2}{f(z)}\right] + R^2 d\Omega_5^2,$$
$$f(z) = 1 - \frac{z^4}{z_0^4}. \tag{15.44}$$

Note that even though we started with AdS in global coordinates to obtain the Witten metric, we have now reached a finite temperature version of AdS in Poincaré coordinates! In this metric, we can follow the same steps as in the case of the Schwarzschild and AdS black holes and find that the temperature of the above metric is given by

$$T = \frac{1}{\pi z_0}. \tag{15.45}$$

This is consistent with (15.38), since the periodicity $\beta_\tau = 4\pi R/n = \pi R$ implies a periodicity $\beta_t = \pi R^2/U_0 = \pi z_0$.

We have already explained that in this AdS black hole metric, supersymmetry is broken. At $r = $ infinity, the fermions are antiperiodic around the Euclidean time direction, thus if we dimensionally reduce the $\mathcal{N} = 4$ SYM theory to three dimensions (i.e., compactify on the Euclidean time) the fermions become massive. The gauge fields are protected by gauge invariance and remain massless under this dimensional reduction. The scalars as we saw remain massless in three dimensions, at least at the classical level. At the quantum level, they also get a mass at one loop, through a fermion loop due to a Yukawa scalar-fermion coupling.

Therefore the 3-dimensional theory obtained by dimensionally reducing $\mathcal{N} = 4$ SYM on the compact Euclidean time is pure QCD glue, only gauge fields A_μ^a and nothing else!

15.5 Application of finite temperature: mass gap

We want to understand the mass gap in pure glue theory from AdS/CFT. Consider the dimensionally reduced AdS/CFT at finite temperature, i.e. the AdS_5 black hole metric reduced to three boundary dimensions. The mass gap means spontaneous appearance of a mass for physical states, in this case for $\mathcal{N} = 4$ SYM, reduced to three boundary dimensions with supersymmetry breaking conditions. This translates into having a classical mass for physical states existing in the gravitational dual (15.37).

Thus we consider massless fields in the bulk of (15.37) and we calculate their 3-dimensional mass. For a scalar field, the equation of motion is $\Box\phi = 0$ on this space, and dimensionally reduced solutions (independent of τ) can be put in a factorized form

$$\phi(\rho, \vec{x}, \tau) = f(\rho)e^{i\vec{k}\cdot\vec{x}}. \tag{15.46}$$

At the horizon $\rho = b$ we need to impose that the solution is smooth, i.e. $df/d\rho = 0$, since the horizon acts like the origin of a plane in spherical coordinates, as we saw. On the other hand, at $\rho \to \infty$ we need to impose that the solution is normalizable, since we are interested in physical states in AdS, which means

$$f \sim \frac{1}{\rho^4}. \tag{15.47}$$

Choosing the normalizable solution at infinity (among the two solutions, one normalizable and one non-normalizable) and continuing it until the horizon in general results in a solution that does not satisfy the boundary condition. Imposing the boundary condition at the horizon results in a quantization condition on the parameters of the solution, $\vec{k}^2$, which equals m^2, the 3-dimensional mass squared. Then the mass squared has a positive discrete spectrum, which is the statement of mass gap. In conclusion, the finite temperature AdS space (15.37) behaves like a quantum mechanical box, with a nonzero ground state energy. This is as expected from the general analysis, since the finite temperature effectively cuts off the interior of AdS space at $z = z_0$, and we argued that AdS space with the middle cut off should function as a quantum mechanical box, since then a light ray goes from the horizon to the boundary in finite time.

15.6 $\mathcal{N} = 4$ SYM plasmas from AdS/CFT

We have seen that we can put AdS/CFT at finite temperature by introducing a black hole in AdS_5, and that breaks supersymmetry, leaving only massless 3-dimensional pure glue and massive states after dimensional reduction on the periodic Euclidean time.

But considered from the 4-dimensional perspective (no dimensional reduction), we still have $\mathcal{N} = 4$ SYM at finite temperature. This is a priori very different from QCD at finite temperature, which would be relevant for the real world. Nevertheless, we will see that various quantities calculated *at finite temperature* within $\mathcal{N} = 4$ SYM give results similar to QCD, hence it is believed that there is a certain universality for gauge theories at finite temperature, which encouraged the use of the original AdS/CFT at finite temperature.

In particular, in heavy ion collisions like Brookhaven's RHIC experiment and LHC's ALICE experiment one obtains a strongly coupled quark–gluon plasma (sQGP), i.e. a system of quarks and gluons at high temperature (above the QCD phase transition), and even though it is a dynamical (time dependent) situation, and spatially bounded, as opposed to the space-filling time independent case in the previous sections, one can still apply the AdS/CFT methods used for $\mathcal{N} = 4$ SYM at finite temperature. Besides the finite temperature, the sQGP has finite density and chemical potential, and also it was recently found that magnetic fields are very important. Therefore, in the next sections we also study how to add finite density, chemical potential, and magnetic fields. By analogy with the real world QCD state at high temperature, we call the $\mathcal{N} = 4$ SYM state plasma as well.

Bulk properties: entropy, energy, pressure

We begin by calculating the bulk thermodynamic properties of the $\mathcal{N} = 4$ SYM plasma.

The entropy of a 5-dimensional black hole is given by the usual Bekenstein–Hawking formula,

$$S = \frac{A}{4G_{N,5}}, \tag{15.48}$$

where A is the area of the horizon. For the metric (15.44), the horizon area (at $z = z_0$) is

$$A = \frac{R^3}{z_0^3} \int dy_1 dy_2 dy_3. \tag{15.49}$$

The entropy S of the horizon of the AdS black hole is identified with the entropy of the dual field theory. It is infinite, but there is a finite 4-dimensional *entropy density*, given using AdS/CFT by

$$s = \frac{S}{\int dy_1 dy_2 dy_3} = \frac{R^3}{4G_{N,5} z_0^3}. \tag{15.50}$$

To calculate this, we note that in 10-dimensional string theory we have (this normalization of the Einstein action can be derived in string theory)

$$2\kappa_N^2 = 16\pi G_{N,10} = (2\pi)^7 g_s^2 \alpha'^4, \tag{15.51}$$

and in $AdS_5 \times S^5$ we have $R^4 = \alpha'^2 g_{YM}^2 N = \alpha'^2 (4\pi g_s) N$, which gives

$$G_{N,10} = \frac{\pi^4}{2N^2} R^8. \tag{15.52}$$

Doing the dimensional reduction on the S^5 of radius R, with the volume of the S^5 of unit radius $\Omega_5 = \pi^3$, we obtain

$$G_{N,5} = \frac{G_{N,10}}{\Omega_5 R^5} = \frac{\pi}{2N^2} R^3. \tag{15.53}$$

Together with (15.45), this gives for the entropy density

$$s_{\lambda=\infty} = \frac{\pi^2}{2} N^2 T^3. \tag{15.54}$$

This is the entropy density at infinite coupling $\lambda \to \infty$ for $\mathcal{N} = 4$ SYM. From the thermodynamic relations $s = \partial P / \partial T$ and $\epsilon = -P + Ts$, where ϵ is the energy density E/V, we find for the pressure and energy density

$$P_{\lambda=\infty} = \frac{\pi^2}{8} N^2 T^4,$$
$$\epsilon_{\lambda=\infty} = \frac{3\pi^2}{8} N^2 T^4, \tag{15.55}$$

at infinite coupling.

On the other hand, at weak coupling we can calculate the entropy density as follows. For a free bosonic degree of freedom, the entropy density is $s = 2\pi^2 T^3/45$, and for a free fermionic degree of freedom, it is $7/8$ of the bosonic result. But in $\mathcal{N} = 4$ we have a gauge field, with two degrees of freedom, and six scalars, for a total of eight bosonic degrees of freedom, and matching them eight fermionic degrees of freedom. All of them are in the adjoint of $SU(N)$, thus each having $N^2 - 1$ components. In total, the entropy density is

$$s_{\lambda=0} = \left(8 + 8\frac{7}{8}\right)(N^2 - 1)\frac{2\pi^2 T^3}{45} \simeq \frac{2\pi^2}{3}N^2 T^3, \tag{15.56}$$

where in the last equality we have dropped the 1 in $(N^2 - 1)$ at large N.

Then, in between the strong and weak coupling we have

$$\frac{s_{\lambda=\infty}}{s_{\lambda=0}} = \frac{3}{4}. \tag{15.57}$$

By using the thermodynamic relations we find the same ratio for the pressure and energy density

$$\frac{P_{\lambda=\infty}}{P_{\lambda=0}} = \frac{\epsilon_{\lambda=\infty}}{\epsilon_{\lambda=0}} = \frac{3}{4}. \tag{15.58}$$

But in lattice QCD, the numerical value for the ratio of the pressure at infinite coupling to the pressure at zero coupling, and the ratio of energy density at infinite coupling to the energy density at zero coupling is found to be about 80%, very close to the result of 75% in $\mathcal{N} = 4$ SYM. This is the first example of a finite temperature calculation for which the $\mathcal{N} = 4$ SYM result is close to the QCD result, and it led to the suggestion of universality among gauge theories at finite temperature as a justification for using $\mathcal{N} = 4$ SYM via AdS/CFT to apply to the real world.

Energy loss: drag on heavy quarks

When a fast heavy quark (a would-be "jet") passes through the sQGP plasma, it loses energy at a high rate, due to the interaction with the medium, a phenomenon known as "jet quenching." This is observed experimentally in the RHIC and ALICE experiments. Due to the fact that the sQGP is strongly coupled, it is an ideal quantity to be calculated using AdS/CFT, and various calculations were done using $\mathcal{N} = 4$ SYM at finite temperature, and were shown to give reasonable results when compared to the experimental results (which are for QCD).

As we saw, a heavy quark corresponds to a long string stretching between the boundary at infinity and a point in the interior of AdS. A moving heavy quark on a straight path then corresponds to a string moving at infinity on this straight path, i.e. to the holographic prescription for a Wilson line. But now the string does not return to infinity, but instead stretches all the way (asymptotically) to the horizon. The string as it moves loses momentum, so one needs a constant force to keep it at constant velocity. One can calculate this momentum loss, or force by calculating the string stationary configuration.

We consider the ansatz for a string with endpoint moving on the boundary, $z = 0$ in (15.44), in the y_1 direction. Therefore we consider $y_1(t, z \rightarrow 0) = vt$ as a boundary condition, and the stationary ansatz $y_1(t, z) = vt + h(z)$. We moreover choose the static gauge

$\tau = t, \sigma = z$ for the worldsheet coordinates (τ, σ) of the string. Substituting the ansatz in the Nambu–Goto action, we obtain

$$S = -\frac{R^2}{2\pi\alpha'} \left(\int dt \right) \int \frac{dz}{z^2} \sqrt{\frac{f(z) - v^2 + f(z)^2 h'^2(z)}{f(z)}}. \tag{15.59}$$

The action is independent of $h(z)$, it only depends on $h'(z)$, which means that the canonical momentum P_z^1 is conserved,

$$\begin{aligned} P_z^1 &= \frac{\delta S}{\delta h'(z)} = \frac{\delta S}{\delta y^{1\prime}(z)} \\ &= -\frac{R^2}{2\pi\alpha'} \frac{1}{z^2} \frac{f^{3/2}(z) h'(z)}{\sqrt{f(z) - v^2 + f(z)^2 h'^2(z)}}. \end{aligned} \tag{15.60}$$

Then we can solve $h'(z)$ as a function of P_z^1, to obtain

$$h'^2(z) = \left(\frac{2\pi\alpha' P_z^1}{R^2} \right)^2 \frac{z^4}{f(z)^2} \frac{f(z) - v^2}{f(z) - \left(\frac{2\pi\alpha' P_z^1}{R^2} \right)^2 z^4}. \tag{15.61}$$

Both the numerator and denominator are positive at the boundary $z = 0$ ($1 - v^2$ and 1, respectively, for the last ratio), and negative at the horizon $z = z_0$ where $f(z) = 0$, and since $h'^2(z) \geq 0$, it means that both must equal zero at the same time. This gives $(2\pi\alpha' P_z^1 / R^2)^2 z_1^4 = v^2 = 1 - z_1^4 / z_0^4$, and eliminating z_1 we obtain

$$P_z^1 = \pm \frac{R^2}{2\pi\alpha' z_0^2} \gamma v, \tag{15.62}$$

where $\gamma = 1/\sqrt{1 - v^2}$ is the Lorentz factor. That means that only for this *momentum flux in the z direction*, P_z^1, can we have a stationary solution. We choose the positive sign, corresponding to momentum flowing from the $z = 0$ boundary to the $z = z_0$ horizon due to the pulling of an external force, which pumps momentum that dissipates into the medium.

Plugging this P_z^1 back into (15.61), we can integrate it for $h(z)$, obtaining

$$h(z) = -\frac{v z_0}{2} \left[\operatorname{arctanh} \left(\frac{z}{z_0} \right) - \arctan \left(\frac{z}{z_0} \right) \right]. \tag{15.63}$$

This corresponds to a string trailing behind the quark, hanging down from the boundary ($h(0) = 0$), and becoming almost parallel to the horizon near $z = z_0$ ($h(z_0) = -\infty$), as in Fig.15.3.

Then P_z^1 is momentum flow into the plasma, thus the momentum *loss* in the plasma is

$$\frac{dp}{dt} = -P_z^1 = -\frac{R^2}{2\pi\alpha' z_0^2} \gamma v. \tag{15.64}$$

Translating into gauge theory variables using $R^2/\alpha' = \sqrt{\lambda}$ and $T = 1/\pi z_0$ ($\lambda = g_{\text{YM}}^2 N$ is the 't Hooft coupling), we obtain

$$\begin{aligned} \frac{dp}{dt} &= -\frac{\pi}{2} \sqrt{\lambda} T^2 \gamma v = -\frac{\pi}{2M} \sqrt{\lambda} T^2 p \\ &\equiv -\eta_D p, \end{aligned} \tag{15.65}$$

where $p = M\gamma v$ is the momentum of the heavy quark and η_D is the drag coefficient.

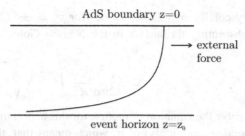

Figure 15.3   A quark being pulled by an external force at the boundary, and a string trailing behind it, hanging down from the boundary.

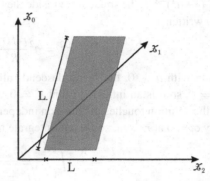

Figure 15.4   Wilson loop contour relevant for jet quenching.

Jet quenching from (semi-) light-like Wilson loop

There is a quantity that defines the amount of energy lost by an ultrarelativistic heavy quark or gluon passing through the plasma; and is called the *jet quenching parameter* $\hat{q}$. It is formally defined as the mean transverse momentum squared gained by the heavy object per distance travelled, $\hat{q} \equiv \langle k_\perp^2 \rangle / L$, but is related to jet quenching (energy loss). It can be calculated from a particular kind of Wilson loop. The fact that the relevant objects are ultrarelativistic heavy quarks uniquely defines this loop: it has to have two long light-like sides of length L_-, and it has to be separated by a spatial distance L, so two sides are short, space-like and of distance L. In the previous case, we considered only one heavy quark moving, but for using Wilson loops, we need a closed contour, so the one above is uniquely defined, as in Fig.15.4.

One can prove that for this Wilson loop, the jet quenching parameter $\hat{q}$ is found from the asymptotic scaling formula

$$\langle W[C] \rangle \simeq \exp\left[-\frac{1}{4\sqrt{2}} \hat{q} L_- L^2 \right], \tag{15.66}$$

though we will not do it here.

We then calculate the Wilson loop defined by the above contour, in the Poincaré AdS black hole background (15.43), with U called r. We choose a static gauge $\tau = y^-$, $\sigma = y^2$ ($y^\pm = (t \pm y^1)/\sqrt{2}$), and consider the static ansatz that is translationally invariant in x^- (ignoring the worldsheet deformation near the short L side of the Wilson loop) $y^\mu = y^\mu(\sigma)$, with $y^3(\sigma) = $ constant, $y^+(\sigma) = $ constant, and the boundary condition $r(\pm L/2) = \infty$ (for $y^2 = \pm L/2$, $r = \infty$, i.e. we reach the boundary). Substituting the ansatz in the Nambu–Goto action, we obtain for the Euclidean action

$$S = \frac{\sqrt{2}r_0^2 L_-}{2\pi\alpha' R^2} \int_0^{L/2} d\sigma \sqrt{1 + \frac{r'^2 R^4}{f(r)r^4}}. \tag{15.67}$$

We can solve the equation of motion for this action in the following way. The Lagrangean is independent (explicitly) of σ, which means that, thinking of it as "time," the "Hamiltonian" $H = r' \partial \mathcal{L}/\partial r' - \mathcal{L}$ is conserved, i.e. constant. Defining the constant a by $H = \frac{\sqrt{2}r_0^2 L_-}{2\pi\alpha' R^2}(1+a^2)^{-1/2}$, the square root inside the Lagrangean is $\sqrt{1+a^2}$ and the equation of motion is written as

$$r'^2 = a^2 \frac{r^4 f(r)}{R^4}. \tag{15.68}$$

For a solution with $a > 0$, the string descends all the way down to the horizon $r = r_0$, where $f(r) = 0$, so it is an inflexion point $r' = 0$, and then goes back up to the boundary. Notice that the solution touches the horizon independent of how small L, or T are. Because of symmetry $\sigma = 0$ at $r = r_0$, so we can integrate the above equation of motion to obtain

$$\frac{L}{2} = \frac{R^2}{a} \int_{r_0}^{\infty} \frac{dr}{\sqrt{r^4 - r_0^4}} = \frac{cR^2}{ar_0}, \tag{15.69}$$

where $c = \sqrt{\pi}\,\Gamma(5/4)/\Gamma(3/4) \simeq 1.311$. Then the integration constant is $a = 2cR^2/r_0 L$, and with $r_0 = \pi R^2 T$ it becomes $a = 2c/(\pi LT)$. Finally we can write the on-shell action as a function of only L_-, L, T as

$$S = \frac{\pi\sqrt{\lambda}L_- LT^2}{2\sqrt{2}} \sqrt{1 + \frac{4c^2}{\pi^2 T^2 L^2}}. \tag{15.70}$$

From this action we must subtract the action of the "W bosons," the perimeter, $\simeq 2L_-$ times the tension $1/2\pi\alpha'$ times the length from the boundary to the horizon (now the space finishes at the horizon), $\int_{r_0}^{\infty} dr \sqrt{g_{--}g_{rr}}$, giving

$$S_0 = \frac{c\sqrt{\lambda}L_- T}{\sqrt{2}}. \tag{15.71}$$

Then we have for the regularized action at small LT

$$S - S_0 \simeq \frac{\pi^2}{8\sqrt{2}c} \sqrt{\lambda}T^3 L_- L^2 + O(T^5 L_- L^4). \tag{15.72}$$

The Wilson loop prescription gives $\langle W[C] \rangle = \exp[S - S_0]$, but the trace in $W[C]$ is in the adjoint representation, Tr_a, which equals twice the trace in the fundamental representation Tr, for which the formula (15.66) was written. Then we obtain for the jet quenching parameter

$$\hat{q}_{\mathrm{SYM}} = \frac{\pi^2}{c}\sqrt{\lambda}T^3 = \frac{\pi^{3/2}\Gamma(3/4)}{\Gamma(5/4)}\sqrt{g_{\mathrm{YM}}^2 N}T^3. \tag{15.73}$$

This again gives a reasonable comparison with QCD, once we replace the constant g_{YM}^2 with the corresponding QCD running coupling, again lending credence to the idea of universality of gauge theories at finite temperature.

15.7 Adding a finite chemical potential $\mu \neq 0$

We now want to learn how to add a finite charge (particle) density or finite chemical potential into AdS/CFT, since in real experiments like heavy ion collisons one needs to consider these cases. First off, charge in the case of $\mathcal{N} = 4$ SYM can only mean R-charge, i.e., a subgroup of the $SU(4) = SO(6)$ global symmetry of the CFT. On the gravity side, this charge is sourced by the gauge field in AdS space.

As we saw, in AdS/CFT, the source coupling for a boundary gauge field a_μ is $\int d^d x J^\mu a_\mu$. In particular, J^0 is a charge density in the gauge theory, and it couples to a_0, the boundary value of A_0 inside AdS space. It follows that the boundary condition corresponding to nonzero charge inside AdS space is

$$A = A_0(z)dt + \ldots, \tag{15.74}$$

as we go to the boundary $z \to 0$. That in turn means that we need to have an electrically charged solution inside AdS space. Then electrical charge in AdS space corresponds to having a charge density on the boundary. Moreover, the source coupling in the action should be $q\mu$, hence $A_0(z = 0) = a_0 = \mu$ is the chemical potential for the R-charge. That means that the boundary condition is $A \to \mu dt$ as $z \to 0$.

To find the corresponding solution, we need to consider the Einstein–Maxwell system in AdS space (with a negative cosmological constant). One can find the Reissner–Nordstrom solution in AdS in Poincaré coordinates in a way similar to the Reissner–Nordstrom solution in flat space. The solution is found as usual by simply replacing the $f(z)$ in (15.44) with a charged expression,

$$ds^2 = \frac{R^2}{z^2}\left(-f(z)dt^2 + d\vec{x}^2 + \frac{dz^2}{f(z)}\right),$$
$$f(z) = 1 - \left(1 + Kz_+^2\mu^2\right)\left(\frac{z}{z_+}\right)^d + Kz_+^2\mu^2\left(\frac{z}{z_+}\right)^{2(d-1)},$$
$$K \equiv \frac{(d-2)\kappa_{N,d+1}^2}{(d-1)g^2R^2}, \tag{15.75}$$

where we have written the solution in $d + 1$ dimensions for completeness. The gauge field corresponding to this solution is

$$A_0 = \mu\left[1 - \left(\frac{z}{z_+}\right)^{d-2}\right]. \tag{15.76}$$

Note that we might think that we could drop the constant part of A_0, but we need to have $A = 0$ at the horizon $z = z_+$. This boundary condition is necessary for A to be well defined, since otherwise A is singular at the horizon.

The temperature of this metric is found in the same way as in the other cases from the periodicity of the Euclidean version of the black hole, as

$$T = \frac{1}{4\pi z_+} \left(d - K(d-2) z_+^2 \mu^2 \right). \tag{15.77}$$

Note that we can have $T = 0$ and $\mu \neq 0$ for this black hole.

To find all the relevant thermodynamic quantities in this case, we need to calculate the thermodynamic potential. But for that, we need to decide whether to consider the case of constant chemical potential μ or constant charge density ρ. From the above analysis, constant μ corresponds to constant a_μ source for the gauge field on the boundary. The thermodynamic potential in this case is the grand canonical potential $\Omega = \Omega(\mu, V, T) = U - TS - \mu N$, with $d\Omega = -SdT - PdV - Nd\mu$. It is found by the usual AdS/CFT relation,

$$Z_{\text{CFT}} = e^{-\beta\Omega} = Z_{\text{sugra}} = e^{-S_{\text{sugra}}} \Rightarrow \Omega = T S_{\text{sugra}}. \tag{15.78}$$

The necessary condition to find this potential is that the on-shell AdS supergravity action S_{sugra} does not have any boundary terms (they vanish) if we take $a_0 = $ constant (constant chemical potential). It is satisfied for the Einstein–Maxwell action, so we have $\Omega = T S_{\text{sugra}}$.

We can calculate the on-shell supergravity action and find

$$\Omega = -\frac{R^{d-1}}{2\kappa_N^2 z_+^d} \left(1 + K z_+^2 \mu^2 \right) V_{d-1}. \tag{15.79}$$

In this case, the charge density VEV is a one-point function, so it is found by differentiating the supergravity action with respect to the source,

$$\rho = \langle J^0 \rangle = \left. \frac{\delta S_{\text{sugra}}}{\delta a_0} \right|_{a_0 = 0}. \tag{15.80}$$

Another possibility is to consider constant charge density ρ, in which case the thermodynamic potential is the free energy (canonical ensemble) $F = \Omega + \mu Q$. We see that it corresponds to adding a term linear in $\mu = a_0$ to the on-shell gravitational action. Indeed, one can add a boundary term to the gravitational action of the form (in the supergravity action we have $-1/4g^2 \int d^{d+1}x \sqrt{-g} F_{\mu\nu}^2$)

$$+ \frac{1}{g^2} \int_{z \to 0} d^d x \sqrt{-h} n^a F_{ab} A^b, \tag{15.81}$$

where h_{ab} is the metric on the boundary. The addition of this term means that we keep fixed $n^a F_{ab}$ instead of A_a on the boundary, and since a_0 is a chemical potential, now we fix instead its conjugate, the charge density.

15.8 Adding a magnetic field $B \neq 0$

To add a magnetic field in the gauge theory we need to add a magnetic field in the bulk gravitational theory as well. The reason is as follows. The $\mathcal{N} = 4$ SYM theory has a global $SU(4) = SO(6)$ R-symmetry, a $U(1)$ subgroup of which we are considering. But in order to talk about magnetic fields we must *gauge* this symmetry by coupling its current to a gauge field, i.e. add a term $\int d^d x J^\mu a_\mu$ to the action. We do not want to alter the dynamics of the theory, so we will not add a kinetic term for a_μ, but instead consider a_μ like an *external* (background) gauge field. But in this case the coupling $\int d^d x J^\mu a_\mu$ in the action takes exactly the form of the current-source coupling needed for AdS/CFT, hence we can identify the external a_μ, giving the magnetic field, with the boundary source for the bulk gauge field.

The case of a magnetic field in AdS_5 is a bit more complicated, so we discuss the AdS_4 case here, which is easier to understand. We also consider the case with both electric and magnetic charges in AdS, so we look to generalize (15.75) and (15.76) for $d = 3$ by introducing magnetic charge, with ansatz $A = A_0(z)dt + B(z)xdy$. The generalization is obtained by once again replacing $f(z)$ with the appropriate electric–magnetic duality invariant result,

$$f(z) = 1 - \left[1 + K\left(z_+^2\mu^2 + z_+^4 B^2\right)\right]\left(\frac{z}{z_+}\right)^3 + K\left(z_+^2\mu^2 + z_+^4 B^2\right)\left(\frac{z}{z_+}\right)^4, \quad (15.82)$$

and the gauge field gets an extra term

$$A = \mu\left[1 - \frac{z}{z_+}\right]dt + Bxdy. \quad (15.83)$$

We can rewrite $f(z)$ in a form making more obvious the electric–magnetic duality as

$$f(z) = 1 - [1 + h^2 + q^2]\left(\frac{z}{z_+}\right)^3 + (h^2 + q^2)\left(\frac{z}{z_+}\right)^4, \quad (15.84)$$

and the field strength coming from A as

$$F = \frac{1}{z_+^2\sqrt{K}}[h\,dx \wedge dy + q\,dt \wedge dz]. \quad (15.85)$$

We see that both $\vec{E}$ and $\vec{B}$ are finite at the boundary $z = 0$, but $\vec{B}$ is interpreted as external magnetic field, and $\vec{E}$ as source for the charge density, with VEV (15.80).

Since only the coefficients of the z^3 and z^4 terms in $f(z)$ change, but not the functional form of the metric, it is clear that the temperature is found by replacing the corresponding coefficients, as

$$T = \frac{1}{4\pi}\left[3 - K(z_+^2\mu^2 + z_+^4 B^2)\right]. \quad (15.86)$$

The thermodynamic potential at constant chemical potential can also be found, generalizing (15.79) for $d = 3$, as

$$\Omega = -\frac{R^2}{2\kappa_N^2 z_+^3}\left[1 + K(z_+^2\mu^2 - 3z_+^4 B^2)\right]V_2. \quad (15.87)$$

In the case of AdS_{d+1}, in particular for AdS_5, one can do a similar analysis, again with the ansatz

$$F = Bdx^1 \wedge dx^2, \qquad (15.88)$$

but unless we are in AdS_4 this ansatz breaks the isotropy of the solution, and one needs to consider a more complicated metric ansatz, and the solutions are numerical and not very illuminating, so we will not describe them here.

Important concepts to remember

- Finite temperature field theory is obtained by having a periodic Euclidean time, with period $\beta = 1/T$. The partition function for such periodic paths gives the thermal partition function, from which we can extract correlators by adding sources, etc.
- The Wick-rotated Schwarzschild black hole has a smooth (non-singular) "horizon" only if the Euclidean time is periodic with period $\beta = 1/T_{BH} = 8\pi M$. Thus black holes Hawking radiate.
- Quantum field theory in the presence of a black hole does not have finite temperature though, since the Schwarzschild black hole is thermodynamically unstable ($C = \partial M/\partial T < 0$).
- Fermions in the Wick-rotated black hole are antiperiodic around the Euclidean time at infinity, thus they are massive if we dimensionally reduce the theory on the periodic time. Since bosons are massless, the black hole (and finite temperature) breaks supersymmetry.
- By putting a black hole in AdS space, the thermodynamics is stable if we are at high enough black hole mass M.
- The Witten prescription for finite temperature AdS/CFT is to put a black hole of mass $M \to \infty$ inside global AdS_5 and to take a certain scaling of coordinates, giving the metric (15.37). It can also be obtained from the near-horizon limit of a non-extremal D3-brane metric, giving a Poincaré metric.
- By dimensionally reducing $d = 4$ $\mathcal{N} = 4$ SYM on the periodic Euclidean time, we get pure Yang–Mills in three dimensions, which has a mass gap (spontaneous appearance of a lowest nonzero mass state in a massless theory).
- The mass gap is obtained in AdS space from solutions of the wave equations in AdS that have a 3-dimensional mass spectrum like the one of a quantum mechanical box with the ground state removed. Thus the Witten metric is similar in terms of eigenmodes to a quantum mechanical finite box.
- The ratio of entropy densities, pressures, and energy densities (and thus of effective degrees of freedom) at infinite coupling to zero coupling is 3/4, close to the 80% of lattice QCD simulations.
- Jet quenching, the energy loss of a heavy quark ("jet") passing through the sQGP medium, can be modelled by the energy loss of a string moving with constant v at the boundary, and trailing behind all the way to the horizon. One finds a drag coefficient $\eta_D \propto \sqrt{\lambda}T^2$.
- One can calculate the jet quenching parameter $\hat{q}$ from a Wilson loop with two long light-like sides of length L_- and two short space-like sides of length L, and AdS/CFT gives $\hat{q}_{SYM} \propto \sqrt{\lambda}T^3$.

- A chemical potential for the R-charge of $\mathcal{N} = 4$ SYM corresponds to a constant source for A_0 on the boundary of AdS_5, i.e. $A \rightarrow \mu dt$ as $z \rightarrow 0$, and gives an electrically charged solution in AdS, namely the Reissner–Nordstrom AdS black hole.
- The grand-canonical ensemble of constant μ is found from the usual AdS action for supergravity, whereas the canonical ensemble for constant charge density is found by adding an extra boundary term to the AdS action for the gauge field.
- Adding an external magnetic field in $\mathcal{N} = 4$ SYM with respect to the R symmetry group is done by adding a magnetic field in AdS. In AdS_4, one simply adds magnetic charge to the AdS black hole.

References and further reading

The prescription for AdS/CFT at finite temperature was done by Witten in [43]. The calculation of the jet quenching parameter from a (partially) light-like Wilson loop was originally done in [73]. More details about the drag on heavy quarks, jet quenching, and $\mathcal{N} = 4$ SYM plasmas in general and how they apply to heavy ion collisions can be found in [72]. The way to add magnetic field in AdS_4 and more details on adding chemical potential can be found in [69, 71]. In [70] how to add magnetic field in AdS_5 is described.

Exercises

1. Parallel the calculation of the Schwarzschild black hole to show that the extremal ($Q = M$) black hole has zero temperature.
2. Check that the rescaling plus the limit given in (15.36) gives the Witten background for finite temperature AdS/CFT.
3. Take a near-horizon nonextremal D3-brane metric,

$$ds^2 = \alpha' \left\{ \frac{U^2}{R^2}[-f(U)dt^2 + d\vec{y}^2] + R^2 \frac{dU^2}{U^2 f(U)} + R^2 d\Omega_5^2 \right\},$$

$$f(U) = 1 - \frac{U_0^4}{U^4}, \tag{15.89}$$

where U_0 is fixed, $U_0 = \pi T R^2$ (T = temperature). Note that for $f(U) = 1$ we get the near-horizon extremal D3 brane, i.e. $AdS_5 \times S^4$. Check that a light ray traveling between the boundary at $U = \infty$ and the horizon at $U = U_0$ takes a finite time (for $U_0 = 0$, it takes an infinite time to reach $U = 0$).
4. Check that the rescaling

$$U = \rho \cdot \frac{U_0}{R}; \quad t = \frac{\tau R}{U_0}; \quad \vec{y} = \vec{x}\frac{R^2}{U_0}, \tag{15.90}$$

where R = AdS radius, takes the above near-horizon nonextremal D3-brane metric to the Witten finite T AdS/CFT metric.

5. Check that the temperature of the AdS–Reissner–Nordstrom solution (15.75) is given by (15.77).
6. Calculate the grand-canonical thermodynamic potential (15.79) by calculating the regularized on-shell action, subtracting the contribution of pure AdS space.
7. Check that for the magnetic solution with (15.82), the temperature is given by (15.86), and the grand-canonical potential by (15.87).

16

Scattering processes and gravitational shockwave limit

In previous chapters we have seen how to describe correlators, states, and Wilson loops. But in QCD we are interested in S matrices that describe the scattering of physical asymptotic states. The LSZ formalism relates the S matrices to correlators, but it assumes the existence of separated asymptotic states.

In a conformal field theory, however, there is no notion of scale, therefore there is no notion of infinity, and no asymptotic states, so we cannot construct S matrices from correlators. We can define via regularization a notion of scattering amplitudes, and we study that in Chapter 26, but S-matrices cannot be consistently defined.

Therefore in order to construct S matrices so that we can study scattering of states as in QCD, we need to break the conformal invariance. In this chapter we describe the simplest possible modification of $AdS_5 \times S^5$ that does the job, leaving more general cases to Part III of the book.

16.1 The "hard-wall" model for QCD

At high energy (in the UV), QCD is approximately conformal since the small mass parameters are irrelevant, therefore the gravity dual background corresponding to QCD (of course, assuming that there is such a thing, which is not obvious) should be approximately like $AdS_5 \times X^5$ at large fifth dimension ρ, since the conformal group $SO(4, 2)$ matches the isometry group of AdS_5. Here X_5 is some compact space. This $AdS_5 \times X_5$ is then modified in some way at small ρ, corresponding to the low energy (IR) behavior of QCD.

The metric for this model is

$$ds^2 = \frac{r^2}{R^2} d\vec{x}^2 + R^2 \frac{dr^2}{r^2} + R^2 ds_X^2$$
$$= e^{-2y/R} d\vec{x}^2 + dy^2 + R^2 ds_X^2. \tag{16.1}$$

The simplest possible model that captures some of the properties of QCD is then obtained by just cutting off $AdS_5 \times X_5$ at a certain value of r called r_{min}. To relate it to parameters in QCD, we note that momenta $p_i = -i\partial_i$ in QCD are related to momenta in ten dimensions $\tilde{p}_i$ by the metric factor, i.e.

$$\tilde{p}_\mu = \frac{R}{r} p_\mu. \tag{16.2}$$

When we have a characteristic momentum in ten dimensions, i.e. of the scale of the metric, $\tilde{p}_\mu \sim 1/R$, we obtain a QCD momentum

$$p \sim \frac{r}{R^2}.\tag{16.3}$$

But in this case, if the cut-off AdS theory has the characteristic momentum of the curvature, we expect the QCD momentum to be of the order of the QCD scale Λ, the scale of the lowest fundamental excitations. Therefore we have

$$r_{\min} = R^2 \Lambda.\tag{16.4}$$

This is called the "hard-wall" model for holographic QCD, and was introduced by Polchinski and Strassler.

16.2 Scattering in QCD and the Polchinski–Strassler scenario

Polchinski and Strassler went on to describe scattering in this hard-wall model for QCD. As in the case of $AdS_5 \times S^5$, fields in the hard-wall metric correspond to 4-dimensional composite operators, which correspond to gauge invariant, composite particles. Examples are nucleons and mesons or glueballs. We saw an example of a glueball operator described by AdS/CFT, Tr $[F_{\mu\nu}F^{\mu\nu}]$, but one can write many.

The wavefunction for a glueball state, for instance $e^{ik\cdot x}$, corresponds via AdS/CFT to a wavefunction Φ for the $AdS_5 \times S_5$ field corresponding to the glueball, times a wavefunction in the extra coordinates, e.g.

$$\Phi = e^{ik\cdot x} \times \Psi(\rho, \vec{\Omega}_{X_5}).\tag{16.5}$$

Assuming that the states on the gravitational side scatter locally according to the flat space string amplitude, Polchinski and Strassler proposed an ansatz for the scattering amplitude of gauge invariant states in QCD. It is given by a convolution of the string amplitude with the wave functions for the extra dimensions, rescaling the momenta between the local string amplitude and the QCD amplitude according to (16.2),

$$\mathcal{A}_{\text{QCD}}(p_i) = \int dr d^5\Omega_{X_5} \sqrt{-g}\, \mathcal{A}_{\text{string}}(\tilde{p}_i) \prod_i \Psi_i\left(r, \vec{\Omega}_{X_5}\right).\tag{16.6}$$

We can define a QCD scale corresponding to the α' scale, the "QCD string scale" $\hat{\alpha}'$, defined by

$$\hat{\alpha}' = (g_{\text{YM}}^2 N)^{-1/2} \Lambda^{-2}.\tag{16.7}$$

This definition is made such that the string momenta in string units are bounded from above by QCD momenta in QCD string units, since using $R^2/\alpha' = (g_{\text{YM}}^2 N)^{1/2}$ we obtain

$$\sqrt{\alpha'}\tilde{p} = \sqrt{\hat{\alpha}'}p\left(\frac{r_{\min}}{r}\right) \le \sqrt{\hat{\alpha}'}p.\tag{16.8}$$

But then we can also define a "QCD Planck scale" $\hat{M}_{\text{Pl}}$. In ten dimensions, the coefficient of the Einstein action is M_{Pl}^8, but in string theory it is, up to numbers, $\sim 1/(g_s^2\alpha'^4)$, the $1/g_s^2$

since it comes from a closed string interaction, with coupling g_s and the $1/\alpha'^4$ for dimensional reasons, since we need to form a dimension eight object. Then $M_{\text{Pl}} \sim g_s^{-1/4}\alpha'^{-1/2}$, which leads to the QCD Planck scale

$$\hat{M}_{\text{Pl}} = g_s^{-1/4}\hat{\alpha}'^{-1/2} = g_{\text{YM}}^{-1/2}\Lambda(g_{\text{YM}}^2 N)^{1/4} = N^{1/4}\Lambda. \tag{16.9}$$

We discuss the relevance of these QCD scales later on in the chapter. Here we just note that in the 't Hooft limit, $N \to \infty$, $g_{\text{YM}}^2 N$ finite, $\hat{M}_{\text{Pl}} \gg 1/\sqrt{\hat{\alpha}'}$.

16.3 Regge behavior

In gauge theories it is expected that when $s \gg -t > 0$ (here s and t are the Mandelstam variables for 2 to 2 scattering), or $s \to \infty$ with $|t|$ fixed, we have *Regge behavior* for the amplitude,

$$\mathcal{A}(s, t) \simeq \beta(t)s^{\alpha(t)},$$
$$\alpha(t) = \alpha_0 + \frac{\hat{\alpha}'}{2}t. \tag{16.10}$$

In string theory, amplitudes also show Regge behavior. In particular, the amplitude for 2 to 2 scalars scattering has the *Virasoro–Shapiro* form (which we will not derive here),

$$\mathcal{A}_{\text{string}} = g_s^2 \alpha'^3 \left[\prod_{x=s,t,u} \frac{\Gamma(-\alpha'\tilde{x}/4)}{\Gamma(1+\alpha'\tilde{x}4)} \right] K(\sqrt{\alpha'}\tilde{p}), \tag{16.11}$$

where the K is a kinematic factor of order $\tilde{p}^8$. Note that for massless external states we always have $s + t + u = \sum_i m_i^2 = 0$. Then in the Regge limit $s \gg -t > 0$, or rather $\alpha's \gg 1$, $\alpha'|t|$ fixed, we obtain

$$\mathcal{A}_{\text{string}}(s, t) \simeq g_s^2 \alpha'^3 [\text{polariz.tensors}](\alpha's)^{\alpha't/2+2} \frac{\Gamma(-\alpha't/4)}{\Gamma(1+\alpha't/4)}. \tag{16.12}$$

The proof of this statement is left as an exercise.

Doing the Polchinski–Strassler integral (16.6), one finds that for small enough $\tilde{t}$ (specifically for $\hat{\alpha}'|t| < (\Delta - 4)/\ln(s/|t|)$), the integral is dominated by the lowest r, i.e. $r = r_{\text{min}}$, because the integrand is larger there. Thus in first approximation, instead of the integral we can use the integrand at r_{min}. That in turn means that the QCD amplitude has Regge behavior,

$$\mathcal{A}_{\text{QCD}}(p) \sim \beta(t)(\hat{\alpha}'s)^{2+\hat{\alpha}'t/2}, \tag{16.13}$$

since (16.8) means that at $r = r_{\text{min}}$ we have $\alpha'\tilde{t} = \hat{\alpha}'t$, so the string theory power law turns into a QCD power law.

The fact that the Polchinski–Strassler integral is concentrated near the cut-off could be guessed on physical grounds. Indeed, it is known that the Regge limit $s \to \infty$, t fixed is governed by the low energy modes of the theory. This is the regime of "soft scattering" which is characterized by the emission of a large number of the smallest mass particles, so even though the total energy is very large, the *per particle* (per "parton") energy is very small, and the physics is governed by the low energy (IR) part of the theory. We discuss this further in the next sections. But low energy corresponds in the gravity dual to r close to the cut-off r_{min} (the fifth dimension r acts roughly as an energy scale, as we saw).

16.4 Gravitational shockwave scattering as a model of QCD high energy scattering

One can now ask, how can we describe high energy scattering from the point of view of the gravitational theory? We have seen that the regimes of interest in QCD correspond to $\alpha's \gg 1$ and $s/M_{\text{Pl}}^2 \gg 1$ in string theory. The answer to how to describe this scattering was formulated by 't Hooft.

At sufficiently high energies, in a gravitational theory only the momentum of the particle matters, and it produces a gravitational shockwave. The solution was described by Aichelburg and Sexl. Then 't Hooft argues that the scattering of two nearly massless particles (with masses much smaller than the Planck scale) with $G_N s \sim 1$, yet still $G_N s < 1$, is described by one particle creating an A–S shockwave, and the other particle moving on a geodesic on this shockwave, which allows one to calculate a scattering amplitude. But at trans-Planckian energies, $G_N s > 1$ ($\sqrt{s} > M_{\text{Pl}}$), both particles create an Aichelburg–Sexl shockwave, and the collision of the two shockwaves creates a black hole. This is as expected since at energies higher than the Planck mass we expect quantum gravity to be important, which leads to the quantum creation of a gravitational object, the black hole. Of course, when $\sqrt{s} \sim M_{\text{Pl}}$ the term black hole for the quantum gravitational object created is not too appropriate, it becomes appropriate in the classical limit, when $\sqrt{s} \gg M_{\text{Pl}}$.

The A–S gravitational shockwave solution is of the general type of a "parallel plane," or "pp," gravitational wave type,

$$ds^2 = 2dx^+ dx^- + H(x^+, x^i)(dx^+)^2 + \sum_i dx_i^2, \tag{16.14}$$

which is a solution of Einstein's equations that corresponds to perturbations moving at the speed of light, having a plane wave front.

The only nonzero component of the Ricci tensor for this metric is

$$R_{++} = -\frac{1}{2}\partial_i^2 H(x^+, x^i). \tag{16.15}$$

This is left as an exercise to check. That means that Einstein's equations are consistent only if T_{++} is the only nonzero component of the energy-momentum tensor, in which case Einstein's equations read

$$R_{++} = -\frac{1}{2}\partial_i^2 H(x^+, x^i) = 8\pi G T_{++}. \tag{16.16}$$

This means that the function H is harmonic in x^i if T_{++} is independent of x^i or with a delta function support.

These *pp wave* solutions are special, since, as we see, the Einstein equation, written in terms of the Ricci tensor R_{ij}, linearizes, so the usual highly nonlinear behavior of gravity is avoided. In particular, this leads to an important result proved by Horowitz and Steif, that in a pp wave background there are no α' corrections to the equations of motion, since all possible R^2 corrections vanish on-shell (by the use of the zero order equations of motion), hence pp waves give *exact string solutions*.

The A–S shockwave is a pp wave created by a massless particle of momentum p (as we said, at sufficiently high energy, all particles look nearly massless and source a gravitational wave), with energy-momentum tensor

$$T_{++} = p\delta^{d-2}(x^i)\delta(x^+). \tag{16.17}$$

Then the harmonic function H factorizes as

$$H(x^+, x^i) = \delta(x^+)\Phi(x^i), \tag{16.18}$$

and Einstein's equation reduces to the Poisson equation for Φ,

$$\partial_i^2 \Phi(x^i) = -16\pi G_{N,d} p\delta^{d-2}(x^i). \tag{16.19}$$

This solution was originally found by infinitely boosting a black hole, whose energy-momentum tensor is a point particle with $T_{00} = M\delta^{d-2}(x^i)\delta(y)$, boosted to

$$T_{00} = \frac{m}{\sqrt{1-v^2}}\delta^{d-2}(x^i)\delta(y - vt), \tag{16.20}$$

and corresponding values for T_{11} and T_{01}. Taking $v \to 1$ while keeping the momentum $p = mv\gamma$ constant leads to the A–S T_{++}, and taking the limit on the black hole solution leads to the A–S shockwave solution.

In $d = 4$ flat spacetime dimensions, we find

$$\Phi = -4G_{N,4} \ln \rho^2, \tag{16.21}$$

where $\rho^2 = x_i x_i$ is the transverse radius squared, and in $d > 4$ flat spacetime dimension we find

$$\Phi = \frac{16\pi G_{N,d}}{\Omega_{d-3}(d-4)} \frac{p}{\rho^{d-4}}. \tag{16.22}$$

Then in flat spacetime, in the trans-Planckian regime $G_N s > 1$, collisions of high energy particles are described by the collision of two gravitational A–S shockwaves, one propagating in the x^+ direction, say, and the other in the x^- direction, with an impact parameter b (spatial separation) in the transverse directions x^i. Since the background metric is flat, and we only have a term proportional to $(dx^{\pm})^2$ nontrivial, the metric *before* the collision is simply the sum of the two metrics,

$$ds^2 = 2dx^+dx^- + dx_i^2 + (dx^+)^2\Phi_1(x^i) + (dx^-)^2\Phi_2(x^i). \tag{16.23}$$

On the other hand, *after* the collision, we expect to form a black hole.

Indeed, it was shown in flat four dimensions first by Penrose at zero impact parameter $b = 0$, and then by Eardley and Giddings at nonzero b, smaller than a b_{max}, that a "marginally trapped surface" forms at the collision point $x^+ = x^- = 0$. For a

Schwarzschild black hole, the horizon surface $r = r_H \equiv 2MG_N$ is a marginally trapped surface, but in general the marginally trapped surface is different from the event horizon. There is a general relativity theorem saying that if there is at some point a marginally trapped surface, there is an event horizon outside it, and thus a black hole, in its future. Since one can prove the existence of these surfaces for $b \leq b_{max}$, it follows that we can put a bound on the cross-section for black hole formation, $\sigma_{BH} \geq \pi b_{max}^2$.

We can apply the same mechanism for high energy scattering inside the hard-wall model, the cut-off $AdS_5 \times S^5$, and use the Polchinski–Strassler formula (16.6) to relate to QCD scattering. The only caveat is that the Polchinski–Strassler formula requires the use of an amplitude, and in the case of black hole formation, in the classical picture above one can only calculate a b_{max}. But given a b_{max} and the intuition that the scattering is almost classical in nature, we can calculate an amplitude using the *black disk eikonal approximation*, that reproduces the classical cross-section.

One considers an S-matrix given by an $S = e^{i\delta}$, where δ satisfies

$$
\begin{aligned}
&\text{Re}[\delta(b, s)] = 0, \\
&\text{Im}[\delta(b, s)] = 0, \quad b > b_{max}, \\
&\text{Im}[\delta(b, s)] = \infty, \quad b \leq b_{max}.
\end{aligned}
\tag{16.24}
$$

Then the amplitude corresponding to it is the Fourier transform of $T = -i(S - 1)$, i.e.

$$
\begin{aligned}
\frac{1}{s}\mathcal{A}(s, t) &= -i \int d^2 b\, e^{i\vec{q}\cdot\vec{b}}(e^{i\delta} - 1) = i \int_0^{b_{max}(s)} b\, db \int_0^{2\pi} d\theta\, e^{iqb\cos\theta} \\
&= 2\pi i \frac{b_{max}(s)}{\sqrt{t}} J_1\left(\sqrt{t} b_{max}(s)\right).
\end{aligned}
\tag{16.25}
$$

The optical theorem says that the total cross-section is given by the imaginary part of the forward elastic amplitude (at $t = 0$), $\text{Im}[1/s\mathcal{A}(s, 0)]$, which is seen to equal πb_{max}^2.

16.5 Black holes in the IR of the gravity dual and the Froissart bound

As an application of this formalism, we describe the maximal asymptotic behavior that can happen in the hard-wall model. First, we need to understand the possible QCD energy scales relevant for $\sqrt{s}$. We saw the QCD string scale $(\hat{\alpha}')^{-1/2} = \Lambda(g_{YM}^2 N)^{1/2}$ and the QCD Planck scale $\hat{M}_P = N^{1/4}\Lambda$. For $\sqrt{s} > \hat{M}_P$, in the gravity dual we start to create small black holes, smaller than the curvature scale of the background. But eventually, when $\sqrt{s}$ increases enough, the black holes will have a scale comparable with the scale of the curvature. This will happen when the horizon size r_H equals R. Since the metric factor vanishing on the horizon is $1 - \#M/M_{Pl}^{d-2}r^{d-3}$, the energy (or mass M) for a given r_H is $E \sim M_{Pl}^8 r_H^7$ in $d = 10$ dimensions. Then for $r_H = R$ we have the energy scale

$$
E_R = M_{Pl}^8 R^7.
\tag{16.26}
$$

The scale R^{-1} corresponds to Λ in QCD. A simple way to see this is to note that $\alpha' = R^2(g_{YM}^2 N)^{-1/2}$, whereas as we saw in QCD $\hat{\alpha}' = \Lambda^{-2}(g_{YM}^2 N)^{-1/2}$. Then substituting M_{Pl} by $\hat{M}_{Pl}$ and R by Λ^{-1} in E_R, we obtain

$$\hat{E}_R = N^2 \Lambda^8 \Lambda^{-7} = N^2 \Lambda. \qquad (16.27)$$

Then at energies higher than $\hat{E}_R$ in QCD, we obtain black holes of a size larger than the scale of AdS R in the gravity dual.

It would seem as if there is no higher energy scale, but in fact there is. One can calculate the behavior of the total cross-section with energy corresponding to $E > \hat{E}_R$, from scattering in AdS space for $r_H > R$, and obtain a power law. But we know that there is a *unitarity bound* in field theories and QCD in particular, the Froissart bound. The total cross-section is bounded by

$$\sigma_{\text{tot}} \le C \ln^2 \frac{s}{s_0}; \quad C \le \frac{\pi}{m_\pi^2}. \qquad (16.28)$$

In theories other than QCD, m_π refers to the lowest energy state of the theory. So if we obtain a power law in the case of $E > \hat{E}_R$, it means that we have not reached the asymptotic Froissart regime above, which has a slower dependence on energy.

The Polchinski–Strassler integral (16.6) is dominated, as we saw, by a scattering region r_{scatt} close to r_{min}, but in reality not exactly at r_{min}, but separated slightly from it. Then the black holes that are created, are created with a center close to r_{scatt}. When the size of the black hole is larger than the AdS size, the fluctuations in the position of the created black hole are smaller than the black hole size, meaning that the black hole becomes more and more classical, as in Fig.16.1a and b. Eventually, at large enough s, the black hole becomes so large that it reaches the cut-off r_{min}, and it starts to look as if it is situated on the IR cut-off itself, as in Fig.16.1b. This happens at some higher energy $\hat{E}_F$, which would correspond to the onset of the maximal behavior, which we guess (and will soon confirm) should be associated with the asymptotic Froissart behavior. Whereas the other scales did not depend on the details of the cut-off (if it was a hard cut-off as in the hard-wall model, or a smooth deformation as in a more physical model), it is clear that $\hat{E}_F$ does, since it is the energy at which value the black hole in the dual reaches the IR brane.

It is more appropriate to consider a symmetrical situation for AdS space with respect to the IR cut-off, i.e. consider the metric

$$ds^2 = e^{\frac{2|y|}{R}} d\vec{x}^2 + dy^2 + R^2 ds_X^2. \qquad (16.29)$$

Then the IR cut-off acts as a brane, with a nonzero tension. Then effectively, asymptotically the black hole is created on this IR brane.

We can write down an A–S gravitational shockwave solution situated on the IR brane, such that the collision of two such waves will create a black hole on the IR brane. The equation it satisfies is a curved space generalization of (16.19),

$$-\frac{1}{2} \left[e^{\frac{2|y|}{R}} \left(\partial_y^2 + \frac{d}{R} \operatorname{sgn}(y) \partial_y \right) + \nabla_x^2 \right] \Phi = 8\pi G_{N,d+1} p \delta^{d-2}(x^i) \delta(y). \qquad (16.30)$$

In Fourier space for $\vec{x}$, i.e. defining $h(\vec{q}, y) = \int d^{d-2}\vec{x} e^{i\vec{q}\cdot\vec{x}} \Phi(\vec{x}, y)$, we get

$$h(\vec{q}, y)'' + \frac{d}{R} \operatorname{sgn}(y) h(\vec{q}, y)' - \vec{q}^2 e^{-\frac{2|y|}{R}} h(\vec{q}, y) = -16\pi G_{N,d+1} p \delta(y). \qquad (16.31)$$

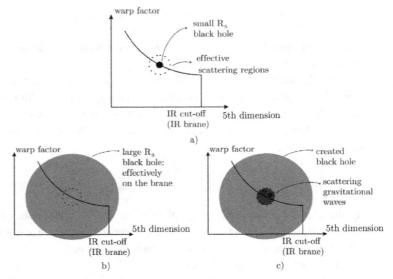

Figure 16.1   a) At small enough energies, the created black hole is small, and fluctuates (is created at a random point) inside a small region of effective scattering; b) At large enough energies, the created black hole is so large, that it is effectively fixed (has small fluctuations) and it looks as if it sits on the IR brane; c) At these large energies, the process is effectively classical: two shockwaves going in opposite directions scatter creating a black hole larger than the scattering region.

Its solution translates into the solution for Φ,

$$\Phi(r,y) = \frac{4G_{N,d+1}p}{(2\pi)^{\frac{d-4}{2}}} \frac{e^{-\frac{d|y|}{2R}}}{r^{\frac{d-4}{2}}} \int_0^\infty dq\, q^{\frac{d-4}{2}} J_{\frac{d-4}{2}}(qr) \frac{I_{d/2}\left(e^{-\frac{|y|}{R}}Rq\right)}{I_{d/2-1}(Rq)}. \tag{16.32}$$

The proof of this is left as an exercise. One needs to impose normalizability at $y = \pm\infty$ of the solutions of the two sides of $y = 0$, and the jump conditions (for Φ') obtained by integrating the equation of motion across $y = 0$. At $r \to \infty$ and $y = 0$, we have (the proof is again left as an exercise)

$$\Phi(r, y = 0) \simeq R_s \sqrt{\frac{2\pi R}{r}} C_1 e^{-M_1 r},$$

$$C_1 = \frac{j_{1,1}^{-1/2} J_2(j_{1,1})}{a_{1,1}}, \tag{16.33}$$

where $M_1 = j_{1,1}/R$ is the first KK mode for the effective theory when we reduce the 5-dimensional theory onto the IR brane, $j_{1,1} \simeq 3.83$ is the first zero of the first Bessel function, i.e.

$$J_1(z) \simeq a_{1,1}(z - j_{1,1}), \quad z \to j_{1,1}, \tag{16.34}$$

$R_s = G_{N,4}\sqrt{s}$ is the Schwarzschild radius corresponding to the energy of one of the colliding particles, $\sqrt{s}/2$. The prefactor in (16.33) could not be guessed, except the proportionality with the input energy $\sqrt{s}$, but the $\sim e^{-M_1 r}$ behavior is to be expected, since if we KK reduce the 5-dimensional theory down to the four dimensions of the IR brane, we expect at large distance the exponential decay of the first KK mode (the lightest mode) to dominate.

One can do an exact analysis of the black hole creation, but a simple argument shows that we will saturate the Froissart bound. The shockwave profile is $\Phi \sim \sqrt{s}e^{-M_1 r}$, so if we have two separated shockwaves at a distance (impact parameter) b, the emitted energy should be proportional to this $\sqrt{s}e^{-M_1 r}$. When the emitted energy equals the minimum, $\hat{M}_{Pl}$, to create a black hole in the gravity dual, we reach the maximum distance. Then

$$b_{max} \sim \frac{1}{M_1}\ln\frac{s}{\hat{M}_{Pl}}. \tag{16.35}$$

Therefore the total cross-section is

$$\sigma_{tot} = \pi b_{max}^2 \sim \frac{\pi}{M_1^2}\ln^2\frac{s}{\hat{M}_{Pl}}, \tag{16.36}$$

indeed saturating the Froissart bound.

Note that in this asymptotic regime, the gravitational wave scatterings in the gravity dual happen on the IR cut-off of the geometry, thus there are no quantum fluctuations in the fifth dimensional position of the black hole (the changes due to the Polchinski–Strassler integration are very small compared to the size of the created black hole), and the scattering is approximately classical, see Fig. 16.1c.

16.6 QCD plasmas and shockwave models for heavy ion collisions

In heavy ion collisions in experiments like RHIC and ALICE one obtains a hot strongly coupled plasma of quarks and gluons, as we discussed in the previous chapter. The collisions are at very high energies, $E \gg \hat{E}_R = N^2 \Lambda_{QCD}$, so it is natural to expect that we are already in the asymptotic regime, which corresponds to saturation of the Froissart unitarity bound in QCD and to classical scattering of gravitational shockwaves on the IR cut-off of the geometry in the gravity dual.

Also from the point of view of the standard description of the heavy ion collision process, we expect that classical scattering of the gravitational shockwaves in the gravity dual would give a good model. Indeed, highly boosted nuclei would look like pancakes because there is a longitudinal Lorentz contraction, but no transverse Lorentz contraction, or shockwaves in the ultrarelativistic limit. Moreover, the effective low energy pion field surrounding the nuclei would also get contracted, so in the limit we would get two colliding pion field shockwaves, see Fig. 16.2. This is in fact part of the 1952 Heisenberg model for saturation of the Froissart bound (note that the model was written before Froissart, and even before QCD, so Heisenberg's foresight is remarkable).

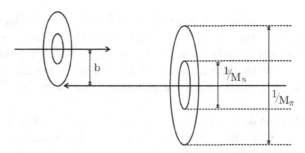

Figure 16.2 Scattering of nuclei or heavy ions at high energy in QCD described by the Heisenberg model. $1/M_N$ is the size of the nucleon, $1/M_\pi$ is the size of the pion field.

The Heisenberg model assumes that the probability to emit energy is proportional to the overlap of the pion wavefunctions of the colliding nuclei. The overlap decays with distance (impact parameter b) as $e^{-m_\pi b}$. Since the total energy of the collision is $\sqrt{s}$, the emitted energy is $\sim \sqrt{s}e^{-m_\pi b}$. The maximum impact parameter $b_{\max}$ corresponds to the case when this energy equals the average *per pion* emitted energy, $\langle E_0 \rangle$, thus

$$\sqrt{s}e^{-m_\pi b_{\max}} = \langle E_0 \rangle \Rightarrow b_{\max} = \frac{1}{m_\pi} \ln \frac{\sqrt{s}}{\langle E_0 \rangle} \Rightarrow$$

$$\sigma_{\text{tot}} = \pi b_{\max}^2 = \frac{\pi}{m_\pi^2} \ln^2 \frac{\sqrt{s}}{\langle E_0 \rangle}. \tag{16.37}$$

For a polynomial action for the scalar pions Heisenberg finds that $\langle E_0 \rangle \propto \sqrt{s}$, so is no good, but for the DBI action for the scalar (as needed in the case of AdS/CFT, where the scalar pion corresponds roughly to the fluctuation of the IR brane in the hard-wall model), Heisenberg finds that $\langle E_0 \rangle \simeq$ constant, leading to the saturation of the Froissart unitarity bound.

A good model for the heavy ion collisions via AdS/CFT is then the collision of two gravitational shockwaves on the IR brane of the hard-wall model, creating a black hole on the IR brane, that is in one-to-one correspondence with the sQGP fireball created in the heavy ion collisions, as proposed by myself in 2005. Moreover, as seen from the above, the picture matches with the Heisenberg model for the saturation of the Froissart bound.

We can obtain a lower bound on the entropy formed in these collisions if we can calculate the area of the marginally trapped surface formed at the point of collision of the shockwaves in the gravity dual, since the event horizon of the black hole being formed in the future of the collision is larger than, or equal to, the area of the marginally trapped surface, so

$$A_{\text{ev.hor.}} \geq A_{\text{marg.trap.}} \Rightarrow$$

$$S_{\text{emitted}} = S_{\text{BH}} = \frac{A_{\text{ev.hor.}}}{4G_{N,5}} \geq \frac{A_{\text{marg.trap.}}}{4G_{N,5}}. \tag{16.38}$$

The picture of a colliding gravitational shockwave in a gravity dual for heavy ion collisions has been employed recently also in the context of $\mathcal{N} = 4$ SYM, having in mind the same universality of gauge theories at finite temperature. For instance, one can calculate entropy production in the context of these gravitational shockwave collisions. The shockwave ansatz considered by Gubser *et al.* [75] is (for AdS in Poincaré coordinates)

$$ds^2 = \frac{R^2}{z^2}\left(2dx^+dx^- + (dx^1)^2 + (dx^2)^2 + dz^2\right) + \frac{R}{z}\Phi(x^1, x^2, z)\delta(x^+)(dx^+)^2, \quad (16.39)$$

and the energy-momentum tensor for the source is

$$T_{++} = E\delta(x^+)\delta(z - R)\delta^{d-2}(x^i). \quad (16.40)$$

From Einstein's equations, one finds

$$\Phi(x^1, x^2, z) = \frac{2G_{N,5}E}{R}\frac{1 + 8q(1+q) - 4\sqrt{q(1+q)}(1+2q)}{\sqrt{q(1+q)}},$$
$$q \equiv \frac{(x^1)^2 + (x^2)^2 + (z - R)^2}{4zR}. \quad (16.41)$$

Considering the extra term as a small perturbation of the AdS metric, it can be related to a VEV of the energy-momentum tensor $\langle T_{ij}\rangle$ on the boundary. Indeed, the metric is a "gauge field for the local diffeomorphism invariance," and it must then correspond to a global current on the boundary, specifically the energy-momentum tensor. This mapping is discussed in greater detail in Chapter 22, but here we will just quote the result,

$$\langle T_{ij}(\vec{x})\rangle = \frac{R^3}{4\pi G_{N,5}}\lim_{z\to 0}\frac{1}{z^4}\delta g_{ij}. \quad (16.42)$$

Substituting $\delta g_{ij} = R/z\Phi(x^1, x^2, z)\delta(x^+)$, we get

$$\langle T_{ij}(\vec{x})\rangle = \frac{2R^4E}{\pi\left[R^2 + (x^1)^2 + (x^2)^2\right]^3}\delta(x^+), \quad (16.43)$$

which should be compared with the shockwaves coming from colliding heavy nuclei. The entropy produced in the collision has a lower bound from (16.38), and Gubser *et al.* find $S \geq S_{\text{trap}} \propto E^{2/3}$, which, however, does not fit the observed data well, so one needs more precise calculations.

Important concepts to remember

- Since $\mathcal{N} = 4$ SYM is conformal, it does not have asymptotic states, so no S matrices. To define scattering, one must modify the duality and introduce a fundamental scale (break scale invariance).
- The simplest model for QCD is then the "hard-wall model," where we just cut off AdS_5 at an $r_{\min} = R^2\Lambda_{\text{QCD}}$, since a gravity dual of QCD would look like AdS_5 at large r (corresponding to the UV, where QCD is almost conformal), and would get cut off in some way at small r (corresponding to the IR, where QCD is confining).
- Gauge invariant scattered states (nucleons, mesons, glueballs) correspond to fields in $AdS_5 \times S^5$.

- The Polchinski–Strassler ansatz for scattering in QCD (or QCD-like models) is to consider the scattering amplitude in AdS space, at rescaled momenta, and convolute it with the wavefunctions in the extra dimensions for the gauge invariant states scattered.

- Corresponding to energy scales in AdS_5, we have energy scales in QCD. Corresponding to $1/R$ we have Λ_{QCD}, to the string scale $1/\sqrt{\alpha'}$ we have the QCD string scale $1/\sqrt{\hat{\alpha}'} = \Lambda(g_{YM}^2 N)^{1/4}$, to the Planck scale M_{Pl} we have the QCD Planck scale $\hat{M}_{Pl} = \Lambda N^{1/4}$.

- Regge behavior $\mathcal{A}(s,t) \sim \beta(t)s^{\alpha(t)}$ for $s \to \infty$, t fixed in string theory leads to Regge behavior for QCD via AdS/CFT.

- High energy scattering on the gravity side, for $\sqrt{s} > M_{Pl}$ (corresponding to $\sqrt{s} > \hat{M}_{Pl}$ in QCD), is described by collision of gravitational shockwaves of Aichelburg–Sexl type, sourced by $T_{++} = p\delta^{d-2}(x^i)\delta(x^+)$, and leading to black hole formation.

- Gravitational shockwaves are solutions for which gravity linearizes, which allows us to write exact solutions, and write a solution for the colliding shockwaves *before* the collision.

- If we find a marginally trapped surface at the collision point, we know there should be an event horizon, and thus a black hole, in the future of the collision, which allows us to calculate a maximum impact parameter b_{max}, and from it a cross section $\sigma_{tot} = \pi b_{max}^2$.

- At an energy $E_R = M_{Pl}^8 R^7$ on the gravity side, black holes created in the collision have size of order the AdS scale R, corresponding in QCD to an energy $\hat{E}_R = \Lambda N^2$.

- At a higher energy scale E_F, we start producing black holes on the IR brane (IR cut-off in a symmetric configuration around it) on the gravity side, and correspondingly at a scale $\hat{E}_F$ in QCD we start saturating the *Froissart unitarity bound* $\sigma_{tot} \leq \pi/m_\pi^2 \ln^2(s/s_0)$.

- In the gravity dual, the saturation of the unitarity bound arises from the exponential decay of the A–S type gravitational shockwave, $\Phi \sim \sqrt{s}e^{-M_1 r}$.

- The Heisenberg model saturates the Froissart bound, and has Lorentz-boosted nuclei giving pion field shockwaves, with profile $\Phi \sim e^{-m_\pi r}$.

- Heavy ion collisions can be modelled in the gravity dual by collisions of gravitational shockwaves with black hole formation, and the black hole corresponds to the sQGP fireball.

- The emitted entropy is bounded from below by the area of the marginally trapped surface formed during the gravitational shockwave collision.

References and further reading

The hard-wall model and the prescription for scattering of gauge invariant states in it was described by Polchinski and Strassler in [44]. The prescription for using gravitational shockwaves collisions in AdS/CFT to describe high energy QCD scattering, and in particular to obtain the Froissart bound, was given in [58, 59]. In [57] it was shown how to calculate b_{max} for gravitational shockwave collisions with black hole formation in flat four dimensions. The identification of the black holes on the IR brane in the gravity dual with the sQGP plasma fireball in heavy ion collisions was done in [74], and in [75] it was shown how to calculate entropy production in heavy ion collisions using gravitational shockwave collisions.

Exercises

1. Show that the Virasoro–Shapiro amplitude (16.11) reduces to (16.12) in the Regge limit.
2. Show that the only nonzero component of the Ricci tensor for the metric (16.14) is given by (16.15).
3. Show that the solution to (16.31) is given by (16.32).
4. Show that at large distances, (16.32) turns into (16.33).
5. Calculate the components of the Ricci tensor for (16.39). Show that Einstein's equations for the energy momentum-tensor given by (16.40), together with the AdS cosmological constant, are satisfied if Φ is given by (16.41).
6. Near the boundary at $r = \infty$, the normalizable solutions (wavefunctions) of the massive AdS Laplacean go as $(x_0^\Delta \sim) r^{-\Delta}$ (where $\Delta = 2h_+ = d/2 + \sqrt{d^2/4 + m^2 R^2}$). Substitute in the Polchinski–Strassler formula to obtain the r dependence of the integral at large r, and using that $r \sim 1/p$, estimate the hard scattering (all momenta of the same order, p) behavior of QCD amplitudes.

The pp wave correspondence

Until now, in AdS/CFT we have seen how to deal with the supergravity limit of string theory in $AdS_5 \times S^5$, but we have not seen any real string theory. Now we show how to obtain string theory in a particular limit, the Penrose limit, in which gravitational spaces turn to pp waves, leading to what has been called the pp wave correspondence.

17.1 The Penrose limit in gravity and pp waves

pp waves

We have already seen in the previous chapter what pp waves are. They are plane fronted gravitational waves ("parallel plane," or "pp"), which are solutions to Einstein's equations, corresponding to perturbations moving at the speed of light. In flat background, they have the form (16.14).

We also saw that the only nontrivial component of the Ricci tensor is R_{++}, satisfying (16.15), which can be guessed as follows. It must be proportional to g_{++}, the only nontrivial component. On dimensional grounds (since $R_{\mu\nu}$ contains two derivatives), we must have $R_{++} \propto \partial_i^2 g_{++}$. Because of general coordinate invariance, it cannot have $\partial_+ \partial_i$ or ∂_+^2, for instance, so ∂_i^2 is the only possibility. Then we obtain (16.15), which actually equals the linearized result. Thus for pp waves, the linearized solution is exact.

pp waves can be defined in pure Einstein gravity, supergravity, or any theory that includes gravity, the only difference being that the Einstein equation contains the energy momentum tensor of the extra matter.

11-dimensional pp waves

In particular, in the maximal 11-dimensional supergravity, we find a solution that has the same metric (16.14), together with

$$F_4 = dx^+ \wedge \phi : \quad F_{(4)+\mu_1\mu_2\mu_3} = \phi_{(3)\mu_1\mu_2\mu_3}, \tag{17.1}$$

where ϕ is a 3-form that satisfies (from the equations of motion of the action (7.127))

$$d\phi = 0: \quad \partial_{[\mu_1}\phi_{\mu_2\mu_3\mu_4]} = 0,$$
$$d*\phi = 0: \quad \partial_{[\mu}\epsilon_{\mu_1...\mu_8]}{}^{\mu_9\mu_{10}\mu_{11}}\phi_{\mu_9\mu_{10}\mu_{11}} = 0 \Leftrightarrow \partial^{\mu_1}\phi_{\mu_1\mu_2\mu_3} = 0,$$
$$-\partial_i^2 H = |\phi|^2: \quad -\frac{1}{2}\partial_i^2 H = \frac{2}{4!}\phi_{\mu\nu\rho}\phi^{\mu\nu\rho}. \tag{17.2}$$

As an observation, for $\phi = 0$ we have a solution with

$$H = \frac{Q}{|x - x_0|^7} \tag{17.3}$$

that corresponds to a D0-brane that is localized in space and time.

On the other hand, if

$$H = \sum_{ij} A_{ij} x^i x^j; \quad -2\text{Tr } A = |\phi|^2, \tag{17.4}$$

we have a solution that is not localized in space and time, since the spacetime is not flat at infinity. For $\phi = 0$ we have purely gravitational solutions that obey Tr $A = 0$. A solution for generic (A, ϕ) preserves 1/2 of the supersymmetry, namely the supersymmetry that satisfies $\Gamma_-\epsilon = 0$, where ϵ is a generic supersymmetry parameter.

There is, however, a very particular case, that was found by Kowalski-Glikman in 1984, that preserves ALL the supersymmetry. It is

$$A_{ij}x^i x^j = -\sum_{i=1,2,3}\frac{\mu^2}{9}x_i^2 - \sum_{i=4}^{9}\frac{\mu^2}{36}x_i^2,$$
$$\phi = \mu\, dx^1 \wedge dx^2 \wedge dx^3. \tag{17.5}$$

He also showed that the only background solutions that preserve all the supersymmetry of 11-dimensional supergravity are 11-dimensional Minkowski space, $AdS_7 \times S_4, AdS_4 \times S_7$ and the maximally supersymmetric wave (17.5).

We also noted in the last chapter that Horowitz and Steif showed in 1990 that pp waves are exact string solutions, since all possible α' corrections vanish on-shell.

10-dimensional pp waves

The case relevant for AdS/CFT, 10-dimensional type IIB string theory, which has $AdS_5 \times S_5$ as a background solution, also contains solutions of the pp wave type, with metric (16.14), together with

$$F_5 = dx^+ \wedge (\omega + *\omega): \quad F_{+\mu_1...\mu_4} = \omega_{\mu_1...\mu_4}; \quad F_{+\mu_5...\mu_8} = \omega_{\mu_5...\mu_8},$$
$$H = \sum_{ij} A_{ij} x^i x^j; \quad \phi = \phi_0, \tag{17.6}$$

satisfying (from the equations of motion of the action (7.116)

$$d\omega = 0: \quad \partial_{[\mu_1}\omega_{\mu_2...\mu_5]} = 0,$$
$$d*\omega = 0: \quad \partial^{\mu_1}\omega_{\mu_1...\mu_5} = 0,$$
$$\partial_i^2 H = -|\omega|^2: \quad -\frac{1}{2}\partial_i^2 H = \frac{1}{48}\omega_{\mu_1...\mu_n}\omega^{\mu_1...\mu_n}. \tag{17.7}$$

As in 11 dimensions, here the general metric preserves 1/2 of the supersymmetry defined by $\Gamma_- \epsilon = 0$. There is also a maximally supersymmetric solution, that has

$$H = -\frac{\mu^2}{64} \sum_i x_i^2; \quad \omega = \frac{\mu}{2} dx^1 \wedge dx^2 \wedge dx^3 \wedge dx^4. \tag{17.8}$$

Penrose limit

Formally, this says that in the neighborhood of a null geodesic, we can always put the metric in the form

$$ds^2 = dV \left(dU + \alpha dV + \sum_i \beta_i dY^i \right) + \sum_{ij} C_{ij} dY^i dY^j, \tag{17.9}$$

and then we can take the limit

$$U = u; \quad V = \frac{v}{R^2}; \quad Y^i = \frac{y^i}{R}; \quad R \to \infty, \tag{17.10}$$

and obtain a pp wave metric in u, v, y^i coordinates.

The interpretation of this procedure is: we boost along a direction, e.g. x, while taking the overall scale of the metric to infinity. The boost

$$t' = \cosh\beta \ t + \sinh\beta \ x; \quad x' = \sinh\beta \ t + \cosh\beta \ x, \tag{17.11}$$

implies

$$x' - t' = e^{-\beta}(x - t); \quad x' + t' = e^{\beta}(x + t), \tag{17.12}$$

so if we scale all coordinates (t, x and the rest, y^i) by $1/R$ and identify $e^{\beta} = R \to \infty$ we obtain (17.10).

We can show that the maximally supersymmetric pp waves are Penrose limits of maximally supersymmetric $AdS_n \times S_m$ spaces. In particular, the maximally supersymmetric IIB solution (17.8) is a Penrose limit of $AdS_5 \times S_5$. This can be seen as follows. We boost along an equator of S_5 and stay in the center of AdS_5, therefore expanding around the null geodesic defined by $\theta = 0$ (equator of S_5) and $\rho = 0$ (center of AdS_5), as in Fig. 17.1.

We obtain

$$ds^2 = R^2 \left(-\cosh^2\rho \ d\tau^2 + d\rho^2 + \sinh^2\rho \ d\Omega_3^2 \right) + R^2 \left(\cos^2\theta \ d\psi^2 + d\theta^2 + \sin^2\theta \ d\Omega_3'^2 \right)$$
$$\simeq R^2 \left[-\left(1 + \rho^2 \right) d\tau^2 + d\rho^2 + \rho^2 d\Omega_3^2 \right] + R^2 \left[\left(1 - \theta^2 \right) d\psi^2 + d\theta^2 + \theta^2 d\Omega_3'^2 \right]. \tag{17.13}$$

We then define the null coordinates $\tilde{x}^{\pm} = (\tau \pm \psi)/\sqrt{2}$, since ψ parameterizes the equator at $\theta = 0$. And we make the rescaling (17.10), i.e.

$$\tilde{x}^+ = x^+; \quad \tilde{x}^- = \frac{x^-}{R^2}; \quad \rho = \frac{r}{R}; \quad \theta = \frac{y}{R}, \tag{17.14}$$

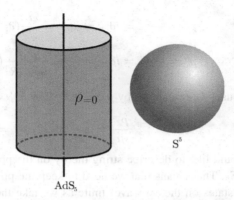

Figure 17.1   Null geodesic in $AdS_5 \times S^5$ for the Penrose limit giving the maximally supersymmetric wave. It is in the center of AdS_5, at $\rho = 0$, and on an equator of S_5, at $\theta = 0$.

and we get

$$ds^2 = -2dx^+dx^- - \mu^2(\vec{r}^2 + \vec{y}^2)(dx^+)^2 + d\vec{y}^2 + d\vec{r}^2. \qquad (17.15)$$

Here we have also introduced a parameter μ by rescaling $x^+ \to \sqrt{2}\mu x^+$, $x^- \to x^-/\mu\sqrt{2}$, for future use. This metric is the maximally supersymmetric wave (17.8), with μ rescaled by a factor of 8. The F_5 field also matches.

One can also similarly show that the maximally supersymmetric 11-dimensional pp wave is the Penrose limit of both the $AdS_4 \times S^7$ and the $AdS_7 \times S^4$ spaces.

17.2 The Penrose limit of AdS/CFT: large R-charge

AdS/CFT relates 4-dimensional $\mathcal{N} = 4$ SYM in flat Minkowski space to string theory in $AdS_5 \times S^5$, but we also saw that the Penrose limit of $AdS_5 \times S^5$ is the maximally supersymmetric pp wave. We want to understand then to what limit of $\mathcal{N} = 4$ SYM does the Penrose limit correspond. This was done by Berenstein, Maldacena, and myself (BMN) in 2002.

The energy is the Noether generator of time translations, τ in the global AdS_5 metric, hence the energy in AdS space is given by $E = i\partial_\tau$, and the Noether generator of rotations, the angular momentum for rotations in the plane of two coordinates X^5, X^6, is $J = -i\partial_\psi$, where ψ is the angle between X^5 and X^6.

But by the AdS/CFT dictionary the energy E corresponds to the conformal dimension Δ in $\mathcal{N} = 4$ SYM, whereas the angular momentum J corresponds to an R-charge, specifically a $U(1)$ subgroup of $SU(4) = SO(6)$ that rotates the scalar fields X^5 and X^6 corresponding in spacetime to the coordinates X^5 and X^6.

After taking the Penrose limit, in the pp wave background we have momenta $p^\pm$ defined as

$$p^- = -p_+ = i\partial_{x^+} = i\partial_{\tilde{x}^+} = \frac{i}{\sqrt{2}}(\partial_\tau + \partial_\psi) = \frac{1}{\sqrt{2}}(\Delta - J),$$

$$p^+ = -p_- = i\partial_{x^-} = i\frac{\partial_{\tilde{x}^-}}{R^2} = \frac{i}{\sqrt{2}R^2}(\partial_\tau - \partial_\psi) = \frac{\Delta + J}{\sqrt{2}R^2}. \tag{17.16}$$

For later use, we rescale p^- by $\mu\sqrt{2}$ and p^+ by $1/\mu\sqrt{2}$, obtaining

$$\frac{p^-}{\mu} = \Delta - J; \quad 2\mu p^\mu = \frac{\Delta + J}{R^2}. \tag{17.17}$$

We would like to describe string theory on the pp wave, which is the Penrose limit of $AdS_5 \times S_5$. That means that we need to keep the pp wave momenta p^+, p^- (momenta of physical states on the pp wave) finite as we take the Penrose limit. That means that we must take to infinity the radius of AdS space, $R \to \infty$, but keep $\Delta - J$ and $(\Delta + J)/R^2$ of $\mathcal{N} = 4$ SYM operators fixed in the limit. Therefore we must consider only SYM operators that have $\Delta \simeq J \sim R^2 \to \infty$, thus only operators with large R-charge!

We then conclude that *the Penrose limit corresponds to a large R-charge limit* in $\mathcal{N} = 4$ SYM.

We have already explained in Chapter 10 that from the supersymmetry algebra we can obtain the bound $\Delta \geq |J|$, in a similar manner to the condition $M \geq |Q|$, the BPS condition, which means that $p^\pm > 0$. Operators that saturate the bound belong in short multiplets. Since $R^2/\alpha' = \sqrt{4\pi g_s N} = \sqrt{g_{YM}^2 N}$, if we keep g_s fixed, $J \sim R^2$ means that $J/\sqrt{N}$ is fixed, or in other words that we consider only operators with R-charge $J \sim \sqrt{N}$.

Thus the Penrose limit corresponds in $\mathcal{N} = 4$ SYM to considering the sector of operators with R-charges $J \sim \sqrt{N}$.

Note, however, that there is a better way to think about the Penrose limit of $AdS_5 \times S^5$ and of AdS/CFT.

The point is that if we choose $\tilde{x}^\pm = (\psi \pm \tau)/\sqrt{2}$, since ψ is periodic with period 2π, the "light-cone time" $x^+ = \tilde{x}^+$ is periodically identified with period $2\pi/\sqrt{2}$, which is not good for a time variable.

A better prescription (though maybe where the physical interpretation of the Penrose limit is not so clear) is to take instead

$$x^+ = \tau, \quad x^- = R^2(\psi - \tau), \quad r = \rho R, \quad y = \theta R, \tag{17.18}$$

and then take $R \to \infty$. Then we again get (17.15), and now we have only x^- periodic, with period $2\pi R^2 \to \infty$, but the lightcone time x^+ is not periodic.

When mapping to the CFT through AdS/CFT, the energy is still

$$p^- = i\partial_{x^+} = i(\partial_\tau + \partial_\psi) = \Delta - J, \tag{17.19}$$

but the momentum is now

$$p^+ = i\partial_{x^-} = -\frac{i}{R^2}\partial_\psi = \frac{J}{R^2}, \tag{17.20}$$

and then J is naturally quantized, because x^- is periodic with period $2\pi R^2$.

We will nevertheless still use (17.14) and (17.16) when no confusion arises.

17.3 String quantization and Hamiltonian on the pp wave

Considering the bosonic Polyakov action for the string in the pp wave background, we obtain

$$S = -\frac{1}{2\pi\alpha'} \int_0^l d\sigma \int d\tau \frac{1}{2}\sqrt{-\gamma}\gamma^{ab}\left[-2\partial_a x^+ \partial_b x^- - \mu^2 x_i^2 \partial_a x^+ \partial_b x^+ + \partial_a x^i \partial_b x^i\right].$$
(17.21)

Here $x^i = (\vec{r}, \vec{y})$. Consider the conformal gauge $\sqrt{-\gamma}\gamma^{ab} = \eta^{ab}$, and the light-cone gauge $x^+(\sigma, \tau) = p^+\tau$. But we want to slightly change the light-cone gauge convention, by choosing $x^+(\sigma, \tau) = \tau$. Then $\eta^{ab}\partial_a x^+ \partial_b x^- = 0$ and $\eta^{ab}\partial_a x^+ \partial_b x^+ = -1$, and we obtain

$$S = -\frac{1}{2\pi\alpha'} \int d\tau \int_0^l d\sigma \left[\frac{1}{2}\eta^{ab}\partial_a x^i \partial_b x^i + \frac{\mu^2}{2}x_i^2\right].$$
(17.22)

The free wave equation in flat space, $\partial^2 x^i/\partial\tau^2 = \partial^2 x^i/\partial\sigma^2$ before, becomes under $\tau \to \tau/p^+$,

$$\frac{\partial^2}{\partial\tau^2}x^i = \frac{1}{(p^+)^2}\frac{\partial^2}{\partial\sigma^2}x^i.$$
(17.23)

That means that we must also rescale the length of the string, and instead of $l = 2\pi$ have arbitrary l. Also introducing α' to match dimensions, we get

$$\frac{\partial^2}{\partial\tau^2}x^i = c^2 \frac{\partial^2}{\partial\sigma^2}x^i,$$
(17.24)

where

$$c = \frac{l}{2\pi\alpha'p^+}.$$
(17.25)

Therefore, in order to have $c = 1$ for convenience, we choose $l = 2\pi\alpha'p^+$, and we then get for the gauge-fixed action in the pp wave background

$$S = -\frac{1}{2\pi\alpha'} \int d\tau \int_0^{2\pi\alpha'p^+} d\sigma \left[\frac{1}{2}(-(\dot{x}^i)^2 + (x'^i)^2) + \frac{\mu^2}{2}x_i^2\right].$$
(17.26)

The equation of motion in the pp wave background is then

$$(-\partial_\tau^2 + \partial_\sigma^2)x^i - \mu^2 x^i = 0.$$
(17.27)

We expand a solution of the equations of motion in plane waves $x^i \propto e^{-i\omega_n\tau + ik_n\sigma}$, which implies

$$\omega_n^2 = k_n^2 + \mu^2.$$
(17.28)

But in flat space, at $\mu = 0$, we had $\omega_n = k_n = n$, since the length of σ was 2π. After our rescalings, we have $k_n = n/\alpha'p^+$, both in flat space and in the pp wave, since k_n is defined simply by Fourier expansion on the circle of length $2\pi\alpha'p^+$. Therefore, we obtain

$$\omega_n = \sqrt{\mu^2 + \frac{n^2}{(\alpha'p^+)^2}}.$$
(17.29)

As we saw in Chapter 7 (Eq. (7.77) for the bosonic string), the light-cone Hamiltonian for the open string in flat space is

$$H_{\text{l.c.}} = p^- = -p_+ = \frac{p^i p^i - 1/\alpha'}{2p^+} + \frac{1}{2} \sum_{n \geq 1} \omega_n N_n. \tag{17.30}$$

Here N_n is the total occupation number,

$$N_n = \sum_i a_n^{i\,\dagger} a_n^i + \sum_\alpha b_n^{\alpha\dagger} b_n^\alpha , \tag{17.31}$$

where $n > 0$ are left-movers and $n < 0$ are right-movers.

But now there are no usual zero modes p^i, since the x^is are massive, rather they become simply modes of $n = 0$.

Then for the closed string on the pp wave we get the light-cone Hamiltonian

$$H = \sum_{n \in \mathbb{Z}} N_n \omega_n. \tag{17.32}$$

Here we have considered $n < 0$ for left-movers and $n > 0$ for right-movers, we have included the zero modes $n = 0$, and there are eight transverse oscillators, as before.

We also have the condition that the total momentum along the closed string should be zero, by translational invariance, the same as for the flat space string, giving

$$P = \sum_{n \in \mathbb{Z}} n N_n = 0. \tag{17.33}$$

A physical state is defined by the n_i and p^+, i.e. it is $|\{n_i\}, p^+\rangle$.

Finally, we can verify that the flat space limit $\mu \to 0$ gives

$$2p^+ p^- = \frac{1}{\alpha'} \sum_n n N_n , \tag{17.34}$$

which is indeed the flat space spectrum in light-cone quantization.

We also can consider a limit of highly curved background, $\mu \alpha' p^+ \gg 1$, in which case we can expand the square root,

$$\omega_n \simeq \mu \left(1 + \frac{1}{2} \frac{n^2}{(\alpha' \mu p^+)^2} \right) , \tag{17.35}$$

that from the point of view of $\mathcal{N} = 4$ SYM is a perturbative expansion.

For comparison with the field theory, we translate the result into $\mathcal{N} = 4$ SYM variables. We use that $E/\mu = (\Delta - J)$, $2\mu p^+ = (\Delta + J)/R^2 \simeq 2J/R^2 = 2J/(\alpha'\sqrt{g_{\text{YM}}^2 N}$, and obtain

$$(\Delta - J)_n = w_n = \sqrt{1 + \frac{g_{\text{YM}}^2 N n^2}{J^2}} , \tag{17.36}$$

where we are in the limit that

$$\frac{g_{\text{YM}}^2 N}{J^2} = \text{fixed}. \tag{17.37}$$

17.4 String states from $\mathcal{N} = 4$ SYM; BMN operators

We have seen that corresponding to string states of given energy on the pp wave we have operators of given $\Delta - J$, which at zero coupling should have $\Delta - J = 1$. The vacuum of course has zero energy, therefore $\Delta - J = 0$. Thus we organize fields according to their $\Delta - J$, and construct operators from them.

The charge J was defined as rotating X^5, X^6 in spacetime, and correspondingly in $\mathcal{N} = 4$ SYM rotating the scalars Φ^5, Φ^6. Then we write

$$Z = \Phi^5 + i\Phi^6 , \tag{17.38}$$

defined to have charge $+1$ under the rotation, thus $\bar{Z}$ has charge -1. The rest of the scalars, Φ^m, $m = 1, \ldots, 4$ are neutral. The gauge fields A_μ are also neutral, which means that

$$D_\mu Z = \partial_\mu Z + [A_\mu, Z] \tag{17.39}$$

also has charge $+1$. However, the fermions are charged under the R symmetry as well, and if we consider χ^I, they are in the spinor representation of $SO(6)$, which means that under a $U(1) = SO(2)$ subgroup they have charge $\pm 1/2$. We write the eight components of positive charge as $\chi^a_{J=+1/2}$ and the eight components of negative charge as $\chi^a_{J+-1/2}$. In terms of dimensions, as usual Z, Φ^m, and A_μ have $\Delta = 1$, and χ have $\Delta = 3/2$.

Then the fields are organized by their $g_{YM} = 0$ values for $\Delta - J$ as follows. The unique combination with $\Delta - J = 0$ is Z, the combinations with $\Delta - J = 1$ are $(D_\mu Z, \Phi^m)$ for bosons and $\chi^a_{J=+1/2}$ for fermions, the combinations with $\Delta - J = 2$ are $\bar{Z}$ and $\chi^a_{J=-1/2}$.

The vacuum state must be composed only of Zs, in order to have $\Delta - J = 0$, and must have momentum p^+ corresponding to charge J, hence we can identify the vacuum with the operator

$$|0, p^+\rangle = \frac{1}{\sqrt{J}N^{J/2}} \text{Tr} \, [Z^J]. \tag{17.40}$$

Note that this vacuum has momentum $2\mu p^+ = 2J/R^2$, i.e. J units of momentum.

The string oscillators (creation operators) on the pp wave at the $n = 0$ level are eight bosons and eight fermions of light-cone energy $p^- = \mu$, therefore should correspond to fields of $\Delta - J = 1$ to be inserted inside the operator corresponding to the vacuum, (17.40). They must be gauge covariant, in order to obtain a gauge invariant operator by taking a trace. It is easy to see then that the unique possibility is the eight $\chi^a_{J=+1/2}$ for the fermions and the four ϕ^is, together with four covariant derivatives $D_\mu Z = \partial_\mu Z + [A_\mu, Z]$ for the bosons. We have replaced the four A_μs with the covariant derivatives $D_\mu Z$ in order to obtain a covariant object.

These fields are to be inserted inside the trace of the vacuum operator (17.40), for example a state with $2 n = 0$ excitations will be ($a^\dagger_{0,r}$ corresponds to Φ^r and $b^\dagger_{0,b}$ to $\psi^b_{J=1/2}$)

$$a^\dagger_{0,r} b^\dagger_{0,b} |0, p^+\rangle = \frac{1}{N^{J/2+1}\sqrt{J}} \sum_{l=1}^{J} \text{Tr} \, [\Phi^r Z^l \psi^b_{J=1/2} Z^{J-l}] , \tag{17.41}$$

where we have put the Φ^r field on the first position in the trace by cyclicity of the trace.

The string oscillators at excited levels $n \geq 1$ are obtained in a similar manner. But now they correspond to excitations that have a momentum wavefunction $e^{2\pi i n x/L}$ around the closed string of length L. Since the closed string is modelled by the vacuum state $\sim \text{Tr}\,[Z^J]$, the appropriate operator corresponding to an $a_{n,4}^\dagger$ insertion is

$$a_{n,4}^\dagger |0, p^+\rangle = \frac{1}{\sqrt{J}} \sum_{l=1}^{J} \frac{1}{\sqrt{J}N^{J/2+1/2}} \text{Tr}\,[Z^l \Phi^4 Z^{J-l}] e^{\frac{2\pi i n l}{J}}. \qquad (17.42)$$

Actually, this operator vanishes by cyclicity of the trace, and the corresponding string state does not satisfy the equivalent zero momentum constraint (cyclicity). In order to obtain a nonzero state, we must introduce at least two such insertions, such that the total momentum vanishes.

An example of a physical state, that corresponds to an operator that does not vanish, is obtained by acting with a bosonic oscillator with momentum n and another with momentum $-n$,

$$a_{n,4}^\dagger a_{-n,3}^\dagger |0, p^+\rangle = \frac{1}{\sqrt{J}} \sum_{l=1}^{J} \frac{1}{N^{J/2+1}} \text{Tr}\,[\Phi^3 Z^l \Phi^4 Z^{J-l}] e^{\frac{2\pi i n l}{J}}. \qquad (17.43)$$

We consider here the "dilute gas approximation," where there are very few "impurities" among the J Zs. These operators have been called BMN operators in the literature.

17.5 The discretized string action from $\mathcal{N} = 4$ SYM

We can also define a Hamiltonian that acts on the BMN operators, corresponding to string states. In the previous subsection, we have simplified things, and we had mapped an operator like

$$\mathcal{O} = \frac{1}{\sqrt{J}N^{J/2+1/2}} \text{Tr}\,[Z^l \Phi Z^{J-l}] \qquad (17.44)$$

to a string state with a single creation operator acting on a vacuum.

At the end of Chapter 10, we saw the operator–state correspondence in four dimensions, and how it relates terms in the Taylor expansion of fields $z_{a_1 \ldots a_m}^{(m)} \sim \partial_{a_1} \ldots \partial_{a_m} Z$ on $\mathbb{R}^4$ with states for the KK expansion on $\mathbb{R}_t \times S^3$. In global AdS/CFT, the $\mathbb{R}_t \times S^3$ exists at the boundary of global AdS, whereas $\mathbb{R}^4$ exists at the boundary of the Poincaré patch of AdS.

Then the constant term for the scalar Z corresponds to a state of energy 1, i.e. a harmonic oscillator of unit frequency, with creation operator $b^\dagger$. We also find a harmonic oscillator, with creation operator $a^\dagger$ for the state corresponding to a Φ scalar. But the $SU(N)$ index structure is also important, which means that the order of creation operators inside a trace is important, and we have a state denoted by

$$|a_l^\dagger\rangle \equiv \text{Tr}\,\left[(b^\dagger)^l a^\dagger (b^\dagger)^{J-l}\right] |0\rangle. \qquad (17.45)$$

In order to consider the order of oscillators in a "word," without the need to use the $SU(N)$ index structure, we need to consider instead of usual harmonic oscillators, "Cuntz oscillators" a_i, $i = 1, \ldots, n$, satisfying

$$a_i|0\rangle = 0, \quad a_i a_j^\dagger = \delta_{ij}, \quad \sum_{i=1}^{n} a_i^\dagger a_j = 1 - |0\rangle\langle 0| , \tag{17.46}$$

and no other relations (in particular, no relations between $a_i^\dagger a_j^\dagger$ and $a_j^\dagger a_i^\dagger$, so the order is important).

For a single Cuntz oscillator, we have

$$a|0\rangle = 0, \quad a a^\dagger = 1, \quad a^\dagger a = 1 - |0\rangle\langle 0|. \tag{17.47}$$

One can consider a further simplification by instead of using J $b^\dagger$s and $a^\dagger$ impurities, using J different *independent* Cuntz oscillators $a_i^\dagger$ at each site (here $j = 1, \ldots, J$ corresponds to sites), therefore satisfying

$$[a_i, a_j] = [a_i^\dagger, a_j] = [a_i^\dagger, a_j^\dagger] = 0, \quad i \neq j$$
$$a_i a_i^\dagger = 1, \quad a_i^\dagger a_i = 1 - (|0\rangle\langle 0|)_i; \quad a_i|0\rangle_i = 0. \tag{17.48}$$

Then the action of the interacting piece of the $\mathcal{N} = 4$ SYM Hamiltonian,

$$H_{\text{int}} = -g_{\text{YM}}^2 \text{Tr} \sum_{I>J} \left\{ [\Phi^I, \Phi^J][\Phi_I, \Phi_J] \right\} , \tag{17.49}$$

which in terms of Z and Φ^m is given by

$$H_{\text{int}} = -g_{\text{YM}}^2 \text{Tr} \left\{ [Z, \Phi^m][\bar{Z}, \Phi^m] \right\} \rightarrow 2g_{\text{YM}}^2 N[b^\dagger, \phi][b, \phi]; \quad \phi = \frac{a + a^\dagger}{\sqrt{2}} , \tag{17.50}$$

on states corresponding to operators $\mathcal{O}$ can be defined through the action of Feynman diagrams for the 2-point functions $\langle \mathcal{O}\mathcal{O} \rangle$. A relevant Feynman diagram for the 2-point function is given in Fig. 17.2.

The interacting Hamiltonian (17.50) becomes, when acting on states corresponding to operators,

$$H_{\text{int}} = 2g_{\text{YM}}^2 N \sum_{j} (\phi_j - \phi_{j+1})^2 , \tag{17.51}$$

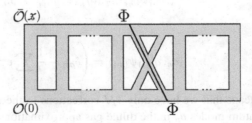

Figure 17.2   Feynman diagram for the 2-point function of $\mathcal{O}(x)$ at one-loop.

where as before $\phi_j = (a_j + a_j^\dagger)/\sqrt{2}$ and $\lambda = g^2 N$. For instance, the term with two $a^\dagger$s, $\sum_j (a_j^\dagger a_{j+1}^\dagger + a_{j+1}^\dagger a_j^\dagger - (a_j^\dagger)^2 - (a_{j+1}^\dagger))^2$ comes from the contraction of the $\bar{Z}$ in $H_{\text{int}} \sim \bar{Z}\Phi Z\Phi + Z\Phi\bar{Z}\Phi - \bar{Z}\Phi^2 Z - \bar{Z}Z\Phi^2$.

As we see, the Feynman diagram coming from this corresponds to a nearest-neighbor interaction, hopping from site l to site $l+1$ or $l-1$.

As we said, the harmonic oscillators $a^\dagger$ have frequency one, meaning we also have a kinetic term $\sum_j a_j^\dagger a_j$, though because the oscillators are Cuntz, we need to write $\sum_j (a_j^\dagger a_j + a_j a_j^\dagger)/2$, for a total Hamiltonian

$$H = \sum_{j=1}^{J} \frac{a_j^\dagger a_j + a_j a_j^\dagger}{2} + \frac{2\lambda}{(2\pi)^2} \sum_{j=1}^{J} (\phi_j - \phi_{j+1})^2. \tag{17.52}$$

The kinetic term is a discrete version of the continuum of the harmonic oscillator chain, $\int dx [\dot{\phi}^2(x) + \phi^2(x)]$, where ϕ_j is a discretized version of the relativistic field $\phi(x,t)$, and $(\phi_{j+1} - \phi_j)^2$ is the discretized version of $\phi'^2(x,t)$.

Then the continuum version of the Hamiltonian is

$$H = \int_0^L d\sigma \frac{1}{2} [\dot{\phi}^2 + \phi'^2 + \phi^2]. \tag{17.53}$$

The length of the chain is (from the fact that $\phi'^2/2$ comes from the last term in (17.52))

$$L = \frac{2\pi J}{\mu\sqrt{\lambda}} = 2\pi\alpha' p^+, \tag{17.54}$$

as expected. This is indeed the Hamiltonian of the light-cone string on a pp wave, and it was derived from the $\mathcal{N} = 4$ SYM theory.

One subtle point is the use of Cuntz oscillators, but a correct treatment in the "dilute gas approximation" of the Hamiltonian (17.52) leads indeed to the frequency (17.36), as obtained from the string Hamiltonian.

One first defines "Fourier modes" (momentum space) for the Cuntz oscillators a_j, by

$$a_j = \frac{1}{\sqrt{J}} \sum_{n=1}^{J} e^{\frac{2\pi i j n}{J}} a_n. \tag{17.55}$$

When acting on states in the dilute gas approximation, of the form

$$|\psi_{\{n_i\}}\rangle = |0\rangle_1 \ldots |n_{i_1}\rangle \ldots |n_{i_k}\rangle \ldots |0\rangle_J, \tag{17.56}$$

we get

$$[a_n, a_m^\dagger]|\psi_{\{n_i\}}\rangle = \left(\delta_{nm} - \frac{1}{J} \sum_k e^{2\pi i i_k \frac{m-n}{J}} \right) |\psi_{\{n_i\}}\rangle. \tag{17.57}$$

That means that we have only $1/J$ corrections to the usual commutation relations for the momentum modes a_n in the dilute gas approximation. We have $[a_n, a_m^\dagger] \simeq \delta_{nm} + O(1/J)$. Further defining

$$a_n = \frac{c_{n,1} + c_{n,2}}{\sqrt{2}},$$

$$a_{J-n} = \frac{c_{n,1} - c_{n,2}}{\sqrt{2}}, \tag{17.58}$$

and performing a Bogoliubov transformation rotating the creation and annihilation operators as

$$\tilde{c}_{n,1} = a_n c_{n_1} + b_n c_{n,1}^\dagger,$$
$$\tilde{c}_{n,2} = a_n c_{n_1} - b_n c_{n,1}^\dagger,$$
$$a_n = \frac{(1+\alpha_n)^{1/4} + (1+\alpha_n)^{-1/4}}{2},$$
$$b_n = \frac{(1+\alpha_n)^{1/4} - (1+\alpha_n)^{-1/4}}{2}, \tag{17.59}$$

where

$$\alpha_n = \frac{\lambda}{8\pi^2}(\cos(2\pi n/J) - 1) = -\frac{\lambda}{(2\pi)^2}\sin^2\frac{\pi n}{J}, \tag{17.60}$$

the Hamiltonian becomes diagonal, with frequency ω_n given by

$$\omega_n = \sqrt{1 + 4|\alpha_n|} = \sqrt{1 + \frac{4\lambda}{(2\pi)^2}\sin^2\frac{\pi n}{J}}. \tag{17.61}$$

The corresponding Fock states are

$$c_{n,1/2}^\dagger|0\rangle = \frac{a_n^\dagger \pm a_{J-n}^\dagger}{\sqrt{2}}|0\rangle = \frac{1}{\sqrt{J}}\sum_{j=1}^{J}\frac{e^{\frac{2\pi ijn}{J}} \pm e^{-\frac{2\pi ijn}{J}}}{\sqrt{2}}a_j^\dagger|0\rangle. \tag{17.62}$$

These are mapped to the BMN operators with $\cos(2\pi nj/J)$ and $i\sin(2\pi nj/J)$.

Note that for $n \ll J$, the result for the frequency reduces to the Hamiltonian of the string on the pp wave.

17.6 Spin chain interpretation

The above result for the Cuntz oscillator eigenfrequencies is exact both in λ and in n/J, as long as $J \to \infty$ and we are in the dilute gas approximation, with a small number M of impurities (excitations) $\ll J$. The calculation in a sense resums the one-loop SYM calculations through the Bogoliubov transformation to obtain the square root form. But one can generalize the case of $M \ll J$ to a *spin chain*, with the number of excitations $M \sim J$.

Indeed, the description of the insertion of Φs and their corresponding $a_l^\dagger$ operators among a loop of Zs and their corresponding $b^\dagger$ operators reminds one of spin chains. A spin chain is a 1-dimensional system of length L of spins with only up $|\uparrow\rangle$ or down $|\downarrow\rangle$ degrees of freedom, as in Fig. 17.3. This equivalence can in fact be made exact, as we discuss in the next chapter.

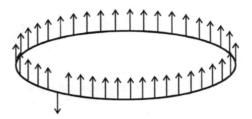

A periodic spin chain of the type that appears in the pp wave string theory. All spins are up, except that one excitation has one spin down.

Then the string on the pp wave is obtained in the dilute gas approximation of the spin chain.

The first order contribution to the energy of the string can be obtained from the one-loop correction to the anomalous dimension of the BMN operator. We are in the 't Hooft planar limit when we consider AdS/CFT, hence the one-loop correction, involving only a single 4-point SYM interaction, can only connect nearest neighboring fields in the operator.

To obtain the one-loop correction to Δ, we consider the one-loop correction to the two point function of operators in x-space, $\langle \mathcal{O}(x)\mathcal{O}(0)\rangle$. Then the interaction term in the Lagrangean,

$$\mathcal{L}_{\text{int}} = 2g_{\text{YM}}^2 \text{Tr} \ [Z, \Phi^m][\bar{Z}, \Phi^m]$$
$$= 2g_{\text{YM}}^2 \left(2\text{Tr} \ [\Phi^m Z \Phi^m \bar{Z}] - \text{Tr} \ [(Z\bar{Z} + \bar{Z}Z)\Phi^m \Phi^m]\right), \qquad (17.63)$$

leads *in the planar limit* only to interactions that change the site l of the impurity Φ^m in the BMN operator $\mathcal{O}$ to $l \pm 1$, as we can see from Fig. 17.4b and c. Diagrams that connect fields further away are nonplanar (cannot be drawn on a plane without crossing lines), as we can easily check. The relevant diagrams are the tree diagram and the two one-loop "hopping" diagrams in Fig. 17.4 a, b and c. The result is easily seen to be

$$\langle \mathcal{O}(x)\mathcal{O}^*(0)\rangle = \frac{\mathcal{N}}{|x|^{2J+2}} \left[1 + g_{\text{YM}}^2 N I(x) \left(e^{\frac{2\pi i n}{J}} + e^{-\frac{2\pi i n}{J}}\right)\right]. \qquad (17.64)$$

Here the N factor comes from doing the index loop contractions due to 't Hooft's double-line notation (there is an extra index loop for the one-loop correction with respect to the tree result), and the exponential factors come from the difference between the $e^{\frac{2\pi i n l}{J}}$ factor in the operator on the top and the $e^{\frac{2\pi i n(l\pm1)}{J}}$ factor in the operator on the bottom, and the integral $I(x)$ is

$$I(x) = \frac{|x|^4}{(4\pi^2)^2} \int d^4 y \frac{1}{y^4(x-y)^4} \sim \frac{1}{4\pi^2} \log(|x|\Lambda) + \text{finite}, \qquad (17.65)$$

where we have dimensionally regularized the integral and extracted the log-divergent piece. The regularization of this integral (cut-off or dimensional regularization) is left as an exercise.

We also need to consider that there are other diagrams, like d and e in Fig. 17.4. As we can see, they do not change the site number l, so do not depend on the momentum factor

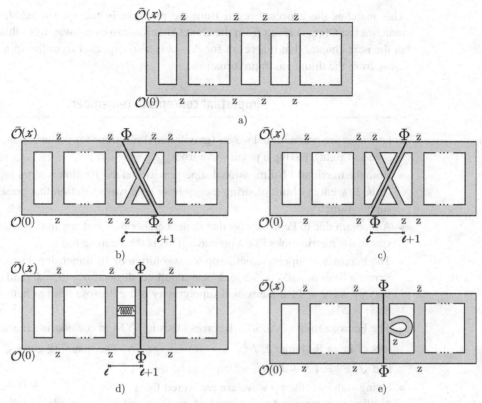

Figure 17.4 Planar Feynman diagrams for the 2-point function of $\mathcal{O}$: a) The planar tree level diagram; b) Planar one-loop Feynman diagram with hopping from $l + 1$ to l; c) Planar one-loop diagram with hopping from l to $l + 1$; d) One-loop planar diagram with gluon exchange; e) One-loop planar diagram with scalar self-energy.

$e^{\frac{2\pi i n l}{J}}$, meaning that they can be calculated for $n = 0$. But at $n = 0$, we have protected operators, which do not get quantum corrections, so the sum of all contributions at $n = 0$ must be 0, which means that in (17.64) we replace the sum of the exponentials with the same, minus 2. Finally, we obtain

$$\langle \mathcal{O}(x)\mathcal{O}^*(0)\rangle = \frac{\mathcal{N}}{|x|^{2J+2}}\left[1 - 2g_{\mathrm{YM}}^2 N\left(\cos\left(\frac{2\pi n}{J}\right) - 1\right)\frac{1}{4\pi^2}\log(|x|\Lambda)\right]. \quad (17.66)$$

But this is to be compared with the expansion of the 2-point function in $g_{\mathrm{YM}}^2 N$ for Δ:

$$\langle \mathcal{O}(x)\mathcal{O}^*(0)\rangle = \frac{\mathcal{N}}{|x|^{2(J+1+(\Delta-J)[g^2N])}} \simeq \frac{\mathcal{N}}{|x|^{2(J+1)}}\left[1 + 2(\Delta - J)[g^2N]\ln(|x|) + \dots\right], \quad (17.67)$$

from which we deduce

$$(\Delta - J)_n = \left[1 + \frac{g_{\mathrm{YM}}^2 N}{2\pi^2}\sin^2\left(\frac{\pi n}{J}\right)\right]. \quad (17.68)$$

This matches the expected result from the pp wave in the limit $n \ll J$, but moreover matches the $O(g_{\text{YM}}^2)$ result from the Cuntz Hamiltonian even away from this limit. We see in the next chapter that this result for $\Delta - J$ is also expected from the spin chain picture, away from the dilute gas approximation.

Important concepts to remember

- pp waves are gravitational waves (gravitational solutions for perturbations moving at the speed of light), having a plane wave front.
- Both the maximal 11-dimensional supergravity and the 10-dimensional supergravity that is the low energy limit of string theory have a pp wave solution that preserves maximal supersymmetry.
- A theorem due to Penrose says that in the neighborhood of any null geodesic in a curved space, the metric looks like a pp wave. This is the Penrose limit.
- The maximally supersymmetric pp wave solution of 10-dimensional supergravity is the Penrose limit of $AdS_5 \times S_5$: look near a null geodesic at $\rho = \theta = 0$ (and the maximally susy pp wave of 11-dimensional supergravity is the Penrose limit of both $AdS_4 \times S^7$ and $AdS_7 \times S^4$).
- The Penrose limit of AdS/CFT corresponds in SYM to considering the sector of operators of large R charge J, $J \simeq \Delta \sim R^2 \sim \sqrt{g_{\text{YM}}^2 N}$. The mapping gives $\sqrt{2}p^- = \Delta - J$ and $\sqrt{2}p^+ = (\Delta + J)/R^2$, which must be fixed.
- String states on the pp wave are recovered from AdS/CFT if $g_{\text{YM}}^2 N/J^2 = $ fixed. String oscillators correspond to insertion of $\phi^i, D_\mu Z$ and $\chi_{J=1/2}^a$ inside Tr $[Z^J]$ (which corresponds to the vacuum $|0, p^+\rangle$ with J units of p^+), with some momentum $e^{2\pi i n/J}$. The modes with $n = 0$ have $\Delta - J = 1$ at all couplings.
- The discretized string action is obtained from the action of $\mathcal{N} = 4$ SYM Feynman diagrams on operators $\mathcal{O}$ in their two-point function $\langle \mathcal{O}(0)\mathcal{O}(x)\rangle$. Thus the long operator acts as a discretized closed string.
- The oscillators appearing from SYM are Cuntz oscillators, but when acting on states in the dilute gas approximation, the creation oscillators for the eigenmodes obey the usual relations.

References and further reading

The pp wave correspondence was defined by Berenstein, Maldacena, and myself in [45] (BMN). For a review of the correspondence, see for instance [46]. The maximally supersymmetric 11-dimensional supergravity pp wave was described by Kowalski-Glikman in [47]. The Aichelburg–Sexl shockwave was described in [48]. Horowitz and Steif [49] proved that pp waves are solutions of string theory exact in α'. The Penrose limit can be found in [50], whereas its physical interpretation is described in [45]. The type IIB maximally supersymmetric plane wave was found in [51] and it was shown to be the Penrose limit of $AdS_5 \times S^5$ in [45, 52].

Exercises

1. An Aichelburg–Sexl shockwave is a gravitational solution given by a massless source of momentum p, i.e. $T_{++} = p\delta(x^+)\delta(x^i)$. Find the function $H(x^+, x^i)$ defining the pp wave on a *UV brane*, with $e^{-\frac{2|y|}{R}}$ instead of $e^{\frac{2|y|}{R}}$ in (16.29).

2. If the null geodesic moves on S^5, one can choose the coordinates such that it moves on an equator, thus the Penrose limit gives the maximally supersymmetric pp wave. Show that if instead the null geodesic moves on AdS_5, the Penrose limit gives 10-dimensional Minkowski space (again choose $\rho = 0$).

3. The Killing spinor equation in 10-dimensional type IIB theory is

$$\mathcal{D}_M\epsilon = \nabla_M\epsilon + \frac{i}{24}F_{ML_1...L_4}\Gamma^{L_1...L_4}\epsilon. \qquad (17.69)$$

   Show that for a generic pp wave, we have 1/2 supersymmetry preserved, but for the maximally supersymmetric pp wave all the susy is preserved, and find the solution $\epsilon(x^+, x^i)$ in terms of an independent constant spinor parameter ψ.

4. Write down *all* the $\mathcal{N} = 4$ SYM fields (including derivatives) with $\Delta - J = 2$.

5. Check that, by cyclicity of the trace, the operator with two insertions of Φ^1, Φ^2 at levels $+n$ and $-n$ equals (up to normalization)

$$Tr[\Phi^1 Z^l \Phi^2 Z^{J-l}]. \qquad (17.70)$$

6. Fill in the details of the diagonalization of the Cuntz Hamiltonian (17.52) to obtain the eigenenergies (17.61).

7. Prove the regularization (cut-off or dimensional regularization) of the integral (17.65).

Spin chains

18.1 The Heisenberg XXX spin chain Hamiltonian, H_{XXX}

A spin chain is a model for magnetic interactions in one dimension, where the only relevant degrees of freedom are the electron spins. The simplest model we can write is a model for the rotationally invariant interaction of a system of objects of spin 1/2 with nearest-neighbor interactions. At the classical level, the model was written (and solved in one dimension) by Ernst Ising, a student of Wilhelm Lenz, in his PhD thesis in 1924, with Hamiltonian

$$H(\sigma) = -\sum_{\langle ij \rangle} J_{ij}\sigma_i\sigma_j - \sum_j h_j\sigma_j, \qquad (18.1)$$

where J_{ij} is a coupling ($J_{ij} > 0$ being a ferromagnetic case and $J_{ij} < 0$ an antiferromagnetic case), $\sigma_i = \pm 1$ is a classical spin (up or down), $\langle ij \rangle$ means nearest-neighbor, and h_j is an external field. Usually one considers $J_{ij} = J$ and $h_j = 0$, giving

$$H(\sigma) = -J\sum_{\langle ij \rangle} \sigma_i\sigma_j. \qquad (18.2)$$

The quantum version of the Ising model is called the Heisenberg model. In the case of a rotationally invariant system of spin 1/2 in one spatial dimension, we call it the Heisenberg $XXX_{1/2}$ model; it was introduced by Heisenberg in 1928. We replace $\sigma_i = \pm 1$ with Pauli matrices, and add a constant to the energy, obtaining the Hamiltonian

$$H = -J\sum_{j=1}^{L} (\vec{\sigma}_j \cdot \vec{\sigma}_{j+1} - 1). \qquad (18.3)$$

Here $\vec{\sigma}_j$ are Pauli matrices (spin 1/2 operators) at site j, with periodic boundary conditions, i.e. $\vec{\sigma}_{L+1} \equiv \vec{\sigma}_1$ and J is a coupling constant.

- If $J > 0$ the system is ferromagnetic, and the interaction of spins is minimized if the spins are parallel, therefore the vacuum is $| \uparrow\uparrow \ldots \uparrow \rangle$.

- If $J < 0$ the system is antiferromagnetic and the interaction is minimized for antiparallel spins, therefore the vacuum is $| \uparrow\downarrow\uparrow\downarrow \ldots \uparrow\downarrow \rangle$.

Note that the non-rotationally invariant case is called the XYZ model, and has Hamiltonian

$$H = -\sum_{j=1}^{L}\left[\sum_{\alpha=1}^{3} J_\alpha \sigma_j^\alpha \cdot \sigma_{j+1}^\alpha - 1\right].$$ (18.4)

The $XXX_{1/2}$ Heisenberg model was "solved" (or rather, reduced to solving a set of algebraic equations, the Bethe ansatz equations) by Bethe in 1931, by the "coordinate Bethe ansatz," that is described further on in this chapter.

We can write the $XXX_{1/2}$ Hamiltonian in a different way by defining the *permutation operator* acting on spin states, P_{ij}, with

$$P_{ij}|\ldots \uparrow_i \ldots \uparrow_j \ldots\rangle = |\ldots \uparrow_i \ldots \uparrow_j\rangle; \quad P_{ij}|\ldots \downarrow_i \ldots \downarrow_j \ldots\rangle = |\ldots \downarrow_i \ldots \downarrow_j\rangle,$$
$$P_{ij}|\ldots \uparrow_i \ldots \downarrow_j \ldots\rangle = |\ldots \downarrow_i \ldots \uparrow_j\rangle; \quad P_{ij}|\ldots \downarrow_i \ldots \uparrow_j \ldots\rangle = |\ldots \uparrow_i \ldots \downarrow_j\rangle. \quad (18.5)$$

We can write P_{ij} in terms of Pauli matrices,

$$P_{ij} = \frac{1}{2} + \frac{1}{2}\vec{\sigma}_i \cdot \vec{\sigma}_j = \frac{1}{2} + \frac{1}{2}\sigma_i^3 \sigma_j^3 + \sigma_i^+ \sigma_j^- + \sigma_i^- \sigma_j^+,$$ (18.6)

valid on spin states, as we can easily check. Then the Heisenberg Hamiltonian without the constant, $H = -\sum_j \vec{\sigma}_j \cdot \vec{\sigma}_{j+1}$, can be rewritten as

$$H = -J\sum_{j=1}^{L}(2P_{j,j+1} - 1).$$ (18.7)

Often one uses (18.3), i.e. subtract the constant $-JL$ from H, to obtain

$$H' = -2J\sum_{j=1}^{L}(P_{j,j+1} - 1).$$ (18.8)

18.2 The $SU(2)$ sector and H_{XXX} from $\mathcal{N} = 4$ SYM

We will see that we can obtain the Heisenberg spin chain model from $\mathcal{N} = 4$ SYM. To obtain that, we need an $SU(2)$ sector that can act like the spin up, spin down states of the Heisenberg model. In the pp wave correspondence from the last chapter, we saw that a string of length J of complex Zs with a few, M, "impurities" Φ^m and no $\bar{Z}$s acted like a spin chain in the "dilute gas approximation," with a discretized string Hamiltonian acting on it. In this chapter, we call the number of sites of the spin chain L, to avoid confusion with J, the coupling constant of the Heisenberg Hamiltonian.

Therefore, in order to construct a full spin chain, we need to consider an $SU(2)$ sector made up of *two* complex scalars

$$Z = \Phi^1 + i\Phi^2; \quad \text{and} \quad W = \Phi^3 + i\Phi^4,$$ (18.9)

and no Φ^5, Φ^6, nor $\bar{Z}$ and $\bar{W}$. To construct the relevant $SU(2)$ sector we consider operators with large, but arbitrary numbers of both Z and W, that is, operators of the type

$$\mathcal{O}_\alpha^{J_1,J_2} = \text{Tr } [Z^{J_1} W^{J_2}] + \ldots (\text{permutations}). \tag{18.10}$$

The $\mathcal{N} = 4$ SYM interaction Hamiltonian is

$$H_{\text{int}} = -g_{\text{YM}}^2 \sum_{I>J} \text{Tr } [\Phi^I, \Phi^J][\Phi_I, \Phi_j], \tag{18.11}$$

and this reduces in the $SU(2)$ sector to

$$H_{\text{int}} = -g_{\text{YM}}^2 [Z, W] \text{Tr } [\bar{Z}, \bar{W}]. \tag{18.12}$$

Considering more general single trace operators made up of Φ^Is, $\text{Tr } [\Phi^{I_1} \ldots \Phi_{I_L}]$, the operator that permutes the order of Φ^{I_i} and Φ^{I_j} is

$$P_{ij} \equiv P_{I_i I_j}^{J_i J_j} = \delta_{I_i}^{J_j} \delta_{I_j}^{J_i}. \tag{18.13}$$

We can then consider the action of *planar* Feynman diagrams (in the large N limit) on operators (18.10). Planar Feynman diagrams only connect nearest-neighbors.

Anomalous dimension matrix

Since energy corresponds to anomalous dimension in AdS/CFT, the Hamiltonian should correspond to the matrix of anomalous dimensions. The renormalized operators $\mathcal{O}_{\text{ren}}$ are written in terms of the bare ones $\mathcal{O}$ via a matrix of renormalization factors,

$$\mathcal{O}_{\text{ren}}^A = Z^A{}_B \mathcal{O}^B. \tag{18.14}$$

Then the matrix of anomalous dimensions is

$$\Gamma \equiv \frac{dZ}{d \ln \Lambda} \cdot Z^{-1}. \tag{18.15}$$

The eigenvectors $\mathcal{O}_n$ of Γ are multiplicatively renormalizable and give

$$\langle (Z \cdot \mathcal{O})_n(x)(Z \cdot \mathcal{O})_n(y) \rangle = \langle \mathcal{O}_n^{\text{ren}}(x) \mathcal{O}_n^{\text{ren}}(y) \rangle = \frac{\text{const.}}{|x - y|^{2(L+\gamma_n)}}. \tag{18.16}$$

Here the renormalization factor Z is $1 + O(\lambda)$ and γ_n is the eigenvalue of Γ. As we see, the eigenvalue of Γ appears in the 2-point function exponent.

In the $SU(2)$ sector, from the Feynman diagrams in the planar limit in Fig. 18.1, we obtain for the renormalization factor Z,

$$Z_{\ldots I_l I_{l+1} \ldots}^{\ldots J_l J_{l+1} \ldots} = \mathbb{1} + \frac{g_{\text{YM}}^2 N}{16\pi^2} \ln \Lambda \, 2 \left(\delta_{I_l}^{J_l} \delta_{I_{l+1}}^{J_{l+1}} - \delta_{I_l}^{J_{l+1}} \delta_{I_{l+1}}^{J_l} \right), \tag{18.17}$$

where the I on Φ^I runs over Z and W. Then the one-loop planar Hamiltonian, i.e. the matrix of anomalous dimensions Γ, is

$$H_{\text{planar}}^{(1)} = \Gamma_{\text{planar}}^{(1)} = \frac{g_{\text{YM}}^2 N}{16\pi^2} \sum_{l+1}^L 2 \left(1 - P_{l,l+1} \right). \tag{18.18}$$

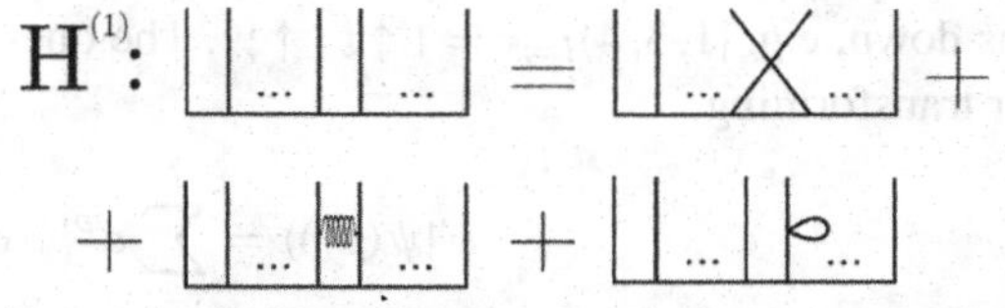

$$\text{H}^{(1)}: \quad \boxed{|\,|\,\ldots\,|\,|\,\ldots|} = \boxed{|\,\ldots\,\times\,\ldots|} +$$

$$+ \boxed{|\,\ldots\,\text{⦚}\,\ldots|} + \boxed{|\,\ldots\,|\,\circ\,\ldots}$$

Figure 18.1 The action of the "Hamiltonian" or anomalous dimension matrix on operators through Feynman diagrams is defined by the renormalization factor. Relevant diagrams are shown.

Dilatation operator

But a more precise concept of Hamiltonian, easily extendable to higher loops, is the concept of a *dilatation operator* $\mathcal{D}$ obtained by attaching the Feynman diagrams to the operators $\mathcal{O}_\alpha^{J_1,J_2}$, giving

$$\mathcal{D} \circ \mathcal{O}_\alpha^{J_1,J_2}(x) = \sum_\beta \mathcal{D}_{\alpha\beta} \mathcal{O}_\beta^{J_1,J_2}(x). \tag{18.19}$$

In perturbation theory, $\mathcal{D}$ is a sum over loops, $\mathcal{D}^{(n)} \sim O(g_{\text{YM}}^{2n})$. As an action on the operators, we can think of it as adding and removing fields in the operator. Defining the operator

$$\check{Z}_{ij} \equiv \frac{d}{dZ_{ji}}, \tag{18.20}$$

the tree level dilatation operator is the identity, which can be written as

$$\mathcal{D}^{(0)} = \text{Tr}\left(Z\check{Z} + W\check{W}\right), \tag{18.21}$$

and the one-loop dilatation operator can be written as

$$\mathcal{D}^{(1)} = -\frac{g_{\text{YM}}^2}{8\pi^2} \text{Tr}\,[Z,W][\check{Z},\check{W}]. \tag{18.22}$$

On the operator as a spin chain, it acts as before,

$$\mathcal{D}^{(1)}_{\text{planar}} = \frac{g_{\text{YM}}^2 N}{8\pi^2} \sum_{l+1}^{L} \left(\mathbb{1}_{l,l+1} - P_{l,l+1}\right), \tag{18.23}$$

i.e., giving the Heisenberg Hamiltonian, with $J = g_{\text{YM}}^2 N/(16\pi^2)$.

18.3 The coordinate Bethe ansatz

One magnon

We now diagonalize the Heisenberg Hamiltonian using the method introduced by Bethe, of the *coordinate Bethe ansatz*. The eigenstates (pseudoparticles) are called *magnons*. To write them, we first denote by $|x_1, \ldots, x_N\rangle$ the state with spins up at sites x_i along the chain of spins down, e.g. $|1, 3, 4\rangle_{L=5} = |\uparrow\downarrow\uparrow\uparrow\downarrow\rangle$. The one-magnon state is obtained by simply Fourier transforming,

$$|\psi(p_1)\rangle = \sum_{x=1}^{L} e^{ip_1 x} |x\rangle, \qquad (18.24)$$

which leads to

$$H|\psi(p_1)\rangle = -2J \sum_{x=1}^{L} e^{ip_1 x} (|x-1\rangle + |x+1\rangle - 2|x\rangle). \qquad (18.25)$$

If we have periodic boundary conditions around the chain, i.e. $|L + 1\rangle \equiv |1\rangle$, then the momentum p_1 is quantized, $p_1 = 2\pi k/L$, and the action of the Hamiltonian on $|\psi(p_1)\rangle$ becomes

$$H|\psi(p_1)\rangle = -2J(e^{ip_1} + e^{-ip_1} - 2) \sum_{x=1}^{L} e^{ip_1 x} |x\rangle = 8J \sin^2(p_1/2) |\psi(p_1)\rangle. \qquad (18.26)$$

Two magnons

A two-magnon state $|\psi(p_1, p_2)\rangle$, eigenstate of the Hamiltonian, is obtained by the ansatz of a superposition of an incoming and an outgoing wave, related by the 2-body S-matrix $S(p_1, p_2)$ for scattering in 1+1 dimensions,

$$\psi(x_1, x_2) = e^{i(p_1 x_1 + p_2 x_2)} + S(p_2, p_1) e^{i(p_2 x_1 + p_1 x_2)}. \qquad (18.27)$$

The first term is an incoming plane wave, and the second a scattered wave, where particles interchange their momenta and are scattered by the 2-body S-matrix. The two-magnon state is then the sum over positions of the above wavefunction, i.e.

$$|\psi(p_1, p_2)\rangle = \sum_{1 \le x_1 < x_2 \le L} \psi(x_1, x_2) |x_1, x_2\rangle. \qquad (18.28)$$

This is the (coordinate space) Bethe ansatz for the two-magnon state. By substituting this ansatz in the Schrödinger equation,

$$H|\psi(p_1, p_2)\rangle = E|\psi(p_1, p_2)\rangle, \qquad (18.29)$$

we obtain

$$E = 8J \left[\sin^2 \frac{p_1}{2} + \sin^2 \frac{p_2}{2} \right], \qquad (18.30)$$

and the form of the 2-point S-matrix

$$S(p_1, p_2) = \frac{\phi(p_1) - \phi(p_2) + i}{\phi(p_1) - \phi(p_2) - i} = S^{-1}(p_2, p_1), \qquad (18.31)$$

where

$$\phi(p) = \frac{1}{2} \cot \frac{p}{2} \equiv u. \qquad (18.32)$$

The proof for E and $S(p_1, p_2)$ is left as an exercise. We note that the energy of the two-magnon state is the sum of the energies of two one-magnon states of momenta p_1 and p_2. This continues to be true in general, for an M-magnon state, with

$$E = \sum_{j=1}^{M} 8J \sin^2 \frac{p_j}{2} = \sum_{j=1}^{M} 2J \frac{1}{u_j^2 + 1/4}. \tag{18.33}$$

Here $\phi(p) = u$ are called rapidities, and when ps are true magnon momenta, they are called the *Bethe roots*. We must find sets of these Bethe roots, or equivalently the momenta p_1, p_2 that solve a set of equations known as the Bethe equations, in the same way as we had $p_1 = 2\pi k/L$ for the one-magnon state.

The 2-body S-matrix has poles at $\phi_{12} \equiv \phi(p_1) - \phi(p_2) = \pm i$, which means that it is a bound state of two magnons (it is imaginary, not real). The Bethe equations come from the condition of periodicity for $x_1 < x_2$,

$$\psi(x_1, x_2) = \psi(x_2, x_1 + L). \tag{18.34}$$

Substituting in the Bethe ansatz for $\psi(x_1, x_2)$, we obtain the equations

$$e^{ip_1 L} = S(p_1, p_2) = \frac{\cot p_1/2 - \cot p_2/2 + 2i}{\cot p_1/2 - \cot p_2/2 - 2i},$$
$$e^{ip_2 L} = S(p_2, p_1) = \frac{\cot p_2/2 - \cot p_1/2 + 2i}{\cot p_2/2 - \cot p_1/2 - 2i}, \tag{18.35}$$

called the *Bethe ansatz equations*. They give solutions for the Bethe roots u_1, u_2 or p_1, p_2. Multiplying the two equations we obtain

$$e^{i(p_1 + p_2)L} = 1 \Rightarrow p_1 + p_2 = \frac{2\pi n}{L}, \quad n = 0, 1, \ldots, L-1. \tag{18.36}$$

In particular, we have the *real* solutions $p_2 = -p_1 \in \mathbb{R}$, for which the Bethe ansatz equations give

$$e^{ip_1 L} = S(p_1, -p_1) = \frac{2\cot p_1/2 + 2i}{2\cot p_1/2 - 2i} = e^{ip_1} \Rightarrow p_1 = \frac{2\pi n}{L-1}. \tag{18.37}$$

Note that in general p_1, p_2 must be different numbers, since $S(p_1, p_1) = -1$, so we get $\psi = 0$, a signature of the fermionic nature of the spin chain.

Substituting $-p_2 = p_1 = 2\pi n/(L-1)$ in the Bethe ansatz for the 2-magnon wavefunction, we obtain

$$|\psi(n)\rangle \equiv |\psi(p_1(n), -p_1(n))\rangle = C_n \sum_{l=1}^{L} \cos\left(\pi n \frac{2l+1}{L-1}\right) |x_2 + l, x_2\rangle,$$
$$C_n = 2e^{-\frac{i\pi n}{L-1}}. \tag{18.38}$$

$\mathcal{N} = 4$ SYM map and pp wave limit

Translating into $\mathcal{N} = 4$ SYM objects, this 2-magnon state corresponds to an operator eigenstate of the dilatation operator,

$$\mathcal{O}_n^{J,2} = C_n \sum_{l=0}^{L-1} \cos\left[\pi n \frac{2l+1}{L-1}\right] \mathrm{Tr}\,[WZ^l WZ^{J-l}]. \tag{18.39}$$

In the limit of $n \ll L, L \to \infty$, we obtain the BMN operator for $\Phi^i \to W$,

$$\mathcal{O}_n^{J,2} \to C_n \sum_{l=0}^{L-1} \cos \frac{2\pi n l}{L} \mathrm{Tr}\left[WZ^l WZ^{L-l}\right]. \tag{18.40}$$

In general, when we have a number M of magnons much less than L and $n \ll L$, the spectrum of states is given by acting with the creation operator

$$a_n^\dagger = \frac{1}{\sqrt{L}} \sum_{l=1}^{L} e^{\frac{2\pi i n l}{L}} \sigma_l^-, \tag{18.41}$$

where σ_l^- are Pauli matrices at site l. Acting with these creation operators, we obtain the BMN operators, as we can easily see. The momenta are $p_i \simeq 2\pi n_k/L$. Translating the energies into $\mathcal{N} = 4$ SYM variables, we get

$$\gamma = \Delta - L - M = \frac{\lambda}{16\pi^2} \sum_{k=1}^{M} 8 \sin^2 \frac{p_k}{2} \simeq \frac{\lambda}{8\pi^2} \sum_{k=1}^{M} p_k^2 = \frac{\lambda}{2L^2} \sum_{k=1}^{M} n_k^2. \tag{18.42}$$

These are indeed the eigenstates and eigenvalues of the independent string oscillators on the pp wave. But now we have that even when $M \sim L$ (outside of the "dilute gas approximation" of the last chapter), we have a Fock spectrum, which should correspond to string excitations in the *full AdS$_5 \times S^5$*, not just its Penrose limit.

General Bethe ansatz equations and integrability

The general ansatz for M magnons (M-body scattering) is obtained from the ansatz for two magnons (two-body scattering). The fundamental reason is the *integrability* of the Heisenberg spin chain. At the classical level, for a system with a finite number of degrees of freedom N, integrability means the existence of N integrals of motion T_i, i.e. independent conserved quantities (constant during the time evolution). In the case of the spin chain, we have L degrees of freedom. At the quantum level, integrability still means the existence of L independent integrals of motion T_i, but now this is expressed by saying that the operators $\hat{T}^i$ obey $[\hat{T}_i, \hat{T}_j] = 0$, and the Hamiltonian $\hat{H}$ is one of the $\hat{T}_i$s. In turn, quantum integrability implies that the dynamics is completely determined by giving the 2-body scattering matrix $S(p_1, p_2)$. In particular, the M-body scattering is also given in terms of the 2-body scattering.

Then, defining the *phase shifts* $\delta_{ij} = -\delta_{ji}$ by

$$S(p_i, p_j) \equiv e^{i\delta_{ij}}, \tag{18.43}$$

the M-body wavefunction (for M magnons) is

$$\psi(x_1,\ldots,x_M) = \sum_{P \in \text{Perm}(M)} \exp\left[i\sum_{i=1}^{M} p_{P(i)}x_i + \frac{i}{2}\sum_{i<j} \delta_{P(i)P(j)}\right]. \tag{18.44}$$

Then the Schrödinger equations imply again that the energies of the magnons,

$$\epsilon_j = 8J \sin^2 \frac{p_j}{2}, \tag{18.45}$$

add up to give (18.33). Also again periodicity of the wavefunction $\psi(x_1,\ldots,x_M)$ gives the *Bethe ansatz equations*

$$e^{ip_k L} = \prod_{i \neq k,\, i=1}^{M} S(p_k, p_i). \tag{18.46}$$

As before, taking products of all the equations, we obtain $e^{i(\sum_k p_k)L} = 1$, which implies

$$\hat{P} \equiv \sum_k p_k = \frac{2\pi n}{L}; \quad n = 0,\ldots,L-1. \tag{18.47}$$

And we still need all the p_is to be different, since if two ps are the same, we get $\psi = 0$ because of the fermionic nature of the chain, as we can verify from the M-body wavefunction ansatz.

Bethe equations have sets of both real and complex solutions for fixed L, M. The sets are $\{p_1,\ldots,p_M\}_n$, where n labels the solution. For $M = 1$ we have $n_1 = L - 1$ sets (since $p_1 = 2\pi k_1/(L-1)$, $k_1 = 0,\ldots,n_1 - 1$), for $M = 2$ we have $n_2 = L(L-3)/2$, etc. The total sum of solution sets is

$$\sum_{M=1}^{L} n_M = 2^L, \tag{18.48}$$

since at each site we can have two possibilities: spin up or down, for a total of 2^L possibilities.

The Bethe ansatz equations (18.46) can be rewritten in terms of the rapidities $u_k = 1/2 \cot(p_k/2)$, using the fact that

$$\frac{u_k + i/2}{u_k - i/2} = \frac{\cos(p_k/2) + i\sin(p_k/2)}{\cos(p_k/2) - i\sin(p_k/2)} = e^{ip_k}, \tag{18.49}$$

to obtain

$$\left(\frac{u_k - i/2}{u_k + i/2}\right)^L = \prod_{j \neq k,\, j=1}^{M} \left(\frac{u_k - u_j - i}{u_k - u_j + i}\right), \quad k = 1,\ldots,M. \tag{18.50}$$

Here u is called rapidity because $u = 1/2 \cot(p/2)$ implies

$$\frac{du}{dp} = -\frac{1 + 4u^2}{4} \Rightarrow \frac{dp}{du} = -\frac{1}{u^2 + 1/4} = \epsilon(u), \tag{18.51}$$

as in the case of a one-dimensional relativistic massive particle, $E^2 - p^2 = m^2$, written in terms of rapidities μ as

$$E = m \cosh \mu; \quad p = m \sinh \mu \Rightarrow E = \frac{dp}{d\mu}. \tag{18.52}$$

Therefore the u_i are the nonrelativistic analogs of rapidities, hence the name.

Taking the log of the Bethe ansatz equations in the original form (18.46), we obtain

$$p_k L = \sum_{i \neq k, i=1}^{M} \delta_{ki} + 2\pi n_k, \tag{18.53}$$

where n_k are integers. Therefore for each solution set $\{p_k\}$ (Bethe roots) we have a set of integers $\{n_k\}$. Note that we do not need real momenta p_k, or correspondingly real rapidities u_k, only real energies, which are given by the sum (18.33). That means that it is enough to have u_ks in pairs: for any Bethe root u_k, we must also have u_k^* as a Bethe root. That means that the Bethe roots are situated in the complex u plane, symmetrically about the real axis $\mathbb{R}$.

18.4 Thermodynamic limit and Bethe strings

We now consider the thermodynamic limit $L \to \infty$, $M \to \infty$ of the Bethe ansatz, since as we see later, it is related to classical strings in the gravity dual.

Taking the log of (18.50), we obtain

$$L \log \left(\frac{u_i + i/2}{u_i - i/2} \right) = \sum_{k \neq i, k=1}^{M} \log \left(\frac{u_i - u_k + i}{u_i - u_k - i} \right) - 2\pi i n_i. \tag{18.54}$$

Here n_i are arbitrary integers for each root u_i, which means that the set $\{n_i\}$ are quantum numbers for the multiparticle system characterized by the roots $\{u_i\}$.

Assuming self-consistently that in the thermodynamic limit the magnon momenta are of order $p_i \sim 1/L$, which means that the rapidities are of order $u_i \sim L$, makes the Bethe equations (18.54) finite in terms of $x_i = u_i/L$,

$$\frac{1}{x_i} + 2\pi n_i = \frac{2}{L} \sum_{k \neq i, k=1}^{M} \frac{1}{x_i - x_k}. \tag{18.55}$$

We saw that the roots u_i can be anywhere in the complex plane, and in order to have a real energy, we need to have for any root u_i, that u_i^* is also a root. Moreover, if $\mathrm{Re}(u_k) = \mathrm{Re}(u_i)$, the roots should have different imaginary parts, due to the fermionic nature of the spin chain.

In the thermodynamic limit $L \to \infty$, the Bethe equations (18.50) have left-hand side 0 or ∞, which means the right-hand side must be the same, i.e. $S(p_i, p_k)$ has a pole. In turn, that means that $u_k = u_i \pm i$, i.e. the two magnons of the same energy form a bound state by splitting their rapidities in the complex plane. Therefore, in the $L \to \infty$ limit, the

Bethe roots with the same real parts are separated in an array with $u_k = \text{Re}(u) + ik$. Then from (18.55) on such an array, n_i must be constant, $n_i = -n_C$ characterizing the array, or contour C, and the quantum numbers of the solution $\{u_k\}$ being the ks. Since $u_{k+1} - u_k \sim 1$, it follows that $x_{k+1} - x_k \sim 1/L \to 0$, even when $M \to \infty$. In this $M \to \infty$ limit, arrays are not vertical lines anymore ($u_k = \text{Re}(u) + ik$), but can curve in the complex plane.

Therefore in the thermodynamic limit, the roots u_i accumulate on smooth contours in the complex u plane called *Bethe strings*, which must be situated symmetrically with respect to the real axis (in order to have a real energy for the system). Then Bethe equations become integral equations.

We define the Bethe root density

$$\rho(x) \equiv \frac{1}{L} \sum_{j=1}^{M} \delta(x - x_j), \tag{18.56}$$

which was normalized such that

$$\int_C dx \, \rho(x) = \frac{M}{L}. \tag{18.57}$$

Then the thermodynamic limit of the Bethe equations (18.55) can be written in the integral form

$$2P \int_C dy \frac{\rho(y)}{y - x} = -\frac{1}{x} + 2\pi n_{C(u)}; \ x \in C, \tag{18.58}$$

where P means the principal part of the integral.

The energy of the system becomes

$$\frac{E}{2J} = \left(-\frac{L}{2}\right) + \sum_{j=1}^{M} \frac{1}{u_j^2 + 1/4} \to \left(-\frac{L}{2}\right) + \frac{1}{L} \int_C \frac{\rho(x)}{x^2}. \tag{18.59}$$

From (18.58) at $x = 0$, we obtain

$$\int_C dx \frac{\rho(x)}{x} = 2\pi m. \tag{18.60}$$

This can also be obtained from the quantization of the total momentum (18.47), which gives in the thermodynamic limit

$$\sum_{i=1}^{M} \frac{1}{u_i} = \frac{1}{L} \sum_{i=1}^{M} \frac{1}{x_i} = 2\pi m, \tag{18.61}$$

thus reducing to the same.

The Bethe ansatz equations in the thermodynamic limit (18.57–18.60) can be written independently for various Bethe strings, i.e. smooth components C_n, corresponding to various macroscopic solutions in the thermodynamic limit.

18.5 Spin chains from AdS space

We now turn to understanding the spin chain from the point of view of the gravity dual to $\mathcal{N} = 4$ SYM. The spin chain corresponds to large operators in the $SU(2)$ sector, i.e. made from Z and W.

Magnon excitations with $M/J \rightarrow 0$ (BMN operators) correspond to the strings on the pp wave, i.e. near a geodesic situated in the center of AdS_5 and rotating around an equator of S^5. On the other hand, at large M, we expect *semiclassical strings* instead of pointlike strings. Since we want the gravity dual to the $SU(2)$ sector, the $SU(2) = SO(3)$ invariance suggests that the strings corresponding to them in the gravity dual must belong in an $S^3 \subset S^5$ in the gravity dual.

There are various matchings of string solutions with solutions of the Bethe equations, but we can treat them together, by looking at the thermodynamic limit with M/J finite, as we show in the next subsection.

The worldsheet of strings belonging to an $S^3 \subset S^5$ has nontrivial values for X^0 and X^i, $i = 1, \ldots, 4$, with $X^i X^i = 1$. We form

$$Z = X^1 + iX^2; \quad W = X^3 + iX^4, \tag{18.62}$$

and we define an $SU(2)$ element

$$g = \begin{pmatrix} Z & W \\ -\bar{W} & \bar{Z} \end{pmatrix} = \begin{pmatrix} X^1 + iX^2 & X^3 + iX^4 \\ -X^3 + iX^4 & X_1 - iX^2 \end{pmatrix} = X^1 \mathbb{1} + iX^4 \sigma_1 + iX^3 \sigma_2 + iX^2 \sigma_3$$
$$= X^i \tilde{\sigma}_i, \quad i = 1, 2, 3, 4. \tag{18.63}$$

Then we can calculate the form of the matrix *currents j_a*

$$j_a = g^{-1} \partial_a g, \tag{18.64}$$

and obtain

$$\mathrm{Tr}\, (j_a)^2 = -2 \sum_{i=1}^{4} (\partial_a X^i)(\partial_a X^i), \tag{18.65}$$

where there is no sum over a. The calculation is left as an exercise.

String action

The action in conformal gauge of a string moving in the flat (Euclidean) embedding space for S^3 is

$$S = \frac{1}{4\pi \alpha'} \int d\tau d\sigma [-(\partial_a \tilde{X}^0)^2 + (\partial_a \tilde{X}^i)^2], \tag{18.66}$$

where $\tilde{X}^i \tilde{X}^i = R^2$ and $\tilde{X}^0$ also is the embedding coordinate for AdS_5 of radius R. Redefining $\tilde{X}^i = RX^i$ and $\tilde{X}^0 = RX^0$, one obtains the action

$$S = \frac{R^2}{4\pi \alpha'} \int d\tau \int_0^{2\pi} d\sigma \left[-(\partial_a X^0)^2 + \sum_{i=1}^{4} (\partial_a X^i)^2 \right], \tag{18.67}$$

where $X^i X^i = 1$ and $R^2/4\pi\alpha' = \sqrt{\lambda}/(4\pi)$. Writing this in terms of the currents j_a calculated above, we obtain

$$S = -\frac{\sqrt{\lambda}}{4\pi} \int_0^{2\pi} d\sigma \int d\tau \left[\frac{\mathrm{Tr}\,(j_a)^2}{2} + (\partial_a X^0)^2 \right]. \qquad (18.68)$$

Equations of motion and constraints

The X^0 equation of motion is

$$\partial_+ \partial_- X^0 = 0, \qquad (18.69)$$

and, using $\delta j_a/\delta(X^i \tilde{\sigma}^i) = \partial_a \delta(\sigma - \sigma_0)$, the equation of motion for $X^i \tilde{\sigma}^i$ is

$$\eta^{ab} \partial_a j_b = 0. \qquad (18.70)$$

In terms of $\partial_\pm = \partial_\tau \pm \partial_\sigma$, we get

$$\partial_+ j_- + \partial_- j_+ = 0. \qquad (18.71)$$

Since we have used the conformal gauge for the string action, we must supplement it with the Virasoro constraint,

$$(\partial_\pm X^A)^2 = 0 \Rightarrow (\partial_\pm X^0)^2 = (\partial_\pm X^i)^2. \qquad (18.72)$$

But (18.69), which means $(\partial_\tau^2 - \partial_\sigma^2)X^0 = 0$, can be solved by

$$X^0 = \kappa\tau. \qquad (18.73)$$

Since as we saw, $(\partial_\pm X^i)^2 = -1/2\mathrm{Tr}\,(j_\pm)^2$, the Virasoro constraint reduces to

$$\frac{1}{2}\mathrm{Tr}\,(j_\pm)^2 = -\kappa^2. \qquad (18.74)$$

Symmetries

The action (18.68) has global $SU(2)_L \times SU(2)_R$ symmetry acting as $g \to gh$ and $g \to hg$, with Noether currents $j_a \equiv \sigma^A j_a^A/2i$ and

$$l_a = g j_a g^{-1} = \partial_a g \cdot g^{-1} \equiv \frac{\sigma^A}{2i} l_a^A, \quad A = 1, 2, 3. \qquad (18.75)$$

The corresponding Noether charges are (see (18.68))

$$Q_R^A = \frac{\sqrt{\lambda}}{4\pi} \int_0^{2\pi} d\sigma j_\tau^A,$$

$$Q_L^A = \frac{\sqrt{\lambda}}{4\pi} \int_0^{2\pi} d\sigma l_\tau^A. \qquad (18.76)$$

For the spin chain operators $\mathcal{O}_\alpha^{L-M,M}$, under the right $SU(2)$ action $g \to gh$, (ZW) acts as a doublet, so $Q_R^3(Z) = +1$ and $Q_R^3(W) = -1$. Under the left $SU(2)$ action, $g \to hg$,

$\left(\begin{array}{c} Z \\ -\bar{W} \end{array}\right)$ and $\left(\begin{array}{c} W \\ \bar{Z} \end{array}\right)$ act as doublets, which means that $Q_L^3(Z) = +1$ and $Q_L^3(-\bar{W}) = -1$, or that $Q_L^3(W) = +1$. Finally, for the spin chain operators, we obtain

$$Q_R^3(\mathcal{O}_\alpha^{L-M,M}) = L - 2M; \quad Q_L^3(\mathcal{O}_\alpha^{L-M,M}) = L. \tag{18.77}$$

These are the charges which we should have for a semiclassical string in AdS space that corresponds to spin chain operators.

18.6 Bethe strings from AdS strings

The semiclassical strings in AdS space correspond to Bethe strings, but instead of checking this for each semiclassical string, we check that the Bethe equations in the thermodynamical limit, (18.57–18.60), are obtained from the equations for strings in AdS space.

Since $j_a = g^{-1}\partial_a g$, by explicit substitution we can check that

$$\partial_+ j_- - \partial_- j_+ + [j_+, j_-] = 0. \tag{18.78}$$

Then, defining

$$J_\pm = \frac{j_\pm}{1 \mp x}, \tag{18.79}$$

where x is an arbitrary variable, we obtain

$$\partial_+ J_- - \partial_- J_+ + [J_+, J_-]$$
$$= \frac{1}{1 - x^2} \left[\partial_+ j_- - \partial_- j_+ + [j_+, j_-] - x(\partial_+ j_- + \partial_- j_+) \right]. \tag{18.80}$$

But then from (18.78) and (18.71) this vanishes for any x. This looks like the condition of a gauge field with zero field strength in two dimensions, $F_{+-} = \partial_+ A_- - \partial_- A_+ + [A_+, A_-] = 0$, so we can consider J_a as a "flat connection" in mathematical language. We can define $J_\sigma = (J_+ - J_-)/2$ and the "monodromy" of this connection (the equivalent of the Wilson loop $W[C]$ for a contour wrapping around the periodic σ coordinate, and giving the parallel transport of objects in the fundamental representation, as we saw in Chapter 14), is defined by

$$\Omega(x) \equiv P \exp\left[-\int_0^{2\pi} d\sigma J_\sigma \right] = P \exp\left[\int_0^{2\pi} d\sigma \frac{1}{2}\left(\frac{j_+}{x-1} + \frac{j_-}{x+1} \right) \right]. \tag{18.81}$$

From its trace, we define the object $p(x)$ by

$$\mathrm{Tr}\ \Omega(x) \equiv 2\cos p(x) = e^{ip(x)} + e^{-ip(x)}. \tag{18.82}$$

From the definition, we can see that $p(x)$ has singularities at $x = \pm 1$.

Near $x = \mp 1$, expanding (18.81) and using (18.74), we find

$$p(x) \simeq -\frac{\pi \kappa}{x \pm 1} + \dots \tag{18.83}$$

Near $x = \infty$, we find by expanding (18.81) that

$$\text{Tr } \Omega \simeq 2 + \frac{1}{x^2}\left(\frac{1}{2}\int_0^{2\pi} d\sigma_1 \int_0^{2\pi} d\sigma_2\right)\text{Tr } (j_\tau(\sigma_1)j_\tau(\sigma_2)) + \ldots$$

$$= 2 - \frac{4\pi^2}{\lambda}\frac{1}{x^2}Q_R^2 + \ldots = 2 - \frac{4\pi^2}{\lambda x^2}(L - 2M)^2 + \ldots, \tag{18.84}$$

where we have used $(j_+ + j_-)/2 = j_\tau$. Then we find that near $x = \infty$,

$$p(x) = -\frac{2\pi(L - 2M)}{\sqrt{\lambda}x} + \ldots \tag{18.85}$$

On the other hand, near $x = 0$, using $J_\pm(x) = j_\pm(1 \pm x) + \ldots$, and $j_a = g^{-1}\partial_a g$, we obtain

$$J_\sigma(x) = \frac{J_+ - J_-}{2} \simeq j_\sigma + xj_\tau + \ldots = g^{-1}(\partial_\sigma + xl_\tau)g \Rightarrow$$

$$\text{Tr } \Omega \simeq 2 + \frac{x^2}{2}\int_0^{2\pi} d\sigma_1 d\sigma_2 \text{Tr } [l_\tau(\sigma_1)l_\tau(\sigma_2)]$$

$$= 2 - \frac{4\pi^2 Q_L^2}{\lambda}x^2 + \ldots = 2 - \frac{4\pi^2 L^2}{\lambda}x^2 + \ldots \tag{18.86}$$

Then for $p(x)$ near $x = 0$ we obtain

$$p(x) \simeq 2\pi m + \frac{2\pi L}{\sqrt{\lambda}}x + \ldots \tag{18.87}$$

We define a function G that excludes the poles at $x = \pm 1$,

$$G(x) \equiv p(x) + \frac{\pi\kappa}{x + 1} + \frac{\pi\kappa}{x - 1}, \tag{18.88}$$

which has only branch cut singularities, and so is completely determined from its discontinuities

$$G(x + i0) - G(x - i0) \equiv 2\pi i\rho(x), \tag{18.89}$$

via a dispersion relation

$$G(x) = \int_{\mathcal{C}} \frac{\rho(y)}{x - y}. \tag{18.90}$$

On the other hand, for $x \in \mathcal{C}_k$, we now have

$$p(x + i0) + p(x - i0) = 2\pi n_k, \tag{18.91}$$

in order to have $e^{ip(x+i0)}e^{ip(x-i0)} = 1$. Then finally we obtain for $x \in \mathcal{C}_k$

$$2P \int dy \frac{\rho(y)}{x - y} = \frac{2\pi\kappa}{x - 1} + \frac{2\pi\kappa}{x + 1} + 2\pi n_k. \tag{18.92}$$

The worldsheet energy of the string in AdS, $\delta S/\delta\dot{X}^0$, should be identified with the anomalous dimension Δ in the CFT,

$$\Delta = \frac{\sqrt{\lambda}}{2\pi}\int_0^{2\pi} d\sigma\, \partial_\tau X^0 = \sqrt{\lambda}\kappa, \tag{18.93}$$

where we have used $X^0 = \kappa\tau$ to obtain a relation between Δ and κ.

Using the behavior at infinity of $p(x)$, we obtain

$$\int dx \rho(x) = \frac{1}{2\pi i} \int dx \left[p(x+i0) - p(x-i0) + \frac{2\pi\kappa}{x+i0} - \frac{2\pi\kappa}{x-i0} \right] = \frac{2\pi}{\sqrt{\lambda}}(\Delta + 2M - L).$$

(18.94)

Similarly, using the behavior at zero of $p(x)$, we obtain (from the constant term)

$$\int dx \frac{\rho(x)}{x} = \frac{1}{2\pi i} \int dx \frac{1}{x} \left[p(x+i0) - p(x-i0) + \frac{2\pi\kappa}{x+i0-1} - \frac{2\pi\kappa}{x-i0-1} \right] = 2\pi m.$$

(18.95)

Finally, using the behavior at zero of $p(x)$, we obtain (from the linear term in x)

$$\int dx \frac{\rho(x)}{x^2} = \frac{1}{2\pi i} \int dx \frac{1}{x^2} \left[p(x+i0) - p(x-i0) + \frac{2\pi\kappa}{x+i0-1} - \frac{2\pi\kappa}{x-i0-1} \right]$$
$$= \frac{2\pi}{\sqrt{\lambda}}(\Delta + L).$$

(18.96)

To match with the Bethe equations from $\mathcal{N} = 4$ SYM in the thermodynamic limit, we rescale $x \to 4\pi L x/\sqrt{\lambda}$, obtaining for $x \in \mathcal{C}$,

$$2P \int dy \frac{\rho(y)}{x-y} = \frac{x}{x^2 - \frac{\lambda}{16\pi^2 L^2}} \frac{\Delta}{L} + 2\pi n_k,$$
$$\int dx \rho(x) = \frac{M}{L} + \frac{\Delta - L}{2L},$$
$$\int dx \frac{\rho(x)}{x} = 2\pi m,$$
$$\frac{\lambda}{8\pi^2 L} \int dx \frac{\rho(x)}{x^2} = \Delta + L = \frac{\lambda}{8\pi^2} H^{(1-\text{loop})}.$$

(18.97)

In the thermodynamic limit of the Bethe equations, $\lambda/L^2 \to 0$, $\frac{\Delta-L}{L} \to 0$, $\frac{\Delta}{L} \to 1$, we obtain the same Bethe equations of $\mathcal{N} = 4$ SYM, (18.57–18.60).

Then each individual Bethe string corresponds to an individual macroscopic string in $AdS_5 \times S^5$. We must calculate the Bethe curve, then calculate the resolvent $G(x)$, the density $\rho(x)$, then the energy, to match against a corresponding classical AdS string. Agreement has been found for various AdS strings (folded strings, circular strings, etc.).

Important concepts to remember

- The Heisenberg $XXX_{1/2}$ spin chain Hamiltonian is the simplest magnetic interaction in 1+1 dimensions, coupling nearest Pauli spins.
- It is diagonalized by a Bethe ansatz, for excitations ("magnons") of spin up propagating in a sea of spin down states.
- The Heisenberg Hamiltonian is reproduced from the $SU(2)$ sector of $\mathcal{N} = 4$ SYM, using $Z = \Phi^1 + i\Phi^2$ and $W = \Phi^3 + i\Phi^4$ to build large R-charge single trace operators, at the one-loop planar level.
- The Hamiltonian can be thought of as an anomalous dimension matrix, or dilatation operator, and is obtained by attaching Feynman diagrams to the large operators.

- In the coordinate Bethe ansatz, the 2-magnon S matrix defines any M-magnon scattering, due to integrability, and the energy of the multimagnon state is the sum of the energies of each magnon.
- The momenta p_k of the magnons or the rapidities $u_k = 1/2 \cot(p_k/2)$ satisfy the Bethe ansatz equations, and sets $\{u_k\}$ of solutions are called Bethe roots, and form a contour in the complex plane, in the thermodynamic limit, called a Bethe string.
- Bethe strings are obtained from classical (macroscopic) AdS strings moving in an $S^3 \subset S^5$. Each Bethe string corresponds to a different AdS string.
- The Bethe ansatz equations in the thermodynamic limit can be obtained from the equations of motion and constraints of strings in AdS.

References and further reading

The identification of the 1-loop SYM Hamiltonian with the Heisenberg spin chain was done in [53]. Good reviews of spin chains in SYM are [54, 55]. However, none of the papers and reviews for spin chains is easy to digest, the papers are at an advanced level.

Exercises

1. Check that the Bethe ansatz for two magnons, with

$$E = E_1 + E_2; \quad S(p_1, p_2) = \frac{\phi(p_1) - \phi(p_2) + i}{\phi(p_1) - \phi(p_2) - i}; \quad \phi(p) = \frac{1}{2} \cot \frac{p}{2}, \quad (18.98)$$

   solves the Schrödinger equation for $H_{XXX1/2}$.
2. Prove the relation (18.65).
3. Check that the Bethe ansatz for three magnons satisfies the Schrödinger equation for $H_{XXX1/2}$ if the Bethe ansatz for two magnons does.
4. Write down explicitly a 3-magnon eigenstate with $p_1 + p_2 + p_3 = 0$, then deduce the corresponding spin chain operator in $\mathcal{N} = 4$ SYM and show that it reduces to a BMN operator in the limit $n \ll J, J \to \infty$.
5. Write down all the eigenstates and eigenenergies of a spin chain with three sites, using the coordinate Bethe ansatz.
6. We have seen that Bethe strings in the $J \to \infty$ limit have Bethe roots lying on vertical lines $u_k = \mathrm{Re}(u) + ik$. Why does the argument fail for $M \to \infty$ as well, with M/J fixed, i.e. in the thermodynamic limit, and why can we have a curved line (Bethe strings)?

PART III

AdS/CFT DEVELOPMENTS AND GAUGE–GRAVITY DUALITIES

In Part II of the book we have been dealing with the original duality of string theory in $AdS_5 \times S^5$ background vs. $\mathcal{N} = 4$ SYM in 3+1-dimensional flat space, since it is the best understood, and we can calculate the largest number of quantities. But since the original paper of Maldacena, a plethora of other cases have been discovered, including theories with less supersymmetry and/or no conformal invariance. In the cases when there is no conformal field theory, and no AdS space, one could use the term AdS/CFT, but the proper name of the correspondence is *gravity–gauge duality*, since it relates a gravitational theory (string theory in a certain background) to a gauge theory in fewer dimensions, via holography. In that case the gravitational background is known as the *gravity dual* of the gauge theory.

In this chapter we start with other cases with conformal invariance, when the proper name of the correspondence is still AdS/CFT.

19.1 $AdS_4 \times S^7$ and $AdS_7 \times S^4$

There are two other cases where the theory living at the boundary of AdS space is maximally supersymmetric and conformal. They are not related to the same 10-dimensional type IIB string theory, since in fact there is a theorem [51], stating that the only maximally supersymmetric backgrounds of type IIB are 10-dimensional Minkowski space, $AdS_5 \times S^5$, and the maximally supersymmetric pp wave, which is just the Penrose limit of $AdS_5 \times S^5$.

Instead, they are related to M theory, which, as we explained in Chapter 7, is a theory in 11 dimensions obtained as type IIA string theory at strong coupling, since the radius of the eleventh dimension is $R_{11} = g_s \sqrt{\alpha'}$. In the low energy limit, M theory becomes the unique 11-dimensional supergravity theory. The relevant maximally supersymmetric backgrounds are $AdS_4 \times S^7$ and $AdS_7 \times S^4$, and a theorem by Kowalski-Glikman [47] states that the only maximally supersymmetric backgrounds of 11-dimensional supergravity are 11d Minkowski space, $AdS_7 \times S^4$, $AdS_4 \times S^7$ and the maximally supersymmetric wave, which can be obtained as a Penrose limit of both $AdS_4 \times S^7$ and $AdS_7 \times S^4$.

As in the case of the 10-dimensional $AdS_5 \times S^5$ background, the $AdS_4 \times S^7$ and $AdS_7 \times S^4$ backgrounds can be obtained as the near-horizon limit of extremal p-brane objects in the theory. Of course in M-theory there are no D-branes, since there are no strings to end on them. But instead we have branes that are the uplift of the D-branes of type IIA string theory. In type IIA string theory there is a fundamental string ("F1-brane") electrically charged with respect to the NS-NS field $B_{\mu\nu}$, as explained in Chapter 7, and a magnetically charged object corresponding to the same $B_{\mu\nu}$, called an NS5-brane. There is also a D2-brane, electrically charged under the $A_{\mu\nu\rho}^{(10)}$ field of type IIA supergravity, and a D4-brane, magnetically charged under $A_{\mu\nu\rho}^{(10)}$. These branes uplift to 11-dimensional M theory into an electrically charged M2-brane with respect to $A_{MNP}^{(11)}$, and an M5-brane magnetically charged under the same $A_{MNP}^{(11)}$. The M2-brane reduces to the fundamental string (F1-brane) in ten dimensions when dimensionally reducing along the brane (when the string coupling is the radius of a direction along the M2-brane), and to the D2-brane when reducing along a direction transverse to the M2-brane (the string coupling is the radius of a transverse direction). The M5-brane reduces to the D4-brane in ten dimensions when dimensionally reducing along the brane, and to the NS5-brane when reducing along a direction transverse to the brane. Thus in M theory the fundamental object can be thought to be the M2-brane, the uplift of the fundamental string, whereas the M5-brane is solitonic in nature. Also, since fundamental strings can end on D4-branes, it means that M2-branes can end on M5-branes.

$AdS_7 \times S^4$ vs. 6-dimensional $(2, 0)$ theory

The metric for N extremal M5-branes is

$$ds^2 = f_5^{-1/3}(r)\left(-dt^2 + d\vec{x}_{(5)}^2\right) + f_5^{2/3}(r)(dr^2 + r^2 d\Omega_4^2),$$

$$f_5(r) = 1 + \frac{\pi N l_P^3}{r^3}. \tag{19.1}$$

Here l_P is the 11-dimensional Planck length. There is also a 4-form field strength,

$$F_{(4)} = *\left(dt \wedge dx^1 \wedge \ldots \wedge dx^5 \wedge d(f_5^{-1})\right); \quad (A_{(6)} = f_5^{-1} dt \wedge dx^1 \wedge \ldots \wedge dx^5), \tag{19.2}$$

where $A_{(6)}$ is the Poincaré dual to the 3-form gauge field $A_{(3)}$. We take the near-horizon limit for the same reason as in the case of the D3-branes, to decouple gravity from field theory modes along the brane, in order to have a *duality* between string theory in a gravitational background and the field theory existing on the brane. We define

$$U^2 \equiv \frac{r}{l_P^3}, \tag{19.3}$$

and we take the gravity decoupling limit $l_P \to 0$, together with the near-horizon limit $r \to 0$, with $U =$ fixed. One way to define what we keep fixed (and thus this is how we define U) is to impose that the near-horizon, decoupling limit leads to an overall l_P^2 factor and everything inside is fixed. Then the metric becomes

$$ds^2 = l_P^2 \left[\frac{U^2}{(\pi N)^{1/3}}(-dt^2 + d\vec{x}_{(5)}^2) + 4(\pi N)^{2/3}\frac{dU^2}{U^2} + (\pi N)^{2/3} d\Omega_4^2 \right], \tag{19.4}$$

which is the metric of $AdS_7 \times S^4$, with radii $R_{S^4} = l_P(\pi N)^{1/3}$ and $R_{AdS_7} = 2l_P(\pi N)^{1/3} = 2R_{S^4}$. The 4-form field strength in the near-horizon limit becomes

$$F_{(7)} = \frac{3r^2}{\pi N l_P^3} dx^0 \wedge \ldots \wedge dx^5 \wedge dr,$$

$$F_{\mu_1 \ldots \mu_4} = \frac{1}{\sqrt{4!\,7!}\sqrt{-g}} \epsilon_{\mu_1 \ldots \mu_4}{}^{\mu_5 \ldots \mu_{11}} F_{\mu_5 \ldots \mu_{11}} = \frac{1}{\sqrt{4!\,7!}\sqrt{-g}} g_{S^4} \epsilon^{\mu_1 \ldots \mu_{11}} F_{\mu_5 \ldots \mu_{11}} \Rightarrow$$

$$F_{S^4} = \frac{3}{\sqrt{4!\,7!}} l_P^3 \pi N \epsilon_{S^4}, \tag{19.5}$$

where $\mu_1, \ldots, \mu_4$ are in the S^4 directions and $\mu_5, \ldots, \mu_{11}$ in the rest and ϵ_{S^4} is the volume form in the S^4, integrating to 1 (so that R_{S^4} cancels in the $\sqrt{g_{S^4}}/R_{S^4}^4$ factor defining it). As in the case of $AdS_5 \times S^5$, in order to have the supergravity approximation valid, we need that $R_{AdS}/l_P, R_S/l_P \gg 1$, which implies $N \to \infty$.

In the decoupling limit above, supergravity in the $AdS_7 \times S^4$ background should be equivalent to the theory on the N M5-branes, so we need to define this theory. The M5-brane is the uplift of the D4-brane of string theory, when the eleventh dimension $R_{11} = g_s \sqrt{\alpha'}$ is large. But for the coupling of a Dp-brane we have from (9.12)

$$g_{Dp}^2 = (2\pi)^{p-2} g_s \alpha'^{\frac{p-3}{2}} \Rightarrow g_{D4}^2 = 4\pi^2 g_s \sqrt{\alpha'} = 4\pi^2 R_{11}. \tag{19.6}$$

On the D4-brane we have 5-dimensional maximally supersymmetric Yang-Mills theory, in the same way as happens in four dimensions on D3-branes, with coupling g_{D4}^2. Thus the coupling in the 6-dimensional theory on the M5-branes reduced on R_{11} is given only in terms of R_{11}, meaning that the original 6-dimensional theory has no intrinsic coupling (dimensionless parameter). It also has no dimensional parameters (since otherwise they would also be evident in the reduced theory, the 5-dimensional maximal SYM), and is in fact a conformal theory. This theory has $(2, 0)$ supersymmetries, i.e. 16 supercharges divided into two minimal Weyl spinors of the same chirality. This superalgebra has an $SO(5) \simeq USp(4)$ R-symmetry, and the spinors can be considered as four spinors ψ^I in the **4** representation of $USp(4)$ obeying a modified Majorana reality condition (as we saw in Chapter 3 we can define such a condition for spinors in representations of $USp(2N)$) $\bar{\chi}_C^i = \chi_j^T \Omega^{ji} C$. The only massless irreducible representation of the $(0, 2)$ algebra is made up of a tensor $B_{\mu\nu}$ with self-dual field strength, five scalars X^m in the fundamental representation of $SO(5)$, corresponding to the five transverse directions to the M5-brane, and the fermions. Then the R-symmetry rotating the scalars and the fermions corresponds to the rotation group on the five directions transverse to the brane, or the symmetry group of the S^4 present in the near-horizon limit.

While the *abelian* theory on a single M5-brane is well understood, on N coincident M5-branes we have a conformal field theory with $(2, 0)$ supersymmetry that is strongly coupled (since there is no dimensionless coupling that can be made small, i.e. perturbative), that is little understood. The AdS/CFT correspondence relates this theory with supergravity in $AdS_7 \times S^4$ background. While we now cannot make many tests of AdS/CFT, we can assume that AdS/CFT is indeed valid, since we have the same heuristic derivation as in the case of $\mathcal{N} = 4$ SYM in four dimensions vs. $AdS_5 \times S^5$, and derive quantities in the 6-dimensional $(2, 0)$ theory via AdS/CFT.

$AdS_4 \times S^7$ vs. 3-dimensional $\mathcal{N} = 8$ theory

The other case of interest is the case of M2-branes. The metric for N extremal M2-branes is

$$ds^2 = f_2^{-2/3}(r)(-dt^2 + dx^2 + dy^2) + f_2^{1/3}(r)(dr^2 + r^2 d\Omega_7^2),$$

$$f_2(r) = 1 + \frac{32\pi^2 N l_P^6}{r^6}, \tag{19.7}$$

and the 4-form field strength is

$$F_{(4)} = d(f_2^{-1}(r)) \wedge dt \wedge dx \wedge dy \quad (A_{(3)} = f_2^{-1}(r)dt \wedge dx \wedge dy). \tag{19.8}$$

We again consider the near-horizon limit $r \to 0$, $l_P \to 0$, but now we keep constant

$$U = \frac{r^2}{l_P^3}, \tag{19.9}$$

again such that the near-horizon metric is fixed except for an l_P^2 overall factor. The metric becomes

$$ds^2 = l_P^2 \left[\frac{U^2}{(32\pi N)^{2/3}}(-dt^2 + dy^2 + dz^2) + (32\pi^2 N)^{1/3} \left(\frac{1}{4} \frac{dU^2}{U^2} + d\Omega_7^2 \right) \right], \tag{19.10}$$

which is the metric of $AdS_4 \times S^7$, with $R_{AdS_4} = l_P(32\pi^2 N)^{1/6}/2$, $R_{S^7} = l_P(32\pi^2 N)^{1/6} = 2R_{AdS_4}$, and the 4-form field strength in the directions of AdS_4 becomes

$$F_{(4)} = \frac{6r^5}{32\pi^2 N l_P^6} dr \wedge dt \wedge dx \wedge dy = \frac{\pi\sqrt{N}}{2} l_P^3 \epsilon_{(4)}, \tag{19.11}$$

where $\epsilon_{(4)}$ is the volume form, $\sqrt{-g}/R^4$ on AdS_4 (note that by rescaling t, y, z by a constant, we can rescale the coefficient of $\epsilon_{(4)}$ to anything we want).

The theory on the M2-branes should be maximally supersymmetric, i.e. with $\mathcal{N} = 8$ supersymmetries. That means that there are eight Majorana fermions in three dimensions (each with two real components, for a total of 16 supercharges), transforming in the fundamental of the $SO(8)$ R-symmetry. The theory should be related to the dimensional reduction of the 4-dimensional $\mathcal{N} = 4$ SYM theory, i.e. 3-dimensional maximal SYM, which is expected to live on the worldvolume of D2-branes. Indeed, we saw that D2-branes are M2-branes reduced on a transverse direction, so for them only the rotational $SO(7)$ symmetry should be manifest, not the full $SO(8)$.

Indeed, for the Lagrangean formulation that exists for D2-branes, we have seven scalars corresponding to the seven transverse coordinates to the brane X^m, and a Yang-Mills gauge field A_μ that in three dimensions has also one degree of freedom, like a scalar does. In fact, it is Poincaré dual to a scalar ($F_{\mu\nu} = 1/2\epsilon_{\mu\nu\rho}\partial^\rho\phi$). The D2-brane theory has a coupling g_{D2}^2 with mass dimension one, which means that it becomes strongly coupled in the IR, since the effective dimensionless coupling will be $g_{eff}^2 = g_{D2}^2/E$. In the strong coupling limit we should obtain the M2-brane theory, which is therefore defined as the IR limit (fixed point) of the D2-brane theory. For a single D2-brane, the action for the corresponding single M2-brane is simple to obtain because of the lack of self-coupling. It is obtained by just dualizing the gauge field to another scalar, to obtain a theory of eight scalars (corresponding to the eight directions transverse to an M2-brane) and fermions. But the theory of multiple coincident M2-branes is again very mysterious, and is hard to define. We could again, as in the case of M5-branes, define it by AdS/CFT. The heuristic derivation of the correspondence follows in the same way, since we have a decoupling limit in which the field theory on the branes decouples from the gravity theory in the bulk of (19.10), and we obtain a duality between the two.

However, we will see in the next chapter that we can modify the duality of the multiple M2-brane theory vs. $AdS_4 \times S^7$ to a well-defined model called the ABJM model, that involves a parameter k. When $k = 1$ we recover the M2-brane theory, but the perturbative expansion, which is in $1/k$, is lost. We will therefore not describe the duality further here.

Observables

We saw that in both cases, the gravitational side of the duality is better defined than the field theory side, so observables are defined there. We can calculate correlation functions of field theory operators in the same way as in the $AdS_5 \times S^5$ case, by calculating Witten diagrams in AdS space. We can also identify field theory solitons from solitonic branes in the gravity dual. We can put the system at finite temperature and calculate transport properties. We can calculate the anomalous dimensions of large operators from the energies of strings in the gravity dual. We can calculate scattering of gauge invariant states. Most of the calculations that can be done in AdS_5 can be extended to these other conformal cases.

One difference appears in the case of the Wilson loop, which was defined for gauge theories. But the M5- and M2-brane theories do not have gauge fields per se in order to define $P \exp\left[i \oint_C A_a dx^a\right]$. The single M5-brane theory has a self-dual antisymmetric tensor field B_{ab} on its worldvolume, and in the multiple M5-brane case we expect to have some tensor field as well. The natural generalization will be something like $P \exp\left[i \oint_\Sigma B_{ab} d\Sigma^{ab}\right]$, which we will call a Wilson surface operator. The correct definition of this operator in field theory is difficult, since there are no "nonabelian tensor fields B_{ab}" in a simple sense, yet we want an interacting version of the abelian operator $\exp\left[\oint_\Sigma B_{ab} d\Sigma^{ab}\right]$. But again there is no problem of defining this operator in the gravitational side, in a similar way to the Wilson loop, as $e^{-S_{M2}[\Sigma]}$, where S_{M2} is the M2-brane on-shell action, and corresponds to a minimal volume for the M2-brane bounded by the *closed* surface Σ on the boundary at infinity.

19.2 $\mathcal{N} = 2$ orientifold of $AdS_5 \times S^5$

We can also consider modifying the $AdS_5 \times S^5$ background. If we still want conformal invariance of the dual field theory, we need to maintain the AdS_5 space untouched, so we can only modify the S^5, to obtain some $AdS_5 \times X_5$ space. Yet if we also want to preserve some supersymmetry, we need also to maintain invariance under a subgroup of the R-symmetry, corresponding as we saw to the isometry group of X_5 (S^5 in the standard case, with isometry $SO(6)$). There is a well-defined procedure that can modify only X_5 in such a way, namely *orbifolding* or *orientifolding*.

Orbifolds and orientifolds in string theory

The concept of orbifold is one familiar from mathematics. It involves turning a base space into a coset space, by dividing by a discrete isometry group Γ, namely identifying the base space under the action of elements of Γ: $x \to \gamma x$. That is, points mapped under the action of Γ are identified, and the coset space is the space of equivalence classes under this relation. The simplest example of Γ is a $\mathbb{Z}_2$ action, for instance the one that interchanges $x \to -x$. In the case of a circle, $\theta \in [-\pi, \pi]$, $\theta \sim \theta + 2\pi$, considering the coset $S^1/\mathbb{Z}_2$ gives $\theta \in [0, \pi]$, i.e. an interval, with the endpoints 0 and π being fixed points of the $\mathbb{Z}_2$ action. In general, under the orbifold identification by Γ we have several fixed points, which are singular points. It is somewhat more obvious that fixed points are singular if we consider the example of the complex plane $z = x + iy$, identified under the spatial reflection $z \to -z$. In this case, as we can easily see, we obtain a cone with deficit angle π, so the origin is singular (has infinite curvature).

A field theory defined on an orbifold space is obviously also singular, but one interesting thing about string theory is that it is actually well defined on the orbifold space. For a field theory, one only projects the Hilbert space to states invariant under Γ, but for a string theory, one needs also to consider *twisted states*. These states are the result of the fact that a closed string in the orbifold can be closed only in the projected space, not in the covering space. Namely, it can start and end at different points in the covering space, that are identified under Γ. For instance, in the case of the identification under spatial reflection R, twisted states would be

$$X^m(\sigma + 2\pi, \tau) = -X^m(\sigma, \tau), \tag{19.12}$$

when going around the string we identify only modulo R. As we see, there are twisted states localized at the fixed point at the origin ($X^m = 0$). In the case of $S^1/\mathbb{Z}_2$, we also have twisted states under the combined action with the circle periodicity P,

$$X(\sigma + 2\pi, \tau) = 2\pi - X(\sigma, \tau), \tag{19.13}$$

i.e. when going around the string, we identify modulo RP. Thus there are twisted states localized at the fixed point $X = \pi$. In general then, twisted states are those that satisfy

$$X(\sigma + 2\pi, \tau) = hX(\sigma, \tau), \tag{19.14}$$

where h is a discrete symmetry of the space of Xs.

Note then that there are twisted states localized at the fixed points of the orbifold.

An orientifold is an object that exists only within string theory. Like an orbifold, it is obtained as a coset of a space, divided by the action of a discrete group. But the discrete group is the product of a discrete spacetime isometry Γ with the worldsheet parity, $\Omega : \sigma \to 2\pi - \sigma$, or $z \to \bar{z}$.

In our example with the circle, the orientifold version would be $S^1/(\mathbb{Z}_2 \times \Omega)$, i.e. identify under

$$X(\sigma, \tau) \leftrightarrow -X(2\pi - \sigma, \tau) \quad (X(z, \bar{z}) \leftrightarrow -X(\bar{z}, z)). \tag{19.15}$$

Identifying the theory under worldsheet parity Ω alone leads to unoriented string theory, whereas identifying under the $\mathbb{Z}_2$ action alone leads to an orbifold, and identifying under the product leads to an orientifold. The orientifold has fixed points, which are the same as those of the orbifold action (since the worldsheet parity does not change the action on spacetime).

The main difference with respect to the orbifold is that there is no analog of twisted states, since already the discrete symmetry involves an action on the worldsheet (one can make a more rigorous argument for this). As a result, there are no states localized at the fixed point of the orientifold. When the orientifold action is only in directions $p+1, \ldots, D-1$, but does not affect directions $0, 1, \ldots, p$, we have an $O(p)$-plane, where the terminology follows that for D-branes.

Consider N D-branes in the presence of an orientifold plane, together with their N images under the orientifold action. The generic gauge group on the D-branes, when they are separated, is $U(1)^N$. But besides the usual strings between the N D-branes, there are now also strings between the D-branes and their images under the orientifold action. If we have m D-branes coincident, we have as usual a $U(m)$ gauge group. But if the m D-branes are also at the orientifold plane, we have more light states coming from strings between the D-branes and their images. In fact, they fit in the adjoint of the group $SO(2m)$ or $USp(2m)$ ($m(2m-1)$ or $m(2m+1)$ dimensional representation). If all the D-branes are on a single orientifold plane, the gauge group is the maximal $SO(2N)$ or $USp(2N)$.

The orientifold $O(p)$ plane has a negative Dp-brane charge and a negative tension, that can be cancelled by adding Dp-branes. On a compact space, this is necessary, since flux lines that start on a source must end on another, so the total charge must be zero. The charge of an $O(p)$-plane is $Q_{Op} = -2^{p-5}Q_{Dp}$, which is cancelled by 2^{p-5} Dp-branes on a compact space (note that we are counting here only independent D-branes, without their images; including their images we need 2^{p-4} Dp-branes to cancel the charge).

$\mathcal{N} = 2$ orientifold of $AdS_5 \times S^5$

We focus on an $\mathcal{N} = 2$ supersymmetric orientifold of $AdS_5 \times S^5$. The orientifold action gives an orientifold $O(7)$ plane singularity, therefore only two transverse coordinates are affected. In terms of the six coordinates transverse to the N D3-branes defining $AdS_5 \times S^5$, $z = x_1 + ix_2, x_3, \ldots, x_6$, the S^5 metric $d\Omega_5^2$ is modified to $d\tilde{\Omega}_5^2$ as

$$dr^2 + r^2 d\tilde{\Omega}_5^2 = \frac{|dz|^2}{|z|} + dx_3^2 + dx_4^2 + dx_5^2 + dx_6^2. \tag{19.16}$$

Specifically, this means that if we write the $AdS_5 \times S^5$ metric in the form

$$ds^2 = R^2(-\cosh^2 \rho \, dt^2 + d\rho^2 + \sinh^2 \rho \, d\Omega_3^2 + \sin^2 \theta \, d\psi^2 + d\theta^2 + \cos^2 \theta \, d\tilde{\Omega}_3^2),$$
$$d\tilde{\Omega}_3^2 = \sin^2 \theta' d\psi'^2 + d\theta'^2 + \cos^2 \theta' d\phi^2, \tag{19.17}$$

then the $\mathbb{Z}_2$ action defining the orientifold is $\psi \to \psi + \pi$. The effect of the orientifold is to restrict to $\psi \in [0, \pi)$, $\theta \in [0, \pi/2)$. The orientifold $O(7)$ plane is situated at $\sin \theta = 0$, i.e. at $\theta = 0$, which is an S^3 inside the S^5, parameterized by $d\tilde{\Omega}_3^2$.

It is also the position of an orbifold-type singularity of the metric, as we can easily verify (we change the periodicity of the angle ψ, so we get a cone). The $O(7)$ has a charge -4, which must be cancelled by the addition of four independent D7-branes, since the $O(7)$ is inside the compact space S^5, so we cannot have un-compensated charge. The gauge group on the N D3-branes (and their images) is $USp(2N)$, and the gauge group on the four D7-branes (and their images) is $SO(8)$. However, if we take the decoupling limit for D3-branes, $g_s \to 0, \alpha' \to 0$, it means that the coupling for the D7-branes, $\sim g_s \alpha'^{(p-3)/2} \to 0$, hence we obtain simply an $SO(8)$ global symmetry.

Because of the singularity, the symmetry $SO(6)$ of the S^5 is broken to the symmetry of $S^3 \times S^1$, i.e. $SO(4) \times SO(2) \simeq SU(2)_L \times SU(2)_R \times U(1)_R$, which suggests that the dual gauge theory must have R-symmetry $SU(2) \times U(1)$, corresponding to an $\mathcal{N} = 2$ susy theory.

The gauge theory

Strings stretching between D3-brane i and D3-brane j have a state $|i\rangle|j\rangle$ as before. In the un-orientifolded case the state corresponds to the adjoint of $SU(N)$, giving the usual $N = 4$ SYM multiplet. Due to the orientifold projection, one instead obtains the $N = 2$ SYM multiplet in the adjoint (symmetric, $N(2N+1)$ dimensional) of $USp(2N)$, which contains an $\mathcal{N} = 1$ chiral multiplet W, coupled to a hypermultiplet in the antisymmetric representation of $USp(2N)$ ($N(2N-1)$ dimensional), made up of two $\mathcal{N} = 1$ chiral multiplets Z, Z'. The complex scalar W (two real scalars) corresponds to the D3-brane motion in the two directions transverse to the $O(7)$ plane, whereas the complex scalars Z, Z' correspond to the D3-brane motion in the four directions transverse to the D3-brane, but parallel to the $O(7)$-plane.

Strings stretching between D7-brane m and D7-brane n have a state $|m\rangle|n\rangle$ in the adjoint (antisymmetric, 28 dimensional) representation of $SO(8)$, but as we mentioned, they decouple from the theory (their coupling to each other and to the rest of the fields becomes zero in the AdS/CFT decoupling limit). Finally, strings stretching between D3-brane i and D7-brane m give a state $|i\rangle|m\rangle$ in the fundamental representation of the gauge group $USp(2N)$ (due to the index i) and the fundamental representation of the (global symmetry) $SO(8)$ (due to the index m), described by hypermultiplets $q^{im}, \tilde{q}^{im}, i = 1, \ldots, 2N; m = 1, \ldots, 4$. These can be thought of as $\mathcal{N} = 2$ "quarks" in the theory, in that they have a fundamental gauge index i ("color"), and a global symmetry index m ("flavor"). Note, however, that the theory is still conformal, as seen by the gravity dual still having an unmodified AdS_5 factor (with $SO(2,4)$ isometry), so one can think of the resulting theory as a supersymmetric and conformal version of QCD.

The symmetries of the theory are $U(1) \times SU(2)_R$ R-symmetry and $SU(2)_L \times SO(8)$ global symmetry. The chiral fields Z, Z' are doublets under $SU(2)_L$, have zero charge under $U(1)_R$, and together with their complex conjugates form a doublet of $SU(2)_R$. The chiral field W is a singlet of $SU(2)_L \times SU(2)_R$ and has charge $+1$ under $U(1)_R$. The fields $q, \tilde{q}$

have zero charge under $U(1)_R$, are singlets under $SU(2)_L$, and together with their complex conjugates, form a doublet of $SU(2)_R$. The fields charged under $SU(2)_R$ can be written as doublets satisfying a reality condition. Z, Z' are written as $Z_{ij}^{AA'}$, where A, B are $SU(2)_R$ indices and A', B' are $SU(2)_L$ indices (and i, j are $USp(2N)$ indices), and satisfy

$$Z_{ij}^{AA'} = \epsilon^{AB} \epsilon^{A'B'} \Omega_{ii'} (Z^\dagger)_{BB'}^{i'j'} \Omega_{jj'}. \tag{19.18}$$

Also $q, \tilde{q}$ are written as q_i^{Am}, satisfying the reality condition

$$q_i^{Am} = \epsilon^{AB} \Omega_{ij} (q^\dagger)_B^{jm}. \tag{19.19}$$

The superpotential of the theory is given by (up to normalizations)

$$\mathcal{W} \sim (W_{ij} q^{im} \tilde{q}^{jm} + W_{ii'} \Omega^{i'j} Z_{jj'} \Omega^{j'k} Z'_{kk'} \Omega^{k'i}). \tag{19.20}$$

Here W_{ij} is the adjoint chiral field, Z_{ij}, Z'_{ij} are the antisymmetric chiral fields and Ω^{ij} is the antisymmetric form defining the $USp(2N)$ gauge group.

From this superpotential, we can derive the F-terms, i.e. the on-shell values for the auxiliary scalars F in the chiral multiplets,

$$F_{\tilde{q}\ i}^{\ m} = W_{ij} q^{jm}; \quad F_{q\ i}^{\ m} = W_{ij} \tilde{q}^{jm},$$
$$F_{Z'\ ij'} = W_{ii'} \Omega^{i'j} Z_{jj'} - (i \leftrightarrow j'),$$
$$F_{Z\ ij'} = W_{ii'} \Omega^{i'j} Z'_{jj'} - (i \leftrightarrow j'),$$
$$F_{W\ ij'} = Z_{ii'} \Omega^{i'j} Z'_{jj'} - (i \leftrightarrow j') + \Omega_{ii'} q^{i'm} \tilde{q}^{jm} \Omega_{jj'}. \tag{19.21}$$

The scalar potential is as usual (as we saw in Chapter 3) the sum of the F-terms and the D-terms, $V = \sum_i |F_i|^2 + g^2 D^a D^a$.

Gravity dual description

The gravity dual metric, obtained in the decoupling limit for the D3-branes, is (19.17), which has as fixed point ("orientifold $O(7)$ plane") an $S^3 \subset S^5$, together with AdS_5, therefore an $AdS_5 \times S^3$ $O(7)$ worldvolume. As we saw, at the same location as the $O(7)$ we have four D7-branes and their images, and the AdS/CFT decoupling limit means that the $SO(8)$ gauge theory on the D7-branes decouples. But as in the case of $AdS_5 \times S^5$, where gravity decouples from the D3-brane gauge theory, the decoupling of the D7-brane gauge theory from the D3-brane gauge theory means that the D7-brane gauge theory is part of the "gravity dual" theory to the D3-brane gauge theory. Therefore, we should supplement the gravity dual with the 7+1-dimensional $SO(8)$ SYM multiplet on $AdS_5 \times S^3$.

Then, to compare with the spectrum of operators in the $\mathcal{N} = 2$ $USp(2N)$ SYM theory with matter, we need to KK expand the 9+1-dimensional bulk supergravity modes on S^5, and the 7+1-dimensional vector multiplet on the S^3, giving towers of fields in AdS_5. This is as expected, since now there is a global $SO(8)$ symmetry for the CFT fields, which should correspond to a symmetry in the "gravity dual" (Note that the term "gravity dual" is now slightly misleading, since it involves also the $SO(8)$ SYM fields in 7+1-dimensions).

Therefore, the bulk supergravity modes KK expanded on the S^5 are as usual boundary sources for CFT operators $\mathcal{O}$ with no $SO(8)$ charges (they can have contributions from the $SO(8)$-charged fields $q^{im}, \bar{q}^{im}$, but with the $SO(8)$ indices contracted), but now we also have 7+1-dimensional $SO(8)$ vector modes reduced on the S^3 corresponding to boundary sources for $SO(8)$-charged CFT operators $\mathcal{O}^{m\cdots}$. These $SO(8)$-charged operators are analogous to pion operators in QCD. Indeed, the simplest $SO(8)$-charged operator would be made up of two $|i\rangle|m\rangle$ fields, in the fundamental of the gauge symmetry $USp(2N)$ and of the global symmetry $SO(8)$, thus $\mathcal{O}^{mn} = \bar{q}_i^m q^{in}$ (summed over gauge indices i, but not over global symmetry indices m). For the pion we have a similar operator, with two quark fields q, summed over $SU(3)_c$ gauge indices i, but not isospin $SU(2)_f$ indices m.

We see that *quarks* are introduced in the field theory for AdS/CFT by the addition of a different kind of *probe* branes (here $D7$ and $O(7)$) in the gravity dual. Here we added only four D7-branes to the background of N D3-branes. Of course, ideally one should consider also the backreaction of the extra branes on the geometry, but if $N_f \ll N_c$ (as here, $4 \ll N$), we can neglect it. In the gravitational description, the addition of the branes implies *adding SYM modes living on these branes, which act as sources for pion-like operators*.

19.3 Other orbifolds and orientifolds

To obtain orbifolds or orientifolds of $AdS_5 \times S^5$, in general we are interested in the near-horizon limit of D3-branes sitting at the origin on $\mathbb{R}^{3,1} \times \mathbb{R}^6/\Gamma$, where Γ is a discrete subgroup of $SO(6) \simeq SU(4)_R$, which corresponds to rotational symmetry transverse to the D3-branes in the background. If $\Gamma \subset SU(3) \subset SU(4)_R$, we have $\mathcal{N} = 1$ supersymmetry, since then the remaining symmetry, commuting with Γ, is $U(1)_R$, whereas if $\Gamma \subset SU(2) \subset SU(4)_R$, we have $\mathcal{N} = 2$ supersymmetry. Since the near-horizon limit commutes with the orbifold procedure, the near-horizon limit produces the space $AdS_5 \times S^5/\Gamma$, and leads to a CFT with the corresponding amount of supersymmetry. The action of Γ on S^5 is the same as the action on the angular coordinate on $\mathbb{R}^6$.

Orbifold example: D3-branes on an $\mathbb{R}^4/\mathbb{Z}_k$ singularity

The field theory on N D3-branes at the $\mathbb{Z}_k$ singularity has gauge group $SU(N)^k$, since we have an $SU(N)$ at each image of the branes. Moreover, we have states $|i\rangle|\bar{j}\rangle$ starting on an image and ending on the next one, therefore we have bifundamental fields in representations $(\mathbf{N}, \bar{\mathbf{N}}, \mathbf{1}, \dots, \mathbf{1}), (\mathbf{1}, \mathbf{N}, \bar{\mathbf{N}}, \mathbf{1}, \dots, \mathbf{1}), \dots, (\mathbf{1}, \dots, \mathbf{1}, \mathbf{N}, \bar{\mathbf{N}}), (\bar{\mathbf{N}}, \mathbf{1}, \dots, \mathbf{1}, \mathbf{N})$.

The gravity dual is obtained in the near horizon of the D3-branes, as $AdS_5 \times S^5/\mathbb{Z}_k$, where the $\mathbb{Z}_k$ action leaves an S^1 fixed inside S^5. The low-energy states are the untwisted sector states, which are just the $\mathbb{Z}_k$ projection of the original $AdS_5 \times S^5$ states, and the twisted states, which are $k - 1$ tensor multiplets existing at the orbifold fixed "plane" $AdS_5 \times S^1$, so reduce on S^1 to $k - 1$ $U(1)$ gauge fields on AdS_5.

Other orientifolds

We can construct more orientifolds of a similar type with the one from the previous section, by considering instead of the $\mathbb{Z}_2$ orbifold action, leading to a D_4 singularity, i.e. a singularity that gives rise to an $SO(8)$ $(= D_4)$ global symmetry from the D7-branes added to cancel the orientifold charge, an E_6, E_7, E_8 singularity, i.e. a singularity that gives rise to the corresponding global symmetry when adding the D7-branes to cancel the charge.

The singular metric is obtained from an orbifold action on the angular (rotational) part of $\mathbb{R}^6$ at infinity, transverse to the D3-branes, giving

$$dr^2 + r^2 d\tilde{\Omega}_5^2 = \frac{|dz|^2}{|z|^\alpha} + dx_3^2 + dx_4^2 + dx_5^2 + dx_6^2, \tag{19.22}$$

where, from $\alpha = 1$ for D_4, we now have $\alpha = 4/3, 3/2, 5/3$ for E_6, E_7, E_8, respectively. By defining

$$w = z^{1-\alpha/2},$$

$$\tan^2 \theta = \frac{|w|^2}{x_3^2 + x_4^2 + x_5^2 + x_6^2}, \tag{19.23}$$

the angular part of (19.22) is written in a form similar to S^5, just with a different periodicity,

$$ds^2 = \sin^2 \theta d\psi^2 + d\theta^2 + \cos^2 \theta d\tilde{\Omega}_3^2, \tag{19.24}$$

where $\theta \in [0, \pi/2]$ and $\psi = \arg(w)$ is periodic with periodicity $2\pi(1 - \alpha/2)$ (corresponding to a full rotation in the z plane). We see that for $\alpha = 0$ we obtain the usual $AdS_5 \times S^5$, and otherwise we get an orbifold action that reduces the periodicity, obtaining a singularity at $\theta = 0$, which is an $AdS_5 \times S^3$ locus for an $O(7)$ orientifold.

19.4 Open strings on pp waves and orientifolds

We can take the $\mathcal{N} = 2$ orientifold projection of the pp wave, or equivalently the Penrose limit of the $\mathcal{N} = 2$ orientifold (19.17). In terms of the oscillators for the eight transverse directions of the maximally supersymmetric pp wave, with the two directions transverse to the orientifold plane being called 7 and 8, we have the action for the bosonic oscillators

$$a_n^a \to a_{-n}^I; \quad I = 1, \ldots, 6; \quad a_n^{7,8} \to -a_{-n}^{7,8}, \tag{19.25}$$

and for the fermionic oscillators $b_n \to i\Gamma^{56} b_{-n}$.

We can also obtain the same result from the Penrose limit of the orientifold metric. In the metric (19.17) we can consider the $\mathbb{Z}_2$ action to be $\psi' \to \psi' + \pi$ instead, giving a fixed plane at $\theta' = 0$ ($\sin \theta' = 0$), and then consider the null geodesic that is fixed in AdS_5 and moves along the S^5 circle parameterized by ψ, at $\theta = 0$. Therefore we consider the rescaling

$$\rho = \frac{r}{R}; \quad \theta = \frac{y}{R}; \quad x^+ = \frac{\psi + t}{\sqrt{2}}; \quad x^- = \frac{t - \psi}{\sqrt{2}}, \tag{19.26}$$

which leads to the momentum and energy

$$p^+ = \frac{\Delta + J}{\sqrt{2}R^2}; \quad p^- = \frac{\Delta - J}{\sqrt{2}}, \tag{19.27}$$

in terms of gauge theory quantities.

The $S^3 \subset S^5$ fixed plane at $\theta' = 0$ has a symmetry $SO(4) = SU(2)_L \times SU(2)_R$, and then the J above, corresponding in the pp wave to rotations in the x^7, x^8 plane, and another charge J' that corresponds to rotations in the x^5, x^6 plane, are defined from the two J^3 generators as

$$J = J^3_{SU(2)_R} + J^3_{SU(2)_L}; \quad J' = J^3_{SU(2)_R} - J^3_{SU(2)_L}. \tag{19.28}$$

To construct operators corresponding to the string states on the pp wave, we have to consider the combinations $(Z\Omega)^j_i = Z_{ik}\Omega^{kj}, (Z'\Omega)^j_i = Z'_{ik}\Omega^{kj}, (W\Omega)^j = W_{ik}\Omega^{kj}$. A natural ground state to be considered, by analogy with the $\mathcal{N} = 4$ SYM case, is the BPS operator

$$\text{Tr } [(Z\Omega)^J]. \tag{19.29}$$

It corresponds to the closed string ground state with $\sqrt{2}p^+ = 2J/R^2$.

Next, we consider BPS operators that correspond to the introduction of zero momentum "impurities," i.e. $n = 0$ string states, or supergravity states. In $\mathcal{N} = 4$ SYM, we had insertions of the 4 D_as acting on Zs, corresponding to the action of the $a_0^{\dagger a}$, $a = 1, \ldots, 4$, and insertions of Φ^m, $m = 1, \ldots, 4$, corresponding to the action of the $a_0^{\dagger m+4}$, together with insertions of half of the fermions, all having $\Delta - J = 1$.

Similarly now, we still have insertions of the 4 D_as on $(Z\Omega)$, with $\Delta - J = 1$, corresponding to the action of the $a_0^{\dagger i}$, $a = 1, \ldots, 4$, exactly as in $\mathcal{N} = 4$ SYM. The difference arises in the introduction of scalar impurities (and fermions, which we will skip) of $\Delta - J = 1$, which are now $(Z'\Omega)$ and $(\Omega\bar{Z}')$, corresponding to $a_0^{\dagger 5,6}$ (inside the $O(7)$ plane) and $(W\Omega)$ and $(\Omega\bar{W})'$, corresponding to $a_0^{\dagger 7,8}$ (transverse to the $O(7)$ plane).

We have BPS states protected against quantum correction by supersymmetry, where we consider insertions of only $(Z'\Omega)$ and $(W\Omega)$. The states should be $\text{Tr } [(Z\Omega)^J(Z'\Omega)^{l_1}(W\Omega)^{l_2}]$, appropriately symmetrized. Indeed, the on-shell conditions $F_Z = 0$ and $F_{Z'} = 0$ in (19.21) mean that we can commute freely $(W\Omega)$ past $(Z\Omega)$ and $(Z'\Omega)$, and the on-shell condition $F_W = 0$ says that we can commute freely $(Z\Omega)$ past $(Z'\Omega)$ (there are $q\tilde{q}$ terms which lead to splitting of the trace in two, but these are subleading in $1/N$). Therefore the protected $n = 0$ closed string states are (the example below is for $l_1 = l_2, j_{l_2-1} < i_{l_1} < j_{l_2}$)

$$\sum_{i_1,\ldots,i_{l_1},j_1,\ldots,j_{l_2}=1}^{J} \text{Tr } \left[((Z\Omega)^{i_1}(Z'\Omega)) ((Z\Omega)^{j_1-i_1}(W\Omega)) \ldots \right.$$
$$\left. \ldots ((Z\Omega)^{i_{l_1}-j_{l_2-1}}(Z'\Omega)) ((Z\Omega)^{j_{l_2}-i_{l_1}}(W\Omega)) (Z\Omega)^{J-j_{l_2}} \right]. \tag{19.30}$$

Note that since Z, Z' are antisymmetric, while W is symmetric, we need l_2 even.

States with $n \neq 0$ are obtained by adding a momentum along the string. Therefore, for instance, the operator corresponding to the orientifold projected closed string state $(a_{-n}^{\dagger 7+i8} a_n^{\dagger 5+i6} - a_n^{\dagger 7+i8} a_{-n}^{\dagger 5+i6})|0, p^+\rangle_{l.c.}$ is

$$\sum_l e^{\frac{2\pi inl}{J}} \text{Tr} \; [(W\Omega)(Z\Omega)^j(Z'\Omega)(Z\Omega)^{J-l}]. \tag{19.31}$$

The string Hamiltonian acting on these states can be derived to be exactly the same as in the $\mathcal{N} = 4$ SYM case.

The interesting new feature of the $\mathcal{N} = 2$ orientifold case, however, is the introduction of open strings. The open strings arise from strings ending on the D7-branes at the orientifold $O(7)$ fixed point, therefore they obey Neumann boundary conditions in the directions $a = 1, \ldots, 6$ parallel to the fixed point and Dirichlet conditions in the directions 7,8 transverse to the $O(7)$.

The open string ground state, with Chan–Patton factors (mn) is

$$q^m \Omega(Z\Omega)^J q^n. \tag{19.32}$$

It has $\Delta - J = 1$, since the qs have $J = 1/2$, and $J' = 1$, since q is an $SU(2)_R$ doublet. The state is antisymmetric, so is in the adjoint (antisymmetric) representation of $SO(8)$, as we expect from the open strings ending on D7-branes.

We have protected BPS operators corresponding to $n = 0$ (massless) open string states at the orientifold fixed point, which have the same quantum numbers as the operator $q^m \Omega(Z\Omega)^J(Z'\Omega)^j q^n$, but are symmetrized over the insertions of $(Z'\Omega)$, i.e.

$$\sum_{k_1, \ldots, k_l = 1}^{J} q^m \Omega(Z\Omega)^{k_1}(Z'\Omega)(Z\Omega)^{k_2 - k_1}(Z'\Omega) \ldots (Z\Omega)^{k_l - k_{l-1}}(Z'\Omega)(Z\Omega)^{J - k_l} q^n. \tag{19.33}$$

Note that we cannot have Ws inside the trace, since by the $F_Z = 0$ and $F_{Z'} = 0$ conditions we can commute the Ws until the end, where by $F_q = 0 = F_{\bar{q}}$ it should be zero.

Open string excited states are again found by acting with the same oscillators as in the closed string case. We are not constrained, however, by the condition of zero total momentum along the string worldsheet, since the string is no longer translationally invariant due to the endpoints. In the gauge theory, there is no longer cyclicity of the trace to impose that modes of nonzero total momentum give zero.

The orientifold action on the oscillators of the *open string* is

$$a_n^I \to (-1)^n a_n^I; \quad I = 1, \ldots, 8; \quad b_n \to (-1)^n b_n, \tag{19.34}$$

where b_n are fermionic oscillators.

As a final comment, we have described here how to construct open strings from an "orientifold" of $\mathcal{N} = 4$ SYM (corresponding to an orientifold of the $AdS_5 \times S^5$ pp wave), but we can also consider open strings in $\mathcal{N} = 4$ SYM, that end on D-branes wrapped on cycles ("giant gravitons"). We will, however, not describe them here.

Important concepts to remember

- The AdS/CFT correspondence can be generalized to other cases; in general, when there is no AdS space, it is called gravity/gauge duality, and the gravitational background is called gravity dual.
- The other two conformal and maximally supersymmetric cases of duality are $AdS_4 \times S^7$ and $AdS_7 \times S^4$ in M-theory.

- The $AdS_7 \times S^4$ case is dual to the $(2, 0)$ supersymmetric and conformal theory on multiple M5-branes, which is not well understood, but is a nonabelian version of an abelian theory with a self-dual 2-form field B_{ab}, and admitting Wilson surface observables.
- The $AdS_4 \times S^7$ case is dual to an $\mathcal{N} = 8$ (maximally supersymmetric) theory in 2+1 dimensions on multiple M2-branes (membranes), and a not well understood theory, the IR (strong coupling) limit of the theory on N D2-branes. It will be better understood from the ABJM theory in the next chapter.
- Orbifolds are defined by dividing (considering equivalence classes under) a space by a discrete symmetry group Γ. In general, the fixed points of the symmetry are singular in the orbifold, which makes field theory on orbifolds singular, but string theory is well-defined on them.
- String theory on an orbifold has twisted states, which are strings starting on a brane and ending on an image of the same brane.
- Orientifolds are defined only in string theory, by dividing a space by a discrete symmetry group Γ (orbifold action), times the worldsheet parity on the string.
- An orientifold fixed plane ($O(p)$-plane) has no twisted states on it, and D-branes on an orientifold lead to $USp(2N)$ or $SO(2N)$ gauge groups, whereas on orbifold fixed planes they lead to products of $SU(N)$ groups.
- The $\mathcal{N} = 2$ orientifold of $AdS_5 \times S^5$ with an $O(7)$ plane needs the addition of four D7-branes at the plane and leads to a theory with an $SO(8)$ global ("flavor") symmetry for decoupled gauge fields, and "quarks" q^{im}, i.e. bifundamental fields $|i\rangle|m\rangle$ with one leg on a D3-brane and one leg on a D7-brane, namely fundamentals of the gauge $USp(2N)$ and fundamentals of the global $SO(8)$.
- In the gravity dual of the $\mathcal{N} = 2$ orientifold we have a 7+1-dimensional SYM vector field on the orientifold fixed plane, giving sources to the $SO(8)$-charged ("pion") operators.
- The orientifold of the pp wave and the Penrose limit of the orientifolded $AdS_5 \times S^5$ give the same result, and permit the introduction of operators corresponding to open strings on the pp wave, with $q, \tilde{q}$ at the endpoints of the operator.

References and further reading

For more on the various orbifolds and orientifolds of $AdS_5 \times S^5$, see the review [26]. The $\mathcal{N} = 2$ orientifolds of $AdS_5 \times S^5$ were studied in [76]. The $\mathcal{N} = 2$ orientifold of the pp wave theory (with open string spin chain) was studied in [77].

Exercises

1. Consider the metric of N extremal M5-branes and make it non-extremal by as usual $-dt^2 \rightarrow -dt^2 f_T(r)$ and $dr^2 \rightarrow dr^2/f_T(r)$. Take the near-horizon limit of the near-extremal metric to find the finite temperature version of the $AdS_7 \times S^4$ vs. $(2, 0)$ theory

duality. If we reduce on the Euclidean time direction of radius R_1 and on another spatial direction of radius R_2, what can you say about the 4-dimensional theory that we obtain?

2. Consider the "Wilson surface" operator in the 6-dimensional $(2, 0)$ theory at finite temperature defined above. Using AdS/CFT, argue for the correct scaling of the VEV of this operator when the surface that defines it bounds a large volume.

3. Consider $\mathbb{C}^n$ with complex coordinates $z_1, \ldots, z_n$ and a $\mathbb{Z}_k$ action on it that acts in the same way on all the coordinates z_i and has $z_i = 0$ as fixed point. Define the orbifold $\mathbb{C}^n / \mathbb{Z}_k$. How do you define string theory twisted states in this background?

4. Consider the 7+1-dimensional vector field A_M existing on the 7-branes at the $\mathcal{N} = 2$ orientifold fixed point in the gravity dual. Reducing it on the S^3, we get a vector field in AdS_5. To what operator in the field theory should it correspond?

5. Consider a D3-brane wrapping the $S^3 \subset S^5$ on the 7-brane in the gravity dual of the $\mathcal{N} = 2$ orientifold. Propose a field theory interpretation for this solitonic object.

6. Consider the closed string oscillator state on the pp wave $a_{-n-m}^{\dagger 7+i8} a_n^{\dagger 5+i6} a_m^{\dagger 5+i6} |0, p^+\rangle_{\text{closed}}$. Construct from it a state invariant under the $\mathcal{N} = 2$ orientifold action, and write explicitly the corresponding field theory operator.

7. Write down explicitly the $\mathcal{N} = 2$ orientifold operator corresponding to the *open string* oscillator state on the pp wave $a_n^{\dagger 7+i8} a_m^{\dagger 5+i6} |0, p^+\rangle_{\text{open}}$, properly symmetrized by the orientifold projection.

In the previous chapter, we saw that one of the other conformal and maximally supersymmetric cases of AdS/CFT is the duality of M theory in $AdS_7 \times S^4$ background vs. the CFT on the IR of N coincident M2-branes. We also mentioned that there is a better definition of the duality where we deform it by a discrete parameter k, and $k = 1$ corresponds to the original case. The conformal field theory is called the ABJM model, after the authors of the original paper, and is described next.

20.1 The ABJM model

We want to consider a model for the theory on coincident (nonabelian) M2-branes. Since there are eight coordinates transverse to the M2-brane of M-theory, we expect a theory with eight scalars. On the other hand, since the M2-brane theory is supposed to be the strong coupling limit of the D2-brane theory, where there are seven scalars and a gauge field, one expects that at least the degrees of freedom do not change. As also mentioned in the previous chapter, what happens in the case of a single M2-brane is that we can dualize the gauge field (with a single degree of freedom) to a scalar as $F_{\mu\nu} = 1/2\epsilon_{\mu\nu\rho}\partial^\rho\phi$, and have a theory with eight scalars and no other bosons.

But at the nonabelian level, we cannot make the duality transformation. Therefore, to describe a theory on M2-branes we would need eight scalars from the start. But we also need nonabelian gauge fields, in order to have a relation to the theory of multiple D2-branes, which is three-dimensional SYM. Yet we cannot have more degrees of freedom, since we already have the required number. However, in three dimensions we can have Chern–Simons gauge fields that have no degrees of freedom. A minimal, Majorana, fermion in three dimensions has two real degrees of freedom, so to have a supersymmetric theory, we need four Majorana fermions. Therefore, the fields of the ABJM model are four complex scalars C^I, $I = 1, \ldots, 4$, four Majorana fermions ψ_I and CS gauge fields A_μ. It turns out that we need a gauge group $SU(N) \times SU(N)$ instead of the usual $SU(N)$, which means that there are CS gauge fields A_μ and $\hat{A}_\mu$ for the two $SU(N)$ factors, and C^I and ψ_I are bifundamental, i.e. transforming as the representation $(N, \bar{N})$ under $SU(N) \times SU(N)$.

In this section we just write the action down and in the next section we show that it reduces to the D2-brane theory, i.e. 3-dimensional $\mathcal{N} = 8$ supersymmetric gauge theory under a certain Higgs procedure. The action for the ABJM model is then

$$S_{\text{ABJM}} = \int d^3x \left[\frac{k}{4\pi} \epsilon^{\mu\nu\lambda} \text{Tr} \left(A_\mu \partial_\nu A_\lambda + \frac{2i}{3} A_\mu A_\nu A_\lambda - A_\mu \partial_\nu A_\lambda - \frac{2i}{3} A_\mu A_\nu A_\lambda \right) \right.$$

$$- \text{Tr} \left(D_\mu C_I^\dagger D^\mu C^I \right) - i\text{Tr} \left(\psi^{I\dagger} \gamma^\mu D_\mu \psi_I \right)$$

$$+ \frac{4\pi^2}{3k^2} \text{Tr} \left(C^I C_I^\dagger C^J C_J^\dagger C^K C_K^\dagger + C_I^\dagger C^I C_J^\dagger C^J C_K^\dagger C^K \right.$$

$$\left. + 4 C^I C_J^\dagger C^K C_I^\dagger C^J C_K^\dagger - 6 C^I C_J^\dagger C^J C_I^\dagger C^K C_K^\dagger \right)$$

$$+ \frac{2\pi i}{k} \text{Tr} \left(C_I^\dagger C^I \psi^{J\dagger} \psi_J - \psi^{\dagger J} C^I C_I^\dagger \psi_J - 2 C_I^\dagger C^J \psi^{\dagger I} \psi_J + 2 \psi^{\dagger J} C^I C_J^\dagger \psi_I \right.$$

$$\left. \left. + \epsilon^{IJKL} \psi_I C_J^\dagger \psi_K C_L^\dagger - \epsilon_{IJKL} \psi^{\dagger I} C^J \psi^{\dagger K} C^L \right) \right]. \tag{20.1}$$

Here the covariant derivative acts like

$$D_\mu C^I = \partial_\mu C^I + i \left(A_\mu C^I - C^I \hat{A}_\mu \right).$$

The action has $\mathcal{N} = 6$ supersymmetry, instead of the maximal $\mathcal{N} = 8$, which means that the R-symmetry is $U(1) \times SU(4)_R$ ($SU(4) = SO(6)$). The scalars C^I and the fermions ψ_I are in the fundamental **4** representation of $SU(4)_R$, and have charge $+1$ under $U(1)_R$. The gauge fields are Chern–Simons type, i.e. $\propto \text{Tr} [A \wedge dA + 2/3A \wedge A \wedge A]$, whose field equation is $F = dA + A \wedge A = 0$, meaning that there is no propagating degree of freedom. The coefficient of the CS form is $k/4\pi$, with k quantized at the quantum level, i.e. $k \in \mathbb{Z}$, for consistency of e^{iS} (so that it is single valued under all gauge transformations). Here the integer k is called *the level*. Then we see that in the ABJM action the two gauge factors have levels k and $-k$, i.e. $SU(N)_k \times SU(N)_{-k}$.

The level parameter k serves as the deformation parameter away from the theory of N coincident M2-branes in flat space (which corresponds to $k = 1$), and perturbation theory is at large k. Indeed, we see that we have $k/(4\pi)$ in front of the action, where we usually put $1/g_{\text{YM}}^2$, so that after the rescaling $A_\mu \to g_{\text{YM}} A_\mu$, g_{YM} is indeed a coupling constant. Therefore, by the rescaling $A_\mu \to A_\mu/\sqrt{k}$, we now have $1/\sqrt{k}$ as the coupling. Moreover, as we saw, in large N gauge theories, the effective coupling appearing from planar Feynman diagrams is the 't Hooft coupling, $\lambda = g_{\text{YM}}^2 N$, which in our case becomes N/k. It follows that perturbation theory is indeed at $k \to \infty$, and the $k = 1$ case is strongly coupled, and hard to describe, as expected (the fact that we have a Lagrangean for $k = 1$ is not of much use, since we cannot use Feynman diagrams for perturbation theory).

The supersymmetry transformation laws are

$$\delta C^I = i\bar{\epsilon}^{IJ} \psi_J,$$

$$\delta \psi_J = \gamma^\mu \epsilon_{IJ} D_\mu C^I + \frac{2\pi}{k} 2 C^K C_J^\dagger C^L \epsilon_{KL} - \frac{2\pi}{k} (C^I C_K^\dagger C^K - C^K C_K^\dagger C^I) \epsilon_{IJ},$$

$$\delta A_\mu = -\frac{2\pi}{k} (\bar{\epsilon}_{IJ} \gamma_\mu C^J \psi^{\dagger I} - \bar{\epsilon}^{IJ} \gamma_\mu \psi_I C_J^\dagger),$$

$$\delta \hat{A}_\mu = -\frac{2\pi}{k} (\bar{\epsilon}_{IJ} \gamma_\mu \psi^{\dagger I} C^J - \bar{\epsilon}^{IJ} \gamma_\mu C_J^\dagger \psi_I), \tag{20.2}$$

where indices are raised and lowered with the $SU(4)$ invariant metric δ_I^J, and the susy parameters ϵ^{IJ} are antisymmetric in IJ and satisfy

$$\epsilon^{IJ} = \frac{1}{2}\epsilon^{IJKL}\epsilon_{KL}, \tag{20.3}$$

which means they live in the **6** representation of $SU(4)$.

We can understand parts of the supersymmetry laws as follows. As we know, the transformation law for a scalar is of the type $\delta\phi = \bar{\epsilon}\psi$, and in this case the indices match as well. For the fermion, the linear part of the transformation law is $\delta\psi = \not{\partial}\phi\epsilon$, and again the indices match. Since it is a nonabelian theory, we need to add the nonlinear part, which is nontrivial and not easy to explain. The transformation law for the gauge field seems unusual, since it is purely nonlinear, but we need to remember that the gauge field is of CS type, with no propagating degrees of freedom, so its linearized transformation law can only be $\delta A_\mu = \delta \hat{A}_\mu = 0$, as is the case. Again the nonlinear part is nontrivial and not easy to explain.

One important piece of information concerns the gauge group. The theory presented here is the $SU(N) \times SU(N)$ version of the ABJM model, but in fact we will see that the theory related to M2-branes has gauge group $U(N) \times U(N)$. In terms of the action formally nothing changes in the $U(N) \times U(N)$ case.

As we mentioned, we will only justify the ABJM action implicitly, through the fact that we can obtain the multiple D2-brane action, 3-dimensional $\mathcal{N} = 8$ SYM, the KK reduction of 4-dimensional $\mathcal{N} = 4$ SYM.

20.2 Reduction of M2 to D2 and Mukhi–Papageorgakis Higgs mechanism

The M2 to D2 transformation happens through a version of the usual 4-dimensional Higgs mechanism specific to three dimensions, and discovered by Mukhi and Papageorgakis. In four dimensions, in the Higgs mechanism one expands the action around a Higgs vacuum $\langle \Phi \rangle = v$ (Φ is complex, but v is real), and then the vector gauge field A_μ "eats" a real scalar degree of freedom θ (the phase of the complex field) and becomes massive, through the redefinition $A'_\mu = A_\mu + \partial_\mu\theta$. The massive field has one more degree of freedom than the massless gauge field, so the number of degrees of freedom is conserved.

In three dimensions, we could have the usual Higgs mechanism, but there is also another version. A Chern–Simons (non-dynamical) gauge field can "eat" a real scalar and become a dynamical Yang-Mills field (with one degree of freedom in three dimensions), through the same redefinition as in the usual 4-dimensional case.

Abelian version of the Mukhi–Papageorgakis Higgs mechanism

We first explain the mechanism in a simple abelian case, where the steps are clearer. More precisely, we start with a theory of two abelian CS fields, $a_\mu^{(1)}$ and $a_\mu^{(2)}$, with levels k and $-k$ respectively, i.e. with action

$$\frac{k}{2\pi}\int \epsilon^{\mu\nu\rho}(a_\mu^{(1)}\partial_\nu a_\rho^{(1)} - a_\mu^{(2)}\partial_\nu a_\rho^{(2)}). \tag{20.4}$$

We can redefine the CS fields by $a_\mu = a_\mu^{(1)} + a_\mu^{(2)}$ and $\tilde{a}_\mu = a_\mu^{(1)} - a_\mu^{(2)}$, which turns the action into $(k/2\pi)\int \epsilon^{\mu\nu\rho}a_\mu\partial_\nu\tilde{a}_\rho$. Therefore consider the starting action for the CS $a_\mu, \tilde{a}_\mu$ and a complex scalar Φ coupled to a_μ,

$$S = -\int d^3x \left[\frac{k}{2\pi}\epsilon^{\mu\nu\rho}a_\mu\partial_\nu\tilde{a}_\rho + \frac{1}{2}|(\partial_\mu - iea_\mu)\Phi|^2 + V(|\Phi|^2)\right], \tag{20.5}$$

and a vacuum $\Phi = b = $ constant (b is real). We can expand the scalar around this vacuum as

$$\Phi = (b + \delta\psi)e^{-i\theta}, \tag{20.6}$$

and substituting back in the action, we obtain

$$S = -\int d^3x \left[\frac{k}{2\pi}\epsilon^{\mu\nu\rho}a_\mu\partial_\nu\tilde{a}_\rho + \frac{1}{2}(\partial_\mu\delta\psi)^2 + \frac{1}{2}(\partial_\mu\theta + ea_\mu)^2 b^2 + \ldots\right]. \tag{20.7}$$

The omitted terms are from $V(|\Psi|^2)$ and the interactions between $\delta\psi$ and θ. Then making the same redefinition as in the usual Higgs mechanism,

$$a_\mu + \frac{1}{e}\partial_\mu\theta \equiv a'_\mu, \tag{20.8}$$

the gauge field a'_μ is now massive, and the scalar θ disappears from the action (is "eaten"). The action then becomes

$$S = -\int d^3x \left[\frac{k}{2\pi}\epsilon^{\mu\nu\rho}a'_\mu\partial_\nu\tilde{a}_\rho + \frac{1}{2}(\partial_\mu\delta\psi)^2 + \frac{1}{2}(ea'_\mu)^2 b^2 + \ldots\right], \tag{20.9}$$

and solving for a'_μ,

$$a'^\mu = -\frac{k}{2\pi b^2}\epsilon^{\mu\nu\rho}\partial_\nu\tilde{a}_\rho, \tag{20.10}$$

and defining $\tilde{f}_{\mu\nu} = \partial_\mu\tilde{a}_\nu - \partial_\nu\tilde{a}_\mu$, we obtain

$$S = \int d^3x \left[-\frac{k^2}{16\pi^2 b^2}(\tilde{f}_{\mu\nu})^2 - \frac{1}{2}(\partial_\mu\delta\psi)^2 + \ldots\right], \tag{20.11}$$

where again we have omitted some interaction terms.

Mukhi–Papageorgakis Higgs mechanism for ABJM: from M2 to D2

Coming back to the case of the ABJM model, we consider the expansion around the Higgs vacuum $C^4 = v$ (with v real) and the rest of the fields equal to 0. That is, one of the eight real scalars takes a nontrivial VEV, and the other seven do not (since they are the ones that will become the D2-brane scalars). Therefore we expand

$$C^I = v\delta^{I4} + z^I, \tag{20.12}$$

or more precisely in terms of real scalars X^I and X^{I+4},

$$C^I = v\delta^{I4}\mathbb{1}_{N\times N} + \frac{1}{\sqrt{2}}X^I + i\frac{1}{\sqrt{2}}X^{I+4}. \tag{20.13}$$

We consider the case of $U(N) \times U(N)$, since it is related to the brane construction to be understood in the next section. This ansatz breaks the $U(N) \times U(N)$ group to the diagonal $U(N)$ (similarly for $SU(N) \times SU(N) \to SU(N)$). Then we also expand in the generators T_a of $SU(N)$, obeying $[T^a, T^b] = if^{ab}{}_c T^c$ and $\text{Tr } [T^a T^b] = \delta^{ab}$ (in this section we consider this normalization), and in $T^0 \equiv \mathbb{1}_{N \times N}$. More precisely, we expand

$$C^I = \left(\frac{X_0^I}{\sqrt{2}} + v\delta^{I,4} \right) T^0 + i\frac{X_0^{I+4}}{\sqrt{2}} T^0 + i\frac{X_a^I}{\sqrt{2}} T^a - \frac{X_a^{I+4}}{\sqrt{2}} T^a ,$$

$$\psi^I = \psi_0^I T^0 + i\psi_a^I T^a + i\psi_0^{I+4} T^0 - \psi_a^{I+4} T^a,$$

$$A_\mu = A_\mu^0 T^0 + A_\mu^a T^a, \qquad \hat{A}_\mu = \hat{A}_\mu^0 T^0 + \hat{A}_\mu^a T^a, \qquad (20.14)$$

and redefine the gauge fields as

$$A_\mu^+ = \frac{A_\mu + \hat{A}_\mu}{2}; \quad A_\mu^- = \frac{A_\mu - \hat{A}_\mu}{2}. \qquad (20.15)$$

In terms of the new gauge fields, the CS part of the ABJM action becomes

$$S_{\text{CS}} = \int d^3x \frac{k}{2\pi} \varepsilon^{\mu\nu\lambda} \text{Tr} \left(A_\mu^- F_{\nu\lambda}^+ + \frac{2i}{3} A_\mu^- A_\nu^- A_\lambda^- \right) . \qquad (20.16)$$

After some algebra, left as an exercise, the CS and scalar kinetic part of the ABJM action becomes

$$S = \int d^3x \left[\frac{k}{2\pi} \varepsilon^{\mu\nu\lambda} \text{Tr} \left(A_\mu^- F_{\nu\lambda}^+ + \frac{2i}{3} A_\mu^- A_\nu^- A_\lambda^- \right) - \text{Tr } |D_\mu C^I|^2 \right]$$

$$= \int d^3x \left[\frac{k}{2\pi} \varepsilon^{\mu\nu\lambda} \left(A_{\mu a}^- + \frac{1}{2v} \frac{1}{\sqrt{2}} (D_\mu X)_a^4 \right) F_{\nu\lambda}^{+a} + \frac{Nk}{2\pi} \varepsilon^{\mu\nu\lambda} \left(A_{\mu 0}^- + \frac{1}{2v} \frac{1}{\sqrt{2}} (\partial_\mu X)_0^8 \right) F_{\nu\lambda}^{+0} \right.$$

$$- \left(2v A_{\mu a}^- + \frac{1}{\sqrt{2}} (D_\mu X)_a^4 \right)^2 - N \left(\frac{1}{\sqrt{2}} \partial_\mu X_0^{I+4} + 2v A_{\mu 0}^- \delta^{I4} \right)^2$$

$$\left. - \frac{1}{2} (D_\mu X)_a^{A'} (D^\mu X)_a^{A'} - \frac{1}{2} N \partial_\mu X_0^I \partial^\mu X_0^I + \text{higher order} \right], \qquad (20.17)$$

where $A' = \{A \neq 4\}$.

After the shift

$$A_{\mu a}^- \to A_{\mu a}^- - \frac{1}{2v} \frac{1}{\sqrt{2}} (D_\mu X)_a^4 \quad \text{and} \quad A_{\mu 0}^- \to A_{\mu 0}^- - \frac{1}{2v} \frac{1}{\sqrt{2}} (\partial_\mu X_0^8), \qquad (20.18)$$

one obtains

$$S = \int d^3x \left(\frac{k}{2\pi} \varepsilon^{\mu\nu\lambda} (A_{\mu a}^- F_{\nu\lambda}^{a+} + N A_{\mu 0}^- F_{\nu\lambda\ 0}^+) - 4v^2 A_{\mu a}^- A_a^{-\mu} - 4N v^2 A_{\mu 0}^- A_0^{-\mu} \right.$$

$$\left. - \frac{1}{2} (D_\mu X)_a^{A'} (D^\mu X)_a^{A'} - \frac{1}{2} N \partial_\mu X_0^{\tilde{A}'} \partial^\mu X_0^{\tilde{A}'} + \text{higher order} \right), \qquad (20.19)$$

where $\tilde{A}' = \{A \neq 8\}$. We can solve for A_μ^- as

$$A_\mu^- = \frac{k}{16\pi v^2} \varepsilon_{\mu\nu\lambda} F^{+\nu\lambda} + \text{higher order}, \qquad (20.20)$$

and after substituting back into the action, we obtain

$$S = \int d^3x \left[\mathrm{Tr} \left(-\frac{k^2}{32\pi^2 v^2} F^{+\mu\nu} F^+_{\mu\nu} \right) - \frac{1}{2} (D_\mu X)^{I'}_a (D^\mu X)^{I'}_a \right.$$
$$\left. - \frac{1}{2} N \partial_\mu X^{I'}_0 \partial^\mu X^{I'}_0 + \text{higher order} \right]. \tag{20.21}$$

We can now define the YM coupling by

$$\frac{k^2}{32\pi^2 v^2} \equiv \frac{1}{4g^2_{\mathrm{YM}}}, \tag{20.22}$$

and taking the limit $k, v \to \infty$, with $\frac{k}{v} = $ fixed, the higher order terms drop out. After combining the traceless part of X^8_a with the trace part of X^4_0, the resulting theory is 3-dimensional $U(N)$ SYM theory, i.e. the low energy theory on N D2-branes.

20.3 Brane construction: the IR limit of M2-branes on $\mathbb{C}^4/\mathbb{Z}_k$ and $AdS_4 \times \mathbb{CP}^3$ gravity dual

We turn next to obtaining the ABJM model from an M2-brane construction and obtaining its gravity dual. We follow the original derivation of Aharony, Bergman, Jafferis, and Maldacena (ABJM).

We show the brane construction, though it is very involved, so we do not explain all its details, and we only use the final result to construct the gravity dual.

Brane construction

The ABJM model is realized on branes as follows. The first step is to start in type IIB string theory with two NS5-branes in the directions 012345, separated in the 6 direction, which is compact. Then consider N D3-branes in the 012 and 6 directions, with the D3-branes wrapping the whole compact direction 6, and thus intersecting both the NS5 and the NS5′, as in Fig. 20.1a. Thus the common worldvolume is in directions 012, i.e. 3-dimensional, and it is an $\mathcal{N} = 4$ susy theory (the system breaks 1/2 susy) with gauge group $U(N) \times U(N)$ (one $U(N)$ factor on each D3–NS5 intersection), with two $\mathcal{N} = 2$ chiral multiplets A_i, $i = 1, 2$ in the $(N, \bar{N})$ representation and two $\mathcal{N} = 2$ chiral multiplets B_j, $j = 1, 2$ in the $(\bar{N}, N)$ representation.

The second step is to add k D5-branes, intersecting at the same point the NS5-brane and the N D3-branes, in the directions 012349, as in Fig. 20.1b. This breaks the susy to $\mathcal{N} = 2$ and adds k massless chiral multiplets in the fundamental N and k massless chiral multiplets in the antifundamental $\bar{N}$ of each $U(N)$ factor.

The third step is to introduce CS gauge fields as follows. One first obtains a mass term for the chiral multiplets by separating the intersection of the NS5-brane and k

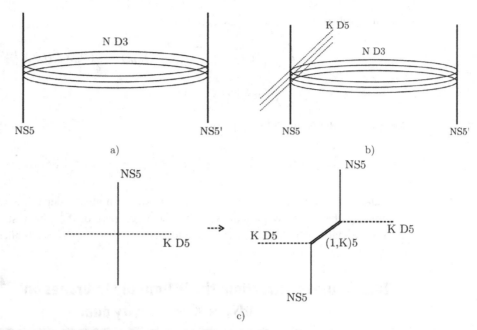

Figure 20.1 a) Brane construction, step 1; b) Brane construction, step 2; c) Brane construction, step 3.

D5-branes through an intermediate bound state, called a $(1, k)$ *5-brane*, as in Fig. 20.1c. Supersymmetry fixed the angle θ in the $(5, 9)$ plane to be given by

$$\theta = \arg(\tau) - \arg(k + \tau); \quad \tau = \frac{i}{g_s} + \chi. \tag{20.23}$$

Then integrating out the fermions in the chiral and antichiral multiplets gives rise to a CS term (as always in three dimensions), with a level $+1/2$ for each Majorana fermion of positive mass and $-1/2$ for each Majorana fermion of negative mass, for a total level of $+k$ for the first $U(N)$ and $-k$ for the second $U(N)$.

The fourth step is to rotate the $(1, k)$ 5-brane in the (37) and (48) directions by the same angle θ above, obtaining a brane in the $012[37]_\theta[48]_\theta[59]_\theta$ directions, for a total 3-dimensional theory with $\mathcal{N} = 3$ supersymmetry.

The final step is to T-dualize to type IIA string theory and lift to M-theory. When T-dualizing on direction 6, we obtain type IIA theory compactified on direction $\tilde{6}$. The D3-brane turns into a D2-brane, since the direction 6 is parallel to the D3-brane. The NS5-brane turns into a "KK monopole" in the directions 012345, associated with the circle $\tilde{6}$. The $(1, k)$ 5-brane turns into a KK monopole in the directions $012[37]_\theta[48]_\theta[59]_\theta$, associated with the circle $\tilde{6}$, and k D6-branes on the same directions. Lifting to M-theory, the D2-brane becomes an M2-brane, the KK monopole remains a KK monopole, just extended in direction 10 also. The D6-brane also turns into a KK monopole, associated with direction 10, which means that the type IIB $(1, k)$ 5-brane turns into a single KK monopole, associated with a direction that is a linear combination of directions $\tilde{6}$ and 10.

All in all, the brane system has lifted to M-theory into a geometry of two intersecting KK monopoles (coming from the NS5 and the $(1, k)$ 5-brane), $\mathbb{R}^{1,2} \times X_8$, probed by N M2-branes, coming from the N D3-branes. Near the core, each monopole looks like $\mathbb{R}^4$, but the intersection of the two looks like $\mathbb{C}^4/\mathbb{Z}_k$. Therefore, the IR limit of the system corresponds to the near-horizon limit of N M2-branes probing the $\mathbb{C}^4/\mathbb{Z}_k$ singularity. On the other hand, the IR limit of the field theory, which is the $\mathcal{N} = 3$ supersymmetric $U(N) \times U(N)$ theory with YM-CS gauge fields described before, is exactly the ABJM model, the $\mathcal{N} = 6$ supersymmetric $U(N) \times U(N)$ CS gauge theory.

Gravity dual

The M2-branes probe the $\mathbb{C}^4/\mathbb{Z}_k$ singularity, where $\mathbb{Z}_k$ acts on the eight Euclidean coordinates of $\mathbb{C}^4$ as

$$Z_i \to e^{\frac{2\pi i}{k}} Z_i, \tag{20.24}$$

and since as we saw in the previous chapter, the gravity dual of N M2-branes in flat space is $AdS_4 \times S^7$, the gravity dual in our case is $AdS_4 \times S^7$, divided by the $\mathbb{Z}_k$ action, i.e. $AdS_4 \times S^7/\mathbb{Z}_k$, since the 7-sphere is defined by $\sum_{i=1}^{4} |Z_i|^2$. But the 7-sphere can be writtten as an S^1 Hopf fibration over $\mathbb{CP}^3$, and then the $\mathbb{Z}_k$ action just makes the S^1 k times smaller, thus leading simply to $\mathbb{CP}^3$ in the $k \to \infty$ limit. The S^1 can be easily identified as the one defined by the common angle multiplying the Z_is, $Z_i = e^{i\alpha} Z_i'$, so Z_i' should define $\mathbb{CP}^3$. We next describe this construction.

As we saw in the previous chapter, the $AdS_4 \times S^7$ metric is

$$ds^2 = \frac{R^2}{4} ds_{AdS_4}^2 + R^2 ds_{S^7}^2, \quad R = l_P (32\pi^2 N')^{1/6}. \tag{20.25}$$

Here $N' = Nk$ is the charge in the covering space, such that in the gravity dual $AdS_4 \times S^7 \mathbb{Z}_k$ we have charge N.

We can easily see that by eliminating the S^1 fiber of S^7 we get $\mathbb{CP}^3$, since $\mathbb{CP}^3$ is defined by coordinates $\zeta_l = Z_l/Z_4$, $l = 1, 2, 3$; and by defining them, we get rid of the $e^{i\alpha}$ fiber and the 7-sphere constraint becomes $\sum_{l=1,2,3} |\zeta_l|^2 + 1 = 1/|Z_4|^2$, which therefore becomes moot (it is a constraint on Z_4, that is outside $\mathbb{CP}^3$). To find the metric on $\mathbb{CP}^3$, we first write the metric on S^7 as a Hopf fibration, by solving the constraint in terms of $|Z_4|$ and substituting in the embedding metric $ds^2 = \sum_{i=1}^{4} |dZ_i|^2$, with $Z_4 = |Z_4| e^{i\tau}$, to obtain

$$ds_{S^7}^2 = (d\tau + \mathcal{A})^2 + \frac{\sum_l |d\zeta_l|^2}{(1 + \sum_l |\zeta_l|^2)} - \frac{\sum_{l,k} \zeta_l \bar{\zeta}_k d\zeta_l d\bar{\zeta}_k}{(1 + \sum_l |\zeta_l|^2)^2},$$

$$\mathcal{A} = \frac{i \sum_l \bar{\zeta}_l d\zeta_l - c.c}{2(1 + \sum_l |\zeta_l|^2)}. \tag{20.26}$$

By dropping the first term, we get the metric on $\mathbb{CP}^3$. Indeed, the $\mathbb{Z}_k$ action amounts to the substitution $\tau \to \tau/k$, which shrinks the first factor to zero, except for the 1-form $\mathcal{A}$.

We can also rewrite the metric in terms of the original Z_i variables as

$$ds^2_{S^7} = (d\tau + \mathcal{A})^2 + ds^2_{\mathbb{CP}^3},$$

$$ds^2_{\mathbb{CP}^3} = \frac{\sum_i dZ_i d\bar{Z}_i}{\sum_j |Z_j|^2} - \frac{|\sum_i Z_i d\bar{Z}_i|^2}{(\sum_j |Z_j|^2)^2},$$

$$d\tau + \mathcal{A} = \frac{i}{2\sum_j |Z_j|^2} \sum_i (Z_i d\bar{Z}_i - \bar{Z}_i dZ_i),$$

$$d\mathcal{A} = i \sum_i d\left(\frac{Z_i}{\sum_j |Z_j|^2}\right) d\left(\frac{d\bar{Z}_i}{\sum_k |Z_k|^2}\right). \tag{20.27}$$

The periodicity of τ is 2π, but it is reduced when we act with $\mathbb{Z}_k$, to obtain

$$ds^2_{S^7/\mathbb{Z}_k} = \left(\frac{1}{k}d\tau + \mathcal{A}\right)^2 + ds^2_{\mathbb{CP}^3}. \tag{20.28}$$

The radius of the S^1 fiber is $R/kl_P \sim (Nk)^{1/6}/k$, so the M theory limit is $k^5 \ll N$. For larger ks, we have a small circle, and we can reduce to type IIA string theory. The reduction to type IIA, using the reduction ansatz in Chapter 7 gives for the string metric, dilaton, and forms

$$ds^2_{\text{string}} = \frac{R^3}{kl_P^3}\left(\frac{1}{4}ds^2_{AdS_4} + ds^2_{\mathbb{CP}^3}\right),$$

$$e^{2\phi} = \frac{R^3}{k^3 l_P^3} \sim \frac{1}{N^2}\left(\frac{N}{k}\right)^{5/2},$$

$$F_4 = \frac{3}{8}R^3 \hat{\epsilon}_4,$$

$$F_2 = kd\mathcal{A}, \tag{20.29}$$

where $\hat{\epsilon}_4$ is the volume form on S^7, normalized to 1. The radius of curvature in string units is given by

$$\frac{R^2_{\text{string}}}{l_s^2} = \frac{R^3}{kl_P^3} = 2^{5/2}\pi\sqrt{\lambda}, \tag{20.30}$$

where we have introduced the 't Hooft coupling $\lambda = N/k$ described before. Note that, as in $\mathcal{N} = 4$ SYM, we have $R^2_{\text{string}} \propto \sqrt{\lambda}$.

In conclusion, the ABJM model is dual under the AdS/CFT correspondence to IIA string theory on $AdS_4 \times \mathbb{CP}^3$. In order for the type IIA supergravity approximation to be valid, we need that $R^2_{\text{string}}/l_s^2 \gg 1$, i.e. $\lambda \gg 1$, and in order to be in type IIA string theory, and not in M-theory, we need $\lambda^{5/2}/N^2 \ll 1$.

Finally, we want to understand from the gravity dual the fact that we proved in the previous section, that the ABJM model reduces, around a Higgs vacuum $C^4 = v$, and in the limit $k, v \to \infty$, with k/v fixed, to the $\mathcal{N} = 8$ $U(N)$ SYM theory. Specifically, this means that the theory on the IR of N M2-brane on $\mathbb{C}^4/\mathbb{Z}_k$ in the above limit, reduces to the theory on N D2-branes in flat space. This is understood as follows.

Consider first as an example the case of $\mathbb{C}/\mathbb{Z}_{k\to\infty}$. Then the action of the $\mathbb{Z}_k$ is

$$Z \to Ze^{\frac{2\pi i}{k}} = Z\left(1 + \frac{2\pi i}{k} + \dots\right) = Z + 2\pi i\frac{Z}{k} + \dots \qquad (20.31)$$

Therefore, expanding around the Higgs vacuum $Z = v$, with v real and $v/k \equiv r$ fixed, we obtain the invariance under

$$Z \to Z + 2\pi i r, \qquad (20.32)$$

or, if we write $Z = X^1 + iX^2$, it means X^2 is compactified with radius r. Moving on to $\mathbb{C}^4/\mathbb{Z}_{k\to\infty}$, with the $\mathbb{Z}_k$ action

$$Z_j \to Z_j e^{\frac{2\pi i}{k}} = Z_j\left(1 + \frac{2\pi i}{k} + \dots\right) = Z_j + 2\pi i\frac{Z_j}{k} + \dots, \qquad (20.33)$$

and expanding around the Higgs vacuum $Z^1 = v$, with v real and $Z^{2,3,4} = 0$, then writing $Z^1 = X^1 + iX^2$, it means X^2 is compactified with radius r.

Therefore, in the limit $k, v \to \infty$ with k/v fixed, the ABJM model, for N M2-branes on $\mathbb{C}^4/\mathbb{Z}_{k\to\infty}$, turns into the theory on N M2-branes in a flat space with a transverse compactified dimension, i.e. N D2-branes in flat 10-dimensional space, as it should.

20.4 The massive deformation of the ABJM model

The ABJM model admits a unique maximally supersymmetric mass deformation, i.e. a mass deformation that preserves $\mathcal{N} = 6$ supersymmetry. This is unlike the case of $\mathcal{N} = 4$ SYM in four dimensions, where any deformation, including a mass deformation, decreases the supersymmetry. However, even though the amount of supersymmetry is the same, which means that the supersymmetry *transformation laws* transform under the same R-symmetry $SU(4) \times U(1)$, in the *Lagrangean* the R-symmetry is broken to $SU(2) \times SU(2) \times U(1)_A \times U(1)_B \times \mathbb{Z}_2$. This is done by splitting the scalars as

$$C^I = (Q^\alpha, R^\alpha), \qquad \alpha = 1, 2, \qquad (20.34)$$

and adding terms to the Lagrangean and supersymmetry transformation rules.

The $SU(2)$ factors act independently on each of the doublets Q^α and R^α, under $U(1)_A$ Q^α has charge $+1$ and R^α has charge -1; under $U(1)_B$ (that was the $U(1)$ multiplying $SU(4)_R$ before the split) Q^α and R^α both have charge $+1$, and $\mathbb{Z}_2$ exchanges Q^α and R^α.

The extra terms in the Lagrangean can be found as follows. We add fermionic mass terms with mass μ, and we also add a deformation to the bosonic potential.[1] The bosonic mass-deformed ABJM Lagrangean is written as

---

[1] These are the only possible terms in $\mathcal{L}$ if the only dimensionful parameter is μ, since from the kinetic terms $\sim \bar{\psi}\partial\!\!\!/\psi$ and $\sim (\partial C)^2$, the fermion ψ has dimension 1, and the scalar C has dimension 1/2. Then the only possible interaction terms of dimension 3 are C^6 and $\bar{\psi}\psi C^2$, already in the pure ABJM action, and $\mu\bar{\psi}\psi$ and $\mu C^4, \mu^2 C^2$ giving the deformation.

$$\mathcal{L}_{\text{bosonic}} = \frac{k}{4\pi}\epsilon^{\mu\nu\lambda}\text{Tr}\left[A_\mu\partial_\nu A_\lambda - \hat{A}_\mu\partial_\nu\hat{A}_\lambda + \frac{2i}{3}\left(A_\mu A_\nu A_\lambda - \hat{A}_\mu\hat{A}_\nu\hat{A}_\lambda\right)\right]$$
$$- \text{Tr} |D_\mu Q^\alpha|^2 - \text{Tr} |D_\mu R^\alpha|^2 - V. \tag{20.35}$$

Here the potential splits as

$$V = \text{Tr}\left(|M^\alpha|^2 + |N^\alpha|^2\right),$$

where

$$M^\alpha = \mu Q^\alpha + \frac{2\pi}{k}\left(2Q^{[\alpha}Q^\dagger_\beta Q^{\beta]} + R^\beta R^\dagger_\beta Q^\alpha - Q^\alpha R^\dagger_\beta R^\beta + 2Q^\beta R^\dagger_\beta R^\alpha - 2R^\alpha R^\dagger_\beta Q^\beta\right),$$
$$N^\alpha = -\mu R^\alpha + \frac{2\pi}{k}\left(2R^{[\alpha}R^\dagger_\beta R^{\beta]} + Q^\beta Q^\dagger_\beta R^\alpha - R^\alpha Q^\dagger_\beta Q^\beta + 2R^\beta Q^\dagger_\beta Q^\alpha - 2Q^\alpha Q^\dagger_\beta R^\beta\right).$$

The mass deformation of the supersymmetry transformation rules is given by[2]

$$\delta^{(\mu)}\psi_I = \frac{1}{2}\epsilon_{JK}C^K\begin{pmatrix} \mu & 0 & 0 & 0 \\ 0 & \mu & 0 & 0 \\ 0 & 0 & -\mu & 0 \\ 0 & 0 & 0 & -\mu \end{pmatrix}^J_I. \tag{20.36}$$

Because of the reality condition (20.3), we can split the susy parameters ϵ^{IJ} in terms of $\alpha, \dot{\alpha}$ independent complex components ($\epsilon^{12}, \epsilon^{1\dot{1}}, \epsilon^{1\dot{2}}$) (thus giving $\mathcal{N} = 6$ real supersymmetries), and dependent components (the ϵs are antisymmetric, so e.g., $\epsilon^{\dot{1}1} = -\epsilon^{1\dot{1}}$, etc.):

$$\epsilon^{\dot{1}\dot{2}} = \epsilon^{12}; \quad \epsilon^{2\dot{2}} = -\epsilon^{1\dot{1}}; \quad \epsilon^{\dot{1}2} = -\epsilon^{1\dot{2}}. \tag{20.37}$$

20.5 The fuzzy sphere ground state

The ground states of the mass-deformed ABJM model are found by setting to zero the potential in (20.36), which in turn implies that $M^\alpha = 0 = N^\alpha$. We can easily find that the two solutions to these equations are

$$R^\alpha = cG^\alpha; \quad Q^\alpha = 0 \quad \text{and} \quad Q^\dagger_\alpha = cG^\alpha; \quad R^\alpha = 0, \tag{20.38}$$

where $c \equiv \sqrt{\frac{\mu k}{2\pi}}$, and the matrices G^α, $\alpha = 1, 2$, satisfy the equations

$$G^\alpha = G^\alpha G^\dagger_\beta G^\beta - G^\beta G^\dagger_\beta G^\alpha. \tag{20.39}$$

Since this is a zero energy ground state of the mass-deformed ABJM model, it preserves all the supersymmetry of the model, i.e. it is $\mathcal{N} = 6$ supersymmetric.

---

[2] The normalization is such that the scalar mass term in the Lagrangean is $-\mu^2\text{Tr} [\bar{C}_I C^I]$.

An explicit solution of the above equations (20.39) can be easily found to be

$$
\begin{aligned}
(G^1)_{m,n} &= \sqrt{m-1}\,\delta_{m,n}\,,\\
(G^2)_{m,n} &= \sqrt{(N-m)}\,\delta_{m+1,n}\,,\\
(G^\dagger_1)_{m,n} &= \sqrt{m-1}\,\delta_{m,n}\,,\\
(G^\dagger_2)_{m,n} &= \sqrt{(N-n)}\,\delta_{n+1,m}\,.
\end{aligned}
\tag{20.40}
$$

In fact, one can show that this is the unique *irreducible* solution, up to $U(N) \times U(N)$ gauge transformations. Note that, since Q^α and R^α are bifundamenal, so are G^α, meaning that $G^\alpha G^\dagger_\beta$ is in the adjoint of the first $U(N)$ and $G^\dagger_\alpha G^\beta$ is the adjoint of the second $U(N)$. One can construct more general reducible solutions as direct sums of these irreducible ones.

In the case of the pure (massless) ABJM model, we see that $c \to 0$, so the ground state solution disappears. In fact, instead of a ground state solution, now we have a *BPS solution*, i.e. a nontrivial, space-dependent solution that has nonzero energy, but preserves 1/2 of the supersymmetry. The solution is formally of the same type, except now the constant c is replaced with the function

$$
c(s) = \sqrt{\frac{k}{4\pi s}},
\tag{20.41}
$$

where s is one of the two spatial directions on the worldvolume of the ABJM model.

The interpretation of the ground state solution of massive ABJM, the solution to (20.39), is of a fuzzy sphere. It was initially thought to be a fuzzy 3-sphere, since the ABJM model lives on M2-branes, and in M-theory M2-branes can end on M5-branes, which would have been the natural interpretation of the BPS solution of pure ABJM: on M2-branes, at $s = 0$ an extra infinite 3-dimensional volume grows, corresponding to the point on which the M2-brane touches the M5-brane. However, it was shown that it is actually a fuzzy 2-sphere, and we can make a precise map to the usual formulation of the fuzzy S^2 described in Section 13.3. Then the interpretation is that we dimensionally reduce to type IIA string theory, so we have actually D2-branes ending on D4-branes, and the fuzzy S^2 sphere at $s = 0$ is the point where the D4-brane grows out of the D2-brane. In fact, this was confirmed by calculating the action of small fluctuations around the fuzzy S^2 in the classical limit $N \to \infty$, finding that the action is the D4-brane action wrapped on S^2 in the gravity dual.

The reason why the fuzzy sphere is actually a fuzzy S^2 at all values of N (finite, except $N = 2$, which is special) is still unclear, but in the classical limit $N \to \infty$, when the fuzzy sphere becomes classical, it can be easily understood. The perturbative parameter of ABJM is the 't Hooft coupling $\lambda = N/k$, and at $N \to \infty$, if we want to have a perturbative expansion, which is needed in order to define the fuzzy sphere and the field theory expansion around it, we need then to have also $k \to \infty$. But as we saw, at $k \to \infty$, the 11-dimensional circle shrinks to zero and we are in 10-dimensional type IIA string theory ($\mathbb{C}^4/\mathbb{Z}_k$ becomes $\mathbb{R}^7$ and $AdS_4 \times S^7/\mathbb{Z}_k$ becomes $AdS_4 \times \mathbb{CP}^3$).

We give some simple arguments why (20.39) gives a fuzzy S^2 to motivate it. A much more complete analysis can be done, but will not be explained here. First, we note that the matrices (20.40) imply the relations

$$\sum_{\alpha=1,2} G^\alpha G_\alpha^\dagger = (N-1)\mathbb{1}; \quad \sum_{\alpha=1,2} (G_\alpha^\dagger G^\alpha)_{mn} = N\delta_{mn} - N\delta_{m1}\delta_{n1}; \quad \Rightarrow$$

$$\sum_{\alpha=1,2} \mathrm{Tr}\,[G^\alpha G_\alpha^\dagger] = N(N-1) = \mathrm{const.} \tag{20.42}$$

This would seem like the definition of a (normalized) unit 3-sphere, using a constraint on two complex coordinates, $\sum_{\alpha=1} |Z^\alpha|^2 = 1$, but the matrices are fixed (given) matrices, so we cannot necessarily deduce this. In fact, we see that $G^1 = G_1^\dagger$, so if these G^α represent coordinates, then G^1 would be interpreted as being real (zero imaginary part), in which case the constraint would truly be for an S^2, $|Z^2|^2 + (\mathrm{Re}Z^1)^2 = 1$.

We can also directly find the usual $SU(2)$ formulation of the fuzzy S^2. The composite matrices

$$J_i = (\sigma_i^T)^\alpha{}_\beta G^\beta G_\alpha^\dagger, \tag{20.43}$$

where G^α are the matrices (20.40) satisfy the usual $SU(2)$ algebra defining the fuzzy S^2,

$$[J_i, J_j] = 2i\epsilon_{ijk}J_k. \tag{20.44}$$

Here σ_i are the Pauli matrices, and σ_i^T are their transposes. In fact, we can directly find the $SU(2)$ commutation relations starting from the algebra (20.39), which we leave as an exercise to prove. But we also find that

$$\bar{J}_i = (\sigma_i^T)^\alpha{}_\beta G_\alpha^\dagger G^\beta \tag{20.45}$$

satisfy the same $SU(2)$ algebra for the fuzzy S^2. It would seem that we have two independent fuzzy S^2s, with symmetry $SU(2) \times SU(2)$, which would equal the $SO(4)$ symmetry of S^3, but in fact the two algebras are not independent, since it is only when acting together on the representation G^α we have a symmetry, namely

$$J_i G^\alpha - G^\alpha \bar{J}_i = (\sigma_i^T)^\alpha{}_\beta G^\beta. \tag{20.46}$$

Finally, since the G^α describe a fuzzy S^2, it follows that the ground state of massive ABJM is the fuzzy S^2, and the BPS state of the massless (pure) ABJM is the fuzzy S^2-funnel, since the radius of the fuzzy sphere varies as $\propto 1/\sqrt{s}$, from zero at $s = \infty$, to infinity at $s = 0$.

20.6 Some comments on applications of the ABJM/$AdS_4 \times$ $\mathbb{CP}^3$ correspondence

For the ABJM/$AdS_4 \times \mathbb{CP}^3$ correspondence, one can calculate almost all the quantities that can be calculated in the 4-dimensional $\mathcal{N} = 4$ SYM case. The ABJM model has been used primarily as a toy model for condensed matter, where most of the interesting behavior occurs in three dimensions. We describe this in Chapter 25, hence we do not do it here. This includes the treatment of finite temperature. Another important application is to spin chains and integrable systems, but since we have described at length how to do this in the case of $\mathcal{N} = 4$ SYM, we also do not describe it. Correlators of gauge invariant operators can also be easily calculated using AdS/CFT, as in the case of $\mathcal{N} = 4$ SYM treated in Chapter 11. One can also calculate scattering of (non-gauge invariant) external states, but since even in the case of $\mathcal{N} = 4$ SYM we will only be describing that in Chapter 26, we do not study it here either.

Important concepts to remember

- The ABJM model is a 3-dimensional conformal $\mathcal{N} = 6$ supersymmetric, $U(N) \times U(N)$ (or $SU(N) \times SU(N)$) CS gauge theory (with levels k and $-k$ respectively) with bifundamental scalars and fermions.

- The Mukhi–Papageorgakis Higgs mechanism in three dimensions means that around the Higgs vacuum, the nondynamical CS gauge field eats a real scalar and becomes dynamical, i.e. Yang–Mills.

- The mechanism can be used to show that the ABJM model (on M2-branes) around a Higgs vacuum reduces to $\mathcal{N} = 8$ SYM theory, the theory on D2-branes.

- The ABJM model arises in the IR limit of N coincident M2-branes on the space $\mathbb{R}^{2,1} \times \mathbb{C}^4/\mathbb{Z}_k$, and its gravity dual is $AdS_4 \times S^7/\mathbb{Z}_k$. In the $k \to \infty$ limit, we have $AdS_4 \times \mathbb{CP}^3$ in type IIA string theory.

- The ABJM model admits a mass deformation that preserves the whole $\mathcal{N} = 6$ supersymmetry.

- The ($\mathcal{N} = 6$ supersymmetric) ground state of mass-deformed ABJM is a fuzzy 2-sphere written in terms of matrices G^α, and the BPS ($\mathcal{N} = 3$) state of pure ABJM is the fuzzy S^2 funnel.

References and further reading

The ABJM model and its gravity dual were defined in [78]. The Mukhi–Papageorgakis Higgs mechanism was defined in [79]; a fuller analysis was done in [82]. The massive deformation of ABJM, and the definition of Eqs. (20.39) and their explicit solutions (20.40) can be found in [83]. The fact that these solutions describe a fuzzy 2-sphere, which can be mapped to the usual description of the fuzzy 2-sphere, was explained in [80, 81].

Exercises

1. Prove the invariance of the part quadratic in fields of the ABJM action (20.1) under the linearized part of the supersymmetry transformation rules (20.2).

2. Prove relation (20.17).

3. Prove that the transformation of coordinates from (20.26) to

$$\zeta_1 = \tan \mu \sin \alpha \sin(\theta/2)\, \epsilon^{i(\psi-\phi)/2}\, \epsilon^{i\chi/2},$$
$$\zeta_2 = \tan \mu \cos \alpha \, \epsilon^{i\chi/2},$$
$$\zeta_3 = \tan \mu \sin \alpha \cos(\theta/2)\, \epsilon^{i(\psi+\phi)/2}\, \epsilon^{i\chi/2}, \qquad (20.47)$$

and reduction on τ, results in the $\mathbb{CP}^3$ metric

$$ds^2_{\mathbb{CP}^3} = d\mu^2 + \sin^2 \mu \left[d\alpha^2 + \frac{1}{4} \sin^2 \alpha \left(\sigma_1^2 + \sigma_2^2 + \cos^2 \alpha\, \sigma_3^2\right) \right.$$
$$\left. + \frac{1}{4} \cos^2 \mu \left(d\chi + \sin^2 \alpha\, \sigma_3\right)^2 \right], \qquad (20.48)$$

where $\sigma_1, \sigma_2, \sigma_3$ are left-invariant 1-forms on S^3, given by

$$\sigma_1 = \cos \psi\, d\theta + \sin \psi \sin \theta\, d\phi,$$
$$\sigma_2 = \sin \psi\, d\theta - \cos \psi \sin \theta\, d\phi,$$
$$\sigma_3 = d\psi + \cos \theta\, d\phi, \qquad (20.49)$$

and the range of the angles is

$$0 \leq \mu, \alpha \leq \frac{\pi}{2}, \quad 0 \leq \theta \leq \pi, \quad 0 \leq \phi \leq 2\pi, \quad 0 \leq \psi, \chi \leq 4\pi. \qquad (20.50)$$

4. Check that (20.40) is a solution of (20.39). Then check that the fuzzy funnel with (20.41) is a solution of the pure ABJM model.

5. Show that we can derive the $SU(2)$ commutation relations for J_i from the G^α matrices algebra (20.39) (independent of any specific representation).

6. Consider the 3-point correlators of $SU(4)_R$ R-symmetry currents J_μ^a in the ABJM model. *From the gravity dual*, argue whether it has an anomalous part, and write down the Witten diagram in AdS_4 for the nonanomalous part. Where does the vertex in the Witten diagram come from when we dimensionally reduce the 10-dimensional type IIA action in (7.113)?

For field theories with no conformal invariance, and less supersymmetry, the *gravity dual* background corresponding to them has no AdS factor, and the compact space is not necessarily a sphere. In order to have a gravity dual background, we need in general to find a brane system on which the field theory lives, and to take a decoupling limit for the corresponding gravity dual, that makes it possible to have a duality between the field theory and gravity in the background.

What we are ultimately interested in is to understand Quantum Chromo Dynamics (QCD), which is a theory that is nonperturbative at low energy, having a coupling constant that runs with energy (asymptotic freedom), is confining, and has a mass gap. Therefore, we look for models that have similar properties, though we will see that up to now there is no model without caveats.

21.1 General properties, map, features

To understand the general properties of gravity duals, in particular ones that could become close to QCD, we first look at an example.

We have in fact already seen one such example in Section 15.5, obtained by putting $\mathcal{N} = 4$ SYM at finite temperature. The finite temperature broke the conformal symmetry by introducing an energy scale, the temperature T, whereas the supersymmetry was broken by the presence of antiperiodic boundary conditions for the fermions around the periodic Euclidean time. We saw that Kaluza–Klein dimensional reduction of $\mathcal{N} = 4$ SYM on the periodic Euclidean time gave us a 3-dimensional theory of pure glue (only 3-dimensional gauge fields A_μ^a), which has a mass gap. We derived the mass gap from the fact that the *gravity dual* (15.37) acts as a 1-dimensional quantum mechanical box, with a nonzero ground state energy.

Unfortunately, in that example, the energy scale M_0 characterizing the mass gap and the masses of the discrete tower of states are proportional to the only other scale available in the theory, the temperature scale T. However, at the scale $T = 1/R$ we start having back the rest of the fields of $\mathcal{N} = 4$ SYM in four dimensions, instead of pure glue in three dimensions. This is so since the masses of Kaluza–Klein states (Fourier modes around periodic time) are given by

$$m^2 e^{\frac{iny}{R}} = -\partial_y^2 e^{\frac{iny}{R}} = \frac{n^2}{R^2} e^{\frac{iny}{R}}, \tag{21.1}$$

so when we reach the energy $E_1 = 1/R$ we can neglect them no longer. The fermions that were dropped out of the theory for being massive also have masses $M \propto T = 1/R$. That means that any quantitative statements about the mass gap or massive states in 3-dimensional QCD are no longer quite valid, since the mass scale involved is the one at which the theory is 3-dimensional QCD no longer. The statements about mass gap and massive states are of course valid in the full *modified* QCD theory.

Unfortunately, this is the situation in all attempts at a QCD (or $\mathcal{N} = 1$ Super QCD) gravity dual analyzed until now. One would hope that there would be a separation of scales between the interesting physics scale M_0 and the cut-off scale of the model (here $T = 1/R$), i.e. $M_0 \ll T$, which could in principle appear due to, e.g., string coupling dependence. However, in all models studied so far, this never happens in a controllable way (such that we can calculate what happens): when there is a separation of scales, we are in a nonperturbative regime and we can calculate no longer.

With this caveat in mind, we now turn to general properties of gravity duals of interesting field theories.

Minimal ingredients of a gravity dual theory that can simulate QCD

- A large N quantum gauge field theory. The gauge group most common is $SU(N)$ (appearing as we saw on D3-branes away from singularities), but $SO(2N)$ or $USp(2N)$ are also possible (appearing when the D3-branes are sitting on orientifold planes). In the gravity theory, N corresponds to a number of branes, and gives a discrete parameter characterizing the curvature of space. We need $N \to \infty$ to have small quantum string (g_s) corrections.

- We usually want the field theory to exist in flat space. This flat space can be identified with the boundary at infinity of the gravity dual. However, often it is better to think that the flat space at some position r in the fifth dimension, corresponding to a field theory energy scale U, is identified with the flat space of the field theory at that scale U.

- Indeed, since we are interested in nonconformal field theories, the energy scale is relevant (the theory does not look the same at all energy scales). The energy scale is identified with the extra (radial) dimension of the gravity dual, whose infinity limit gives the boundary of space. In the case of $\mathcal{N} = 4$ SYM (or more precisely, for the finite temperature case which is nonconformal), the energy scale is $U = r/\alpha'$, as we have already argued. We obtain the so-called UV–IR correspondence: the UV of the field theory (high energies $E = U$) corresponds to the IR of the gravity dual (large distances, or $r \to \infty$), and vice versa.

- The gravity dual is then defined by the $d + 1$-dimensional space obtained from the field theory space (boundary at infinity) and the energy scale, together with a compact space X_m whose symmetries give global symmetries of the field theory.

- The nonconformal theories we are interested in are asymptotically free, which means they are defined in the UV. This is the correct meaning of the statement that the field theory exists at infinity ($U \to \infty$) in the dual: it is there that the theory is perturbative. However, since U is an energy scale, it means that the physics at different energy scales in the field theory correspond to physics at different radial coordinates U in the gravity dual. This means that motion in U corresponds to RG flow in the field theory. In particular, the low energy physics (IR) of interest (since it is hard to calculate in field theory) is situated at small values of U in the gravity dual.

Map between field theory and gravity dual

Next we want to define the general properties of the map between field theory and gravity dual, i.e. the relations that are model-independent.

- The *gauge group* of the field theory has no correspondent in the gravity dual and only quantities that do not involve color indices in a nontrivial way can be calculated (e.g. correlators of gauge invariant states).
- *Global symmetries* in the d-dimensional field theory correspond to gauge symmetries in the $d + 1$-dimensional space (of the gravity dual reduced over the compact space X_m), which themselves correspond to isometries of the compact space X_m. Noether currents J^a_μ for the global symmetries couple to (i.e., correspond to) gauge fields A^a_μ for those gauge symmetries in the gravity dual.
- An important special case of symmetry is translational invariance, with generator P_μ in the field theory, whose local version is general coordinate invariance of the gravity dual. Correspondingly, the energy-momentum tensor $T_{\mu\nu}$ (Noether current of P_μ) couples to the graviton $g_{\mu\nu}$ ("gauge field of local translations") in the gravity dual.
- As in the $AdS_5 \times S^5$ case, the string coupling is $g_s = g^2_{\rm YM}/(4\pi)$, which comes from the fact that $g_{\rm closed} = g^2_{\rm open}$.
- Gauge invariant operators in the dual couple to (are sourced by) $d + 1$-dimensional fields in the gravity dual.
- Supergravity fields in the $d + 1$-dimensional space (dual reduced on X_m) couple to SYM operators (made of adjoint fields) in field theory ("supergravity $\leftrightarrow$ gauge field glueballs").
- To introduce "quarks" in the field theory (fields in the fundamental of the gauge group and some representation of a global symmetry group G), we need to introduce SYM fields for the group G in the gravity dual, which couple to G-charged, pion-like operators (made of "quarks") in the field theory ("SYM $\leftrightarrow$ pion fields"). We have seen this in the case of the $\mathcal{N} = 2$ orientifold in Chapter 19.
- Thus glueballs $\leftrightarrow$ supergravity modes and mesons $\leftrightarrow$ SYM modes.
- The mass spectrum M_n of the tower of glueballs is found as the mass spectrum for the wave equation of the corresponding supergravity mode in the gravity dual, as in the example of $AdS_5 \times S^5$ at finite temperature studied in Section 15.5. A similar statement holds for mesons. We explore this statement in more detail later in the chapter.
- As we explained in Chapter 13, baryons are operators that have more than two fundamental fields (in QCD, there are three, in an $SU(N)$ gauge theory there are N). They have a solitonic character in field theory: for example, they can be obtained as topological solitons in the Skyrme model. The same is true in AdS/CFT: in the original $AdS_5 \times S^5$ AdS/CFT, we saw that the baryon vertex, understood as a bound state that couples to N

external quarks, corresponds to a D5-brane wrapped on the S^5. The same is true in more general models, the baryons being obtained in soliton-like fashion by wrapping branes on nontrivial cycles.

- Wavefunctions of states in field theory, for instance $e^{ik \cdot x}$, correspond to wavefunctions

$$\Phi(x, U, X_m) = e^{ik \cdot x} \Psi(U, X_m) \qquad (21.2)$$

of states in the gravity dual, i.e. obtained by multiplying with a wavefunction for the extra coordinates.

General features of gravity duals for QCD-like or SQCD-like theories

We have seen the minimal ingredients needed to construct a gravity dual, and how to map quantities between field theory and gravity dual. But we are specifically interested in gravity duals of theories that look like QCD, or perhaps like supersymmetric QCD (SQCD), so we want to describe what features we expect from such gravity duals, and are common to attempts at such a solution.

- At high energies, QCD or the QCD-like theories look conformal, since all mass scales become irrelevant. Therefore in the field theory UV, mapped to the region close to the boundary of the dual, $U \to \infty$, the gravitational space looks like $AdS_5 \times X_5$, with X_5 some compact space, maybe with some subleading corrections to the metric.
- At low energies, QCD or the QCD-like theories are nontrivial, and have a mass gap. For $AdS_5 \times S^5$, the wave equation does not give a mass gap (a discrete tower of mass states), a statement mapped to the fact that there is no mass gap in a CFT. Therefore, for the gravity dual of the QCD-like theory, *space must terminate in a certain manner before* $U = 0$, *such that the "warp factor" U^2 in front of $d\vec{x}^2$ remains finite.* In fact, since a singularity would not be good for the field theory, the space must terminate in a smooth manner.

  We have already explained a more rigorous, though not always applicable, way to see this. It takes a finite time for a light ray to go to the boundary of AdS space, since $ds^2 = 0$ implies $\int dt = \int^{\infty} dU/U^2 =$ finite. However, it takes light an infinite time to go to the center of AdS space, since $\int dt = \int_0 dU/U^2 = \infty$. Therefore, if AdS space cuts off before $U = 0$, as far as light is concerned, AdS space acts as a finite box, with a discrete spectrum and a mass gap (like a quantum mechanical 1-dimensional box). We come back to this later in the chapter, when we study mass spectra.
- If fundamental quarks are introduced in the theory, open string modes (forming a gauge theory) living on a certain brane in the gravity dual must be introduced. These will couple to the meson-like (pion-like) operators. We have seen this in the case of the $\mathcal{N} = 2$ orientifold, where the brane was fixed at the orientifold point. A similar way to introduce quarks is to introduce free probe branes, that can move at different positions U in the gravity dual, and thus probe the field theory physics at different energy scales; this is the case of the Sakai–Sugimoto model studied later in the chapter.

- If the QCD-like theory has global symmetries (like flavor symmetries or R symmetries), they are reflected in local symmetries in the $d + 1$-dimensional dual theory, or equivalently in symmetries of the extra dimensional space X_m.

21.2 Finite temperature and cut-off AdS_5 solutions

We now start to look at examples of the general description defined in the last subsection, beginning with examples we have already started to analyze.

Finite temperature redux

We start with another look at the case of $AdS_5 \times S^5$ at finite temperature, since it is the simplest toy model for QCD in three dimensions. The $n + 1$-dimensional metric obtained by Witten from scaling the AdS_{n+1} black hole was, as we saw,

$$ds^2 = \left(\frac{\rho^2}{R^2} - \frac{R^{n-2}}{\rho^{n-2}}\right) d\tau^2 + \frac{d\rho^2}{\frac{\rho^2}{R^2} - \frac{R^{n-2}}{\rho^{n-2}}} + \rho^2 \sum_{i=1}^{n-1} dx_i^2. \tag{21.3}$$

We also saw that the rescaling $\rho/R = r/r_0$, $\tau = tr_0/R$, $\vec{x} = \vec{y}r_0/R^2$ leads to the Poincaré version of the finite temperature AdS_{n+1},

$$ds^2 = \frac{r^2}{R^2}\left[-dt^2\left(1 - \frac{r_0^n}{r^n}\right) + d\vec{y}_{(n-1)}^2\right] + R^2 \frac{dr^2}{r^2\left(1 - \frac{r_0^n}{r^n}\right)}, \tag{21.4}$$

which for $n = 3$, adding the S^5 metric to the above, is the same as the metric of the near-horizon near-extremal D3-branes.

Now we want to verify the general features of gravity duals of QCD-like theories explained in the previous subsection. We see that indeed at large r the metric goes over to $AdS_5 \times S^5$ (in Poincaré coordinates), corresponding to the fact that in the UV the field theory is still $\mathcal{N} = 4$ SYM, whereas at small r the space terminates smoothly at $r = r_0$. The $SO(6)$ symmetry of the S^5 corresponds as at $T = 0$ with the global R-symmetry of the UV theory, i.e. of $\mathcal{N} = 4$ SYM. Since there are no quarks in this gravity dual, this matches all the general features described in the last subsection.

Cut-off AdS_5 redux: modified hard-wall

The simplest possible model for QCD is obtained just by cutting the AdS_5 at an $r_{min} = R^2\Lambda$. This is the "hard-wall" model for QCD, described in Section 16.1, and used for high energy QCD scattering of colorless states via the Polchinski–Strassler prescription.

Then trivially one has that the large r metric is $AdS_5 \times S^5$, and at $r = r_{min}$ the space terminates (though not smoothly). Also trivially, light travels a finite time between $r = r_{min}$

and $r = \infty$, and the spectrum will be discrete and have the mass gap $\propto \Lambda$, the only scale in the theory. Therefore Λ can be phenomenologically fixed to be related to the mass gap. Again we have $SO(6)$ R-symmetry in the UV, corresponding to the symmetry of the S^5, and we have no quarks in the theory, so again we match all the general features of gravity duals of QCD-like theories.

But one can improve this simple model, by making the cut-off dynamical, i.e. considering it as an added D-brane living at $r = r_{min}$ in the gravity dual. By the arguments given before, the modes on this extra D-brane should source pion-like operators. In fact, it turns out that the fluctuation in the position of this D-brane is a good enough model for (a scalar singlet version of) the QCD pion. One can use this to obtain a more precise version of the saturation of the Froissart bound described in Section 16.5, since now we have a gravity dual for the pion, not just for the glueball (in Section 16.5 we implicitly considered the Froissart bound for a theory with only glueballs, no mesons).

21.3 The Polchinski–Strassler and Klebanov–Strassler solutions

We next consider two gravity duals of $\mathcal{N} = 1$ supersymmetric models.

Polchinski–Strassler solution

The Polchinski–Strassler solution gives the gravity dual of $\mathcal{N} = 1^*$ SYM, which is a certain massive deformation of $\mathcal{N} = 1$ SYM (there are couplings to massive modes), but was found in an attempt to describe the gravity dual of the undeformed theory, so we just consider it as a toy model for $\mathcal{N} = 1$ SYM and QCD. The brane configuration giving it is of D3-branes "polarizing" (puffing up, by the appearance of an extra space) into D5-branes due to the presence of a nonzero flux. It exhibits a mass gap through a phenomenon similar to the finite temperature $AdS_5 \times S^5$ case (i.e., near-horizon near-extremal D3-branes). The metric and dilaton are given by

$$ds^2_{\text{string}} = Z_x^{-1/2} d\vec{x}^2_{3+1} + Z_y^{1/2}(dy^2 + y^2 d\Omega_y^2 + dw^2) + Z_\Omega^{1/2} w^2 d\Omega_w^2,$$

$$Z_x = Z_y = Z_0 = \frac{R^4}{\rho_+^2 \rho_-^2}; \quad Z_\Omega = Z_0 \left[\frac{\rho_-^2}{\rho_-^2 + \rho_c^2} \right]^2,$$

$$\rho_\pm = (y^2 + (w \pm r_0)^2)^{1/2}; \quad R^4 = 4\pi g_s N; \quad \rho_c = \frac{2g_s r_0 \alpha'}{R^2}; \quad r_0 = \pi \alpha' m N,$$

$$e^{2\Phi} = g_s^2 \frac{\rho_-^2}{\rho_-^2 + \rho_c^2}, \tag{21.5}$$

and m is the mass parameter of the deformation. The metric goes over to $AdS_5 \times S^5$ at large $\rho = \rho_- \simeq \rho_+$, as it should. The "near-core" region is $\rho \sim r_0$, which, however, is quite complicated (depends on y and w separately), but we see that the typical warp factor $(Z^{1/2})$ is finite in this region, thus we do have the situation we advocated. At particular points, there are still AdS-like throats that are not regulated though. There are no quarks in the field theory, and no nonabelian global symmetries, so we again match all the expected features of gravity duals of QCD-like and SQCD-like theories.

Klebanov–Strassler solution

The Klebanov–Strassler solution describes an $\mathcal{N} = 1$ supersymmetric $SU(N+M) \times SU(N)$ gauge theory, with two bifundamental chiral superfields A_1, A_2, i.e. in the $(N + M, \bar{N})$ representation, and two chiral superfields B_1, B_2 in the conjugate representation $(\overline{N + M}, N)$. This gauge theory is of the "cascading" type, meaning that at consecutive energy scales the relevant degrees of freedom are reduced by "Seiberg duality" transformations, which change the description to one in terms of smaller gauge groups. The Seiberg duality transformation is for an $SU(N_c)$ gauge theory with N_f flavors to go into an $SU(N_f - N_c)$ gauge theory with N_f flavors; when one description is strongly coupled, the other is weakly coupled. In our case, from the point of view of the $SU(N + M)$ group, we have $2N$ flavors, so after the duality it turns into an $SU(2N - (N+M)) = SU(N-M)$ group, with N flavors, i.e. bifundamentals $(N - M, \bar{N})$ and its conjugate. We still have the $SU(N)$ group intact. We see that after the Seiberg duality we have the same theory, but with $N \to N - M$. This process can then continue, $N \to N - M \to N - 2M$, etc., until we reach a minimum group. This is known as the duality cascade, and it means that at different energy scales, the effective (perturbative) degrees of freedom cascade, going from groups with N to $N - M$ to $N - 2M$, etc.

The metric, obtained from a configuration of M "fractional 3-branes" on a "conifold point" (we will not explain these definitions here) in the near horizon limit, is

$$ds_{10}^2 = h^{-1/2}(\tau)d\vec{x}^2 + h^{1/2}(\tau)ds_6^2,$$

$$ds_6^2 = \frac{1}{2}\epsilon^{4/3}K(\tau)\left[\frac{1}{3K^3(\tau)}(d\tau^2 + (g_5)^2) + \cosh^2\left(\frac{\tau}{2}\right)((g_3)^2 + (g_4)^2)\right.$$
$$\left. + \sinh^2\left(\frac{\tau}{2}\right)((g_1)^2 + (g_2)^2)\right],$$

$$K(\tau) = \frac{(\sinh(2\tau) - 2\tau)^{1/3}}{2^{1/3}\sinh\tau},$$

$$h(\tau) = \alpha\frac{2^{2/3}}{4}\int_\tau^\infty dx\frac{x\coth x - 1}{\sinh^2 x}(\sinh(2x) - 2x)^{1/3},$$ (21.6)

and $\alpha \propto (g_s M)^2$ is a normalization factor. If one sets $\epsilon = 12^{1/4}$, the prefactor $\frac{1}{2}\epsilon^{4/3}$ of ds_6^2 turns into $(3/2)^{1/3}$. At large τ it becomes a log-corrected $AdS_5 \times T^{1,1}$ metric, in terms of $r \sim [\epsilon^2 e^\tau]^{1/3}$, given by

$$ds^2 = h^{-1/2}(r)d\vec{x}^2 + h^{1/2}(r)(dr^2 + r^2 ds_{T^{1,1}}^2),$$

$$h(r) \sim \frac{(g_s M)^2 \ln(r/r_s)}{r^4},$$

$$ds_{T^{11}}^2 = \frac{1}{9}\left(d\psi + \sum_{i=1,2}\cos\theta_i d\phi_i\right)^2 + \frac{1}{6}\sum_{i=1,2}(d\theta_i^2 + \sin^2\theta_i d\phi_i^2)$$

$$= \frac{1}{9}(g_5)^2 + \sum_{i=1}^4(g_i)^2,$$ (21.7)

and $g_1, \ldots, g_5$ are some independent basis of 1-forms. Here the dilaton is approximately constant, $\phi = \phi_0$. Thus again, up to the log correction, we have an $AdS_5 \times X_5$ metric. The log correction is in fact related in this model to the log renormalization of the field theory, the running coupling constant. In the field theory IR, i.e. at small τ in the gravity dual, the metric looks like

$$ds^2 = a_0^{-1/2} d\vec{x}^2 + a_0^{1/2} \left(\frac{d\tau^2}{2} + d\Omega_3^2 + \frac{\tau^2}{4}((g_1)^2 + (g_2)^2) \right), \tag{21.8}$$

thus here also the space terminates smoothly and the warp factor $a_0^{1/2}$ remains finite (a_0 is a constant).

21.4 The Maldacena–Núñez and Maldacena–Năstase solutions

We next look for two solutions giving $\mathcal{N} = 1$ supersymmetric theories, obtained from NS5-branes in type IIB theory wrapped on spheres. For wrapping on S^2, we obtain a 4-dimensional $\mathcal{N} = 1$ theory, corresponding to the Maldacena–Núñez solution, and for wrapping on S^3, we obtain a 3-dimensional $\mathcal{N} = 1$ theory, corresponding to the Maldacena–Năstase solution.

Maldacena–Núñez solution

The solution of Maldacena and Núñez involves type IIB NS5-branes wrapped on S_2. Note that in type IIB theory, NS5-branes can be thought of as the S-dual, i.e. $g_s \to 1/g_s$, or $\phi \to -\phi$, transformed D5-branes. The field theory that is obtained is 4-dimensional $\mathcal{N} = 1$ SYM, coupled to other (massive) modes. It has the string frame metric and dilaton

$$
\begin{aligned}
ds_{10}^2 &= ds_{7,\text{string}}^2 + \alpha' N \frac{1}{4}(\tilde{w}^a - A^a)^2, \\
H &= N\left[-\frac{1}{4}\frac{1}{6}\epsilon_{abc}(\tilde{w}^a - A^a) \wedge (\tilde{w}^b - A^b) \wedge (\tilde{w}^c - A^c) + \frac{1}{4}F^a \wedge (\tilde{w}^a - A^a) \right], \\
ds_{7,\text{string}}^2 &= d\vec{x}_{3+1}^2 + \alpha' N[d\rho^2 + R^2(\rho)d\Omega_2^2], \\
A &= \frac{1}{2}\left[\sigma^1 a(\rho)d\theta + \sigma^2 a(\rho)\sin\theta\, d\phi + \sigma^3 \cos\theta\, d\phi \right]; \quad a(\rho) = \frac{2\rho}{\sinh 2\rho}, \\
R^2(\rho) &= \rho \coth(2\rho) - \frac{\rho^2}{\sinh^2(2\rho)} - \frac{1}{4}, \\
e^{2\phi} &= e^{2\phi_0}\frac{2R(\rho)}{\sinh(2\rho)}.
\end{aligned}
\tag{21.9}
$$

Here the first two lines represent the ansatz for uplifting of a solution of 7-dimensional $\mathcal{N} = 1$ supergravity into ten dimensions on an S^3, in our case the S^3 transverse to the 5-branes. The rest of the formulas above represent the 7-dimensional solution of $\mathcal{N} = 1$ supergravity; $\tilde{w}^a$ are left-invariant one-forms on S^3, parameterized as

$$g = e^{\frac{i\psi\sigma^3}{2}} e^{\frac{i\tilde{\theta}\sigma^1}{2}} e^{\frac{i\tilde{\phi}\sigma^3}{2}}, \tag{21.10}$$

defined by

$$\tilde{\omega}^1 + i\tilde{\omega}^2 = e^{-i\psi}(d\tilde{\theta} + i\sin\tilde{\theta}\, d\tilde{\phi}); \quad \tilde{\omega}^3 = d\psi + \cos\tilde{\theta}\, d\tilde{\phi}, \tag{21.11}$$

and σ_i, $i = 1, 2, 3$ are the Pauli matrices for $SU(2)$.

However, the decoupling of other modes, like Kaluza–Klein modes, cannot be done in a controllable way, as usual. To do so, one would need to switch to a D5-brane description, that is highly nonperturbative.

Indeed, we can make an S-duality, which in type IIB is $\phi \to \phi_D = -\phi$, and the Einstein metric is invariant under it, $g_{\mu\nu}^E \to g_{\mu\nu}^E$ (the relation between the Einstein and string metric is $g_{\mu\nu}^S = e^{\phi/2}g_{\mu\nu}^E$), and this leads from the NS5-brane to the D5-brane. The string metric and dilaton for the D5-brane are

$$ds^2_{\text{string}} = e^{\phi_D}\left[dx^2_{(4)} + \alpha'N\left(d\rho^2 + R^2(\rho)d\Omega_2^2 + \frac{1}{4}\sum_a(\tilde{w}^a - A^a)^2\right)\right],$$

$$e^{2\phi_D} = e^{2\phi_{D,0}}\frac{\sinh(2\rho)}{2R(\rho)}. \tag{21.12}$$

The QCD string tension coming from this gravity background is found as follows. The quark–antiquark potential defining it is given by the Wilson loop, itself given by a fundamental string coming down from infinity in the gravity background. From the string tension we get a factor $1/(2\pi\alpha')$ and from the background (21.12) we get the overall factor $e^{\phi_{D,0}}$, for a tension

$$T_s = \frac{e^{\phi_{D,0}}}{2\pi\alpha'}. \tag{21.13}$$

On the other hand, the KK states, which are of the order of the glueball masses, are

$$M^2_{\text{glueballs}} \sim M^2_{\text{KK}} \sim \frac{1}{R^2} \sim \frac{1}{N\alpha'}, \tag{21.14}$$

where $R^2 \sim N\alpha'$, and R, the radius of the 2-sphere Ω_2, is inferred from (21.12). Then decoupling of the KK states from the QCD string tension, which gives the sought-after physics of confinement at low energy, implies

$$T_s \ll M^2_{\text{KK}} \Rightarrow e^{\phi_{D,0}}N \ll 1. \tag{21.15}$$

However, the condition for supergravity to be valid is for the curvature of the background (21.12) to be small in string units, and we can check that the relevant parameter is $e^{\phi_{D,0}}\alpha'N/\alpha'$, and has to be large, i.e.

$$e^{\phi_{D,0}}N \gg 1, \tag{21.16}$$

so the opposite limit to the one needed for decoupling of KK modes.

Next, we check the behavior of the metric in the UV and IR.

The behavior at $\rho \to \infty$, corresponding to the usual boundary of space, i.e. the UV of the field theory, is

$$R^2 \simeq \rho; \quad a \simeq 2\rho e^{-2\rho}; \quad \phi \simeq \phi_0 - \rho + \frac{\log\rho}{4}. \tag{21.17}$$

At first, it seems this is not good, since we do not obtain a log-corrected $AdS_5 \times X_5$ space, but rather

$$ds^2 = d\vec{x}_{3+1}^2 + \alpha' N \left[d\rho^2 + \rho d\Omega_2^2 + \frac{1}{4}(\tilde{w}^a - A^a)^2 \right]$$

$$= d\vec{x}_{3+1}^2 + \alpha' N \left[\frac{dz^2}{z^2} + (-\log z)d\Omega_2^2 + \frac{1}{4}(\tilde{w}^a - A^a)^2 \right], \qquad (21.18)$$

where $\rho = -\log z$. However, now the dilaton is nontrivial, unlike the previous cases (and unlike for pure $AdS_5 \times S^5$). But in fact, this is good, since all we need is that the 5-dimensional KK reduced supergravity action is the same.

In the presence of a dilaton (and using a "string frame" metric as above), the relevant action is

$$S = \frac{1}{2\kappa_N^2} \int d^5x \sqrt{-g_5} \left(\int_{X_5} \sqrt{g_{X_5}} \right) e^{-2\phi} [\mathcal{R} - (\partial X)^2 + \ldots]$$

$$= \frac{1}{2\kappa_N^2} \int d^5x \sqrt{-g_5} \left(\int_{X_5} \sqrt{g_{X_5}} \right) g^{\mu\nu} e^{-2\phi} [R_{\mu\nu} - \partial_\mu X \partial_\nu X + \ldots], \qquad (21.19)$$

where X is a generic scalar. For a metric

$$ds^2 = e^{2A(\rho)} d\vec{x}_{3+1}^2 + d\rho^2 + ds_{X_5}^2 = e^{2A(z)} d\vec{x}_{3+1}^2 + \frac{dz^2}{z^2} + ds_{X_5}^2, \qquad (21.20)$$

the bracket $[R_{\mu\nu} + \partial_\mu X \partial_\nu X + \ldots]$ does not contain e^{2A} factors and we get

$$\frac{1}{2\kappa_N^2} \int d^4x \, d\rho \left(\int_{X_5} \sqrt{g_{X_5}} \right) e^{2(A-\phi)} \delta^{\mu\nu} [R_{\mu\nu} - \partial_\mu X \partial_\nu X + \ldots], \qquad (21.21)$$

thus in fact we have the condition

$$\phi - \phi_0 - A \overset{\rho \to \infty}{\to} -\rho(+\log \text{ corrections}) = +\log z(+ \text{corrections}), \qquad (21.22)$$

which before was satisfied by $\phi = \phi_0$ and $A \to -\rho + \ldots = +\log z + \ldots$, but now is satisfied by $A = 0$ and $\phi = \phi_0 - \rho + \ldots = \phi_0 + \log z + \ldots$. Also note that now there is some ρ dependence in $\sqrt{g_{X_5}}$ as well.

So the behavior in the UV matches the expectations, though in a nontrivial way.

The behavior in the IR of the field theory, i.e. at $\rho \to 0$ is

$$R^2 = \rho^2 + \mathcal{O}(\rho^4); \quad a = 1 + \mathcal{O}(\rho^2); \quad \phi = \phi_0 + \mathcal{O}(\rho^2), \qquad (21.23)$$

which means that the effective warp factor $e^{2(A-\phi)}$ is constant as in the previous examples.

So the behavior in the UV and IR matches the expectations, there are no fundamental quarks, and there is no global nonabelian symmetry, therefore all the expected features of gravity dual to SQCD are satisfied.

The Maldacena–Năstase solution

The Maldacena–Năstase solution is the analog of the Maldacena–Núñez solution (which is for NS5-brane wrapped on S^2) for NS5-branes wrapped on S^3, and gives $\mathcal{N} = 1$ SYM in three dimensions, with a Chern–Simons coupling, and coupled to other modes as in four dimensions. The Chern–Simons coupling is quantized as always by a quantum number k. The gravity dual corresponds to $k = N/2$, where N is the number of 5-branes, and in this case an index computation shows that there is a unique vacuum, that confines.

The solution is

$$ds^2_{10} = ds^2_{7,\text{string}} + \alpha' N \frac{1}{4}(\tilde{w}^a - A^a)^2,$$

$$H = N\left[-\frac{1}{4}\frac{1}{6}\epsilon_{abc}(\tilde{w}^a - A^a) \wedge (\tilde{w}^b - A^b) \wedge (\tilde{w}^c - A^c) + \frac{1}{4}F^a \wedge (\tilde{w}^a - A^a)\right] + h,$$

$$ds^2_{7,\text{string}} = d\vec{x}^2_{2+1} + \alpha' N[d\rho^2 + R^2(\rho)d\Omega^2_3],$$

$$A = \frac{w(\rho)+1}{2}w^a_L,$$

$$h = N[w^3(\rho) - 3w(\rho) + 2]\frac{1}{16}\frac{1}{6}\epsilon_{abc}w^a \wedge w^b \wedge w^c, \tag{21.24}$$

where w^a are left-invariant forms on the S^3 described by $d\Omega^2_3$, $\tilde{w}^a$ are the same left-invariant forms on the transverse $\tilde{S}^3$, and $w(\rho), R(\rho), \phi(\rho)$ have some complicated form that can be evaluated numerically.

At large ρ they become

$$R^2(\rho) \sim 2\rho; \quad w(\rho) \sim \frac{1}{4\rho}; \quad \phi = -\rho + \frac{3}{8}\log\rho, \tag{21.25}$$

thus, as for the Maldacena–Núñez solution,

$$\phi - \phi_0 - A \to -\rho + \log \text{ corrections}, \tag{21.26}$$

where the metric is

$$ds^2 = e^{2A(\rho)}d\vec{x}^2_{2+1} + d\rho^2 + ds^2_{X_6} = e^{2A(z)}d\vec{x}^2_{2+1} + \frac{dz^2}{z^2} + ds^2_{X_6}, \tag{21.27}$$

implying the same dual supergravity action as in a log-corrected $AdS_4 \times X_6$ background. Thus the UV behavior is the expected one.

The behavior at $\rho \to 0$, i.e. for the IR of the field theory, is

$$R^2(\rho) = \rho^2 + \mathcal{O}(\rho^4); \quad w(\rho) = 1 + \mathcal{O}(\rho^2); \quad \phi = \phi_0 + \mathcal{O}(\rho^2), \tag{21.28}$$

giving a finite effective warp factor, $e^{2(A-\phi)} = e^{-2\phi_0 + \cdots}$.

Thus the behavior of the metric in the UV and in the IR is the expected one for the gravity dual of SQCD-like theories. And again there are no fundamental quarks or global nonabelian symmetries, so we have all the needed features.

But we have more interesting features in this geometry. If we wrap a small number n of branes ($n \ll N/2$) on a noncontractible S^3 in the geometry, the metric is unmodified, but the dual field theory is modified by having $k = N/2 + n$, that still preserves supersymmetry. By adding $|n|$ ($n < 0$) antibranes instead, we get $k = N/2 + n$ which breaks supersymmetry dynamically. Thus this dynamical (nonperturbative) breaking of supersymmetry becomes a spontaneous (classical) effect in field theory, another good example of the power of AdS/CFT.

21.5 The Sakai–Sugimoto model

The Sakai–Sugimoto model is a model that incorporates quarks (i.e. fermions in the fundamental representation of the gauge group), but does so in a probe approximation, i.e. without incorporating the backreaction of the modes dual to quark operators to the geometry. As we mentioned, we can introduce quarks by either introducing fixed D-branes in the gravity dual, as in the case of the $\mathcal{N} = 2$ orientifold (where the branes are at the orientifold fixed point), or by introducing moving D-branes that can probe the geometry. The Sakai–Sugimoto model is an example of the latter. The general idea of introducing quarks on probe D-branes, without need of orientifolds to cancel the charge, was originally put forward by Karch and Katz.

Specifically, the Sakai–Sugimoto model involves a large number N_c of Wick-rotated D4-branes at finite temperature (Euclidean periodic time) giving a gravity dual similar to the one studied by Witten for D3-branes, namely

$$ds^2 = \left(\frac{U}{R}\right)^{3/2} (f(U)d\tau^2 + d\vec{x}_{(4)}^2) + \left(\frac{R}{U}\right)^{3/2} \left(\frac{dU^2}{f(U)} + U^2 d\Omega_4^2\right),$$

$$e^\phi = g_s \left(\frac{U}{R}\right)^{3/4}; \quad F_4 = \frac{2\pi N_c}{V_4}\epsilon_4; \quad f(U) = 1 - \frac{U_{KK}^3}{U^3}. \tag{21.29}$$

Inside this background, one considers N_f D8-brane probes, whose single transverse coordinate is U, and its value depends on the worldvolume coordinate τ, i.e. $U = U(\tau)$. The D8-branes form a U-shape, starting and ending at infinity, as in Fig. 21.1. Since the difference between a $D8$-brane and a $\overline{D8}$-brane is only its orientation (charge in one dimension only means the direction, left or right, of its flux lines), this probe is interpreted as a susy-breaking $D8 - \overline{D8}$-brane pair, joined in the bulk.

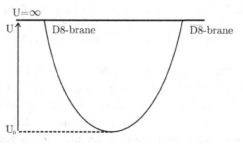

Figure 21.1 The Sakai–Sugimoto model has a probe D8-brane in the gravity dual, starting from infinity and returning to it. At infinity, it looks like a D8-brane/anti-D8-brane pair (parallel branes of opposite orientation).

The solution for $U(\tau)$ is given via its inverse,

$$\tau(U) = U_0^4 f(U_0)^{1/2} \int_{U_0}^{U} \frac{dU}{(\frac{U}{R})^{3/2} f(U) \sqrt{U^8 f(U) - U_0^8 f(U_0)}}. \qquad (21.30)$$

The modes living on this brane are dual to (i.e., couple to) mesonic operators, i.e. pion-like operators, involving quarks and charged under the global symmetry.

Note that the backreaction of this mode on the geometry has not been included, i.e. the background metric is not modified by the presence of the D8-branes. This means that the model is valid only perturbatively if $N_f \ll N_c$, otherwise one has corrections to this solution.

The D4-brane background at finite temperature is similar to the Witten construction for D3-branes at finite temperature, giving $\mathcal{N} = 4$ SYM at finite temperature. In fact, Witten related the D4-brane background to a construction of pure 4-dimensional $SU(N)$ Yang–Mills via (some transformations and) compactification on periodic Euclidean time. We can easily see that up to an overall conformal factor, the metric has the right behavior, by changing variables to $\rho = U^{1/2}$, which gives the metric

$$ds^2 \sim \rho \left[\rho^2(f(\rho)d\tau^2 + d\vec{x}^2) + \frac{d\rho^2}{f(\rho)\rho^2} + d\Omega_4^2 \right]. \qquad (21.31)$$

So up to a conformal factor, we get $AdS_6 \times S^4$ at large U. Compactification on τ then gives the correct behavior for the dual of the 4-dimensional theory.

Then the D4-brane metric has the right properties expected of a gravity dual of a QCD-like theory: smooth cut-off in the IR at finite $U = U_{KK}$ and $AdS_6 \times S^4$ in the UV. The coordinate τ is compact, and moreover we have $\tau = \tau(U)$ (evolves with τ), which means that the field theory probed by the D8-brane is 4-dimensional, as we require.

21.6 Mass spectra in gravity duals from field eigenmodes; examples

As we have already explained, mass spectra for the tower of KK modes in AdS space, which come in increasing representations of the global symmetry group, $SO(6)$ in the case of $AdS_5 \times S^5$ and its related incarnations, correspond to the tower of anomalous dimensions of operators.

But for a given representation of the global symmetry group of the field theory, for instance the singlet, we still have towers of *field theory* states associated with the same operator, dual to a given field in AdS space. For instance, the operator Tr $[F_{\mu\nu}F^{\mu\nu}]$, dual to the massless dilaton (scalar) in AdS, corresponds to a tower of glueball states (excited states of a given glueball, 0^{++}). One way to calculate these masses in field theory is from the 2-point function of the operator. For instance, from the behavior at large distances

$|x - y| \to \infty$ we can calculate the mass m_1 of the lowest state (the mass of the other states appearing in subleading corrections),

$$\langle \mathcal{O}(x)\mathcal{O}(y) \rangle \propto e^{-m_1|x-y|}(+\# e^{-m_2|x-y|} + \ldots). \tag{21.32}$$

A better definition would be to work in momentum space, where the correlator should behave like

$$\langle \mathcal{O}(p)\mathcal{O}(-p) \rangle \sim \sum_j \frac{A_j}{p^2 + m_j^2}, \tag{21.33}$$

i.e., we should be able to find the masses of states from the poles of the correlator.

Then the masses of the tower of states corresponding to the operator $\mathcal{O}$, that couples to the field Φ in AdS space, can be found in the gravity dual from the eigenvalues of $\vec{k}^2 = -m^2$ for solutions of the free equation of motion for Φ, where $\vec{k}$ is the 4-dimensional momentum. For similarity with QCD, we need the spectrum of this tower of states to be discrete (and have a mass gap), which means that the equation of motion in the gravity dual reduces to a case like the quantum mechanical 1-dimensional box.

We have argued that in the case of cut-off AdS space, the reason we obtain a quantum mechanical box is because the time of flight between the center and the boundary of space is finite.

Indeed, in general the simplest way to obtain an infinite discrete spectrum (with no accumulation points) is if the time it takes a light ray to traverse the space from boundary to boundary is finite. This effectively places the system in a box. In cases of the type $AdS_{n+1} \times S^m$, the metric looks like

$$ds^2 = u^\alpha(-dt^2 + d\vec{x}_{n-1}^2) + \frac{du^2}{u^2} + d\Omega_m^2, \tag{21.34}$$

where $\alpha > 0$ and the boundaries of AdS in these Poincaré coordinates are at $u = 0$ and ∞. The time a light ray takes in between the boundary at infinity and the coordinate patch boundary at $u = 0$ is $t = \int_0^\infty \frac{du}{u^{1+\alpha/2}}$ which is finite at infinity, but infinite at zero.

By making the brane nonextremal (introducing a temperature), one regulates this infinity, by placing the new "boundary" a finite time away:

$$ds^2 = u^\alpha \left[\left(1 - \frac{u_T^m}{u^m}\right) d\tau^2 - dt^2 + d\vec{x}_{n-2}^2 \right] + \frac{du^2}{u^2 \left(1 - \frac{u_T^m}{u^m}\right)} + d\Omega_m^2. \tag{21.35}$$

Note that for a D-p-brane in string frame in ten dimensions, the near horizon metric is

$$ds^2_{\text{string}} = u^{\frac{7-p}{2}}(-dt^2 + d\vec{x}_{n-1}^2) + \frac{1}{u^{\frac{7-p}{2}}}(du^2 + u^2 d\Omega_m^2), \tag{21.36}$$

and so the time of travel in between the boundaries is $t = \int du/u^{(7-p)/2}$. For $p \leq 4$, if one regulates the divergence at zero, the time is finite; for $p \geq 6$, the divergence at infinity cannot be regulated, whereas there is none at zero. The case $p = 5$ is special, since both boundaries have divergencies, though only logarithmic.

At the beginning of Chapter 10, we saw that if light takes a finite time between the boundaries, in terms of the *time of flight* variable x defined by $dx = d\rho\sqrt{g_{\rho\rho}/g_{tt}}$, the massless KG equation $\Box \Psi = 0$ becomes a 1-dimensional quantum mechanical problem in terms of $E \equiv m^2 = -k^2$, with $\vec{k}$ being a 4-dimensional momentum,

$$\left[-\frac{d^2}{dx^2} + (V(x) - E) \right] \tilde{\Psi}(x) = 0. \qquad (21.37)$$

Then if the time of flight is finite, i.e. $x_{\max}$ is finite, the potential $V(x)$ of the equivalent quantum mechanical problem has finite support, which means that the energy levels $E_n \equiv m_n^2$ are indeed discrete, infinite in number, and with no accumulation points.

The general procedure for going to an equivalent quantum mechanical problem is thus: change variables to time of flight x, after which the equation becomes ($\Psi \rightarrow \chi(x)$ in $\vec{k}$ space),

$$\chi''(x) + f(x)\chi'(x) + m^2 \chi(x) = 0. \qquad (21.38)$$

After the change of variables $\chi(x) = g(x)h(x)$, with $h(x)$ satisfying $h'/h = -f/2$, the resulting Schrödinger equation has the "potential" $V(x) = 1/4 f^2(x) + 1/2 f'(x)$ and "energy" $E = m^2$.

The case of finite support described above is not the only way to obtain a mass gap though. For instance, the NS5-brane solution in flat space has a string frame metric and dilaton

$$ds_{10,\text{string}}^2 = d\vec{x}_{5+1}^2 + d\rho^2 + d\Omega_3^2, \quad \phi = -\rho. \qquad (21.39)$$

The $\Box$ operator, however, will contain the Einstein metric, so that $e^{\phi/2}\Box = m^2 + e^{2\phi}\partial_\rho e^{-2\phi}\partial_\rho$, and after the procedure above, the Schrödinger potential is $V(x) = 1$, with x ranging from 0 to infinity. Therefore, the spectrum is continuous above a mass gap, though it is not discrete.

Another important case is the case of Witten's construction of 3-dimensional pure glue (QCD_3), obtained from the Witten metric (15.37) (itself obtained either by scaling the AdS_5 black hole, or by the near-horizon limit of the near-extremal D3-branes) by KK reducing on periodic Euclidean time. The analysis of the resulting spectra (for the 0^{++} tower of states of QCD_3, coming from Tr $[F_{\mu\nu}F^{\mu\nu}]$, dual to the massless dilaton) was done in [84], and here we present a simplified analysis. One obtains the reduced 1-dimensional wave equation from the massless scalar KG equation, for $\Psi = \chi(\tilde{\rho})e^{i\vec{k}\cdot\vec{x}}$, $\chi' = d\chi/d\tilde{\rho}$, and $\tilde{\rho} = \rho^2/R^2$,

$$\tilde{\rho}(\tilde{\rho}^2 - 1)\chi'' + (3\tilde{\rho}^2 - 1)\chi' + \frac{m^2}{4}\chi = 0, \qquad (21.40)$$

and so the time of flight x and function $f(x)$ are

$$x = \int d\tilde{\rho} \frac{1}{\sqrt{\tilde{\rho}(\tilde{\rho}^2 - 1)}} = \frac{1}{\sqrt{x}} \, _2F_1(1/4, 1/2, 5/4; 1/x^2),$$

$$f(x) = \frac{3\tilde{\rho}^2 - 1}{2\sqrt{\tilde{\rho}(\tilde{\rho}^2 - 1)}}. \qquad (21.41)$$

Near $\tilde{\rho} = 1$, we obtain for the time of flight $x \simeq \sqrt{2(\rho - 1)}$, and for the potential $V(x) \simeq -1/(4x^2) = -1/(8(\rho - 1))$, whereas near $\tilde{\rho} = \infty$, we obtain for the time of flight $x = -2/\sqrt{\tilde{\rho}} + C$, and for the potential $V(x) = 15/(4(x - C)^2) = 15\tilde{\rho}/16$. So in this case the potential is going to $-\infty$ at $x = 0$, and to $+\infty$ at $x = x_{\max} = C$. The energy levels are found to be all positive, and follow approximately the law $m^2 R^2 = 6n(n + 1)$. Here R is the radius of AdS, the only parameter in the Witten metric (15.37). Note that at large n, $E_n = m_n^2 R^2 \propto n^2$, as for a particle in a box, as it should be. So in this case we have a finite time of flight and a discrete spectrum.

Next, consider the case of the Polchinski–Strassler solution (21.5). The metric goes over to $AdS_5 \times S^5$ at large ρ, so again the time of flight is finite near infinity ($dt^2 = (dy^2 + dw^2)/(y^2 + w^2)^2$). Near the core of the solution, one has a near-shell approximation, with the metric developing a usual 5-brane throat (as in the D3-brane $AdS_5 \times S^5$ case), so strictly speaking one cannot regulate the divergence.

For the case of the Klebanov–Strassler solution (21.6), the time of flight is finite near infinity. As we have seen, at small τ, the metric looks like

$$ds^2 = a_0^{-1/2} d\vec{x}^2 + a_0^{1/2}\left(\frac{d\tau^2}{2} + d\Omega_3^2 + \frac{\tau^2}{4}((g_1)^2 + (g_2)^2)\right), \tag{21.42}$$

which is a smooth cut-off (the metric is similar to that of a plane in spherical coordinates), so the divergence in the time of flight is fully regulated. The spectrum of the scalar is again discrete in this case.

For the case of the Maldacena–Núñez (21.9) and Maldacena–Năstase (21.24) solutions, there is a 7-dimensional metric in the string frame of the type

$$ds_{7,\text{string}}^2 = d\vec{x}^2 + d\rho^2 + R^2(\rho)d\Omega_n^2. \tag{21.43}$$

In this 7-dimensional space, the wave equation becomes

$$\left[-k^2 + e^{2\phi}R^{-n}\partial_\rho e^{-2\phi}R^n\partial_\rho\right]\chi = 0, \tag{21.44}$$

with $n = 2$ or 3 and the factor of $e^{-2\phi}$ because in the string frame the action has such a prefactor. Now the function f in the 1-dimensional quantum mechanical problem is $f(\rho) = d/d\rho(log(R^n) - 2\phi)$, and the time of flight is just ρ, since $|g_{tt}| = g_{\rho\rho}$.

For the S_2 case (Maldacena–Núñez), in the IR, we have $R^2 \simeq \rho^2 - (4/9)\rho^4$ and find $f(\rho) = 2/\rho + (8/9)\rho + \ldots$ and the potential $V(\rho) = 4/3 + \ldots$ In the UV, $R^2 \sim \rho$ and $\phi - \phi_0 = -\rho + 1/4\log\rho$, so that $f(\rho) = 2 + 1/(2\rho)$ and the potential $V(\rho) = 1 + 1/(2\rho)$.

For the S_3 case (Maldacena–Năstase), in the IR, $R^2 \sim \rho^2$ and $\phi \simeq \phi_0$, so $f(\rho) = 3/\rho$ and the potential $V(\rho) = 3/(4\rho^2)$. In the UV, $R^2 \sim \rho$ and $\phi - \phi_0 = -\rho + 3/8\log\rho$, so $f(\rho) = 2 + 3/(4\rho) + \ldots$ and $V(\rho) = 1 + 3/(4\rho) + \ldots$

So we see that in these two cases, the potential does not have finite support (i.e., the time of flight between the boundaries of space is infinite), the potential starts at a higher value and ends up at $V = 1$ at $\rho = \infty$, so we have a continuum of states at high energies, though perhaps also discrete states, if there is a potential well at finite ρ. This is a modification in the IR of the flat space 5-brane spectrum, which as we saw had a continuum of states above a mass gap at $m^2 b^2 = 1$.

What about cases with fundamental quarks, when we can define mesonic states? We have seen that meson operators couple to fields charged under the global symmetry group in the dual ($SO(8)$ in the case of the $\mathcal{N} = 2$ orientifold). These fields belong to a vector multiplet and live on a D-brane in the dual, and they are part of KK towers obtained from

KK expanding down to AdS space (on S^3 in the case of the $\mathcal{N} = 2$ orientifold), in the case the D-brane is not dynamical. In AdS space, these fields can be distinguished from the supergravity fields coupling to nonmeson operators by their quantum numbers, i.e. by being charged under the gauge group. But from the point of view of their equation of motion, there is degeneracy, since the quantum numbers under the global group do not matter: the equation for a massless scalar is still $\Box \Phi = 0$, regardless of global quantum numbers. This means that from AdS/CFT we get a degeneracy in the masses of the towers of states of mesonic and nonmesonic operators.

A way to ensure that does not happen is to consider a different implementation of fundamental quarks, as modes living on D-brane probes in the geometry. Then the fields corresponding to mesonic and nonmesonic operators have different origins and couple differently to the background, giving different equations of motion, hence a more realistic scenario with respect to QCD. We analyze them next in the case of the Sakai–Sugimoto model.

21.7 Mass spectra in gravity duals from mode expansion on probe branes; Sakai–Sugimoto example

In the Sakai–Sugimoto model and in general in models with probe branes moving in a gravity dual geometry, the spectrum of the states coming from nonmesonic (glueball) operators can be obtained as usual from the spectrum of supergravity fields, just that now we have to impose that the fields are restricted (pulled back) to live on the probe brane. So we look at the equation of motion for the pulled back supergravity fields, and we want to obtain a discrete spectrum of $m^2 = -\vec{k}^2$ from them. Since this is a priori no different from the cases in the previous section, we will not describe them here.

Instead, we describe the new feature, the spectrum of states corresponding to mesonic operators. In general, if there is some compact space K with a symmetry group G corresponding to a global symmetry of the mesonic operators, we need to first KK expand the D-brane fields on K, as each reduced field corresponds to a different mesonic operator. Then the discrete spectrum of states coming from a mesonic operator comes from the eigenvalues of the equation of motion of the reduced D-brane field, and usually the quantization condition comes from the behavior in the radial coordinate U associated with the field theory energy.

In the case of the Sakai–Sugimoto model, in the gravity dual metric (21.29) we have an S^4 factor, with an $SO(5)$ gauge group, so we have to KK expand the D-brane fields on it. We are interested in the gauge field, present on any D-brane, and we are interested in particular in not having any $SO(5)$ charges, so we want only the zero mode on S^5 (the KK reduced mode). Then we have the coordinates τ and U, but they are related because of the fact that there is only one coordinate transverse to the D8-brane, so we have $\tau = \tau(U)$. Therefore, after the KK reduction on S^4 and taking into account that $\tau = \tau(U)$, we have a worldvolume action for the gauge field A_M, $M = 0, 1, \ldots, 4$ on the $\vec{x}, U$ coordinates that was calculated by Sakai and Sugimoto. It is best described in terms of coordinates where the D8-brane is flat,

$$(y, z) = (\sqrt{U^3 - 1} \cos \tau, \sqrt{U^3 - 1} \sin \tau), \tag{21.45}$$

such that the D8-brane is localized at $y = 0$ and extends in the z direction. Then the action for the gauge field $A_M = (A_\mu(x^\nu, z), A_z(x^\nu, z))$ becomes

$$S = S_{D8}^{DBI} + S_{D8}^{CS},$$

$$S_{D8}^{DBI} = \frac{\lambda N_c}{216\pi^3} \int d^4x \, dz \text{Tr} \left[\frac{1}{2} K^{-1/3} F_{\mu\nu}^2 + K F_{\mu z}^2 \right],$$

$$S_{D8}^{CS} = \frac{N_c}{24\pi^2} \int_{M^4 \times \mathbb{R}} \omega_{5,CS}(A),$$

$$\omega_{5,CS}(A) = \text{Tr} \left[A \wedge F \wedge F - \frac{1}{2} A^{\wedge 3} \wedge F + \frac{1}{10} A^{\wedge 5} \right],$$

$$K(z) \equiv 1 + z^2, \tag{21.46}$$

where $A = A_\mu dx^\mu + A_z dz$. The tower of meson states (fields) coming from operators coupling to the gauge fields A_μ and A_z correspond to eigenmodes of the equation of motion for A_μ and A_z, that we will call a_n and ϕ_n. Then formally we can expand the gauge fields in these eigenmodes, multiplied by functions identified with the meson states, i.e. write

$$A_\mu(x^\nu, z) = \sum_{n=1}^{\infty} A_\mu^{(n)}(x^\nu) a_n(z),$$

$$A_z(x^\nu, z) = \varphi^{(0)}(x^\nu) \phi_0(z) + \sum_{n=1}^{\infty} \varphi^{(n)}(x^\nu) \phi_n(z). \tag{21.47}$$

The sets of functions $\{a_n(z)\}_{n \geq 1}$ and $\{\phi_n(z)\}_{n \geq 0}$ are complete and orthonormal. The $a_n(z)$ satisfy

$$- K^{1/3} \partial_z(K \partial_z a_n) = \mu_n^2 a_n, \tag{21.48}$$

and $\phi_n(z)$ are $\phi_n \propto \partial_z \phi_n(z)$, $n \geq 1$ and $\phi_0(z) \propto K(z)$. The orthonormality conditions are defined by the action (21.46) as

$$\frac{\lambda N_c}{216\pi^3} \int dz K^{-1/3} a_n(z) a_m(z) = \delta_{nm},$$

$$\frac{\lambda N_c}{216\pi^3} \int dz K \phi_n(z) \phi_m(z) = \delta_{nm}. \tag{21.49}$$

Then the DBI action in (21.46), after integrating over z, becomes

$$S_{D8}^{DBI} = \int d^4x \text{Tr} \left[(\partial_\mu \varphi^{(0)})^2 \right.$$

$$\left. + \sum_{n=1}^{\infty} \left(\frac{1}{2} (\partial_\mu A_\nu^{(n)} - \partial_\nu A_\mu^{(n)})^2 + \mu_n^2 (A_\mu^{(n)} - \mu_n^{-1} \partial_\mu \varphi^{(n)})^2 \right) \right] + \text{int.} \tag{21.50}$$

The massless scalar $\varphi^{(0)}$ can then be identified with the pion in the dual field theory, and we see that the vector fields $A_\mu^{(n)}$ eat the massive scalars $\varphi^{(n)}$ and become massive via the redefinition $A_\mu'^{(n)} = A_\mu^{(n)} - \mu_n^{-1}\partial_\mu\varphi^{(n)}$. After the redefinition we have only $\varphi^{(0)}$ and the massive vectors $A_\mu'^{(n)}$, which can be interpreted as vector meson fields in the dual field theory.

21.8 Finite *N*?

Throughout this book we have analyzed only large N ($N \to \infty$) gauge theories, and one can treat perturbations away from $N = \infty$ by including string g_s corrections. But for the case of interest, namely real QCD, $N = 3$, which is far from infinity. However, it is known that for most quantities of interest, corrections are actually of order $1/N^2 \simeq 0.1$, which can be argued to be small. But one would like to know whether we can calculate anything at finite N, and perhaps at finite $\lambda = g_{\text{YM}}^2 N$, which corresponds to string worldsheet (α') corrections. For generic quantities, the answer (at this point) is no.

However, there is one process for which this is possible, namely forward (small angle) scattering of gauge invariant particles. It was shown in [58, 59] that string corrections (both worldsheet α' and quantum g_s corrections) to the high energy, small angle (forward, or in the case of 4-point scattering $s \to \infty$ and t fixed) scattering cross-section in a gravitational theory are exponentially small in the energy, specifically of order

$$\exp\left(-\frac{G_4^2 s}{8\alpha' \log(\alpha' s)}\right). \tag{21.51}$$

From this we can calculate, using the Polchinski–Strassler formalism described in Section 16.2, that $1/N$ and $1/(g^2 N)$ corrections to the high energy small angle scattering cross-section are exponentially suppressed in energy.

Important concepts to remember

- The noncompact extra dimension acts as an energy scale, and motion in it corresponds to RG flow.
- We introduce quarks in the field theory by introducing branes in the gravity dual, with vector fields on them, that couple to operators charged under the global symmetry corresponding to the brane symmetry.
- In the gravity dual of QCD-like or SQCD-like theories, the space in the UV looks conformal, i.e. it is like $AdS_5 \times X_5$, possibly log-corrected. In the IR, the space terminates in some way, corresponding to the mass gap.
- Examples of QCD-like theories that satisfy these conditions are $AdS_5 \times S^5$ at finite temperature, and cut-off $AdS_5 \times S^5$ (the hard-wall model), which can be modified by introducing a dynamical brane at the cut-off, whose fluctuations correspond to the pion.

Examples of SQCD-like theories that satisfy the conditions are the Polchinski–Strassler and Klebanov–Strassler solutions.

- The Maldacena–Núñez and Maldacena–Năstase solutions correspond to 5-branes wrapped on spheres, giving $\mathcal{N} = 1$ theories in four and three dimensions, respectively, and having a metric with constant scale factor in front of $d\vec{x}^2$, but nontrivial dilaton $\phi - \phi_0 \sim \log z + \ldots$, since the relevant combination is $\phi - \phi_0 - A$, where the scale factor is e^{2A}.

- The Sakai–Sugimoto model is a probe D8-brane (with the U-shaped topology of a D8-brane-anti-D8-brane system) in a Euclidean finite temperature D4-brane. It is a model for QCD with quarks.

- The masses of towers of states corresponding to a certain operator (like a glueball operator and meson operator) are mapped to the eigenstates of the gravity dual field sourcing the operator.

- The easiest way to have a discrete spectrum for the gravity dual field corresponding to the tower of states is a finite time of flight between the boundaries of the gravity dual. Another is a potential that goes to $-\infty$ at $r = 0$ and to $+\infty$ at $r = +\infty$, as in the case of Witten's construction of QCD_3 from compactified near-extremal D3-branes.

- Meson masses in theories with quarks obtained on probe D-branes, as in the case of the Sakai–Sugimoto model, can be calculated from the spectrum of fields living on the probe brane. In the Sakai–Sugimoto model, the gauge field leads to the pion and a tower of massive vector mesons.

References and further reading

The Polchinski–Strassler solution was found in [60], the Klebanov–Strassler solution in [61], the Maldacena–Núñez solution in [62], the Maldacena–Năstase solution in [63], and the Sakai–Sugimoto solution in [64]. The idea of introducing quarks on probe D-branes and how to obtain mesons in these models was introduced by Karch and Katz in [85].

Exercises

1. Consider the modified hard-wall model and for the dynamical brane at $r_{\min}$, the D-brane action in symmetrized cut-off AdS_5, with warp factor $\frac{|r|^2}{R^2} = e^{2k|y|}$ for $|r| \geq r_{\min}$. Write the resulting few terms in the derivative expansion of the scalar pion action.

2. Consider the Klebanov–Strassler solution. Argue that the effective coupling of the gauge theory dual to it runs with the energy as $\alpha \sim \alpha_0 \sqrt{\ln E/\Lambda_0}$. What is the global symmetry of that gauge theory?

3. Consider Wilson loops in the gauge theory dual to the Maldacena–Năstase solution, described by string worldsheets in the gravitational background. Consider the S-dual background, of D5-branes on S^3. Show that the theory is confining, so the potential $V(L) = -\frac{1}{T} \ln\langle W[C] \rangle$ is linear in L at large L, and calculate the string tension.

4. Consider the Sakai–Sugimoto model. Calculate $U(\tau)$ at large U and find there the metric induced on the worldvolume of the D8-brane.

5. Calculate the (maximum) time of flight and equivalent 1-dimensional quantum mechanical potential $V(x)$ for the hard-wall model.

6. Consider a harmonic oscillator potential in 1-dimensional quantum mechanics, $V(x) = ax^2$. Engineer a function $f(x)$, and from it a "gravity dual" metric that would give Eq. (21.40).

7. In the Sakai–Sugimoto model, consider a fluctuation also in the D8-brane worldvolume scalar U. Write down an action for small fluctuations on it.

Holographic renormalization

We have seen in Chapter 11 how to calculate correlation functions in AdS/CFT. But for correlation functions of massive scalar fields, there was a problem that was implicit, so we postponed their treatment until now. Correlation functions arose in AdS from the on-shell supergravity action, but the latter contains divergences coming from near the boundary of AdS. Near the boundary is the IR of the gravitational theory, but by the *UV-IR correspondence* of AdS/CFT, this corresponds to the UV of the field theory. Hence the treatment of these divergences corresponds to the treatment of UV divergences in field theory, subject to regularization, renormalization, and RG flow. Learning how to deal with these divergences is the subject of this chapter. We start with a general treatment, geared towards the application for the metric fluctuations, which was historically the first to be considered, as there the need for a systematic treatment of infinities was clearer. We then apply the formalism for the simplest case, the massive scalar. We also start the treatment of RG flow with a general discussion in this chapter, leaving for the next chapter the important application of RG flow *between fixed points*.

22.1 Statement of the problem and expected results: renormalization of infinities

As we saw in Chapter 11, the Euclidean signature correlators of operators $\mathcal{O}$ coupling to boundary sources $\phi_{(0)}$ for the AdS field ϕ are found by taking derivatives of the on-shell supergravity action, written as a function of those boundary sources,

$$\langle \mathcal{O}(x) \rangle = - \left. \frac{\delta S_{\text{sugra,on-shell}}}{\delta \phi_{(0)}(x)} \right|_{\phi_{(0)}=0} \; ; \quad \langle \mathcal{O}(x_1)\mathcal{O}(x_2) \rangle = \left. \frac{\delta^2 S_{\text{sugra,on-shell}}}{\delta \phi_{(0)}(x_1)\delta \phi_{(0)}(x_2)} \right|_{\phi_{(0)}=0} , \ldots$$

$$\langle \mathcal{O}(x_1) \ldots \mathcal{O}(x_n) \rangle = (-1)^n \left. \frac{\delta^n S_{\text{sugra,on-shell}}}{\delta \phi_{(0)}(x_1) \ldots \delta \phi_{(0)}(x_n)} \right|_{\phi_{(0)}=0} . \tag{22.1}$$

But as we said, the on-shell supergravity action has divergences near the boundary of AdS due to the infinite volume of space there ($\int dz_0 \sqrt{g} \propto \int dz_0/z_0^{d+1}|_{z_0 \sim 0} \to \infty$), not compensated by the behavior of the fields in the Lagrangean. Therefore we need to regularize and renormalize the supergravity action. Note that we have obtained finite results for the n-point functions, also in the case of the 2-point functions having the expected $\sim 1/|\vec{x} - \vec{y}|^{2\Delta}$ behavior, but that corresponded only to taking a certain prescription for regularization, it does not guarantee that we have the right result, unless we derive a general formalism to deal with it. Indeed, the naive extension of the method used in the massless scalar case to the massive case leads to a result with the wrong coefficient, off by a factor of $(2\Delta - d)/\Delta$ with respect to the correct result. One can do a correct calculation, using a bulk-to-epsilon propagator (regulated bulk to boundary propagator) and a careful treatment, as was done in [38], but then the general formalism is obscured.

Instead, we regularize the divergent on-shell action, and get rid of the divergent terms by adding counterterms S_{ct} to the action to obtain the finite renormalized action $S_{ren} = S_{on-shell,sugra} + S_{ct}$. Now taking derivatives of S_{ren} with respect to the boundary sources of the fields, we are able to unambiguously define the finite correlation functions of CFT operators. Moreover, the general renormalization formalism also regulates the 2-point function divergence at $\vec{x} = \vec{y}$.

The most important cases of fields that one can consider are the scalar ϕ, the gauge field A_μ, and the metric $g_{\mu\nu}$. We will see that the *exact one-point function* for the operator corresponding to the scalar ϕ is

$$\langle \mathcal{O}(x) \rangle_{\phi_{(0)}} = -\frac{1}{\sqrt{g_{(0)}}} \frac{\delta S_{ren}}{\delta \phi_{(0)}(x)} \sim \phi_{(2\Delta-d)}(x), \tag{22.2}$$

where $\phi_{(2\Delta-d)}$ is a coefficient appearing in the expansion near the boundary of the solution to the ϕ equation of motion, which is in general not completely determined by the boundary source via the *near-boundary expansion* of the equation of motion, though they are of course determined by an *exact* solution. For the operators J_i and T_{ij} coupling to the bulk fields: vector A_μ and the metric $g_{\mu\nu}$, we similarly obtain (whereas $\mu, \nu = 0, \ldots, d$ is a bulk index, $i, j = 0, 1, \ldots, d - 1$ is a boundary index):

$$\langle J_i(x) \rangle_{A_{(0)i}} = -\frac{1}{\sqrt{g_{(0)}}} \frac{\delta S_{ren}}{\delta A_{(0)i}(x)} \sim A_{(n)i}(x),$$

$$\langle T_{ij}(x) \rangle_{g_{(0)ij}} = -\frac{1}{\sqrt{g_{(0)}}} \frac{\delta S_{ren}}{\delta g_{(0)ij}} \sim g_{(d)ij}(x). \tag{22.3}$$

Here again $A_{(n)ij}$ and $g_{(d)ij}$ are coefficients in the expansion near the boundary of the solution to the corresponding bulk equations of motion, and again in general they are not determined by the boundary sources via the *near-boundary expansion* of the equations of motion, but are determined from an *exact* solution. Note that the boundary operator corresponding to the fluctuations of the bulk metric $g_{\mu\nu}$ is the *energy-momentum tensor* $T_{ij}(x)$ of the CFT. The reason is that the energy-momentum tensor is defined as the variation of the action with respect to the metric, $T_{ij} = -2\delta S_{matter}/\sqrt{g}\delta g^{ij}$, so the source term added in the CFT action by the bulk metric fluctuation $\delta g_{\mu\nu}$ is $1/2 \int \sqrt{g} T_{ij} \delta g^{ij}$, where g_{ij} is the metric on the boundary.

The above exact one-point functions were written *in the presence of nonzero sources*, which means that if we know the *exact* solutions for $\phi_{(2\Delta-d)}$, $A_{(n)i}$, and $g_{(d)ij}$ we can extract all the higher n-point functions via derivatives from it, e.g.

$$\langle \mathcal{O}(x_1) \ldots \mathcal{O}(x_n) \rangle \sim (-1)^{n-1} \frac{\delta^{n-1} \phi_{(2\Delta-d)}(x_1)}{\delta \phi_{(0)}(x_2) \ldots \delta \phi_{(0)}(x_n)} \bigg|_{\phi_{(0)}=0}. \tag{22.4}$$

The exact one-point function also allows us to write holographic Ward identities. Turning on the $g_{ij(0)}$ source (perturbation), one finds the diffeomorphism and conformal (Weyl) Ward identities,

$$\nabla^i \langle T_{ij} \rangle g_{(0)ij} = 0; \quad \langle T^i{}_i \rangle g_{(0)ij} = \mathcal{A}, \tag{22.5}$$

where $\mathcal{A}$ is the (holographic) Weyl anomaly. The above is understood as follows: diffeomorphism invariance, being a gauge invariance has no anomaly, whereas Weyl invariance has a well-defined anomaly.

22.2 Asymptotically AdS spaces and asymptotic expansion of the fields

We saw at the end of Chapter 2 that the Riemann tensor of AdS space is

$$R_{\mu\nu\rho\sigma} = \frac{1}{R^2}(g_{\mu\sigma}g_{\nu\rho} - g_{\mu\rho}g_{\nu\sigma}), \tag{22.6}$$

with the cosmological constant $\Lambda = -d(d-1)M_{\text{Pl},d}^{d-2}/2R^2$. The AdS metric blows up at the boundary, so we do not have a well-defined boundary metric, but rather a well-defined *conformal structure*, i.e. a metric up to conformal transformations. This matches with the idea that we can define the boundary conformal field theory in any coordinates related to the flat ones by a conformal transformation.

A set of well-defined coordinates is the coordinates on the Poincaré patch. We can define *asymptotically AdS spacetimes* in a more formal way, but a simple definition is that the metric near the boundary at $z = 0$ can be put in the form

$$ds^2 = \frac{1}{z^2}(dz^2 + g_{ij}dx^i dx^j), \tag{22.7}$$

where the metric $g_{ij}(\vec{x}, z)$ has a smooth limit as $z \to 0$, and satisfies Einstein's equations. Then near the boundary we can expand

$$g_{ij}(\vec{x}, z) = g_{(0)ij}(\vec{x}) + zg_{(1)ij}(\vec{x}) + z^2 g_{(2)ij} + \ldots \tag{22.8}$$

Einstein's equations then allow one to calculate all the coefficients $g_{(n)ij}(\vec{x})$ with $k > 0$ from $g_{(0)ij}(\vec{x})$. In particular, an explicit computation shows that in pure gravity all coefficients multiplying odd powers of z vanish up to order z^d. If d is even, we can also have a logarithmic term at order z^d,

$$g_{ij}(\vec{x}, z) = g_{(0)ij}(\vec{x}) + z^2 g_{(2)ij}(\vec{x}) + \ldots + z^d(g_{(d)ij}(\vec{x}) + h_{(d)ij}(\vec{x})\log z^2) + \ldots \tag{22.9}$$

In exact solutions the relations among coefficients can be complicated, and even non-local, but when the equations are expanded in z, the solutions relating the higher order coefficients in terms of $g_{(0)ij}$ (and $g_{(d)ij}$ and $h_{(d)ij}$) are *algebraic*. This is a general feature for all the fields, and we see it explicitly for the scalars later on in the chapter. Here $h_{(d)ij}$ equals the metric variation of the conformal anomaly, whereas $g_{(d)ij}$ is undetermined by $g_{(0)ij}$, except for its trace and covariant divergence.

Now moving on to a general field $\Phi(\vec{x}, z)$ (which can be scalar ϕ, gauge field A_μ, metric $g_{\mu\nu}$, etc., but we will suppress the spacetime indices), it has in general an expansion near the boundary as

$$\Phi(\vec{x}, z) = z^m(\Phi_{(0)}(\vec{x}) + z^2\Phi_{(2)}(\vec{x}) + \ldots + z^{2n}(\Phi_{(2n)}(\vec{x}) + \log z^2\tilde{\Phi}_{(2n)}(\vec{x})) + \ldots). \quad (22.10)$$

Except for the case where Φ is the metric itself, we consider the metric to be the (Euclidean) AdS metric in Poincaré coordinates, $g_{(0)ij} = \delta_{ij}, g_{(k)ij} = 0, k > 0$. We should strictly speaking consider at least the coupled Φ-metric system, but we consider a fixed AdS background in the case that Φ is not the metric. If n is not an integer in the above expansion, the $\tilde{\Phi}_{(2n)}$ term is absent.

Since the equation of motion of the field Φ will be second order in derivative, there will be two independent solutions. For the behavior near the boundary, the solutions are z^m and z^{m+2n}. As we have seen in Chapters 11 and 12, the coefficient of z^m, $\Phi_{(0)}$, gives the source for the dual (boundary) operator corresponding to Φ, whereas the coefficient of z^{m+2n} is a linearly independent solution (which in linear analysis does not depend on $\Phi_{(0)}$, but for a specific full nonlinear solution it does) associated with the VEV of the operator. In fact, we already indicated that the exact one-point function $\langle \mathcal{O}(x) \rangle$ will be proportional to the exact solution $\phi_{(2\Delta-d)}$ for the scalar, $g_{(d)ij}$ for the metric, etc. The equations of motion will fix the coefficients of z^{m+2k} for $k < n$, $\Phi_{(2k)}$, algebraically in terms of $\Phi_{(0)}$, but will leave $\Phi_{(2n)}$ unfixed. Then $\tilde{\Phi}_{(2n)}$ will be related to conformal anomalies.

22.3 Method

Once we have the expansion of the fields Φ near the boundary $z = 0$ where the divergences are located, we can isolate the divergent pieces in the action by regularizing the boundary, and add counterterms to the action to make it finite. The resulting finite renormalized action is used to define n-point functions, and to find renormalization group (RG) transformations on them.

Regularization and counterterms

The boundary is regularized, being considered at $z = \epsilon$ instead of $z = 0$, thus we only integrate down to $z = \epsilon$. That means that the regularized action for the field Φ, depending on the boundary value $\Phi_{(0)}$, after doing the integration on z is generally of the type

$$S_{\text{reg}}[\Phi_{(0)}, \epsilon] = \int_{(z=\epsilon)} d^d x \sqrt{g_{(0)}} [\epsilon^{-2\nu} s_{(0)}[\Phi_{(0)}]$$

$$+ \epsilon^{-2\nu+2} s_{(2)}[\Phi_{(0)}] + \ldots - \log \epsilon \, s_{(2\nu)}[\Phi_{(0)}] + \text{finite}]. \quad (22.11)$$

Here, e.g., for a scalar $2\nu = 2\Delta - d$, since a quadratic term (e.g. a mass term) in the action gives $\int dz/z^{d+1}\phi^2$, and we have $\phi \sim z^{d-\Delta}$ near the boundary, leading to $\int_\epsilon dz z^{d-2\Delta-1} \sim \epsilon^{d-2\Delta}$.

The counterterm action should be defined such as to cancel the above divergent terms. The minimal subtraction scheme would correspond to adding exactly minus the divergent terms, with no finite parts, i.e.

$$S_{ct}[\Phi(\vec{x}, \epsilon); \epsilon] = -\text{div.terms in } S_{reg}[\Phi_{(0)}(\Phi(\vec{x}, \epsilon)); \epsilon]. \tag{22.12}$$

Of course, we have to write the action in a general form, so we need to invert the relation $\Phi[\Phi_{(0)}]$ and write the sources $\Phi_{(0)}$ as a function of the general fields living at the regulated boundary $z = \epsilon$, $\Phi_{(0)}[\Phi(\vec{x}, \epsilon)]$, where the metric is $\gamma_{ij} = g_{ij}(\vec{x}, \epsilon)/\epsilon^2$. This is the general quantum field theory procedure, to add counterterms to the action that depend on the same fields in the action, in addition to the regulator ϵ.

Renormalized on-shell action

There is one more subtlety, which is that the subtraction defines the *subtracted action*,

$$S_{sub}[\Phi(\vec{x}, \epsilon); \epsilon] = S_{reg}[\Phi_{(0)}; \epsilon] + S_{ct.}[\Phi(\vec{x}, \epsilon); \epsilon], \tag{22.13}$$

and it is this action that is varied in order to obtain correlation functions, taking $\epsilon \to 0$ only at the end of the calculation.

The renormalized on-shell action is the $\epsilon \to 0$ limit of the subtracted action,

$$S_{ren}[\Phi_{(0)}] = \lim_{\epsilon \to 0} S_{sub}[\Phi(\vec{x}, \epsilon); \epsilon], \tag{22.14}$$

and is finite, but should not be used when variations are taken.

n-point function and RG transformation

The exact one-point function is obtained from the variation of the renormalized action with respect to the boundary source. In the case of a nontrival background metric $g_{(0)ij}(\vec{x})$ we should divide by $\sqrt{g_{(0)}}$, to obtain

$$\langle \mathcal{O}(\vec{x}) \rangle_{\Phi_{(0)}} = \text{``} \frac{1}{\sqrt{g_{(0)}}} \frac{\delta S_{ren}}{\delta \Phi_{(0)}(\vec{x})} \text{''}. \tag{22.15}$$

We put the result in quotation marks, since we really need to take the variations at finite ϵ and take $\epsilon \to 0$ only at the end. Then we must rewrite everything in terms of objects on the $z = \epsilon$ surface: the source $\Phi_{(0)}$ as a function of $\Phi(\vec{x}, \epsilon) = \epsilon^m \Phi_{(0)} + \ldots$ and $g_{(0)ij}$ as a function of $\gamma_{ij} = g_{(0)ij}/\epsilon^2 + \ldots$, obtaining

$$\langle \mathcal{O}(\vec{x}) \rangle_{\Phi_{(0)}} = \lim_{\epsilon \to 0} \frac{1}{\epsilon^{d-m}} \frac{1}{\sqrt{\gamma}} \frac{\delta S_{sub}}{\delta \Phi(\vec{x}, \epsilon)}. \tag{22.16}$$

As we mentioned, the result is proportional to the linearly independent coefficient $\Phi_{(2n)}(\vec{x})$, but there is also a possible local function of the source $\Phi_{(0)}$ that leads to contact terms in the higher *n*-point functions (obtained from it via derivatives), and that is scheme dependent,

$$\langle \mathcal{O}(\vec{x}) \rangle_{\Phi_{(0)}} \sim \Phi_{(2n)}(\vec{x}) + F(\Phi_{(0)}). \tag{22.17}$$

Higher n-point functions arise from further functional derivatives with respect to $\Phi_{(0)}$ according to (22.4).

RG (renormalization group) transformations in the field theory arise from bulk diffeomorphisms that induce Weyl transformations on the boundary, in particular

$$x^i = \mu x^{i'}; \quad z = \mu z'. \tag{22.18}$$

22.4 Example: massive scalar

We now exemplify the general procedure in the case of the massive scalar that we also call Φ as in the general case (though the terms in the expansion are called $\phi_{(k)}$), satisfying the massive KG equation in AdS space.

Perturbative solution

The expansion starts at $z^{d-\Delta}$, as we saw in Chapter 11, i.e. we have

$$\Phi(\vec{x}, z) = z^{d-\Delta} \phi(\vec{x}, z), \tag{22.19}$$

with $\phi(\vec{x}, z)$ having a finite value at the boundary. The KG equation in AdS space in Poincaré coordinates becomes

$$(m^2 R^2 - \Delta(\Delta - d))\phi(\vec{x}, z) - z^2 \partial_i \partial_i \phi(\vec{x}, z) - (d - 2\Delta + 1) z \partial_z \phi(\vec{x}, z) - z^2 \partial_z^2 \phi(\vec{x}, z) = 0. \tag{22.20}$$

Putting $z = 0$ in this equation we get $m^2 R^2 = \Delta(\Delta - d)$, as we knew it should be (from Chapter 11). Next, from the odd powers of z of the equation, we show that the odd powers of z in the expansion of $\phi(\vec{x}, z)$ have coefficient zero, i.e. the expansion is

$$\phi(\vec{x}, z) = \phi_{(0)} + z^2 \phi_{(2)} + z^4 \phi_{(4)} + \dots \tag{22.21}$$

Then, from the coefficient of z^2 in the equation of motion we obtain

$$\phi_{(2)}(\vec{x}) = \frac{1}{2(2\Delta - d - 2)} \partial_i \partial_i \phi_{(0)}. \tag{22.22}$$

Continuing, from the coefficients of $z^4, \dots, z^{2n}$, we get

$$\phi_{(4)}(\vec{x}) = \frac{1}{4(2\Delta - d - 4)} \partial_i \partial_i \phi_{(2)}, \dots, \phi_{(2n)} = \frac{1}{2n(2\Delta - d - 2n)} \partial_i \partial_i \phi_{(2n-2)}. \tag{22.23}$$

This series ends when $2\Delta - d - 2n = 0$, when we need to introduce a term $z^\Delta \log z^2$, i.e.

$$\phi(\vec{x}, z) = \phi_{(0)} + z^2 \phi_{(2)} + \dots + z^{2\Delta - d}(\phi_{(2\Delta - d)} + (\log z^2)\tilde{\phi}_{(2\Delta - d)}) + \dots \tag{22.24}$$

Then $\phi_{(2\Delta - d)}$ is not determined from the equation of motion in the z expansion, and instead we can determine $\tilde{\phi}_{(2\Delta - d)}$ as

$$\tilde{\phi}_{(2\Delta - d)} = -\frac{1}{2^{2\Delta - d} \Gamma\left(\Delta - \frac{d}{2}\right)\left(\Delta - \frac{d-2}{2}\right)} (\partial_i \partial_i)^{\Delta - \frac{d}{2}} \phi_{(0)}. \tag{22.25}$$

Regularization and counterterms

The regularized kinetic on-shell action for the massive scalar is given by,

$$
\begin{aligned}
S_{\text{reg.}} &= \frac{1}{2}\int_{z\geq\epsilon} d^{d+1}x\sqrt{g_{d+1}}(g^{\mu\nu}\partial_\mu\Phi\partial_\nu\Phi + m^2\Phi^2)\\
&= \frac{1}{2}\int_{z\geq\epsilon} d^{d+1}x\sqrt{g_{d+1}}\Phi(-\Box_{g_{\mu\nu}} + m^2)\Phi - \frac{1}{2}\int_{z=\epsilon} d^dx\sqrt{g_{d+1}}g^{zz}\Phi\partial_z\Phi, \quad (22.26)
\end{aligned}
$$

(the minus sign in the second term arises from the convention that $z = \epsilon$ is the lower end of the z integration) and by the equation of motion for Φ the first term is zero, and substituting $\Phi = z^{d-\Delta}\phi(\vec{x}, z)$ in this equation we obtain

$$
S_{\text{reg.}} = -\int_{z=\epsilon} d^dx\, \epsilon^{-2\Delta+d}\left(\frac{1}{2}(d-\Delta)\phi(\vec{x},\epsilon)^2 + \frac{1}{2}\epsilon\phi(\vec{x},\epsilon)\partial_\epsilon\phi(\vec{x},\epsilon)\right). \quad (22.27)
$$

This is of the form of (22.11), with

$$
s_{(0)} = -\frac{1}{2}(d-\Delta)\phi_{(0)}^2,
$$

$$
s_{(2)} = -(d-\Delta+1)\phi_{(0)}\phi_{(2)} = -\frac{d-\Delta+1}{2(2\Delta-d-2)}\phi_{(0)}\partial_i\partial_i\phi_{(0)}, \quad \dots,
$$

$$
s_{(2\Delta-d)} = d\phi_{(0)}\tilde{\phi}_{(2\Delta-d)} = -\frac{d}{2^{2\Delta-d}\Gamma\left(\Delta-\frac{d}{2}\right)\left(\Delta-\frac{d-2}{2}\right)}\phi_{(0)}(\partial_i\partial_i)^{\Delta-\frac{d}{2}}\phi_{(0)}. \quad (22.28)
$$

The counterterm action should cancel these divergences, but should be expressed in terms of the fields at the ϵ-boundary, i.e. we need to solve for $\phi_{(2k)}$ as a function of $\Phi(\vec{x},\epsilon)$. Inverting (22.19),(22.21) to second order in ϵ^2, we obtain

$$
\phi_{(0)} = \epsilon^{-(d-\Delta)}\left(\Phi(\vec{x},\epsilon) - \frac{1}{2(2\Delta-d-2)}\Box_\gamma\Phi(\vec{x},\epsilon)\right),
$$

$$
\phi_{(2)} = \epsilon^{-(d-\Delta)-2}\frac{1}{2(2\Delta-d-2)}\Box_\gamma\Phi(\vec{x},\epsilon), \quad (22.29)
$$

where $\Box_\gamma$ is the Laplacean of the induced metric at $z = \epsilon$, $\gamma_{ij} = 1/\epsilon^2\delta_{ij}$.

Finally we can rewrite the counterterms, i.e. minus the divergent terms in the regularized action, as

$$
S_{\text{ct.}} = \int_{\text{boundary}} d^dx\sqrt{\gamma}\left(\frac{d-\Delta}{2}\Phi^2 + \frac{1}{2(2\Delta-d-2)}\Phi\Box_\gamma\Phi\right) + \mathcal{O}(\Box_\gamma^2), \quad (22.30)
$$

up to first order in $\Box_\gamma$ (corresponding to first order in ϵ^2). The expansion is an expansion in powers of $\Box_\gamma$, or equivalently an expansion in ϵ^2. When $\Delta = d/2 + 1$, the coefficient of $\Phi\Box_\gamma\Phi$ is replaced by $-\frac{1}{2}\log\epsilon$, and in general when $\Delta = d/2 + k$, the coefficient of $\Phi\Box_\gamma^k\Phi$ has a $\log\epsilon$.

Renormalized on-shell action and 1-point and 2-point functions

As in the general case, we write the subtracted action appearing in n-point functions as $S_{\text{sub}}[\Phi(\vec{x},\epsilon)] = S_{\text{reg.}} + S_{\text{ct.}}$, and the renormalized on-shell action as $S_{\text{ren.}} = \lim_{\epsilon\to 0} S_{\text{sub}}$.

Then the 1-point function is (since $m = d - \Delta$ for the scalar)

$$\langle \mathcal{O} \rangle_{\Phi_{(0)}} = \lim_{\epsilon \to 0} \left(\frac{1}{\epsilon^\Delta} \frac{1}{\sqrt{\gamma}} \frac{\delta S_{\text{sub.}}}{\delta \Phi(\vec{x}, \epsilon)} \right). \tag{22.31}$$

In the simplest case $\Delta = d/2 + 1$, the variation of the subtracted action is (note the $\sqrt{g}$ in the first term, which has a bulk origin, vs. the $\sqrt{\gamma}$ in the second, which is truly a boundary term):

$$\delta S_{\text{sub.}} = -\frac{1}{2} \delta \int_{z=\epsilon} d^d x \sqrt{g} g^{zz} \Phi \partial_z \Phi + \delta \int_{z=\epsilon} d^d x \sqrt{\gamma} \left(\frac{d - \Delta}{2} \Phi^2 - \frac{1}{2} \log z \Phi \Box_\gamma \Phi \right) + \dots$$

$$= \int_{z=\epsilon} d^d x \sqrt{\gamma} \delta \Phi \left(-\epsilon \partial_\epsilon \Phi + (d - \Delta) \Phi - \log \epsilon \Box_\gamma \Phi \right). \tag{22.32}$$

Then we obtain

$$\frac{1}{\sqrt{\gamma}} \frac{\delta S_{\text{sub.}}}{\delta \Phi} = -\epsilon \partial_\epsilon \Phi + (d - \Delta) \Phi - \log \epsilon \Box_\gamma \Phi, \tag{22.33}$$

and after substituting the expansion of Φ, $\Phi = z^{d-\Delta} (\phi_{(0)} + z^2 (\phi_{(2)} + \log z^2 \tilde{\phi}_{(2)}) + \dots$, the 1-point function becomes

$$\langle \mathcal{O} \rangle_{\phi_{(0)}} = \lim_{\epsilon \to 0} \left(\frac{1}{\epsilon^\Delta} \frac{1}{\sqrt{\gamma}} \frac{\delta S_{\text{sub.}}}{\delta \Phi} \right) = -2(\phi_{(2)} + \tilde{\phi}_{(2)}). \tag{22.34}$$

Note that we have used $\Delta = d/2 + 1$, $\Box_\gamma = \epsilon^2 \Box_0$, $\Box_0 \phi_{(0)} = -4 \tilde{\phi}_{(2)}$ for the asymptotic solution, log-divergent terms of type $\log \epsilon^2 \tilde{\phi}_{(2)}$ have cancelled, as they should, and the above result is the finite part (before the limit, there were $\mathcal{O}(\epsilon^2 \log \epsilon^2)$ corrections, which vanished).

Thus the 1-point is indeed of the general form that we indicated, $\langle \mathcal{O} \rangle_{\phi_{(0)}} \sim \phi_{(2\Delta - d)} + F(\phi_{(0)})$, since $\tilde{\phi}_{(2)}$ is a function of $\phi_{(0)}$, in this case $-1/4 \Box_0 \phi_{(0)}$. Moreover, this function F is scheme dependent: it can be removed by the addition of a finite counterterm.

In general, the 1-point function for the scalar can be shown to be

$$\langle \mathcal{O} \rangle_{\phi_{(0)}} = -(2\Delta - d) \phi_{(2\Delta - d)} + F(\phi_{(0)}). \tag{22.35}$$

To calculate the 2-point function, we need to find $\phi_{(2\Delta - d)}$ and F, which as we argued can be extracted from the exact solution. To see how this works, we consider the exact solution in the case $d = 4$ and $\Delta = d/2 + 1 = 3$, Fourier transformed on $\vec{x}$ to momenta $\vec{k}$. Then the KG equation after the redefinition $\Phi = z^2 \chi$ is

$$z^2 \partial_z^2 \chi + z \partial_z \chi - (k^2 z^2 + 1) \chi = 0, \tag{22.36}$$

which is the modified Bessel equation, with the unique solution that is *regular in the interior of AdS*,

$$\chi = K_1(kz) \Rightarrow \Phi = z^2 K_1(kz). \tag{22.37}$$

Its expansion near the boundary $z = 0$ gives

$$\Phi(k, z) = \frac{1}{k} z \left[1 + k^2 z^2 \left(\frac{1}{4} (2\gamma - 1) - \frac{1}{2} \log 2 + \frac{1}{2} \log(kz) \right) \right] + \dots \tag{22.38}$$

We replace the overall $1/k$ with an overall normalization $\phi_{(0)}(k)$, and from the expansion we identify $\phi_{(2)}$ and $\psi_{(2)}$,

$$\tilde{\phi}_{(2)}(k) = \frac{k^2}{4}\phi_{(0)}(k) \Rightarrow \tilde{\phi}_{(2)}(x) = -\frac{1}{4}\Box_0\phi_{(0)},$$

$$\phi_{(2)}(k) = \phi_{(0)}(k)k^2\left[\frac{1}{4}(2\gamma - 1) + \frac{1}{2}\log\frac{k}{2}\right]. \tag{22.39}$$

Then

$$\langle\mathcal{O}\rangle_{\phi_{(0)}} = -2(\phi_{(2)} + \tilde{\phi}_{(2)}) = -2\phi_{(0)}(k)\left[k^2\left(\frac{1}{4}(2\gamma - 1) - \frac{1}{2}\log 2 + \frac{1}{4}\right) + \frac{k^2}{4}\log k^2\right], \tag{22.40}$$

and the 2-point function is obtained by derivation,

$$\langle\mathcal{O}(k)\mathcal{O}(-k)\rangle = -\frac{\delta\phi_{(2)}(k)}{\delta\phi_{(0)}(-k)} = \frac{k^2}{2}\log k^2 + \text{contact terms}. \tag{22.41}$$

By going to x space we obtain a regularized version of the naive 2-point function,

$$\langle\mathcal{O}(x)\mathcal{O}(0)\rangle = \frac{4}{\pi^4}\mathcal{R}\frac{1}{x^6}, \tag{22.42}$$

where $\mathcal{R}1/x^6$ equals $1/x^6$ away from $x = 0$. In general, for $\Delta = d/2 + k$, one obtains

$$\langle\mathcal{O}(x)\mathcal{O}(0)\rangle = (2\Delta - d)\frac{\Gamma(\Delta)}{\pi^{d/2}\Gamma(\Delta - d/2)}\mathcal{R}\frac{1}{x^{2\Delta}}. \tag{22.43}$$

The (unregularized) result was first obtained in [38] using a careful application of the Witten method, and it differs from a naive application of the method by a factor of $(2\Delta - d)/\Delta$.

As we see, in the case of the massless scalar, $\Delta = d$, and the naive result was correct (in fact, we can already see that in (22.30) the first leading counterterm, Φ^2, has vanishing coefficient in this case), hence the analysis in Chapter 11 was valid.

RG transformations; operator and VEV deformations

We saw that the RG transformations were $\vec{x} = \vec{x}'\mu$, $z = z'\mu$. But $\Phi(\vec{x}, z)$ is a scalar, so $\Phi'(\vec{x}', z') = \Phi(\vec{x}, z)$, meaning that

$$\phi'_{(2k)}(\vec{x}') = \mu^{d-\Delta+2k}\phi_{(2k)}(\vec{x}'\mu), \quad 2k < 2\Delta - d,$$

$$\tilde{\phi}'_{(2\Delta-d)}(\vec{x}') = \mu^\Delta\tilde{\phi}_{(2\Delta-d)}(\vec{x}'\mu),$$

$$\phi'_{(2\Delta-d)}(\vec{x}') = \mu^\Delta[\phi_{(2\Delta-d)}(\vec{x}'\mu) + \log\mu^2\tilde{\phi}_{(2\Delta-d)}(\vec{x}'\mu)]. \tag{22.44}$$

In particular,

$$\mu\frac{\partial}{\partial\mu}\phi_{(0)}(\vec{x}\mu) = (\Delta - d)\phi_{(0)}(\vec{x}'\mu), \tag{22.45}$$

consistent with a source of an operator of dimension Δ. Moreover, for the general 1-point function (22.35), we obtain

$$\langle \mathcal{O}(\vec{x}') \rangle'_{\phi_{(0)}} = \mu^{\Delta} \left(\langle \mathcal{O}(\vec{x}'\mu) \rangle_{\phi_{(0)}} - (2\Delta - d) \log \mu^2 \tilde{\phi}_{(2\Delta-d)}(\vec{x}'\mu) \right), \tag{22.46}$$

consistent with a $\tilde{\phi}_{(2\Delta-d)}$ giving a conformal anomaly, i.e. an anomalous transformation law under scaling for quantum VEVs.

Then, as we also saw in Chapter 12, in the expansion

$$\Phi = z^{d-\Delta}\phi_{(0)} + \ldots + z^{\Delta}(\phi_{(2\Delta-d)}(+\tilde{\phi}_{(2\Delta-d)} \log z^2)) + \ldots, \tag{22.47}$$

for the two independent linearized solutions in AdS space, $\phi_{(0)}$ corresponds to an operator deformation of the theory, i.e. a source term $\int \mathcal{O}\phi_{(0)}$ in the CFT action, whereas $\phi_{(2\Delta-d)}$ corresponds to a deformation of the VEV of the operator because of (22.35), with the $\tilde{\phi}_{(2\Delta-d)}$ term giving the anomalous contribution.

Important concepts to remember

- The classical on-shell supergravity action in AdS space has divergences near the boundary that correspond to UV divergences in the field theory, so need to be regularized and renormalized by adding counterterms to the action.
- The 1-point function is given by the normalizable mode $\phi_{(2\Delta-d)}$ from the expansion near the boundary, which is linearly independent in the near-boundary expansion, but is *dependent* on the non-normalizable mode $\phi_{(0)}$ in an exact solution (there exists a unique regular solution in Euclidean AdS).
- Higher n-point functions can be derived from further differentiations of the *exact* mode $\phi_{(2\Delta-d)}$, viewed as a function of $\phi_{(0)}$.
- In the near-boundary expansion, we generically have an expansion in z^2, and the coefficients $\Phi_{(2k)}$ with $2k < 2\Delta - d$ are algebraically defined in terms of $\Phi_{(0)}$.
- In the holographic renormalization method, it is crucial to perform all calculations at a finite distance ϵ from the boundary. We integrate only down to ϵ, and write counterterms in terms of fields on a boundary at ϵ.

References and further reading

For more details on the method of holographic renormalization, see the review [86]. The method was first used in [87], where the holographic Weyl anomaly was calculated (we mention this anomaly in the next chapter).

Exercises

1. Calculate the asymptotic expansion (perturbative solution) for a gauge field (with action $\int \sqrt{g} F_{\mu\nu}^2/4$) in AdS_{d+1}, for $d > 3$, as a function of boundary values.
2. Do the same for AdS_4. What changes?
3. Calculate the 1-point function (operator VEV) for a scalar with $\Delta = d/2 + 2$.

4. Calculate the exact solution for a scalar with $\Delta = d/2 + 2$, and from it find $\phi_{(4)}$ as a function of $\phi_{(0)}$.

5. How are the 3-point function calculations affected by the addition of the counterterm action (22.30)?

6. Do the Fourier transform from (22.41) to (22.42) and find the explicit form for $\mathcal{R}1/x^6$.

In the previous chapter we saw that RG transformations in the field theory correspond to motion in the radial coordinate, $x^i = \mu x^{i'}$, $z = \mu z'$. In this chapter we study a particular motion in the radial coordinate, i.e. particular supergravity solutions depending on the radial coordinate, such that we start and end at *fixed points*, with zero field theory beta function $\beta = 0$. That means that in the field theory we have an RG flow between fixed points, whereas in the gravity dual we have a solution that interpolates between points with AdS symmetry.

But such an RG flow in field theory is initiated by a small *relevant* deformation of a conformal field theory, i.e. by a relevant operator. Relevant means that the deformation leads to a flow *away* from the UV fixed point. In the supergravity dual, the RG flow should correspond to a solution of the same supergravity theory that starts in r, approximately like the un-modified AdS vacuum, and ends in r at another AdS vacuum.

23.1 $\mathcal{N} = 1$ supersymmetric mass deformation of $\mathcal{N} = 4$ SYM and an IR fixed point

The cases we are interested in are modifications of the $\mathcal{N} = 4$ SYM dual to $AdS_5 \times S^5$.

In $\mathcal{N} = 4$ SYM, the deformation that is easiest to describe is an $\mathcal{N} = 1$ supersymmetric one. In $\mathcal{N} = 1$ language, we have three chiral multiplets in the theory, Φ_1, Φ_2, Φ_3, with the superpotential

$$W = \text{Tr}\ (\Phi_3[\Phi_1, \Phi_2]). \tag{23.1}$$

The deformation corresponds to adding a mass term to the superpotential

$$\delta W = \frac{m}{2}\text{Tr}\ (\Phi_3^2), \tag{23.2}$$

which breaks $\mathcal{N} = 4$ to $\mathcal{N} = 1$ (we break $\mathcal{N} = 4$, since the $\mathcal{N} = 4$ theory is unique, and we still have $\mathcal{N} = 1$ since we have a superpotential). The deformed theory flows towards a nontrivial IR fixed point (with an exactly zero beta function).

There are two important coefficients that characterize fixed points. They are related to anomalies in the conformal invariance, given by the VEV of $T^\mu{}_\mu$ (the trace of the energy-momentum tensor is classically zero) and in the R-symmetry, given by the VEV of the divergence of the R-symmetry current J^μ. In the presence of a metric $g_{\mu\nu}$ and a source A_μ

for the R-current J_μ, the anomalies are given in terms of the *central charges c and a* (the central charge c is the same one considered in the context of 2-dimensional conformal field theories) by

$$\langle T^\mu{}_\mu \rangle_{g_{\mu\nu}, A_\mu} = \frac{c}{16\pi^2} C^2_{\mu\nu\rho\sigma} - \frac{a}{16\pi^2} \tilde{R}_{\mu\nu\rho\sigma} R^{\mu\nu\rho\sigma} + \frac{c}{6\pi^2} F^2_{\mu\nu},$$

$$\langle \partial_\mu \sqrt{g} J^\mu \rangle_{g_{\mu\nu}, A_\mu} = -\frac{a-c}{24\pi^2} R_{\mu\nu\rho\sigma} \tilde{R}^{\mu\nu\rho\sigma} + \frac{5a-3c}{9\pi^2} F_{\mu\nu} \tilde{F}^{\mu\nu}. \tag{23.3}$$

Here $F_{\mu\nu}$ is the field strength of A_μ, $\tilde{F}^{\mu\nu}$ is its dual, $\tilde{F}^{\mu\nu} = \frac{1}{2}\epsilon^{\mu\nu\rho\sigma} F_{\rho\sigma}$, $\tilde{R}^{\mu\nu\rho\sigma}$ is the dual of the Riemann tensor, $\tilde{R}^{\mu\nu}{}_{\rho\sigma} = \frac{1}{2}\epsilon^{\mu\nu\lambda\tau} R_{\lambda\tau\rho\sigma}$ and $C_{\mu\nu\rho\sigma}$ is the conformal invariant Weyl tensor,

$$C_{\mu\nu\rho\sigma} = R_{\mu\nu\rho\sigma} - \frac{2}{d-2}(g_{\mu[\rho} R_{\sigma]\nu} - g_{\nu[\rho} R_{\sigma]\mu}) + \frac{2}{(d-1)(d-2)} R g_{\mu[\rho} g_{\sigma]\nu}. \tag{23.4}$$

In four dimensions,

$$\tilde{R}_{\mu\nu\rho\sigma} \tilde{R}^{\mu\nu\rho\sigma} = R_{\mu\nu\rho\sigma} R^{\mu\nu\rho\sigma} - 4 R_{\mu\nu} R^{\mu\nu} + R^2 = E_4 \tag{23.5}$$

is proportional to the Euler density (that integrates to a topological invariant), and

$$C_{\mu\nu\rho\sigma} C^{\mu\nu\rho\sigma} = R_{\mu\nu\rho\sigma} R^{\mu\nu\rho\sigma} - 2 R_{\mu\nu} R^{\mu\nu} + \frac{R^2}{3} = I_4 \tag{23.6}$$

is a conformal invariant combination.

The central charge c counts perturbative massless degrees of freedom in the conformal field theory, up to a normalization. Here the normalization is chosen such that $c = (N_c^2 - 1)/4$ for $SU(N_c)$ SYM.

The anomaly in the R-current J^μ has an $a-c$ contribution coming from triangle diagrams with two vertices being energy-momentum tensors and the third $\partial_\mu J^\mu$, and a $5a - 3c$ contribution coming from triangle diagrams with two vertices being J_μ and the third $\partial_\mu J^\mu$, as in Fig. 23.1a and b. Both are given by the chiral fermions running in the loop, as explained in Chapter 1. For $a - c$, since we have a single J_μ, the anomaly is proportional to $\sum_\chi R(\chi)$, where χ is a chiral fermion and $R(\chi)$ is its charge; for $5a - 3c$ the anomaly is proportional to $\sum_\chi R(\chi)^3$, since we have three J_μs. In the UV, the assignment of charge

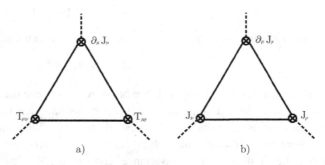

a)   b)

Figure 23.1   a) Anomalous diagram contributing to $a - c$; b) Anomalous diagram contributing to $5a - 3c$.

is $R(\lambda) = 1$ for the gaugino λ (fermion in the $\mathcal{N} = 1$ vector multiplet) and $R(\psi_i) = -1/3$ for the three fermions in the chiral multiplets Φ_i. This means that $\sum_{\chi=\lambda,\psi_i} R(\chi) = 0$ and $\sum_{\chi=\lambda,\psi_i} R(\chi)^3 = 8/9(N_c^2 - 1)$, giving

$$a_{\mathrm{UV}} - c_{\mathrm{UV}} = 0; \quad 5a_{\mathrm{UV}} - 3c_{\mathrm{UV}} \propto \frac{8}{9}(N_c^2 - 1). \tag{23.7}$$

In the IR, an effective charge assignment can also be calculated, with $R(\lambda) = 1$, $R(\psi_1, \psi_2) = -1/2$ and $R(\psi_3) = 0$, leading in a similar way to

$$a_{\mathrm{IR}} - c_{\mathrm{IR}} = 0; \quad 5a_{\mathrm{IR}} - 3c_{\mathrm{IR}} \propto \frac{3}{4}(N_c^2 - 1). \tag{23.8}$$

Finally, this means that

$$\frac{a_{\mathrm{IR}}}{a_{\mathrm{UV}}} = \frac{c_{\mathrm{IR}}}{c_{\mathrm{UV}}} = \frac{27}{32}. \tag{23.9}$$

23.2 c-theorem

We saw in the previous section that the central charges c and a in the IR are smaller than the ones in the UV. This is a general property, since the central charge c counts the number of massless degrees of freedom in the CFT (the number of ways that energy can be transmitted), but RG flow from UV to IR corresponds to coarse graining, when degrees of freedom can only get lost.

In two dimensions, there is a theorem due to Zamolodchikov called the c-theorem, which says that if we have an RG flow between two fixed points, there exists a *monotonically decreasing* function defined along the RG flow, that takes the value of c_{UV} in the UV and of c_{IR} in the IR, called the c-function. Here c appears in the trace anomaly (anomaly in conformal invariance) as

$$\langle T^{\mu}{}_{\mu} \rangle = -\frac{c}{12} R. \tag{23.10}$$

In four dimensions there is a similar statement proved recently by Komargodski and Schwimmer, based on earlier work of Myers and Sinha, and conjectured originally by Cardy, that goes under the name of a-theorem, since as we saw in four dimensions we have a and c charges appearing in the trace anomaly (23.3), and the statement applies to a. Moreover it can be proved constructively case by case using AdS/CFT. We will define an RG flow and explicitly construct a c-function.

The statement of Cardy applies in general dimension to the coefficient A of the term proportional to the Euler density $E_d = \tilde{R}_{\mu\nu\rho\sigma} \tilde{R}^{\mu\nu\rho\sigma}$ in

$$\langle T^{\mu}{}_{\mu} \rangle = -2(-)^{d/2} A E_d + \dots \tag{23.11}$$

23.3 Holographic RG flow and c-theorem; kink ansatz

Holographic Weyl anomaly and central charge

One can calculate the central charge of the conformal field theory from the holographic anomaly. One applies the formalism of holographic renormalization for the VEV (one-point function) of the trace of the energy-momentum tensor, using (22.16), to show that the VEV of the energy-momentum tensor is given by the variation of the action with respect to a Weyl transformation of the metric. In this case the field Φ is the metric variation itself, more precisely a Weyl transformation of the metric, $\delta g_{(0)} = 2\delta\sigma g_{(0)}$, and one considers the renormalized action with divergent part given by (22.11), normalized as

$$S_{\text{reg}} = (16\pi G_N^{(d+1)})^{-1} \int d^d x \sqrt{g_{(0)}}[\ldots + (-\log \epsilon)s_{(d)}] + S_{\text{finite}}, \tag{23.12}$$

with the divergence cancelled by the counterterms. Then the anomaly is given by the variation of the finite part,

$$\delta S_{\text{finite}} = -\int d^d x \sqrt{g_{(0)}}\delta\sigma \mathcal{A}, \tag{23.13}$$

and can be shown to be given by the coefficient of the log divergence (since the regulated action is invariant under constant $\delta\sigma$ for $\delta g_{(0)} = 2\sigma g_{(0)}; \delta\epsilon = 2\delta\sigma\epsilon$), leading to

$$\mathcal{A} = \frac{1}{16\pi G_N^{(d+1)}}(-2s_{(d)}). \tag{23.14}$$

Then using (22.16), we see that the conformal anomaly $\mathcal{A}$ is indeed related to the VEV of the energy-momentum tensor, leading to the 4-dimensional anomaly (for $a = c$)

$$\mathcal{A} = -\frac{a}{16\pi^2}(E_4 + I_4). \tag{23.15}$$

A holographic calculation of $s_{(d)}$ in (23.12) leading to a calculation of $a = c$ via (23.14), that will not be reproduced here, gives

$$a = c = \frac{\pi^2 R^3}{l_{P,5}^3} = \frac{\pi R^3}{8 G_N^{(5)}}. \tag{23.16}$$

Here $l_{P,5}^3 = 8\pi G_N^{(5)}$ (in general $l_{P,d}^{d-2} = 8\pi G_N^{(d)}$), and KK compactification of the gravitational action gives $G_{N,5} = G_N^{(10)}/R^5\Omega_5$ (the coefficient $1/(16\pi G_N)$ of the Einstein action gets multiplied by the volume of the compact space), where $\Omega_5 = \pi^3$ is the volume of the 5-sphere.

Kink ansatz for holographic RG flow and c-function

A solution corresponding to a holographic RG flow has to be of the kink type, with metric

$$ds^2 = e^{2A(r)}(-dt^2 + d\vec{x}_{d-1}^2) + dr^2 = e^{2A(r)}\eta_{\mu\nu}dx^\mu dx^\nu + dr^2, \tag{23.17}$$

and scalars $\phi_i = \phi_i(r)$, such that $A(r) = r/R, \phi_i = 0$ corresponds to AdS space (where we have renamed the coordinate usually called y as r). If the holographic flow is between two field theory fixed points, it should correspond to two approximately AdS points, so $A(r) \simeq r/R_1$ at the UV end (large r), and $A(r) \simeq r/R_2$ at the IR end (small r), with $R_2 < R_1$.

For this metric, the nonzero components of the Ricci tensor are

$$R_{\mu\nu} = e^{2A(r)}[A'' + d(A')^2]\eta_{\mu\nu}; \quad R_{rr} = -d[A'' + (A')^2]. \tag{23.18}$$

For a perfect fluid, the energy-momentum tensor is diagonal, and is $T_{\mu\nu} = \text{diag}(\rho, p_1, \ldots, p_{d-1})$, where p_i are pressures (all equal in a rotationally invariant system). But there is a condition believed to be satisfied by all quantum field theories, called the weakest energy condition, $\rho + p_i \geq 0$, which via the Einstein equations $R_{\mu\nu} - 1/2g_{\mu\nu}R = 8\pi G T_{\mu\nu}$ translates into

$$A'' \leq 0. \tag{23.19}$$

This means that the quantity

$$C(r) = a(r) = \frac{a_0}{(A')^{d-1}} \tag{23.20}$$

is monotonically decreasing (or rather, nonincreasing, since it can be constant) along the flow to the IR. The appropriate normalization in AdS_{d+1} that gives rise to (23.16) at the endpoints is

$$C(r) = a(r) = \frac{\pi^{d/2}}{\Gamma(d/2)(l_P^{(d+1)}A'(r))^{d-1}}, \tag{23.21}$$

which gives the c-function, or rather a-function, which takes the value of the central charge c, or rather a, at the UV and IR fixed points.

Note that we have defined the c-function only in the context of Einstein gravity in the dual, but one can define it with more generality.

23.4 Supersymmetric flow

In $\mathcal{N} = 4$ SYM, one can find an RG flow originating from the addition of the mass term described at the beginning of the chapter, that preserves $\mathcal{N} = 1$ supersymmetry along it and interpolates between the original $\mathcal{N} = 4$ SYM fixed point and the $\mathcal{N} = 1$ supersymmetric IR fixed point.

In AdS_5 supergravity, it should correspond to a kink solution interpolating between the original $\mathcal{N} = 8$ supersymmetric AdS_5 vacuum and another $\mathcal{N} = 2$ supersymmetric (1/4 susy) AdS_5 vacuum. Then the kink must be supersymmetric, i.e. invariant under supersymmetry, which means that the variation of the fermion $\delta_{\text{susy}}\psi$ must vanish (the variation of the bosons is fermionic, which is automatically zero classically, since fermions do not have VEVs because of Lorentz invariance).

The potential for the scalars is derived from a superpotential W. Since we are in a supergravity theory, we also have a term proportional to $|W|^2$, with the result

$$V = \frac{9}{8} \sum_j \left| \frac{\partial W}{\partial \phi_j} \right|^2 - 3l_P^2 |W|^2. \tag{23.22}$$

From the gravitino variation $\delta \psi_\mu^a = 0$ and the spin 1/2 variation $\delta \chi^{abc} = 0$, one finds the condition

$$A' = -l_P^2 W; \quad \frac{d\phi_i}{dr} = \frac{3}{2} l_P^2 \frac{\partial W}{\partial \phi_i}. \tag{23.23}$$

Then the c-function (or a-function) along the RG flow is (substituting $A' = W$ in the general formula)

$$C(r) = a(r) = \frac{\pi^2}{(l_P^{(5)})^9 W^3}, \tag{23.24}$$

i.e. it is given by the value of W. At the AdS vacua, one needs only to compute W to find the central charges $a = c$. For the particular flow of interest, one can analyze the $\mathcal{N} = 8$ supergravity potential minima, and from them find W at the corresponding $\mathcal{N} = 2$ minimum, obtaining indeed that $a(0)/a(\infty) = 27/32$, as expected from the field theory.

Important concepts to remember

- RG flows between fixed points in field theory correspond to kink solutions depending on the radial coordinate r in supergravity, interpolating between different AdS vacua.
- In four dimensions, the anomaly structure is governed by two central charges a and c, which are equal in $\mathcal{N} = 4$ SYM.
- An $\mathcal{N} = 1$ mass deformation of $\mathcal{N} = 4$ SYM triggers an $\mathcal{N} = 1$ RG flow that ends in an IR fixed point.
- The charge a always decreases along the RG flow from UV to IR, similarly to c decreasing in two dimensions.
- In $d = 2$ there is a c-function along the RG flow that is monotonically decreasing and takes the values c_{UV} in the UV and c_{IR} in the IR. In $d = 4$ there is a similar theorem for a.
- One can write a kink ansatz $ds^2 = e^{2A(r)}(-dt^2 + d\vec{x}^2) + dr^2$, and then $A'' \leq 0$, meaning that the c-function is proportional to $1/(A')^{d-1}$ for AdS_{d+1}.

References and further reading

The RG flow described here was found in [88]. The a-theorem was proven in [90], based on earlier work in [89]. The central charge c was calculated in [87]. For more on the holographic c-theorem (i.e., from AdS/CFT) in more general gravity contexts see, e.g. [91].

Exercises

1. Show that the Weyl tensor for AdS is zero.
2. Calculate $\langle T^\mu_\mu \rangle$ for AdS_{d+1} in Poincaré coordinates with radius R.
3. Show that the integral of E_4 in four dimensions is topologically invariant (independent of the metric).
4. Calculate $a_{UV} - c_{UV}$ and $5a_{UV} - 3c_{UV}$ for the conformal field theory corresponding to the $\mathcal{N} = 2$ orientifold of $AdS_5 \times S^5$ in Section 19.2.
5. Prove that the components of the Ricci tensor for the metric (23.17) are given by (23.18).
6. Consider the superpotential

$$W = \frac{1}{4\rho^2} \left[\cos 2\phi_1 (\rho^6 - 2) - (3\rho^6 + 2) \right], \tag{23.25}$$

and $\rho = e^\alpha$. Find the extrema of W in the (α, ϕ_1) plane.

24

Phenomenological gauge–gravity duality I: AdS/QCD

Until now we have dealt with AdS/CFT, or gauge–gravity duality, derived from string theory, even if the derivation did not have mathematical rigor. We had a system of branes in the decoupling limit giving rise to a gravitational background dual to a well-defined field theory. In the language of fundamental string theory (string theory as a unified theory) this is a "top-down" approach: we start with a well-defined string theory system and we see what kind of gravity dual pair we get, looking for something close to QCD.

But another approach is possible, a phenomenological one or "bottom-up." That is, we *assume* that there should be a gravity dual of QCD and see if we can construct it using the properties of QCD. We need to make educated guesses about which properties should appear in the gravity dual, and there is no guarantee that there actually exists a system of branes leading to this dual pair. But the approach has the advantage that it can be taken to be a phenomenological model for QCD that encodes some of its properties in a gravitational theory, and is only approximate, on a par with other phenomenological models for QCD.

In this chapter we follow a natural evolution for three models: first an extension of the "hard-wall" model of Polchinski and Strassler, then a version with a smooth cut-off called the "soft-wall" model, and then a perturbative method to find gravity duals called improved holographic QCD.

24.1 Extended "hard-wall" model for QCD

We have already described the basic "hard-wall" model for QCD of Polchinski and Strassler. It is an AdS_5 space cut off in the IR at $r = r_{\min} = R^2\Lambda$, or in terms of $z = R^2/r$, at $z_m = 1/\Lambda$. An improved version for it was presented, where the IR cut-off is considered as a fluctuating brane, whose scalar mode (brane position) corresponds to a pion.

But one can extend the model further, as was done by Erlich, Katz, Son, and Stephanov (EKSS). We can add two gauge fields $A_{L\mu}^a$, $A_{R\mu}^a$, coupling to the left- and right-handed currents for an $SU(N_f)_L \times SU(N_f)_R$ chiral flavor symmetry, $\bar{q}_L\gamma^\mu T^a q_L$ and $\bar{q}_R\gamma^\mu R^a q_R$, as well as a bifundamental tachyonic scalar field $X^{\alpha\beta}$ of $m^2 R^2 = -3$ (which is OK in AdS space, since it is above the Breitenlohner–Freedman bound, which is the only requirement) coupling to the chiral order parameter $\bar{q}_R^\alpha q_L^\beta$ (α is in the fundamental of $SU(N_f)_R$ and β in the fundamental of $SU(N_f)_L$).

The relevant action is

$$S = \int d^5 x \sqrt{-g} \mathrm{Tr} \left[-|D_\mu X|^2 + 3|X|^2 - \frac{1}{4g_5^2}(F_{L\mu\nu}^2 + F_{R\mu\nu}^2) \right], \qquad (24.1)$$

where $D_\mu X = \partial_\mu - iA_{L\mu}X + iXA_{R\mu}$, $F_{\mu\nu} = \partial_\mu A_\nu - \partial_\nu A_\mu - i[A_\mu, A_\nu]$, $A_{L,R\mu} = A_{L,R\mu}^a T^a$.

At the IR brane one needs to impose boundary conditions. For $A_{L\mu}, A_{R\mu}$, we need gauge invariant boundary conditions, the simplest being $(F_L)_{z\mu} = (F_R)_{z\mu} = 0$, but others being possible (EKSS proved that they have little effect on physical results). As we saw in Chapter 11, we can choose the gauge $A_z = 0$, which implies that the boundary condition is simply Neumann, $\partial_z A_{L,R\mu} = 0$. In the UV, $A_{L\mu}^a$ and $A_{R\mu}^a$ go to the sources $a_{L,R\mu}^a$ for the chiral currents $J_{L,R}^a\mu$.

For the scalar X, there should be a boundary condition in the IR too, which will fully determine the solution for X together with the UV boundary condition. Since the chiral order parameter $\mathcal{O}_X = \bar{q}_R^\alpha a_L^\beta$ has dimension $\Delta = 3$ in $d = 4$, by the general discussion from Chapters 11, 12, and 22, we expect that the scalar boundary condition at $z = 0$ is

$$X \to z^{d-\Delta}(X_0 + z^{2\Delta - d}X_{(2\Delta - d)}) = zX_0 + z^3 X_{(2)}, \qquad (24.2)$$

where X_0 is the source for the chiral order parameter operator $\mathcal{O}_X$, interpreted as a quark mass matrix $M^{\alpha\beta}/2$ (adding the term $X_0^{\alpha\beta}\bar{q}_R^\alpha q_L^\beta$ to the Lagrangean corresponds indeed to adding a quark mass matrix). The simplest case is to assume that $M = m_q \mathbb{1}$. On the other hand, $X_{(2)} \equiv \Sigma/2$ gives the VEV of the operator (1-point function), i.e. $\Sigma^{\alpha\beta} = \langle \bar{q}_R^\alpha q_L^\beta \rangle$, and again the simplest assumption is $\Sigma = \sigma \mathbb{1}$. The boundary condition is thus

$$X(z) \to \frac{1}{2}Mz + \frac{1}{2}\Sigma z^3. \qquad (24.3)$$

This extended hard-wall model then has four parameters m_q, σ, z_m, and g_5 and three fields, $A_{L\mu}, A_{R\mu}$, and X. One can introduce a vector field $V_\mu = (A_{L\mu} + A_{R\mu})/2$ and an axial vector field $A_\mu = (A_{L\mu} - A_{R\mu})/2$, coupling to the vector current $\bar{q}\gamma_\mu T^a q$ and axial vector current, respectively.

2-point function of currents and gauge coupling

The vector field in the gauge $V_z(\vec{x}, z) = 0$ from the action (24.1) has the equation of motion for the transverse Fourier components $V_\mu^a(\vec{q}, z)$

$$\left[\partial_z \left(\frac{1}{z} \partial_z V_\mu^a(\vec{q}, z) \right) + \frac{\vec{q}^2}{z} V_\mu^a(\vec{q}, z) \right] = 0. \qquad (24.4)$$

Denoting the Fourier transform of the boundary source coupling to the vector current by $V_{0\mu}(\vec{q})$, we have

$$V_\mu(\vec{q}, z) = V(\vec{q}, z)V_{0\mu}(\vec{q}); \quad V(\vec{q}, z = \epsilon) = 1. \qquad (24.5)$$

Plugging in the equation of motion and expanding around $z = 0$, we obtain in the UV

$$V(Q, z) = 1 + \frac{Q^2 z^2}{4} \ln(Q^2 z^2) + \dots, \qquad (24.6)$$

where $Q^2 = \vec{q}^2$. Of course, the full solutions will be quantized, since they need also to satisfy the IR boundary condition (Neumann), and as we argued in previous chapters, a cut-off in the IR region leads to a discrete spectrum for the solution to the equation of motion in AdS.

One can calculate the 2-point function of vector currents as in Chapter 11: for the on-shell action, only a boundary term remains, and the 2-point function is the second derivative of the action with respect to the source $V_{0\mu}$,

$$\langle J_\mu^a(x) J_\nu^b(0) \rangle = \frac{\delta^2 S_{\text{sugra}}}{\delta V_{0\mu}^a(x) \delta V_{0\nu}^b(0)}$$

$$= -\frac{1}{2g_5^2} \frac{\delta^2}{\delta V_{0\mu}^a(x) \delta V_{0\nu}^b(0)} \int_{z=\epsilon} d^4x \left(\frac{1}{z} V_\mu^a \partial_z V^{\mu a} \right) \Rightarrow$$

$$\int d^4x e^{i\vec{q}\cdot\vec{x}} \langle J_\mu^a(x) J_\nu^b(0) \rangle = \delta^{ab} (q_\mu q_\nu - \vec{q}^2 g_{\mu\nu}) \Pi_V(Q^2),$$

$$\Pi_V(Q^2) = -\frac{1}{g_5^2 Q^2} \left. \frac{\partial_z V(\vec{q}, z)}{z} \right|_{z=\epsilon}. \tag{24.7}$$

From the solution for $V(Q, z)$ above, we get

$$\Pi_V(Q^2) = -\frac{1}{2g_5^2} \ln Q^2, \tag{24.8}$$

and by comparing with the perturbative QCD result $\Pi_V(Q^2) = -\frac{N_c}{24\pi^2} \ln Q^2$, one can fix

$$g_5^2 = \frac{12\pi^2}{N_c}. \tag{24.9}$$

Decay constants

The solutions to (24.4) with boundary condition $\psi_n(z = \epsilon) = 0$ and $\partial_z \psi_n(z_m) = 0$ are quantized (discrete), and correspond to a tower of meson states associated with the vector current operator, as explained in Chapter 21. Since the field is a vector, the tower is of vector mesons, specifically of ρ vector mesons. Consider the wavefunctions $\psi_n(z)$ for the tower of solutions (labelled by n, with $\vec{q}^2 = -m_n^2$) corresponding to some component μ of V_μ. As usual, the Green's function for that component μ is written in terms of the eigenmodes as

$$G(\vec{q}; z, z') = \sum_n \frac{\psi_n(z) \psi_n(z')}{\vec{q}^2 - m_n^2 + i\epsilon}. \tag{24.10}$$

Since the solution to (24.5) is written in terms of the scalar Green's function G as $V_\mu(\vec{q}, z) = -1/z' \partial_{z'} G(q; z, z')_{z'=\epsilon} V_{0\mu}(\vec{q})$, we can obtain the function $V(\vec{q}, z)$, and from it we get

$$\Pi_V(\vec{q}^2) = -\frac{1}{g_5^2} \sum_n \frac{|\psi'(\epsilon)/\epsilon|^2}{(\vec{q}^2 - m_n^2 + i\epsilon) m_n^2}. \tag{24.11}$$

The decay constants of the vector meson ρ_n with polarization vector ϵ_μ, defined by

$$\langle 0|J_\mu^a|\rho_n^b\rangle = F_n \delta^{ab}\epsilon_\mu, \qquad (24.12)$$

are calculated from the 2-point function of currents. Indeed, considering

$$\sum_n \langle 0|J_\mu^a(q)|\rho_n^b\rangle \frac{1}{m_n^2(\vec{q}^2 - m_n^2 + i\epsilon)} \langle \rho_n^b|J_\mu^a(-q)|0\rangle$$

$$= \frac{1}{q^2}\langle 0|J_\mu^a(q)J_\mu^a(-q)|0\rangle = -3\delta_a^a \Pi_v(q^2)$$

$$= \delta_a^a \epsilon_\mu \epsilon^\mu \sum_n \frac{F_n^2}{m_n^2(\vec{q}^2 - m_n^2 + i\epsilon)}, \qquad (24.13)$$

and equating the two sides, we obtain

$$F_n^2 = \frac{1}{g_5^2}[\psi'(\epsilon)/\epsilon]^2 \rightarrow \frac{1}{g_5^2}[\psi_n''(0)]^2. \qquad (24.14)$$

One can calculate various other quantities like masses, couplings, and decay constants of fields (we will not show them here), like m_ρ, m_π, and f_π, and fit the three remaining parameters, z_m, σ, m_q, after which we get predictions for all the other quantities.

We have seen the method to calculate masses (from eigenmodes of equations of motion of fields in the gravity dual) and decay constants (from current 2-point functions). Couplings between fields are found from the nonlinear terms in the bulk action, integrating over z the wavefunctions in z of the corresponding fields (for example, the ρ_n particle has the wavefunction $\psi_n(z)$).

24.2 "Soft-wall" model for QCD

Next, we consider modifying the background geometry so as to obtain expected properties of the QCD mass spectrum, in order to obtain the so-called "soft-wall" model. In the cases similar to the hard-wall model, the spectrum of modes with high excitation number, $n \gg 1$, is $m_n^2 \propto n^2$ (for instance, we saw in Section 21.6 that the Witten model has $m^2 R^2 = 6n(n+1) \rightarrow 6n^2$). But from QCD data we want $m_n^2 \propto n$, or more precisely $m_n^2 \sim \sigma n$ (with σ the QCD string tension) instead. Similarly, for high spin $S \gg 1$, the picture of a semiclassically rotating relativistic QCD string predicts $m_S^2 = 2\pi\sigma S$.

Therefore, one wants to modify the gravity dual in order to have

$$m_n^2 \sim \sigma n; \quad m_S^2 \sim \sigma S. \qquad (24.15)$$

But the modification is not a UV one, despite the energy (mass) of the state being high, but rather it is from modification of the IR. One way to understand this is that the length of the excited meson is $L \sim m_n/\sigma$, since the energy (mass) of the state with a flux tube (QCD string) between quarks is a linear function of length, $E = \sigma L$. This means that the high mass states are extended, i.e. governed by the IR. Another way to see this is from the analysis of the gravity dual. With just AdS space, there is a continuum of eigenmodes, since the time of flight between the boundary at infinity and the center of AdS is finite. The discrete nature of the spectrum, as in the case of the 1-dimensional quantum mechanical box, comes from the fact that we need also to impose boundary conditions at a finite IR cut-off. Then the precise nature of the way the space is cut off in the IR leads to the discretization of the spectrum, i.e. the behavior of $m_n^2 = m_n^2(n)$.

Then we look for a gravity dual that cuts off smoothly in the IR (hence the name "soft-wall"), and we consider nontrivial metric g_{MN} and dilaton Φ (since, as we saw in all the top-down examples, these were the relevant fields). The space depends only on the radial coordinate for simplicity, as in the domain wall, or "kink" ansatz in the last chapter. To parameterize the solution, we can use a "conformal" coordinate system, with $\Phi(z)$ and metric

$$ds^2 = g_{MN}dx^M dx^N = e^{2A(z)}(\eta_{\mu\nu}dx^\mu dx^\nu + dz^2), \qquad (24.16)$$

but we can also use a "domain wall" (or kink) coordinate system as in the previous chapter,

$$ds^2 = e^{2A(u)}\eta_{\mu\nu}dx^\mu dx^\nu + du^2, \qquad (24.17)$$

the two being related by $du = e^{A(z)}dz$.

Similarly to the case of the top-down models of Chapter 21, we see soon that for the mass spectrum coming from the KG equation (excited states of a scalar) the relevant combination is $\Phi(z) - A(z)$, and not $A(z)$ by itself. To have an asymptotically AdS space in the UV ($z \to 0$), as needed, we would expect $-A(z) \sim \log z$, but we rather need $\Phi(z) - A(z) \sim \log z$, as we saw in the examples of Chapter 21. On the other hand, in the IR (large z), we shortly see that we need $\Phi(z) - A(z) \sim z^2$. The simplest solution to both these conditions is $\Phi(z) - A(z) = z^2 + \log z$.

The action we take in the bulk of the gravity theory needs to have the same fields $A_{L\mu}, A_{R\mu}, X$ as in the extended hard-wall model, and with the same action, but now we need to introduce Φ in the background also, so we write

$$S = \int d^5x \sqrt{-g} e^{-\Phi(z)} \text{Tr} \left[-|D_\mu X|^2 + 3|X|^2 - \frac{1}{4g_5^2}(F_{L\mu\nu}^2 + F_{R\mu\nu}^2) \right] \qquad (24.18)$$

instead of (24.1). The calculation of the gauge coupling from the 2-point function of currents in the previous subsection depended only on the UV behavior of the gravity dual, hence is unmodified. Therefore we can still identify g_5 as in (24.9), i.e. $g_5^2 = 12\pi^2/N_c$. The boundary conditions in the UV are the same (as needed for holography), and the boundary conditions in the IR are that the on-shell action is finite. The ambiguity in the choice of IR boundary conditions noted in the case of the hard wall is lifted in the case of the soft wall.

For the vector field V_μ, in the axial gauge $V_z = 0$ as in the previous section, the transverse components $\partial^\mu V_\mu = 0$ have normalizable solutions $V_n(z)$ only for discrete values of $\vec{q}^2 = -m_n^2$, satisfying a generalization of (24.4), namely

$$\partial_z(e^{-B}\partial_z V_n) + m_n^2 e^{-B} V_n = 0, \qquad (24.19)$$

where $B(z) = \Phi(z) - A(z)$. Therefore, as expected, only the combination B matters for the spectrum m_n. With the field redefinition

$$V_n = e^{B/2}\psi_n, \qquad (24.20)$$

we obtain the Schrödinger equation

$$-\psi_n'' + V(z)\psi_n = m_n^2\psi_n, \qquad (24.21)$$

where the "potential" is

$$V(z) = \frac{1}{4}(B')^2 - \frac{1}{2}B''. \qquad (24.22)$$

To obtain energies $E_n \equiv m_n^2 \propto n$ at large n, as in the case of a harmonic oscillator, we need that an approximately harmonic oscillator potential, $V(z) \propto z^2$ at large z. This in turn requires that $B = \Phi(z) - A(z) \sim z^2$ at large z, as we have indicated.

In the case of $B = z^2/z_m^2 + \log z$, we have the potential $V(z) = z^2/z_m^4 + 3/4z^2$, and the Schrödinger equation

$$-\psi'' + \left[\frac{z^2}{z_m^4} + \frac{3}{4z^2}\right] = E\psi \qquad (24.23)$$

has eigenvalues

$$m_n^2 z_m^2 = E_n = 4(n+1). \qquad (24.24)$$

We see then that the QCD string tension is $\sigma \propto 1/z_m^2$. The wave function corresponding to these eigenvalues is

$$V_n(z) = z^2\sqrt{\frac{2n!}{(n+1)!}}L_n^1(z^2), \qquad (24.25)$$

where L_n^m are associated Laguerre polynomials.

One can also calculate the decay constants through (24.14), which now reads

$$F_n^2 = \frac{1}{g_5^2}[V_n''(0)]^2 = \frac{8(n+1)}{g_5^2}. \qquad (24.26)$$

But until now only $\Phi - A$ was fixed, not Φ and A individually. For that, one needs to look at the spectrum of mesons with higher spin ($S > 2$). For theories with higher spin, interactions are hard to describe. But the kinetic terms of higher spin fields in AdS space, the only thing needed for calculation of the mass spectrum, are easy to write down. The fields are $\phi_{M_1...M_S}$, totally symmetric of rank S, with gauge invariance

$$\delta\phi_{M_1..M_S} = D_{(M_1}\xi_{M_2...M_S)}. \qquad (24.27)$$

Choosing the gauge $\phi_{z...} = 0$, there is still the residual gauge invariance with ξ satisfying $\xi_{z...} = 0$ and

$$\delta\phi_{z...} = D_z\xi_{...} + D_{(.}\xi_{...z)} = \xi_{...}' - 2(S-1)A'\xi_{...} = 0, \qquad (24.28)$$

which implies

$$\xi_{\mu_1...\mu_{S-1}}(z, x^\mu) = e^{2(S-1)A(z)}\tilde{\xi}_{\mu_1...\mu_{S-1}}. \qquad (24.29)$$

Redefining the field as

$$\phi_{...} = e^{2(S-1)A(z)}\tilde{\phi}_{...}, \tag{24.30}$$

the residual gauge invariance is

$$\delta\tilde{\phi}_{...} = \partial_{(.}\tilde{\xi}_{...)}. \tag{24.31}$$

The kinetic action

$$S = -\frac{1}{2}\int d^5x\sqrt{-g}\,e^{-\Phi}[D_N\phi_{M_1...M_S}D^N\phi^{M_1...M_S} + M^2(z)\phi_{M_1...M_S}\phi^{M_1...M_S}], \tag{24.32}$$

where $M(z)$ is a fixed function allowing for gauge invariance, reduces now to (replacing ϕ and $\sqrt{-g}$)

$$S = -\frac{1}{2}\int d^5x\,e^{-\Phi}[e^{(2S-1)A}\partial_N\tilde{\phi}_{\mu_1..\mu_S}\partial^N\tilde{\phi}_{\mu_1...\mu_S}], \tag{24.33}$$

which is explicitly gauge invariant (and without the need to explicitly define $M(z)$).

From this, the equation of motion for $\tilde{\phi}_n$, a transverse traceless mode of $\tilde{\phi}_{...}$, is the same as (24.19), just with

$$B = \Phi - (2S-1)A. \tag{24.34}$$

Again we need $m_n^2 \propto n$ at large n, which means that $V(z) \propto z^2$ at large z, which in turn means that $B \propto z^2$ at large z. But now moreover this has to happen independently of S, which restricts the relations to $\Phi \propto z^2$ at large z and $A \propto -\log z$ at small z. The simplest choice is $A = -\log z$ and $\Phi = z^2/z_m^2$, for which the "potential" is

$$V(z) = \frac{z^2}{z_m^4} + \frac{2(S-1)}{z_m^2} + \frac{S^2 - 1/4}{z^2}, \tag{24.35}$$

which has eigenenergies

$$E_n \equiv m_{n,S}^2 z_m^2 = 4(n+S). \tag{24.36}$$

In conclusion, by imposing that the spectrum of vector mesons has the QCD behavior $m_n^2 \propto n$ we have found $\Phi - A = z^2/z_m^2 + \log z$, and by also imposing the same for the higher spin mesons we have also fixed $\Phi \sim z^2/z_m^2$ and $A \sim -\log z$.

24.3 Improved holographic QCD

The hard-wall and soft-wall models phenomenologically introduced fields corresponding to important operators; a QCD scale, quark mass, string tension, and chiral condensate, and correct behavior for the tower of mesonic states.

But one would also like to obtain a running coupling constant that agrees with the beta function of QCD. We can do this by engineering a scalar potential in the gravity dual that holographically gives the desired running beta function. Note, however, that we want the running beta function at strong coupling (which is not known theoretically), so we need an *ansatz* for the exact form we want to reproduce.

In the conformal coordinate system of the last subsection, in the UV, where $A(z) \simeq -\log z$, z is related to the energy scale as usual by $z = 1/E$. Also in the UV, $du = e^{A(z)}dz$ means that $u \simeq \log z$, so $du = -d \log E$. The 't Hooft coupling constant is $\lambda = g_{YM}^2 N_c = N_c e^{\Phi}$. Its beta function is then generically

$$\mu \frac{d\lambda}{d\mu} = -\frac{d\lambda}{d \log z} = \beta(\lambda) = -b_0 \lambda^2 + b_1 \lambda^3 + b_2 \lambda^4 + \dots \qquad (24.37)$$

From this, one obtains by integrating that

$$\frac{1}{\lambda} \equiv \alpha_s = L - \frac{b_1}{b_0} \log L + \frac{b_1^2}{b_0^2} \frac{\log L}{L} + \mathcal{O}\left(\frac{1}{L^2}\right), \qquad (24.38)$$

where

$$L \equiv -b_0 \log(z\Lambda), \qquad (24.39)$$

and Λ is the RG invariant scale of QCD. We will denote by prime a derivative with respect to $-d(\log z)$, corresponding to $d \log \mu$.

From this beta function we want to obtain a bulk potential for the dilaton, or rather for $\lambda = N_c e^{\Phi}$, as the Taylor expansion

$$V(\lambda) = \sum_{n=0}^{\infty} V_n \lambda^n, \qquad (24.40)$$

corresponding to a perturbative SYM expansion. Substituting λ from (24.38) and expanding in L, one obtains

$$V = V_0 + \frac{V_1}{L} + \frac{V_2}{L^2} + \frac{b_1}{b_0} V_1 \frac{\log L}{L^2} + \mathcal{O}\left(\frac{1}{L^3}\right). \qquad (24.41)$$

Now to fix the coefficients V_i of the potential in terms of the coefficients b_i of the beta function, we need to use a certain form of Einstein's equations. Starting from the string theory action for the supergravity fields, coupled to sources coming from N_f effective $D4 - \bar{D}4$-brane pairs (higher Dp-branes, for instance $D7 - \bar{D}7$-branes, wrapped on the compact space we are reducing on) giving flavors ("quarks") as in the Sakai–Sugimoto model, we can integrate out the 5-form field strength F_5 and put the axion a to zero, along with other fields, to obtain an action for the gravity plus dilaton system in the Einstein frame

$$S = M_{Pl,5}^3 N_c^2 \int d^5 x \sqrt{-g} \left[R - \frac{4}{3} \frac{(\partial_\mu \lambda)^2}{\lambda^2} - V(\lambda) \right]. \qquad (24.42)$$

The exact form of $V(\lambda)$ is not important, since we want to substitute it with the parameterization that will match the beta function. The rest of the action is easy to understand. The $M_{Pl,5}^3$ factor is the standard gravitational coupling. In the string frame, we have normally $e^{-2\Phi} = N_c^2/\lambda^2$ in front of R, and when going to the Einstein frame only the λ^{-2} factor is removed, leaving an N_c^2. The factor of 4/3 arises when going from the string frame Lagrangean $\sqrt{-g}e^{-2\Phi}(R + 4(\partial_\mu \Phi)^2)$ to the Einstein frame by the transformation $g_{\mu\nu} = \lambda^{4/3} g_{\mu\nu}^E$.

The equations of motion of this action, in terms of Φ, are found to be

$$R_{\mu\nu} - \frac{1}{2}g_{\mu\nu}R - \frac{4}{3}\left[\partial_\mu\Phi\partial^\mu\Phi - \frac{1}{2}(\partial_\rho\Phi)^2 g_{\mu\nu}\right] - \frac{1}{2}g_{\mu\nu}V = 0$$

$$\Box\Phi + \frac{3}{8}\frac{dV(\Phi)}{d\Phi} = 0, \qquad (24.43)$$

and for our ansatz in conformal coordinates they become

$$12\dot{A}^2 - \frac{4}{3}\dot{\Phi}^2 - e^{2A}V = 0; \qquad 6\ddot{A} + 6\dot{A}^2 + \frac{4}{3}\dot{\Phi}^2 - e^{2A}V = 0,$$

$$\ddot{\Phi} + 3\dot{A}\dot{\Phi} + \frac{3}{8}e^{2A}\frac{dV}{d\phi} = 0, \qquad (24.44)$$

where dot refers to d/dz. These equations are rewritten in terms of $\lambda = N_c e^\Phi$, in which we impose that it has the expansion (24.38) in L. Also V has an expansion in L (24.41), and the compatibility of the equation of motion expanded in L leads (after some very long algebra) to the identification of the potential coefficients V_i and of the metric coefficient e^{2A} expanded in the same L as

$$V_1 = \frac{8}{9}b_0 V_0;$$

$$V_2 = \frac{23b_0^2 - 36b_1}{3^4}V_0 \Rightarrow V = V_0\left(1 + \frac{8}{9}b_0\lambda + \frac{23b_0^2 - 36b_1}{3^4}\lambda^2\right) + \mathcal{O}(\lambda^3),$$

$$ds^2 = \left[1 + \frac{8}{3^2\log(z\Lambda)} + \frac{4\left(26 + 9\frac{b_1}{b_0^2} - 18\frac{b_1}{b_0^2}\log(b_0\log\frac{1}{z\Lambda})\right)}{3^4\log^2(z\Lambda)}\right.$$

$$\left. + \mathcal{O}\left(\frac{\log^2\log(z\Lambda)}{\log^3(z\Lambda)}\right)\right]\frac{R^2}{z^2}(dz^2 + d\vec{x}^2). \qquad (24.45)$$

Note that V_0, the cosmological constant, must be related to the AdS radius as usual, so $V_0 = 12/R^2$. This solution is given as an expansion in the UV of the theory (small z), as it should, since the beta function was defined perturbatively in the UV.

Therefore, in this improved holographic QCD method we have obtained a metric and dilaton potential for the gravity dual by imposing matching with the beta function of QCD. This was enough for the UV asymptotics, but what we are mostly interested in, in order to obtain the relevant nonperturbative, low energy physics of QCD, is the IR of the solution. But as we have already mentioned, the exact beta function of QCD (needed in order to apply it at large 't Hooft coupling λ as is needed for AdS/CFT in the IR) is of course not known, so we can only propose various forms for $\beta(\lambda)$ that agree with some given order in the expansion in the UV (in the above, up to b_1), and use Einstein's equations in the bulk to solve for the whole solution. This, however, imposes an additional layer of phenomenology, besides the AdS/CFT phenomenology, since we now use phenomenology for QCD in order to define $\beta(\lambda)$. Therefore this method is more useful for the UV expansion of the solution, which we explained above.

Important concepts to remember

- AdS/QCD is a "bottom-up" approach, of engineering a would-be gravity dual using properties of QCD, encoded in a gravitational theory.

- In the extended hard-wall model, one introduces gauge fields $A_{L\mu}, A_{R\mu}$ coupling to the chiral currents of a chiral $SU(N_f)_L \times SU(N_f)_R$ flavor symmetry, and a scalar coupling to the bifundamental chiral order parameter $\bar{q}^\alpha q^\beta$ in the bulk of the cut-off AdS_5 with a fluctuating IR cut-off ("IR brane").

- Masses of QCD states are calculated from eigenmodes of the equations of motion of fields in the gravity dual, decay constants from the 2-point function of currents, and couplings from the nonlinear terms in the bulk action, integrating over the wavefunctions in the fifth dimension z.

- One fixes the free parameters of the model (g_5, z_m, σ, m_q) from four quantities, then the other calculations give predictions.

- The "soft-wall" model modifies the IR of the gravity dual to match the observed spectrum of QCD excited states, which have $m_n^2 \propto n$ at large n for $S \leq 1$ mesons and for higher spin mesons $m_S^2 \propto S$ at large spin S, with the constant of proportionality being (up to a number) the QCD string tension.

- One takes a "domain wall" or "kink" ansatz for the dilaton and metric, depending only on a factor $e^{2A(z)}$, with z being the fifth coordinate.

- The requirement to have $m_n^2 \propto n$ at large n implies that $\Phi - A \propto z^2$ at large z (IR) (and $\Phi - A \propto -\log z$ at small z, in the UV).

- The requirement that $m_S^2 \propto S$ at large S also implies that $\Phi \propto z^2$ at large z and $A \sim -\log z$ at small z, solved by $\Phi = z^2/z_m^2$ and $A = -\log z$, when $m_{n,S}^2 = 4(n + S)/z_m^2$.

- In improved holographic QCD, from the known perturbative beta function of QCD, written as $\beta(\lambda)$, where $\lambda = N_c e^\Phi$, one extracts the perturbative form of the dilaton potential in the gravity dual $V(\lambda) = \sum_n V_n \lambda^n$, and of a corresponding expansion of the metric e^{2A} as an expansion in the UV in $\log(z\Lambda)$.

- If we also want an expansion in the IR, in order to describe low energy physics, we need to make an ansatz for the exact beta function $\beta(\lambda)$ in QCD, and use it, together with Einstein's equations, to determine the gravity dual.

References and further reading

The extended "hard-wall" model was introduced in [92] and the "soft-wall model" of AdS/QCD was introduced in [93]. The improved holographic QCD model was introduced in [94].

Exercises

1. Calculate the mass of the pion, m_π, in the hard-wall model.
2. Calculate the $\rho - \pi - \pi$ coupling in the hard-wall model.
3. For the soft-wall model with $A = -\log z$ and $\Phi = z^2/z_m^2$, calculate the $\rho - \rho - \pi - \pi$ coupling.
4. For the soft-wall model with $A = -\log z$ and $\Phi = z^2/z_m^2$, calculate the mass of the pion.
5. Prove that the equations of motion (24.43) reduce on the conformal coordinates (24.16) to (24.44).
6. Propose an ansatz for $\beta(\lambda)$ that agrees with b_0 and b_1 and can be used at large λ.

25

Phenomenological gauge–gravity duality II: AdS/CMT

In this chapter we study another example of a phenomenological approach to AdS/CFT, namely for condensed matter systems. Strongly coupled condensed matter systems are hard to analyze, and there are few models for them, even phenomenological ones, hence gravity dual phenomenological models are very useful. Contrary to the application of the previous chapter, where AdS/CFT relates in principle a gravity dual to a gauge theory, even though we do not know how to obtain QCD exactly on a system of branes with a gravity dual, in the present case it is not even clear why there should be a nonabelian gauge theory. What is certain is that we expect to have certain operators and symmetries, and based on that we want to construct a would-be gravity dual with fields dual to the operators and with the local versions of the symmetries.

We can start from a well-defined gravity dual in string theory, and try to find a reason to apply it for a condensed matter system, even though the dual field theory is not necessarily related to the system of interest (but operators in it might describe relevant physics). Or we can engineer a would-be gravity dual, with some symmetry and fields corresponding to the operators and symmetries of interest in the condensed matter system, in which case it is not clear that AdS/CFT should be applicable, as in the last chapter.

25.1 Lifshitz, Galilean, and Schrödinger symmetries and their gravity duals

Gravitational dual of Lifshitz points

The first example we start with is of the last type, looking to engineer a would-be gravity dual with the right properties. Most condensed matter systems are nonrelativistic, so in this section we learn how to deal with some nonrelativistic systems.

In condensed matter systems, near a phase transition we have fixed points that sometimes can exhibit so-called "dynamical scaling", or "Lifshitz scaling," in which case we call them Lifshitz points. Instead of the usual relativistic scaling $t \to \lambda t$, $\vec{x} \to \lambda \vec{x}$, at the Lifshitz points we have the scaling

$$t \to \lambda^z t; \quad \vec{x} \to \lambda \vec{x}. \tag{25.1}$$

Here z is called the *dynamical critical exponent*.

A model example for $z = 2$ is the Lifshitz field theory, with Lagrangean

$$\mathcal{L} = \int d^2x \, dt \left[(\partial_t \phi)^2 - k(\vec{\nabla}^2 \phi)^2 \right]. \tag{25.2}$$

It arises in some multicritical points of known materials.

To describe Lifshitz points, we use a phenomenological AdS/CFT approach, namely we try to realize the symmetry group geometrically. In the case of AdS_{d+1}, the symmetry group $SO(d, 2)$ is the same as the conformal group of $Mink_d$, which suggests it is dual to a CFT_d. Assuming that AdS/CFT still holds in general gravity backgrounds, for which there is some indication, we are led to a $d + 1$-dimensional gravitational background dual to the Lifshitz point

$$ds_{d+1}^2 = R^2 \left(-\frac{dt^2}{u^{2z}} + \frac{d\vec{x}^2}{u^2} + \frac{du^2}{u^2} \right). \tag{25.3}$$

Here $d\vec{x}^2 = dx_1^2 + \ldots + dx_{d-1}^2$ and $0 < u < \infty$.

The metric is invariant under the scale transformation,

$$t \to \lambda^z t, \quad \vec{x} \to \lambda \vec{x}, \quad u \to \lambda u, \tag{25.4}$$

with generator (Killing vector)

$$D = -i(zt\partial_t + x^i \partial_i + u\partial_u). \tag{25.5}$$

The metric (25.3) is not geodesically complete: $u = \infty$ is a pp curvature singularity unless $z = 1$ (even though the Ricci scalar is $R = -2/R^2(z^2 + 2z + 3)$, so is constant on the space).

The other Killing vectors generating the algebra are the Lorentz generators M_{ij}, the momentum P_i (generator of space translations), the energy H (generator of time translations), given by

$$M_{ij} = -i(x^i \partial_j - x^j \partial_i); \quad P_i = -i\partial_i; \quad H = -i\partial_t. \tag{25.6}$$

The symmetry algebra generated by $\{M_{ij}, P_i, H, D\}$ is easily found to be

$$[D, H] = z\partial_t = izH,$$
$$[D, P_i] = \partial_i = iP_i; \quad [D, M_{ij}] = 0,$$
$$[M_{ij}, P_k] = \delta_k^i \partial_j - \delta_k^i \partial_i = i(\delta_k^i P_j - \delta_k^j P_i),$$
$$[M_{ij}, M_{kl}] = i(\delta_{ik} M_{jl} - \delta_{jk} M_{il} - \delta_{il} M_{jk} + \delta_{jl} M_{ik}),$$
$$[P_i, P_j] = 0. \tag{25.7}$$

This is the symmetry group of Lifshitz invariance, therefore this lends support to the idea that the background (25.3) is gravity dual to the Lifshitz point.

But in which gravitational theory is the background (25.3) a solution? An example of an action that has this solution for $d + 1 = 4$ dimensions is

$$S = \int d^4x \sqrt{-g}(R - 2\Lambda) - \frac{1}{2} \int \left(\frac{1}{e^2} F_{(2)} \wedge *F_{(2)} + F_{(3)} \wedge *F_{(3)} \right) - c \int B_{(2)} \wedge F_{(2)}, \tag{25.8}$$

where $F_{(2)} = dA_{(1)}$, $F_{(3)} = dA_{(2)}$ are field strength of the 1-form $A_{(1)}$ and 2-form $A_{(2)}$, and where the cosmological constant is

$$\Lambda = -\frac{z^2 + z + 4}{2R^2}. \tag{25.9}$$

There is other matter one can add to gravity with a cosmological constant in order to obtain the metric (25.3). Another example with relativistic gravity is

$$S = \frac{1}{2\kappa_N^2} \int dt d^D x dr \sqrt{-g} \left[R - 2\Lambda - \frac{1}{4} F_{\mu\nu} F^{\mu\nu} - \frac{1}{2} m^2 A_\mu A^\mu \right], \tag{25.10}$$

with Λ defined by the parameters of the theory. Another way to obtain it as a solution is in nonrelativistic gravity, the so-called "Horava gravity" model, that is approximately relativistic in the IR, but has Lifshitz scaling in the UV. We will, however, not discuss this here.

Gravitational dual to Galilean and Schrödinger symmetries

We can realize larger algebras geometrically using the same phenomenological AdS/CFT approach. In particular, there is an algebra relevant for the study of cold atoms and fermions at unitarity. It still has the same generators, $\{M_{ij}, P_i, H, D\}$, but also the generators K_i, called Galilean boosts, for the symmetry

$$t \to t, \quad x_i \to x_i - v_i t, \tag{25.11}$$

together with a conserved rest mass, or *particle number* N. The algebra they generate is a *conformal Galilean algebra*.

In the case of $z = 2$, there is an extra generator C, called a special conformal generator, and the algebra becomes the *Schrödinger algebra* (note that sometimes the conformal Galilean algebra is also called the Schrödinger algebra). This is in fact the symmetry of the Schrödinger equation of a free particle.

In order to realize the algebra geometrically, we need to generalize AdS/CFT to a candidate gravity dual in $d + 2$ dimensions instead of $d + 1$. Specifically, the geometry is

$$ds^2 = R^2 \left(-\frac{dt^2}{u^{2z}} + \frac{d\vec{x}^2}{u^2} + \frac{du^2}{u^2} + \frac{2dt\, d\xi}{u^2} \right). \tag{25.12}$$

Here, as usual, u is the radial coordinate, but we also have an extra coordinate ξ. Note that this metric is *not time reversal invariant* ($t \to -t$), unlike the Lifshitz metric. Note also that the metric is now nonsingular, unlike the Lifshitz metric, since it is conformal to a pp wave spacetime:

$$ds^2 = \frac{R^2}{u^2} \left(-dt^2 u^{2(1-z)} + 2dt\, d\xi + d\vec{x}^2 + du^2 \right). \tag{25.13}$$

The metric is invariant under the scaling symmetry

$$t' = \lambda^z t, \quad \vec{x}' = \lambda \vec{x}, \quad u' = \lambda u, \quad \xi' = \lambda^{2-z} \xi, \tag{25.14}$$

with generator

$$D = -i(zt\partial_t + x^i\partial_i + u\partial_u + (2 - z)\xi\partial_\xi), \tag{25.15}$$

and M_{ij}, P_i, H having the same expressions as before.

The extra symmetry Galilean boost K_i, with transformation

$$\vec{x}' = \vec{x} - \vec{v}t, \quad \xi' = \xi + \frac{1}{2}(2\vec{v} \cdot \vec{x} - v^2 t), \tag{25.16}$$

that implies the transformations

$$(d\vec{x}')^2 = (d\vec{x})^2 + v^2 dt^2 - 2\vec{v} \cdot \vec{x}dt, \quad 2dtd\xi' = 2dtd\xi + 2dt\vec{v} \cdot \vec{x} - v^2 t \, dt, \tag{25.17}$$

has the generator

$$K_i = -i(x^i\partial_\xi - t\partial_i). \tag{25.18}$$

Note that the term $\delta\vec{x} = -\vec{v}t$ leads to $t\partial_i$ in K_i, $\delta\xi = \vec{v} \cdot \vec{x}$ leads to $x^i\partial_\xi$, and the term $\delta\xi = -v^2 t/2$ is nonlinear in the parameter v, so we do not consider it for the calculation of K_i, which corresponds only to the linearized transformations.

Finally, the particle number (or rest mass) N corresponds to the translation symmetry in ξ (the metric is independent of ξ), thus the generator is

$$N = -i\partial_\xi. \tag{25.19}$$

Again we can calculate the algebra of the new generators, obtaining

$$\begin{aligned}
&[K_i, P_j] = \delta_{ij}\partial_\xi = i\delta_{ij}N, \\
&[D, K_i] = zt\partial_i - x^i\partial_\xi + (2 - z)x^i\partial_\xi - t\partial_i = (1 - z)iK_i, \\
&[K_k, M_{ij}] = t(\delta_{ik}\partial_j - \delta_{jk}\partial_i) + \delta_{jk}x^i\partial_\xi - \delta_{ik}x^j\partial_\xi = i(\delta_{jk}K_i - \delta_{ik}K_j), \\
&[K_i, H] = -\partial_i = -iP_i, \\
&[D, N] = (2 - z)\partial_\xi = (2 - z)iN, \\
&[K_i, N] = [H, N] = [P_i, N] = [M_{ij}, N] = 0.
\end{aligned} \tag{25.20}$$

This is the conformal Galilean algebra.

For $z = 2$, it also has a special conformal generator C, corresponding to the symmetry

$$\begin{aligned}
u &\to (1 - at)u, \quad x^i \to (1 - at)x^i, \\
t &\to (1 - at)t, \quad \xi \to \xi - \frac{a}{2}(\vec{x}^2 + u^2).
\end{aligned} \tag{25.21}$$

We can find the commutation relations of C with the other generators,

$$[D, C] = -2iC, \quad [H, C] = -iD, \quad [M_{ij}, C] = 0 = [K_i, C]. \tag{25.22}$$

This is the Schrödinger algebra.

String theory realization

The metric (25.12), dual to Galilean (or Schrödinger) symmetry can be realized in string theory. There is a procedure that maps between string theory solutions called "null Melvin twist," but this will not be explained here. Applying it to $AdS_5 \times S^5$ modified by a nonzero temperature T, we obtain the metric

$$ds^2 = r^2 \left[-\frac{\beta^2 r^2 f(r)}{k(r)} (dt + dy)^2 - \frac{f(r)}{k(r)} dt^2 + \frac{dy^2}{k(r)} + d\vec{x}^2 \right] + \frac{dr^2}{r^2 f(r)} + \frac{(d\psi + A)^2}{k(r)} + d\Sigma_4^2,$$

$$(25.23)$$

where

$$f(r) = 1 - \frac{r_+^4}{r^4} \quad k(r) = 1 + \frac{\beta^2 r_+^4}{r^2}, \qquad (25.24)$$

and the temperature is

$$T = \frac{r_+}{\pi \beta}. \qquad (25.25)$$

At $T = 0$, $k = f = 1$, and then when KK reducing the solution on the coordinates ψ and Σ_4 we get the $z = 2$ metric.

25.2 Spectral functions

Useful quantities that can be calculated using AdS/CFT and are used to describe transport (as we show in the next subsection) are *retarded Green's functions*.

The retarded Green's functions for observables $\mathcal{O}_A$ and $\mathcal{O}_B$ are

$$G^R_{\mathcal{O}_A \mathcal{O}_B}(\omega, k) = -i \int d^{d-1}x \, dt \, e^{i\omega t - i \vec{k} \cdot \vec{x}} \theta(t) \langle [\mathcal{O}_A(t, x), \mathcal{O}_B(0, 0)] \rangle. \qquad (25.26)$$

Here $\theta(t)$ is the Heaviside step function. These Green's functions describe the evolution of small (x, t)-dependent perturbations about equilibrium, in linear response theory.

Indeed,

$$\delta \langle \mathcal{O}_A \rangle (\omega, k) = G^R_{\mathcal{O}_A \mathcal{O}_B}(\omega, k) \delta \phi_{B(0)}(\omega, k). \qquad (25.27)$$

Proof:

Consider a time dependent perturbation to the Hamiltonian arising from the operator $\mathcal{O}_B$,

$$\delta H(t) = \int d^{d-1}x \delta \phi_{B(0)}(t, x) \mathcal{O}_B(x), \qquad (25.28)$$

and compute the resulting VEV of $\mathcal{O}_A$,

$$\langle \mathcal{O}_A \rangle (t, x) = \text{Tr} \, [\rho(t) \mathcal{O}_A(x)], \qquad (25.29)$$

where the density matrix ρ evolves according to

$$i \partial_t \rho = [H_0 + \delta H, \rho]. \qquad (25.30)$$

We go to the interaction picture, so that the time dependence due to H_0 is absorbed into $\mathcal{O}_A$. We then obtain for the VEV,

$$\langle \mathcal{O}_A \rangle (t, x) = \text{Tr} \, [\rho_0 U^{-1}(t) \mathcal{O}_A(t, x) U(t)], \qquad (25.31)$$

where the evolution operator is

$$U(t) = T \exp\left[-i \int^{t} \delta H(t')dt'\right],$$ (25.32)

and $\rho_0 = e^{-iH_0/T}$. Then to first order in perturbation theory (in linear response theory), the variation of the VEV is

$$\delta\langle\mathcal{O}_A\rangle(t,x) = -i\mathrm{Tr}\left\{\rho_0 \int^{t} dt'[\mathcal{O}_A(t,x),\delta H(t')]\right\}$$

$$= -i\int^{t} dt'd^{d-1}x\langle[\mathcal{O}_A(t,x),\mathcal{O}_B(t',x')]\rangle\delta\phi_{B(0)}(t',x').$$ (25.33)

Then Fourier transforming, we obtain

$$\delta\langle\mathcal{O}_A\rangle(\omega,k) = G^R_{\mathcal{O}_A\mathcal{O}_B}(\omega,k)\delta\phi_{B(0)}.$$ (25.34)

q.e.d.

Note that because of causality we must obtain zero for $t < 0$, which means that indeed there is a multiplication by a factor of $\theta(t)$ in the definition of the Green's function $G^R_{\mathcal{O}_A\mathcal{O}_B}(t,x)$, i.e. the Green's function is retarded. We also consider

$$G^R_{\mathcal{O}_A\mathcal{O}_B}(t,k) = \int \frac{d\omega}{2\pi} e^{-i\omega t} G^R_{\mathcal{O}_A\mathcal{O}_B}(\omega,k).$$ (25.35)

Since the Green's function is retarded (has a factor $\theta(t)$ in t space), we must evaluate it by closing the ω contour of integration in the upper-half complex plane.

This leads to two conditions:

1) $G^R_{\mathcal{O}_A\mathcal{O}_B}(\omega,k)$ *is analytic in the complex ω plane for* $\mathrm{Im}(\omega) > 0$.

2) *We must also have* $G^R_{\mathcal{O}_A\mathcal{O}_B}(\omega,k) \to 0$ *for* $|\omega| \to 0$ (in order for the semicircle contour integral at infinity to be zero, so that it can be added for free).

In turn, these two conditions lead to the Kramers–Kronig relations, which would be valid for any complex function satisfying the same two conditions,

$$\mathrm{Re}G^R(\omega) = P\int_{-\infty}^{+\infty} \frac{d\omega'}{\pi} \frac{\mathrm{Im}G^R(\omega')}{\omega'-\omega},$$

$$\mathrm{Im}G^R(\omega) = -P\int_{-\infty}^{+\infty} \frac{d\omega'}{\pi} \frac{\mathrm{Re}G^R(\omega')}{\omega'-\omega},$$ (25.36)

(here P is the principal part) which follow from the relation

$$G^R(z) = \oint_\Gamma \frac{d\zeta}{2\pi i} \frac{G^R}{\zeta - x},$$ (25.37)

where Γ is the contour of integration, the real line in ζ closed by a semicircle at infinity in the upper-half plane, as in Fig. 25.1.

In the $\omega \to 0$ limit of the Kramers–Kronig relation for Re $G^R(\omega)$, we obtain

$$\chi \equiv \lim_{\omega\to 0+i0} G^R_{\mathcal{O}_A\mathcal{O}_B}(\omega,x) = \int_{-\infty}^{+\infty} \frac{d\omega'}{\pi} \frac{\mathrm{Im}G^R_{\mathcal{O}_A\mathcal{O}_B}(\omega',x)}{\omega'}.$$ (25.38)

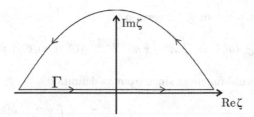

Figure 25.1 Contour of integration for G^R.

Here $\chi \in \mathbb{R}$ is the static thermodynamic susceptibility, since it is

$$\chi = \frac{\partial \langle \mathcal{O}_A \rangle}{\partial \phi_{B(0)}}. \tag{25.39}$$

The relation above is then called a thermodynamic sum rule.

Finally, there is a spectral representation for G^R that follows from its definition by inserting a complete set $\sum_n |n\rangle \langle n|$ between $\mathcal{O}_A$ and $\mathcal{O}_B$. Using the canonical ensemble, which means that the density matrix is $\rho_0 = e^{-H_0/T}$, we obtain

$$G^R_{\mathcal{O}_A \mathcal{O}_B}(\omega, k) = \sum_{mn} e^{-\frac{E_n}{T}} \left(\frac{A_{nm} B_{mn} \delta^{(d)}(k_{nm} - k)}{E_n - E_m + \omega + i0} - \frac{A_{mn} B_{nm} \delta^{(d)}(k_{mn} - k)}{E_m - E_n + \omega + i0} \right), \tag{25.40}$$

where

$$H_0|m\rangle = E_m|m\rangle, \quad k_{nm} = k_n - k_m,$$
$$A_{mn} = \langle m|\mathcal{O}_A(0,0)|n\rangle, \quad B_{mn} = \langle m|\mathcal{O}_B(0,0)|n\rangle. \tag{25.41}$$

Note that $\frac{1}{x \pm i0} = P\frac{1}{x} \mp i\pi \delta(x)$.

The *spectral function* for χ is

$$\chi_A(\omega, k) = -\mathrm{Im} G^R_{\mathcal{O}_A \mathcal{O}_B}(\omega, k), \tag{25.42}$$

since it appears in the integral giving χ, and satisfies $\omega \chi_A(\omega, k) \geq 0$.

Finally, just as we defined the retarded Green's function in x space by

$$G^R_{\mathcal{O}_A \mathcal{O}_B}(t, x) = -i\theta(+t) \langle [\mathcal{O}_A(t,x), \mathcal{O}_B(0,0)] \rangle, \tag{25.43}$$

we can define the *advanced Green's function*

$$G^A_{\mathcal{O}_A \mathcal{O}_B}(t, x) = +i\theta(-t) \langle [\mathcal{O}_A(t,x), \mathcal{O}_B(0,0)] \rangle, \tag{25.44}$$

with the momentum space representation

$$G^R_{\mathcal{O}_A \mathcal{O}_B}(\omega, k) = +i \int d^{d-1}x \, dt e^{i\omega t - i\vec{k}\cdot\vec{x}} \theta(-t) \langle [\mathcal{O}_A(t,x), \mathcal{O}_B(0,0)] \rangle, \tag{25.45}$$

as well as the function

$$\rho_{\mathcal{O}_A \mathcal{O}_B}(t, x) = \langle [\mathcal{O}_A(t,x), \mathcal{O}_B(0,0)] \rangle = i(G^R - G^A). \tag{25.46}$$

Its momentum space version,

$$\rho_{\mathcal{O}_A \mathcal{O}_B}(\omega, k) = \int d^{d-1}x \, dt e^{i\omega t - i\vec{k}\cdot\vec{x}} \langle [\mathcal{O}_A(t, x), \mathcal{O}_B(0, 0)] \rangle = i(G^R - G^A)(\omega, k), \quad (25.47)$$

is a spectral function, since from its definition,

$$G^{R,A}(\omega, \vec{k}) = \int \frac{d\omega'}{2\pi} \frac{\rho(\omega', \vec{k})}{\omega - \omega' \pm i\epsilon}. \quad (25.48)$$

If $\mathcal{O}_A, \mathcal{O}_B$ are Hermitian, then ρ is Hermitian, so it has real diagonal elements, which means that

$$\mathrm{Re}G^R(\omega, \vec{k}) = \mathrm{Re}G^A(\omega, \vec{k}) = P \int \frac{d\omega'}{2\pi} \frac{\rho(\omega', \vec{k})}{\omega - \omega'},$$

$$\mathrm{Im}G^R(\omega, \vec{k}) = -\mathrm{Im}G^A(\omega, \vec{k}) = -\frac{1}{2}\rho(\omega, \vec{k}), \quad (25.49)$$

since $\rho = i(G^R - G^A)$. Then,

$$\rho = -2\mathrm{Im}G^R(\omega, \vec{k}), \quad (25.50)$$

so it is indeed a spectral function (since $\mathrm{Im}G^R$ was a spectral function for χ).

25.3 Transport properties

In the previous subsection we saw that static thermodynamic susceptibilities $\chi = \partial\langle\mathcal{O}\rangle/\partial\phi_{B(0)}$ come from the spectral functions = imaginary parts of retarded Green's functions.

We now describe two important applications of the formalism.

1. Electrical conductivity

Consider electromagnetism on the boundary in the axial gauge $A_0 = 0$, and gauge field source fluctuations $\delta A_{j(0)}$. Then the electric field source is

$$E_j = F_{0j} = -\partial_t \delta A_j = -i\omega \delta A_{j(0)}. \quad (25.51)$$

The induced current 1-point function (VEV) is

$$\langle J_x \rangle = \sigma E_x = -i\omega\sigma \, \delta A_{x(0)}, \quad (25.52)$$

and on the other hand, in accordance with the linear response theory from the last section, it equals $G^R_{J_x J_x}(\omega, \vec{k})\delta A_{x(0)}$, which leads to the conclusion

$$\sigma(\omega, \vec{k}) = \frac{iG^R_{J_x J_x}(\omega, \vec{k})}{\omega}. \quad (25.53)$$

In fact, usually there is no real part involved in the formula, so that the usual *Kubo formula for the electrical conductivity* is

$$\sigma(\omega, \vec{k}) = -\frac{\mathrm{Im}G^R_{J_x J_x}(\omega, \vec{k})}{\omega}.$$

(25.54)

The DC conductivity is obtained as the $\omega \to 0$ limit of the above.

2. Shear viscosity η

Consider a fluid at rest, with velocity field $u^\mu = (1, 0, 0, 0)$, and a perturbation only in h_{xy}, at linear order.

In a general relativistic theory, the viscosity is defined from the expansion of the energy-momentum tensor. The equivalent of the Navier–Stokes equations arises from solving the conservation equation $\nabla_\mu T^{\mu\nu} = 0$ as an expansion in derivatives. For an ideal fluid, we have

$$T^{\mu\nu} = \rho u^\mu u^\nu + P(g^{\mu\nu} + u^\mu u^\nu),$$

(25.55)

where ρ is the density and P is the pressure. Here $P^{\mu\nu} = g^{\mu\nu} + u^\mu u^\nu$ is a projector, since $u_\mu P^{\mu\nu} = 0$, $P^{\mu\rho} P_{\rho\nu} = P^{\mu\rho} g_{\rho\nu}$. For a dissipative fluid, we add a next term in the expansion in derivatives

$$T^{\mu\nu} = \rho u^\mu u^\nu + P P^{\mu\nu} + \Pi^{\mu\nu}_{(1)},$$

(25.56)

where $\Pi^{\mu\nu}_{(1)}$ is linear in ∂u^μ, and we define the *Landau frame* by the condition $\Pi^{\mu\nu}_{(1)} u_\mu = 0$. Then decomposing $\nabla^\nu u^\mu$ in a part parallel to u^μ and a part perpendicular to it, itself expanded in an antisymmetric piece $\omega^{\mu\nu}$, a symmetric traceless piece $\sigma^{\mu\nu}$ and a trace θ,

$$\nabla^\nu u^\mu = -a^\mu u^\nu + \sigma^{\mu\nu} + \omega^{\mu\nu} + \frac{1}{d-1}\theta P^{\mu\nu},$$
$$a^\mu = u^\nu \nabla_\nu u^\mu,$$
$$\theta = \nabla_\mu u^\mu = P^{\mu\nu}\nabla_\mu u_\nu,$$
$$\sigma^{\mu\nu} = \nabla^\mu u^\nu + \nabla^\nu u^\mu - \frac{1}{d-1}\theta P^{\mu\nu},$$
$$\omega^{\mu\nu} = \nabla^{[\mu} u^{\nu]} + \nabla u^{[\mu} a^{\nu]},$$

(25.57)

the first order part of the energy-momentum tensor has a $\sigma^{\mu\nu}$ part with coefficient defining the *shear viscosity* η, and a θ term defining the *bulk viscosity* ζ,

$$\pi^{\mu\nu}_{(1)} = 2\eta\sigma^{\mu\nu} - \zeta\theta P^{\mu\nu}.$$

(25.58)

If we consider only an h_{xy} perturbation to the fluid at rest, we can calculate the resulting energy-momentum tensor substituting in the above expansion, to obtain

$$T_{xy} = -P h_{xy} - \eta \partial_t h_{xy} + \mathcal{O}(h^2) + \mathcal{O}(\partial^2 h).$$

(25.59)

In momentum space, we get

$$T_{xy} = -\eta i\omega h_{xy},$$

(25.60)

which should be compared to the linear perturbation theory, which says we should get $G^R_{T_{xy}T_{xy}} h_{xy}$, leading to

$$\eta = i\frac{G^R_{T_{xy}T_{xy}}(\omega,\vec{0})}{\omega}. \tag{25.61}$$

In fact, again usually there is no contribution from the real part, so the usual *Kubo formula for the shear viscosity* is

$$\eta = -\lim_{\omega\to0}\frac{1}{\omega}\mathrm{Im}G^R_{T_{xy}T_{xy}}(\omega,\vec{0}). \tag{25.62}$$

But since $\mathrm{Im}\, G^R = -\rho/2$, we get also

$$\eta = \lim_{\omega\to0}\frac{1}{2\omega}\int dt\, d^3x e^{i\omega t}\langle[T_{xy}(t,\vec{x}),T_{xy}(0,0)]\rangle. \tag{25.63}$$

Then, to calculate σ and η from AdS/CFT, we need a prescription for how to calculate the retarded Green's functions.

AdS/CFT in Minkowski space at finite temperature

Son and Starinets proposed a way to calculate retarded Green's functions in Minkowski space at finite temperature that generalizes the result at $T = 0$.

If we write the on-shell kinetic action in asymptotically AdS space as a function of the boundary value $\phi_{(0)}$, in momentum space, as the boundary term

$$S = \int\frac{d^d k}{(2\pi)^d}\phi_{(0)}(-k)\mathcal{F}(k,z)\phi_{(0)}(k)\bigg|^{z=z_H}_{z=z_B}, \tag{25.64}$$

where H stands for horizon (since in the case of finite temperature we have a black hole horizon inside the gravity dual) and B for boundary (usually at $z_B = 0$), then the Minkowski space prescription for the retarded Green's function is

$$G^R(k) = -2\mathcal{F}(k,z)|_{z_B}. \tag{25.65}$$

Another equivalent way to calculate it is via (a generalization of) the holographic renormalization method we have already discussed in Chapter 22. The retarded Green's function is

$$G^R_{\mathcal{O}_A\mathcal{O}_B} = \frac{\delta\langle\mathcal{O}_A\rangle}{\delta\phi_{B(0)}}\bigg|_{\delta\phi_{(0)}=0}. \tag{25.66}$$

If the on-shell renormalized action in asymptotically AdS space is

$$S_{\mathrm{ren.}} = S[\partial_\mu\phi] + S_{\mathrm{boundary}}, \tag{25.67}$$

then the 1-point function for the operator corresponding to ϕ is (see (22.31))

$$\langle\mathcal{O}\rangle = \lim_{z\to0}\left(\frac{R}{z}\right)^\Delta\frac{1}{\sqrt{\gamma}}\left(-\frac{\delta S[\phi_{(0)}]}{\delta\partial_z\phi_{(0)}(z)} - \frac{\delta S_{\mathrm{boundary}}}{\delta\phi_{(0)}(z)}\right). \tag{25.68}$$

The on-shell renormalized boundary action is, as we saw in Chapter 22 (see (22.30)),

$$S_{\text{boundary}} = \frac{\Delta - d}{2R} \int_{z \to 0} d^d x \sqrt{\gamma} \phi^2, \tag{25.69}$$

and the field expansion in terms of the independent variations (the non-normalizable and normalizable modes, $\phi_{(0)}$ and $\phi_{(2\Delta - d)}$ is

$$\phi(z) = \left(\frac{z}{R}\right)^{d-\Delta} \phi_{(0)} + \left(\frac{z}{R}\right)^{\Delta} \phi_{(2\Delta - d)} + \cdots \tag{25.70}$$

Thus the 1-point function is

$$\langle \mathcal{O} \rangle = -\lim_{z \to 0} \left(\frac{R}{z}\right)^{\Delta} \left[\frac{z}{R} \partial_z \phi|_{\phi_{(0)} = 0} + \frac{\Delta - d}{2R} 2\phi|_{\phi_{(0)} = 0}\right]$$
$$= -\frac{2\Delta - d}{R} \phi_{(2\Delta - d)}, \tag{25.71}$$

as we have already mentioned in Chapter 22.

Finally, the retarded Green's function is given by the variation of the normalizable mode with respect to the non-normalizable mode,

$$G^R_{\mathcal{O}_A \mathcal{O}_B} = \frac{\delta \langle \mathcal{O}_A \rangle}{\delta \phi_{B(0)}}\bigg|_{\delta \phi_{(0)} = 0} = -\frac{2\Delta_A - d}{R} \frac{\delta \phi_{A(2\Delta - d)}}{\delta \phi_{B(0)}}. \tag{25.72}$$

This formulation is actually equivalent to the one in (25.65).

Other transport properties

In the case of nonzero chemical potential μ, which means nonzero charge density, the heat (energy) and electric currents mix, so

$$\begin{pmatrix} \langle J_x \rangle \\ \langle Q_x \rangle \end{pmatrix} = \begin{pmatrix} \sigma & \alpha T \\ \alpha T & \bar{\kappa} T \end{pmatrix} \begin{pmatrix} E_x \\ -\frac{\nabla_x T}{T} \end{pmatrix}. \tag{25.73}$$

Similarly to the other cases, we find

$$\alpha(\omega) T = i \frac{G^R_{Q_x J_x}(\omega)}{\omega},$$
$$\bar{\kappa}(\omega) T = i \frac{G^R_{Q_x Q_x}(\omega)}{\omega}. \tag{25.74}$$

And again, the usual formulas involve just the imaginary parts,

$$\alpha(\omega) T = -\frac{\text{Im} G^R_{Q_x J_x}(\omega)}{\omega},$$
$$\bar{\kappa}(\omega) T = -\frac{\text{Im} G^R_{Q_x Q_x}(\omega)}{\omega}. \tag{25.75}$$

25.4 Viscosity over entropy density from dual black holes

As an important application of the formalism from the previous section, we calculate the shear viscosity for a system at finite temperature, with a gravity dual having a black hole in asymptotically AdS space. An important result is that the shear viscosity over entropy density, η/s, has the value $1/(4\pi)$ (for a while, it was thought that there was a lower bound, $\eta/s \geq 1/(4\pi)$, saturated by the dual black holes, but it turns out that we can have lower values) for a large class of (almost every) gravity dual with black holes, though we will not prove that here.

We consider just the Witten metric, or rather the black hole in Poincaré coordinates,

$$ds^2 = \frac{r^2}{R^2}\left(-\left(1 - \frac{r_0^4}{r^4}\right)dt^2 + d\vec{x}_3^2\right) + \frac{R^2}{r^2}\frac{dr^2}{1 - \frac{r_0^4}{r^4}}, \tag{25.76}$$

dual to $\mathcal{N} = 4$ SYM at finite temperature. The change of coordinates $u = r_0^2/r^2$ leads to the metric

$$ds^2 = \frac{r_0^2}{R^2}\frac{1}{u}(-f(u)dt^2 + d\vec{x}_3^2) + \frac{R^2}{4}\frac{du^2}{u^2 f(u)}; \quad f(u) = 1 - u^2. \tag{25.77}$$

In Chapter 15 we calculated the entropy density from the area of the horizon of the black hole as $s = \pi^2/2N^2 T^3$.

To calculate the shear viscosity, we need to calculate the retarded Green's function, using the relation (25.72) for a perturbation h_{xy}. On the other hand, writing

$$h_{xy}(\vec{x}, u) = \frac{r_0^2}{R^2 u}e^{-i\omega t + i\vec{q}\cdot\vec{x}}\phi_q(u), \tag{25.78}$$

where $\vec{q} \cdot \vec{x} = qz$ (propagation in the direction z, transverse to x, y) the quadratic action for $\phi_q(u)$ is the same as for a massless scalar field with the gravitational factor $1/(16\pi G_{N,5})$ in front (this is a general property of gravitons, excitations propagating in a direction transverse to the metric polarization). For a massless scalar, $\Delta = d = 4$, so $2\Delta - d = 4$.

We should calculate the exact scalar solution in the black hole background, and calculate $\delta\phi_{(4)}/\omega\delta\phi_{(0)}$. However, that turns out to be difficult.[1] Instead, we can take advantage of the fact that the shear viscosity calculated at the boundary equals the shear viscosity at the horizon. One can prove that the ratio defining η does not change with u if $\omega = \vec{q} = 0$. In fact, in that case, the equation of motion is $\partial_u\phi = 0$, so $\phi(u) = $ constant is an *exact* solution, meaning that $\delta\phi_{(4)}/\omega\delta\phi_{(0)}$ is a constant of u.

At the horizon $u \simeq 1$, the equation of motion is approximately

$$\partial_u^2\phi_q + \frac{1}{1 - u}\partial_u\phi_q + \left(\frac{\omega}{4\pi T}\right)^2\frac{\phi_q}{(1 - u)^2} = 0, \tag{25.79}$$

---

[1] We can calculate approximate solutions on patches instead, and connect the patches, to find the correct solution near the boundary that satisfies the correct boundary condition at the horizon, as in the original calculation of Policastro, Son, and Starinets.

with solutions

$$\phi_q^{\pm} = \phi_0(1 - u)^{\pm i\frac{\omega}{4\pi T}}. \tag{25.80}$$

The boundary condition at the horizon should be infalling, which selects ϕ^-. Then we can calculate at the horizon $u = 1$, using (25.65) and its u-independence, giving

$$G^R(\omega, \vec{k}) = -2\mathcal{F}(\omega, \vec{k}, u)_{u=1}, \tag{25.81}$$

so that we obtain

$$\eta = -\lim_{\omega \to 0} \frac{\text{Im}\, G^R(\omega, \vec{0})}{\omega} = \frac{r_0^3/R^3}{16\pi G_{N,5}} = \frac{\pi}{8}N^2 T^3. \tag{25.82}$$

The details are left as an exercise. Dividing by the entropy density, we obtain

$$\frac{\eta}{s} = \frac{1}{4\pi}, \tag{25.83}$$

as expected. Note that $\eta \propto N^2$, so η itself is not small, but rather η/s is.

As we mentioned, this result is valid for a large class of gravity duals with black holes. In particular, we do not need to have a string theory embedding for the gravity dual. Assuming AdS/CFT is still valid, for any phenomenologically motivated gravity dual we can calculate η/s, and for most we still find the value $1/(4\pi)$. For a while, it was thought that the lower bound $\eta/s \geq 1/(4\pi)$ was valid, but it was proven that in fact quantum corrections, in the form of certain R^2 terms in the gravitational action, as well as anisotropy, can lower the value.

25.5 Gauge fields, complex scalars, and fermions in AdS space vs. CFTs

At the beginning of the chapter we showed how to obtain gravity duals for nonrelativistically invariant condensed matter systems. Indeed, most condensed matter systems are nonrelativistic. However, AdS/CFT is best described when we have relativistically invariant systems, and we can find such relevant systems in condensed matter, so from now until the end of the chapter we deal with relativistic systems only. Moreover, we want to have perturbed conformal field theories, since conformality is mapped to AdS space, and as we saw before, there are many reasons to believe physics in AdS space is always holographic (the boundary is a finite time away, the isometries of AdS match conformal isometries on the boundary, etc.), independently of whether we can embed in string theory via a system of branes. In order to have a weakly coupled gravitational description, where we can calculate things, we need to have a strongly coupled field theory. This is also when the gravity dual description is most useful, since strongly coupled systems are hard to describe in condensed matter, even using phenomenological models. Therefore, from now on we consider relativistically invariant, strongly coupled (possibly perturbed) conformal field theories, dual to a weakly coupled gravitational theory in asymptotically AdS space.

The field theories we are interested in have (global) currents J_μ^a, corresponding to transport of charges, or just transport of particle number (particle number is a global charge). Therefore, by the AdS/CFT dictionary, we want to have at least gauge fields A_μ^a coupling to these currents. One special set of such currents forms the energy-momentum tensor $T_{\mu\nu}$, that couples to the bulk metric $g_{\mu\nu}$, so generically we need to consider the asymptotically AdS background to be dynamical as well. We are also generically interested in charged operators $\mathcal{O}$, i.e. operators that transform under the global symmetries, that can in principle acquire VEVs, so they need to be bosonic, usually scalar in the simplest applications. Therefore, we need to have scalar (or other) fields in the gravity dual that are charged under the gauge symmetry of A_μ^a. In the simplest case of abelian symmetries, the scalar fields would be complex and charged under the $U(1)$ symmetries.

Finally, we are sometimes interested in describing the fermions in the field theory, so we should have fermions in the gravity dual as well. Of course, a fermion χ_α would couple to a fermionic operator $\mathcal{O}_\alpha$, and it is generically difficult to consider the *backreaction* of the fermion on the geometry, so we usually have the gravity dual fermions as probes. In this case, the fermion is a source for the operator $\mathcal{O}_\alpha$, which can be thought of also as an effective *external* fermion introduced in the theory.

Since usually we are interested in the properties of the condensed matter system at finite temperature, we want to have a black hole in the gravity dual.

25.6 The holographic superconductor

We now describe the first, and most common application of this type, the holographic description of the superconductor. It has been developed by many people over the years, but the essential features described here were first described by Gubser, and later developed in a more coherent model by Hartnoll, Herzog, and Horowitz.

Ingredients of a holographic superconductor

The necessary ingredients are:

a) We need an AdS background, to describe a CFT (fixed point) near the phase transition. But since the superconductors of interest, because they are poorly understood, are high T_c superconductors, which are non-Fermi liquids (they have Fermi surfaces, but do not follow the standard Fermi theory), and many of them, mostly cuprates and organics, are layered, meaning that they are effectively $2 + 1$-dimensional, we are interested in an AdS_4 background, dual to a CFT_3.

b) We need to describe charge transport through a $U(1)$ current J_μ, so in the gravity dual we want to have a $U(1)$ bulk gauge field A_μ.

c) We want to describe symmetry breaking, which means that there should be operators (Cooper pair-like) that should condense, i.e. $\langle \mathcal{O} \rangle \neq 0$. We want at least one $\mathcal{O}$, corresponding to a complex field, charged under the $U(1)$ of the current J_μ. In order to describe *s-wave superconductors* (with spherical symmetry, i.e. angular momentum $l = 0$), we want a *charged scalar* ψ. If we wish instead to describe *p-wave superconductors* (with $l = 1$) or *d-wave superconductors* (with $l = 2$), we would need tensor fields.

d) We want to consider temperature, so we want a black hole in AdS_4.

The Lagrangean we consider then is

$$\mathcal{L} = \frac{1}{2\kappa^2}\left(R + \frac{d(d-1)}{R^2}\right) - \frac{1}{4g^2}F_{\mu\nu}^2 - |(\partial_\mu - iqA_\mu)\psi|^2 - m^2\psi^2 - V(\psi). \quad (25.84)$$

For AdS_4, we have $d = 3$.

We further consider $V = 0$ and m^2 satisfying the Breitenloher–Freedman bound in AdS_4, i.e. $m^2R^2 \geq -9/4$, which means that the scalar field ψ will be stable near the boundary at infinity, where the metric is approximately AdS_4.

Since we want to describe superconductivity, it means that the solution with $\psi = 0$ must be *thermodynamically unstable* towards decay to an AdS black hole solution with "scalar hair," i.e. a solution with a nontrivial scalar field profile (note that in asymptotically flat $3 + 1$-dimensional space, there is no such solution, but in the asymptotically AdS there can be). This solution must have $\psi \neq 0$ near the black hole horizon, but only for the temperature T less than a critical temperature T_c, $T < T_c$. For $T > T_c$, we must have $\psi = 0$ as the thermodynamically stable solution. Then T_c corresponds to the temperature of a superconducting phase transition. For dimensional reasons, to have a nonzero T_c, we need another scale, which can be either $\mu \neq 0$, i.e. a nonzero chemical potential, so $T_c \propto \mu$, or a nonzero Q, i.e. a nonzero charge density.

As we saw in Chapter 15, by the AdS/CFT dictionary, at the boundary $r \to \infty$, $A_0 \to \mu$. More precisely, at the AdS_4 boundary

$$A_0 \to \mu - \frac{Q}{r}. \quad (25.85)$$

Here the charge Q is related to the VEV of the current J_μ on the boundary (charge density). So either way, with either $\mu \neq 0$, or $Q \neq 0$, we need A_0 turned on.

Superconducting ansatz

We need to consider a superconducting solution, with metric

$$ds^2 = g_{tt}(r)dt^2 + g_{rr}(r)dr^2 + ds_2^2(r), \quad (25.86)$$

and gauge field and scalar field

$$A_\mu dx^\mu = \Phi(r)dt, \quad \psi = \psi(r). \quad (25.87)$$

On this ansatz, the scalar field terms in the Lagrangean become

$$-|(\partial_\mu - iqA_\mu)\psi|^2 - m^2|\psi|^2 \to -g^{tt}q^2\Phi^2|\psi|^2 - g^{rr}|\partial_r\psi|^2 - m^2|\psi|^2$$
$$= -g^{rr}|\partial_r\psi|^2 - m_{\text{eff}}^2|\psi|^2, \quad (25.88)$$

where the *effective mass* of the scalar fields is

$$m_{\text{eff}}^2 = m^2 + g^{tt}q^2\Phi^2. \quad (25.89)$$

But $g^{tt} < 0$ outside the black hole horizon, and $|g^{tt}| \to \infty$ close to the horizon ($|g_{tt}| \to 0$). We need $\Phi = 0$ at the horizon for it to be well defined. But if Φ drops to zero slower than g^{tt} rises to infinity, we can have $m_{\text{eff}}^2 < 0$ just outside the horizon, and below the Breitenlohner–Freedman bound, which means that it becomes unstable near the horizon. That in turn leads to scalar condensation, since

$$\langle \mathcal{O} \rangle = -i \frac{\delta Z_{\text{bulk}}[\psi_{(0)}]}{\delta \psi_{(0)}} = \frac{\delta S_{\text{bulk}}[\psi_{(0)}]}{\delta \psi_{(0)}} \neq 0. \qquad (25.90)$$

Indeed, as we have seen already, the boundary behavior of the field ψ in AdS_{d+1} follows the form (25.70), i.e.

$$\psi(z) = \left(\frac{z}{R}\right)^{d-\Delta} \psi_{(0)} + \left(\frac{z}{R}\right)^{\Delta} \psi_{(2\Delta-d)} + \ldots, \qquad (25.91)$$

and leads to (25.71), i.e.

$$\langle \mathcal{O} \rangle = \frac{2\Delta - d}{R} \psi_{(2\Delta-d)}. \qquad (25.92)$$

Here in $d = 3$, we have $m^2 R^2 = \Delta(\Delta - 3)$.

Thus we need to consider a solution that is a perturbation by a scalar field of a (possible Reissner–Nordstrom) AdS_4 black hole.

Solutions and operator condensates

We can consider as a background either:

1. The AdS–Reissner–Nordstrom (charged) black hole, as considered by Gubser. The solution is

$$ds^2 = -f(r)dt^2 + \frac{dr^2}{f(r)} + r^2 d\Omega_{2,k}^2,$$
$$f(r) = k - \frac{2M}{r} + \frac{Q^2}{4r^2} + \frac{r^2}{R^2},$$
$$\Phi(r) = \frac{Q}{r} - \frac{Q}{r_H},$$
$$\psi = 0, \qquad (25.93)$$

where for $k = 0$, $d\Omega_2^2 = dx^2 + dy^2$, and for $k = 1$, $d\Omega_2^2$ is the 2-sphere metric.

Or:

2. A neutral AdS black hole, as considered by Hartnoll, Herzog, and Horowitz. In the above charged solution, we consider $k = 0$, so $d\Omega_{2,k}^2 = dx^2 + dy^2$, and we put $Q = 0$, so

$$f(r) = \frac{r^2}{R^2} - \frac{2M}{r}, \quad \Phi = \psi = 0. \qquad (25.94)$$

But in these cases, since we do not have a scalar in the background, ψ must be treated in the probe approximation, i.e. without a backreaction on the metric solution. Note that the backreaction has been treated also, but is not easy to analyze, and one has to resort to numerics.

The equation of motion for ψ in the probe approximation is

$$\frac{1}{\sqrt{-g}}\partial_r\sqrt{-g}g^{rr}\partial_r\psi + m_{\text{eff}}^2\psi = 0 \Rightarrow$$
$$\psi'' + \left(\frac{f'}{f} + \frac{2}{r}\right)\psi' + \frac{\Phi^2}{f^2}\psi - \frac{m^2}{f}\psi = 0, \tag{25.95}$$

and the equation of motion for Φ is the equation of motion for a static massive vector field,

$$\Phi'' + \frac{2}{3}\Phi' - \frac{2\psi^2}{f}\Phi = 0. \tag{25.96}$$

At the horizon, $\Phi = 0$ is needed for normalizability of the solution.

Considering a scalar with mass $m^2 = -2/R^2$, i.e. corresponding to an operator with $\Delta = 1$ or 2 ($m^2R^2 = \Delta(\Delta - 3)$), the solution has the expansion near the boundary

$$\psi = \frac{\psi^{(1)}}{r} + \frac{\psi^{(2)}}{r^2} + \dots,$$
$$\Phi = \mu - \frac{\rho}{r} + \dots \tag{25.97}$$

But in this particular case, we note that *both $\psi^{(1)}$ and $\psi^{(2)}$ are normalizable*, which by the general dictionary means that they should give us two nonzero condensates (1-point functions, i.e. VEVs),

$$\langle \mathcal{O}_i \rangle = \sqrt{2}\psi^{(i)}, \quad i = 1, 2. \tag{25.98}$$

We can choose one of them to vanish, and calculate the other. Numerically, we find for $\langle \mathcal{O}_i \rangle(T/T_c)$ the solutions in Fig. 25.2a and b. Near $T = T_c$, we find we can approximate

$$\langle \mathcal{O}_1 \rangle \simeq 9.3T_c(1 - T/T_c)^{1/2},$$
$$\langle \mathcal{O}_2 \rangle \simeq 144T_c^2(1 - T/T_c)^{1/2},$$
$$T \simeq 0.118\sqrt{\rho}. \tag{25.99}$$

We can also calculate the instability of the Reissner–Nordstrom AdS_4 background under a perturbation with frequency ω, i.e. $\psi = \psi(r)e^{-i\omega t}$. One finds that for $T < T_c$ there is a normalizable mode with ingoing boundary conditions at the black hole horizon, and

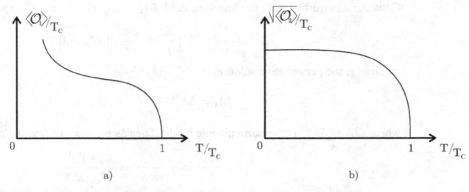

a) Condensate 1 as a function of temperature; b) Condensate 2 as a function of temperature.

Figure 25.2

$\text{Im}(\omega) > 0$, which means that there is an exponentially growing mode, signalling an instability towards a solution with $\psi = \psi(r)$. One can find the backreacted solution numerically with a nontrivial scalar profile, or "charged hairy black hole."

Effective mass and stability

One can calculate the effective mass m_{eff}^2 at the horizon in the AdS_4–RN black hole background, and find that indeed

$$m_{\text{eff}}^2 = m^2 - \frac{\gamma^2 q^2}{2R^2} < m^2, \tag{25.100}$$

where

$$\gamma^2 = \frac{g^2 2R^2}{\kappa_{N,4}^2}. \tag{25.101}$$

On the other hand, the horizon of the AdS_4–RN black hole is $AdS_2 \times S^2$. But for AdS_2, the Breitenlohner–Freedman bound is stronger than for AdS_4, namely $m^2 R_2^2 \geq -1/4$, where R_2 is the radius of AdS_2, different from the radius of the AdS_4 at infinity, more precisely $R_2^2 = R^2/6$. Then there is an instability if

$$-\frac{1}{4} \geq m_{\text{eff}}^2 R_2^2 = \frac{R^2}{6}\left(m^2 - \frac{\gamma^2 q^2}{2R^2}\right), \tag{25.102}$$

i.e. if $q^2\gamma^2 \geq 3 + 2\Delta(\Delta - 3)$.

That means that at low temperature, there is an AdS_2 throat at the horizon in which an asymptotically stable m^2 (i.e., satisfying the AdS_4 BF bound, $m^2 R^2 \geq -9/4$) can become unstable.

This is made possible by two facts: that the BF bound at infinity is different from the BF bound at the horizon (the bound at the horizon is stronger), and that $m_{\text{eff}}^2 < m^2$ due to the coupling to Φ.

Conductivity

Consider a perturbation in the Maxwell field, $\delta A_x = \delta A_x(r)e^{-i\omega t}$, satisfying

$$\delta A_x'' + \frac{f'}{f}\delta A_x' + \left(\frac{\omega^2}{f^2} - \frac{2\psi^2}{f}\right)\delta A_x = 0. \tag{25.103}$$

At large r, the perturbative solution is

$$\delta A_x = \delta A_x^{(0)} + \frac{\langle J_x\rangle}{r} + \ldots, \tag{25.104}$$

where $\langle J_x\rangle = \delta A_x^{(1)}$ is the normalizable mode. Then as we saw, we find the conductivity by

$$\sigma(\omega) = \frac{\langle J_x\rangle}{E_x} = -i\frac{\langle J_x\rangle}{\omega\delta A_x} = -i\frac{\delta A_x^{(1)}}{\omega\delta A_x^{(0)}}. \tag{25.105}$$

Numerically, one obtains $\text{Re}(\sigma)$ as a function of ω/T, for various temperatures $T \geq T_c$.

One finds a gap in frequency: $\sigma = 0$ for $\omega < \omega_g$, and numerically we have the approximate solution

$$\omega_g \approx (q\langle\mathcal{O}\rangle)^{\frac{1}{\Delta}}, \tag{25.106}$$

exact in the probe approximation.

Moreover, one finds $\omega_g/T_c \simeq 8.4$ in this case, and finds in any case a value close to 8 for holographic superconductors, though a range of values is possible. For low T_c superconductors, one usually has $\omega/T_c \simeq 3.5$, and in BCS theory one has $\omega/T_c \simeq 3.54$. However, for high T_c cuprate superconductors, one experimentally has $\omega/T_c \sim 8$ also, which means there is a good chance that we are indeed describing the physics of high T_c superconductors, something that is very hard to obtain by other means.

In weakly coupled superconductors, $\omega_g = 2E_g$, where E_g is the energy gap in the charged spectrum (the spectrum of electronic states). This is so, since Cooper pairs have negligible binding energy, so the energy of the pair is just twice the energy of a single electron.

On the other hand, at strong coupling, this need not happen. We can find the energy gap E_g from the Boltzmann suppression factor

$$\mathrm{Re}[\sigma(\omega \to 0)] = e^{-E_g/T}, \tag{25.107}$$

for $E_g/T \gg 1$. One finds for the holographic superconductor that in general $E_g \neq \omega_g/2$, except for the case of $\Delta = 2$ or 1 ($m^2R^2 = -2$), treated before. It is not clear whether in these cases, $\Delta = 2$ or 1, we have a weakly coupled pairing mechanism, although in general we have a correct strongly coupled behavior.

25.7 The ABJM model, quantum critical systems, and compressible quantum matter

The ABJM model is a 2+1-dimensional Chern–Simons gauge theory with $\mathcal{N} = 6$ super-symmetry, as we saw. It has been used as a primer for condensed matter models in 2+1 dimensions. Its gravity dual is string theory on $AdS_4 \times \mathbb{CP}^3$, which can be dimensionally reduced to AdS_4, and the resulting theory contains a lot of fields, including the ones that were described in previous subsections, i.e. gauge fields, scalars, and fermions charged under the same group. Therefore, we can actually think of the phenomenological AdS_4 models described before as being embedded in the ABJM/$AdS_4 \times \mathbb{CP}^3$ duality, and so of the ABJM model as a toy model in which to embed the physics of the phenomenological models of interest.

There are two types of condensed matter systems for which we can use the ABJM model and its gravity dual as a primer. The systems need to be strongly coupled, conformal, and relativistic, at least in an effective way. The two important such applications are quantum critical systems, and compressible quantum matter in non-Fermi liquids.

Quantum critical systems

One set of interesting phase transitions contains so-called quantum phase transitions, that arise when we tune a parameter ("coupling" g), as opposed to a thermodynamic quantity. Then at $T = 0$, close to the critical value g_c, we have a phase transition, between phases that are different topologically, or some type II (second order) phase transition.

Therefore, at $T = 0$, the transition is driven by quantum mechanical fluctuations, not temperature fluctuations, hence the name quantum phase transitions.

An important example is the *transition between an insulator and a type II superconductor*, as we change the chemical doping g. Then as we increase the temperature, $T > 0$, the phase diagram around $g = g_c$ opens up, and a new state appears, bounded by a phase transition line called the *Kosterlitz–Thouless phase transition* on the low g side, and another line on the high g side, as in Fig. 25.3. The new state is a strongly coupled state called a *quantum critical state*, which is poorly understood and is related via $T \neq 0$ with the $T = 0$ transition at $g = g_c$, hence it is related to a conformal field theory at finite temperature, exactly as the gravity dual of the ABJM model.

One example of such a model is the $1 + 1$-dimensional Ising model in a magnetic field along the direction x, perpendicular to the direction z of the spins, i.e. with Hamiltonian

$$H = -J \sum_i (g\sigma_i^x + \sigma_j^z \sigma_{j+1}^z). \tag{25.108}$$

Then at $g = 0$ we have the usual ferromagnetic state, and at $g = \infty$ we have all the spins along the x direction, so in between there must be a phase transition, which happens in fact at $g = 1$.

Insulator–superconducting phase transition

Another example is an array of superconducting islands on a 2-dimensional lattice: electrons in grains are locked into Cooper pairs with the same phase, meaning that the jth grain has amplitude $|\psi|$ and phase θ_j, with

$$\psi_j = |\psi| e^{i\theta_j}. \tag{25.109}$$

In this case, the Cooper pairs act as bosons ψ.

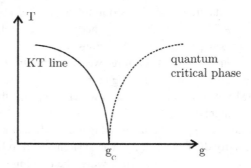

Figure 25.3   Quantum critical phase and Kosterlitz–Thouless phase transition.

Yet another example is provided by actual bosons, namely ^{87}Rb cold atoms, on an optical lattice. In this case, we have a superfluid–normal state phase transition, but the difference between the superfluid and the superconductor is only the charge, which in most applications is not relevant, so it is of the same type as the above.

Bosonic Hubbard model

Both the type II superconductor phase transition and the superfluid phase transition of ^{87}Rb cold atoms are described by the bosonic Hubbard model. It has a Hamiltonian composed of a term with the energy E_c required to remove a boson or Cooper pair to infinity, and a hopping term w between nearest neighbors (between optical sites or grains),

$$H = E_c \sum_i (\hat{n}_i - n_0)^2 - w \sum_{<ij>} (b_i^\dagger b_j + b_j^\dagger b_i), \qquad (25.110)$$

where $\hat{n}_i = b_i^\dagger b_i$. Another way to write this Hamiltonian is as

$$H = \frac{U}{2} \sum_i n_i(n_i - 1) - w \sum_{<ij>} (b_i^\dagger b_j + b_j^\dagger b_i) - \mu \sum_i n_i, \qquad (25.111)$$

where $E_c = U/2$ and $2E_c n_0 = U/2 + \mu$, and $<ij>$ represents the nearest neighbor interaction.

We can also write this in another way, by representing operators in terms of coordinates as

$$\hat{n}_i \equiv \frac{\partial}{i\partial\theta_j}, \quad b_i^\dagger = \sqrt{n_0}e^{i\theta_i}, \quad b_i = \sqrt{n_0}e^{-i\theta_i}, \qquad (25.112)$$

to obtain the *quantum rotor model*,

$$H = E_c \sum_i \left(\frac{1}{i}\frac{\partial}{\partial\theta_j}\right)^2 - J \sum_{<ij>} \cos(\theta_i - \theta_j). \qquad (25.113)$$

The mean field theory of this model generates the qualitative Kosterlitz–Thouless phase transition behavior. More precisely, the ordered (superconducting) phase corresponds to having bound vortex–antivortex pairs, and the Cooper pairs condense with rigid phases. The disordered (insulator) phase corresponds to condensed vortices and localized Cooper pairs.

The KT phase transition is characterized by: 1) unbinding of vortices at criticality; and 2) the duality of charges (particles) and vortices.

Landau–Ginzburg theory

From the bosonic Hubbard model, one can extract an effective field theory. We consider the ground state of the system as $n_i = n_0$, i.e. an equal number of bosons at each site. Next, we consider creation operators $a_i^\dagger$ to produce extra particles at each site and creation operators $h_i^\dagger$ to produce extra "holes" at each site (to remove a particle; this corresponds to an antiparticle in quantum field theory). Then one can construct a discretized *relativistic* field

$$\phi_i \sim \alpha_i a_i + \beta_i h_i^\dagger, \tag{25.114}$$

as in usual quantum field theory. Then time appears from temperature, by analytical continuation to Euclidean time, as usual. One can prove that the effective field theory of the continuum version of the field ϕ_i is the *relativistic* Landau–Ginzburg theory with action

$$S = \int d^3x \left[-(\partial_t \phi)^2 + v^2 |\vec{\nabla}\phi|^2 + (g - g_c)|\phi|^2 + u|\phi|^4 \right]. \tag{25.115}$$

Note that time comes from the temperature of the system, and the system is "relativistic" with "velocity of light" $v \neq 1$.

Therefore the quantum critical state is to be described by the ABJM model and its gravity dual, and it should reduce to an effective Landau–Ginzburg model.

Compressible quantum matter in non-Fermi liquids

Another application of the ABJM model, describing a strongly coupled relativistic conformal field theory, is to compressible quantum matter. This refers to matter that, when in the ground state (at $T = 0$) and coupled to a globally conserved charge Q with chemical potential μ, has $d\langle Q\rangle/d\mu$ smooth and nonzero. A new and important class of such materials is "non-Fermi liquids," with Fermi surfaces that satisfy the compressibility requirement above.

One example is the case of graphene, close to the symmetric point. There, for $\mu > 0$ we have a particle-like Fermi surface, and for $\mu < 0$ a hole-like Fermi surface, both giving locally a close to relativistic effective dispersion relation $d\omega/dk = \text{const.} = v$, but with $v \ll c$. The surface is coupled to an *emergent* $U(1)$ gauge field B_μ (not related to the Maxwell field of electromagnetism) in the material, also coupled to fermionic "spinons" ψ_α, with $\alpha = +, -$ being a spin state, and with charged bosons. The theory should have a *global* symmetry current as well, which corresponds to electric charge.

A nonrelativistic model of this type that can be simulated by ABJM has the action

$$
\begin{aligned}
S = \int d^3x \bigg[& f_+^\dagger \left((\partial_\tau - iA_\tau) - \frac{(\vec{\nabla} - i\vec{A})^2}{2m_f} - \mu \right) f_+ + f_-^\dagger \left((\partial_\tau + iA_\tau) - \frac{(\vec{\nabla} + i\vec{A})^2}{2m_f} - \mu \right) f_- \\
& + b_+^\dagger \left((\partial_\tau - iA_\tau) - \frac{(\vec{\nabla} - i\vec{A})^2}{2m_b} + \epsilon_1 - \mu \right) b_+ \\
& + b_-^\dagger \left((\partial_\tau + iA_\tau) - \frac{(\vec{\nabla} + i\vec{A})^2}{2m_b} + \epsilon_1 - \mu \right) b_- \\
& + \frac{u}{2} (b_+^\dagger b_+ + b_-^\dagger b_-)^2 + v b_+^\dagger b_-^\dagger b_- b_+ - g_1(b_+^\dagger b_-^\dagger f_- f_+ + h.c.) \\
& + c^\dagger \left(\partial_\tau - \frac{(\vec{\nabla})^2}{2m_c} + \epsilon_2 - \mu \right) c - g_2(c^\dagger(f_+ b_- + f_- b_+) + h.c.) \bigg], \tag{25.116}
\end{aligned}
$$

and a $U(1)$ global symmetry current corresponding to electric charge,

$$Q = f_+^\dagger f_+ + f_-^\dagger f_- + b_+^\dagger b_+ + b_-^\dagger b_- + 2c^\dagger c, \tag{25.117}$$

and both fundamental charged bosons $b_\pm$ and fermions $\phi_\pm$ coupled to the emergent gauge field as well as a neutral fermion c.

A relativistic version of this can be obtained by the substitution

$$(\partial_\tau - iqA_\tau) - \frac{(\vec{\nabla} - iq\vec{A})^2}{2m} - \mu \to (\partial_\mu - iqA_\mu)^2 - m^2, \tag{25.118}$$

for bosons, and for fermions, by the square root of the Klein–Gordon operator above.

To describe this in the gravity dual of ABJM, we should have a nonzero chemical potential, therefore nonzero A_0 on the boundary of AdS_4.

25.8 Reducing the ABJM model to the Landau–Ginzburg model

Both the systems described in the previous section are abelian, and should correspond to an effective field theory. Therefore, it is important to embed the abelian effective field theory in the ABJM model. In particular, here we show how to embed the Landau–Ginzburg model in the ABJM model, in an effective field theory manner, thus giving a justification for the application of the ABJM/$AdS_4 \times \mathbb{CP}^3$ duality to quantum critical systems.

We have an embedding of the Landau–Ginzburg model in the massively deformed ABJM model, via the matrices G^α defining the (fuzzy sphere) vacuum. The massive deformation of the ABJM model has bosonic fields, the bifundamental scalars $C^I = (Q^\alpha, R^\alpha)$, and the gauge fields for the two $U(N)$ group factors, A_μ and $\hat{A}_\mu$. Consider the ansatz for the reduction

$$\begin{aligned}
A_\mu &= a_\mu^{(2)} G^1 G_1^\dagger + a_\mu^{(1)} G^2 G_2^\dagger, \\
\hat{A}_\mu &= a_\mu^{(2)} G_1^\dagger G^1 + a_\mu^{(1)} G_2^\dagger G^2, \\
Q^\alpha &= \phi_\alpha G^\alpha, \\
R^\alpha &= \chi_\alpha G^\alpha,
\end{aligned} \tag{25.119}$$

with no summation for the index α in the ansatz for Q^α, R^α. Here the gauge fields $a_\mu^{(1)}, a_\mu^{(2)}$ are real and the scalars ϕ_α, χ_α are complex. From the above ansatz, we find that the field strengths reduce in a simple manner,

$$F_{\mu\nu} = f_{\mu\nu}^{(2)} G^1 G_1^\dagger + f_{\mu\nu}^{(1)} G^2 G_2^\dagger, \quad \hat{F}_{\mu\nu} = f_{\mu\nu}^{(2)} G_1^\dagger G^1 + f_{\mu\nu}^{(1)} G_2^\dagger G^2. \tag{25.120}$$

After some algebra, one finds that the massive ABJM model reduces to

$$S = -\frac{N(N-1)}{2} \int d^3x \left[\frac{k}{4\pi} \epsilon^{\mu\nu\lambda} \left(a_\mu^{(2)} f_{\nu\lambda}^{(1)} + a_\mu^{(1)} f_{\nu\lambda}^{(2)} \right) + |D_\mu \phi_i|^2 + |D_\mu \chi_i|^2 + U(|\phi_i|, |\chi_i|) \right],$$

$$U = \frac{4\pi^2}{k^2} \Big[(|\phi_1|^2 + |\chi_1|^2)(|\chi_2|^2 - |\phi_2|^2 - c^2)^2 + (|\phi_2|^2 + |\chi_2|^2)(|\chi_1|^2 - |\phi_1|^2 - c^2)^2$$

$$+ 4|\phi_1|^2 |\phi_2|^2 (|\chi_1|^2 + |\chi_2|^2) + 4|\chi_1|^2 |\chi_2|^2 (|\phi_1|^2 + |\phi_2|^2) \Big], \tag{25.121}$$

where $c^2 = \mu k/(2\pi)$. One can show that this truncation is consistent, i.e. its equations of motion solve the equations of motion of the full theory.

A further truncation is possible, considering $\phi_1 = \phi_2 = 0$ and $\chi_1 = b$, leading to (after some algebra)

$$S = -\frac{N(N-1)}{2} \int d^3x \left[\frac{k}{2\pi} \epsilon^{\mu\nu\lambda} a_\mu^{(1)} f_{\nu\lambda}^{(2)} + \left(a_\mu^{(1)}\right)^2 |b|^2 + |D_\mu \chi_2|^2 + V \right],$$
$$V = \frac{4\pi^2}{k^2} [|b|^2 |\chi_2|^4 + |\chi_2|^2 ((|b|^2 - c^2)^2 - 2|b|^2 c^2) + c^4 |b|^2]. \tag{25.122}$$

We note that the field $a_\mu^{(1)}$ is auxiliary, so it can be eliminated by its equations of motion, giving

$$a_\mu^{(1)} = -\frac{k}{4\pi |b|^2} \epsilon^{\mu\nu\lambda} f_{\nu\lambda}^{(2)}. \tag{25.123}$$

Substituting this back in the action, we obtain

$$S = -\frac{N(N-1)}{2} \int d^3x \left[\frac{k^2}{8\pi^2 |b|^2} \left(f_{\mu\nu}^{(2)}\right)^2 + |D_\mu \chi_2|^2 + V \right], \tag{25.124}$$

which is the action of the sought-for relativistic Landau–Ginzburg model, with $g \sim (|b|^2 - c^2)^2$ and $g_c \sim 2|b|^2 c^2$ being both tunable couplings. We can tune g to be near zero mass ($g \sim g_c$), in which case $|m^2| \ll \mu^2$. In that case, a careful analysis reveals that all the "nonzero modes" (modes dropped in the consistent truncation) are heavy (masses of order μ), and the only light modes are the ones we have kept. We can then show that the couplings of these massive modes to the light modes are negligible, so that quantum loops of the former are decoupled, and we have a *consistent truncation even at the quantum level*. That means that one can indeed think of the reduced model (Landau–Ginzburg) as an effective theory, *exactly as in the condensed matter case*.

We can write the LG theory in canonical normalization by redefining

$$a^{(2)} = \frac{2\pi b}{Nk} \tilde{a}^{(2)}, \quad \chi_2 = \frac{\tilde{\chi}_2}{N}, \tag{25.125}$$

in terms of which the action becomes

$$S = \int d^3x \left[-\frac{1}{4} \left(\tilde{f}_{\mu\nu}^{(2)}\right)^2 - |D_\mu \tilde{\chi}_2|^2 - V \right], \tag{25.126}$$

where $D_\mu = \partial_\mu - ig\tilde{a}_\mu^{(2)}$ and $g = \frac{2\pi |b|}{Nk}$ and the potential is

$$V = \frac{g^2}{2} \left[|\tilde{\chi}_2|^4 + \frac{\mu^2 k^2 N^4}{4\pi^2} + |\tilde{\chi}_2|^2 N^2 \left(-\frac{4\mu k}{2\pi} + |b|^2 + \frac{\mu^2 k^2}{4\pi^2 |b|^2} \right) \right]. \tag{25.127}$$

Note that in order for the further truncation of (25.121) to the Landau–Ginzburg model to be consistent, one needs to satisfy an extra equation, which turns out to be satisfied if the BPS condition is satisfied.

Important concepts to remember

- Lifshitz scaling is $t \to \lambda^z t$, $\vec{x} \to \lambda \vec{x}$, and can be represented holographically by a metric with the same scaling, together with $u \to \lambda u$ for the radial coordinate.

- The conformal Galilean and Schrödinger algebras can also be realized holographically in a gravity dual, and the resulting metrics can be embedded in string theory (even though the general models are AdS/CFT phenomenological).

- Retarded Green's functions give the linear response theory of a VEV $\langle \mathcal{O} \rangle$ to a source perturbation, $\delta \langle \mathcal{O} \rangle = G^R \phi_{(0)}$, and its imaginary part is a spectral function for $\chi = \partial \langle \mathcal{O} \rangle / \partial \phi$.

- Transport properties can be found from the imaginary parts of Green's functions. The conductivity is given by $\sigma = -\mathrm{Im} G^R_{J_x J_x} / \omega$ and the shear viscosity is $\eta = -\lim_{\omega \to 0} \mathrm{Im} G^R_{T_{xy} T_{xy}} (\omega, \vec{0}) / \omega$.

- In AdS/CFT in Minkowski space at finite temperature, retarded Green's functions are obtained from the on-shell kinetic action $\int \phi_0(-k) \mathcal{F}(k, z) \phi_0(k)|^{z_H}_{z_B}$ as $G^R = -2\mathcal{F}|_{z_B}$, or from the exact solution ϕ expanded near the boundary, which can be written as $G^R = -[(2\Delta - d)/R] \, \delta\phi_{(2\Delta-d)} / \delta\phi_{(0)}$.

- The viscosity over entropy density from plasmas corresponding to most dual black holes, thus including the case of $\mathcal{N} = 4$ SYM at finite temperature, is $\eta/s = 1/(4\pi)$.

- Phenomenological approaches to $2 + 1$-dimensional condensed matter, conformal and strongly coupled, involve asymptotically AdS_4 black holes, with gauge fields and charged scalars, and possibly fermions.

- In the holographic superconductors, the VEV of a scalar operator should condense, $\langle \mathcal{O} \rangle$, for a $T < T_c$. This happens due to the effective mass of a scalar in the gravity dual being lowered by the coupling to an electrostatic potential for a gauge field, Φ, and due to the fact that the Breitenlohner–Freedman bound in the AdS_2 factor near the horizon of the black hole is stronger than the BF bound in the AdS_4 at infinity, allowing for a scalar that is stable at infinity, but unstable at the horizon.

- The ABJM model is a primer for $2 + 1$-dimensional strongly coupled, conformal, relativistic states, that can be used for the quantum critical phase and compressible quantum matter in non-Fermi liquids.

- The massive deformation of the ABJM model can be reduced to an effective Landau–Ginzburg model, which is also the effective model for the quantum critical phase in condensed matter.

References and further reading

For more details on the holographic approach to condensed matter systems (AdS/CMT), see the review by Hartnoll [71]. Other useful reviews for specific areas of AdS/CMT include [112], [113] and [111]. The gravity dual Galilean and Schrödinger symmetry was described by Son [96] and Balasubramanian and McGreevy [97]. The gravity dual of Lifshitz symmetry was described by Kachru, Liu, and Mulligan [95]. The Lifshitz gravity dual was obtained in Horava–Lifshitz gravity in [98]. The string theory embedding was obtained in [99–101]. The recipe for calculation of correlation functions in AdS/CFT at finite temperature was proposed in [104]. The calculation of viscosity over entropy density is described in the review [103]. It was first found in [105] for $\mathcal{N} = 4$ SYM, and in [106] it was shown to be valid for a large class of models with gravity duals involving black holes. The calculation of transport properties in the ABJM model (2+1 dimensions) from dual black holes was described in [108]. The model of a holographic superconductor described here was first proposed by Gubser in [107], and developed by Hartnoll, Herzog, and Horowitz [102]. The applications of the ABJM model to condensed matter physics are reviewed in [109]. The reduction of the ABJM to the Landau–Ginzburg model was found in [110].

Exercises

1. Prove explicitly the invariance of the metric in (25.12) under the (finite) symmetries generated by $D, P_i, H, M_{ij}, K_i, N$, and C.
2. Calculate holographically the spectral function for the Green's function for a scalar operator with $\Delta = 3$ in $d = 4$.
3. Find the Kubo formula for the bulk viscosity ζ.
4. Fill in the details leading to (25.82).
5. Calculate the entropy and the temperature of the black hole solution in (25.93).
6. Consider the further reduction $\chi_1 = \phi_2 = 0$ in (25.121). Find the reduced action and show that the reduction is consistent.

Gluon scattering: the Alday–Maldacena prescription

In this chapter we describe a way to calculate scattering amplitudes of gluons at strong coupling using AdS/CFT, that was proposed by Alday and Maldacena. But in order to describe it, we need first to describe an important symmetry of string theory, that was just touched upon before, called T-duality.

26.1 T-duality of closed strings and supergravity fields

We consider bosonic strings on compact spaces, more precisely the simplest case possible, of a circle S^1 in the direction X^{25}, i.e. we identify

$$X^{25} \sim X^{25} + 2\pi R. \tag{26.1}$$

But then a string can now also *wind around this direction n* times,

$$X^{25}(\tau, \sigma = 2\pi) - X^{25}(\tau, \sigma = 0) = 2\pi mR \sim 0. \tag{26.2}$$

Define the *winding* $w = mR/\alpha'$, so the boundary condition is now

$$X^{25}(\tau, \sigma + 2\pi) = X^{25}(\tau, \sigma) + 2\pi \alpha' w. \tag{26.3}$$

But X^{25} satisfies the wave equation for a closed string, so we have the solution

$$X^{25}(\tau, \sigma) = X_L^{25}(\tau + \sigma) + X_R^{25}(\tau - \sigma) \equiv X_L^{25}(u) + X_R^{25}(v),$$
$$X_L(u) = x_0^L + \sqrt{\frac{\alpha'}{2}} \tilde{\alpha}_0 u + i\sqrt{\frac{\alpha'}{2}} \sum_{n \neq 0} \frac{\tilde{\alpha}_n}{n} e^{-inu},$$
$$X_R(v) = x_0^R + \sqrt{\frac{\alpha'}{2}} \alpha_0 v + i\sqrt{\frac{\alpha'}{2}} \sum_{n \neq 0} \frac{\alpha_n}{n} e^{-inv}. \tag{26.4}$$

The boundary condition, rewritten now as

$$X_L(u + 2\pi) - X_L(u) = X_R(v) - X_R(v - 2\pi) + 2\pi \alpha' w, \tag{26.5}$$

implies that $\tilde{\alpha}_0 - \alpha_0 = \sqrt{2\alpha'}w$. Then the momentum is

$$p = \frac{1}{2\pi\alpha'} \int_0^{2\pi} (\dot{X}_L + \dot{X}_R) d\sigma = \frac{1}{\sqrt{2\alpha'}}(\alpha_0 + \tilde{\alpha}_0), \tag{26.6}$$

so that finally

$$p = \frac{1}{\sqrt{2\alpha'}}(\alpha_0 + \tilde{\alpha}_0), \quad w = \frac{1}{\sqrt{2\alpha'}}(\tilde{\alpha}_0 - \alpha_0) \Rightarrow$$

$$\alpha_0 = \sqrt{\frac{\alpha'}{2}}(p - w), \quad \tilde{\alpha}_0 = \sqrt{\frac{\alpha'}{2}}(p + w). \tag{26.7}$$

Then the classical solution is

$$X^{25}(\tau, \sigma) = X_L + X_R = x_0 + \alpha' p\tau + \alpha' w\sigma + i\sqrt{\frac{\alpha'}{2}} \sum_{n \neq 0} \frac{e^{-in\tau}}{n}(\alpha_n e^{in\sigma} + \tilde{\alpha}_n e^{-in\sigma}). \tag{26.8}$$

Since the momentum in a compact direction is quantized, $p = n/R$, then together with $w = mR/\alpha'$, they are both discrete.

When quantizing, we must consider the constraints $L_0, \tilde{L}_0$, and impose $L_0 - \tilde{L}_0 = 0$,

$$L_0 = \frac{\alpha_0^I \alpha_0^I}{2} + N^\perp = \frac{\alpha'}{4} p^i p^i + \frac{\alpha_0^{25} \alpha_0^{25}}{2} + N^\perp,$$

$$\tilde{L}_0 = \frac{\alpha'}{4} p^i p^i + \frac{1}{2} \tilde{\alpha}_0^{25} \tilde{\alpha}_0^{25} + \tilde{N}^\perp \Rightarrow$$

$$L_0 - \tilde{L}_0 = \alpha' pw + N^\perp - \tilde{N}^\perp. \tag{26.9}$$

The constraint $L_0 - \tilde{L}_0 = 0$ therefore gives

$$N^\perp - \tilde{N}^\perp = \alpha' pw = nm. \tag{26.10}$$

On the other hand, the mass-shell constraint gives

$$M_{\text{compact}}^2 = -k^\mu k_\mu = 2p^+ p^- - p^i p^i = \frac{2}{\alpha'}(L_0 + \tilde{L}_0 - 2) - p^i p^i$$

$$= p^2 + w^2 + \frac{2}{\alpha'}(N^\perp + \tilde{N}^\perp - 2). \tag{26.11}$$

Note that the length of a string winding around the circle m times is $2\pi Rm$, which means that the energy is $2\pi |m| R/(2\pi\alpha') = |m| R/\alpha' = |w|$, which explains the w^2 term. Finally then, writing the spectrum in terms of n, m, and R, we get

$$M_{\text{compact}}^2 = \left(\frac{n}{R}\right)^2 + \left(\frac{mR}{\alpha'}\right)^2 + \frac{2}{\alpha'}(N^\perp + \tilde{N}^\perp - 2). \tag{26.12}$$

At this moment, we note a symmetry in the spectrum, called *"T-duality."* Consider $\tilde{R} = \alpha'/R$, then

$$M^2(R; n, m) = M^2(\tilde{R}; m, n). \tag{26.13}$$

This symmetry then exchanges R with $\tilde{R}$, and m with n (momentum with winding), and leaves the spectrum invariant.

Since R is an adjustable parameter, characterizing the vacuum, i.e. a modulus, this duality relates different vacua. It relates a small compactification circle with a large circle, and happens only in string theory (not in particle theory), since only strings can wind around compact directions.

Exchanging n with m means that we exchange

$$\tilde{\alpha}_0^{25} \leftrightarrow \tilde{\alpha}_0^{25}, \quad \alpha_0^{25} \leftrightarrow -\alpha_0^{25}, \tag{26.14}$$

which can be extended to a symmetry acting on $X^{25}(\tau,\sigma)$, namely exchanging it with

$$X'^{25}(\tau,\sigma) = X_L(\tau+\sigma) - X_R(\tau-\sigma)$$

$$= q_0 + \alpha' w \tau + \alpha' p \sigma + i\sqrt{\frac{\alpha'}{2}} \sum_{n\neq 0} \frac{e^{-in\tau}}{n} (\tilde{\alpha}_n e^{-in\sigma} - \alpha_n e^{in\sigma}), \tag{26.15}$$

where $q_0 = x_0^L - x_0^R$. Thus the extended symmetry exchanges

$$x_0 \leftrightarrow q_0; \quad p \leftrightarrow w; \quad \alpha_n \leftrightarrow -\alpha_n; \quad \tilde{\alpha}_n \leftrightarrow \tilde{\alpha}_n, \tag{26.16}$$

and we can check that again M^2 is invariant under this symmetry. We can in fact extend it to a full quantum symmetry, not only at the free field level, but also at the level of interactions.

We have described here T-duality for bosonic strings, but this generalizes easily to supersymmetric strings.

Buscher rules

In fact, we can extend the action of T-duality to action on the full massless fields of the superstring, i.e. the supergravity fields. The resulting transformation rules are known as the Buscher rules, and their action on $g_{\mu\nu}, B_{\mu\nu}$, and ϕ for duality on direction 0 is given by

$$\tilde{g}_{00} = \frac{1}{g_{00}}; \quad \tilde{g}_{0i} = \frac{B_{0i}}{g_{00}},$$

$$\tilde{g}_{ij} = g_{ij} - \frac{g_{0i}g_{0j} - B_{0i}B_{0j}}{g_{00}},$$

$$\tilde{B}_{0i} = \frac{g_{0i}}{g_{00}}; \quad \tilde{B}_{ij} = B_{ij} + \frac{g_{0i}B_{0j} - B_{0i}g_{0j}}{g_{00}},$$

$$\tilde{\phi} = \phi - \frac{1}{2}\log(g_{00}). \tag{26.17}$$

One can extend the Buscher rules to an action on the antisymmetric tensor fields (coupling to D-branes), and they can be proven from a transformation of the string path integral, but we will not show it here.

26.2 T-duality of open strings and D-branes

We now consider T-duality of bosonic open strings. We start with an open string with Neumann boundary conditions in the compact direction X^{25}. Then the solution to the equations of motion is

$$X^{25}(\tau,\sigma) = x_0^{25} + \sqrt{2\alpha'}\alpha_0^{25}\tau + i\sqrt{2\alpha'}\sum_{n\neq 0}\frac{\alpha_n^{25}}{n}\cos n\sigma\, e^{-in\tau}$$
$$= X_L^{25}(\tau+\sigma) + X_R^{25}(\tau-\sigma),$$

$$X_L^{25} = \frac{x_0^{25} + q_0^{25}}{2} + \sqrt{\frac{\alpha'}{2}}\alpha_0^{25}(\tau+\sigma) + i\sqrt{\frac{\alpha'}{2}}\sum_{n\neq 0}\frac{\alpha_n^{25}}{n}e^{-in(\tau+\sigma)},$$

$$X_R^{25} = \frac{x_0^{25} - q_0^{25}}{2} + \sqrt{\frac{\alpha'}{2}}\alpha_0^{25}(\tau-\sigma) + i\sqrt{\frac{\alpha'}{2}}\sum_{n\neq 0}\frac{\alpha_n^{25}}{n}e^{-in(\tau-\sigma)}, \tag{26.18}$$

where $\alpha_0^{25} = \sqrt{2\alpha'}p^{25} = \sqrt{2\alpha'}n/R$. Under T-duality, $X^{25}(\tau,\sigma)$ is exchanged with

$$X'^{25}(\tau,\sigma) = X_L^{25}(\tau+\sigma) - X_R^{25}(\tau-\sigma) = q_0^{25} + \sqrt{2\alpha'}\alpha_0^{25}\sigma + \sqrt{2\alpha'}\sum_{n\neq 0}\frac{\alpha_n^{25}}{n}e^{-in\tau}\sin n\sigma, \tag{26.19}$$

which is the same expansion as for a string between two D-branes, with

$$\alpha_0^{25} = \frac{1}{\sqrt{2\alpha'}}\frac{x_2^{25} - x_1^{25}}{\pi}. \tag{26.20}$$

Here $x_{1,2}^{25}$ are the positions of the two D-branes in the X^{25} direction. Then

$$X'^{25}(\tau,\pi) - X'^{25}(\tau,0) = 2\alpha'p^{25}\pi = 2\pi\alpha'\frac{n}{R} = 2\pi\tilde{R}n \tag{26.21}$$

can be identified with $x_2'^{25} - x_1'^{25}$ and $X'^{25}(\tau,0) = q_0^{25}$ can be identified with $x_1'^{25}$, which means that there is a D-brane at $X'^{25} = q_0^{25}$, and strings wind n times around the circle and return to it.

Moreover,

$$\partial_\sigma X^{25}(\tau,\sigma) = \frac{dX_L}{du}(u=\tau+\sigma) - \frac{dX_R}{du}(u=\tau-\sigma) = \partial_\tau X'^{25}(\tau,\sigma),$$

$$\partial_\tau X^{25}(\tau,\sigma) = \frac{dX_L}{du}(u=\tau+\sigma) + \frac{dX_R}{du}(u=\tau-\sigma) = \partial_\sigma X'^{25}(\tau,\sigma). \tag{26.22}$$

These can be written together as

$$\partial_\alpha X^{25} = \epsilon_{\alpha\beta}\partial_\beta X'^{25}. \tag{26.23}$$

This means that under T-duality, the Neumann boundary condition in X^{25} is exchanged with the Dirichlet boundary condition in X'^{25} and vice versa, Dirichlet for X^{25} with Neumann for X'^{25},

$$\partial_\sigma X^{25} = 0 \to \partial_\tau X'^{25} = 0 \ (\Rightarrow \delta X'^{25} = 0),$$
$$\partial_\tau X^{25} = 0 \to \partial_\sigma X'^{25} = 0. \tag{26.24}$$

The interpretation is that the space-filling D25-brane we started with is exchanged with a D24-brane under T-duality. That means that T-duality changes the dimensionality of the D-brane: T-duality on a direction parallel to the Dp-brane leads to a D$(p-1)$-brane, and T-duality on a direction transverse to the Dp-brane leads to a D$(p+1)$-brane.

From this fact, we can understand that T-duality takes us from type IIA to type IIB string theory and vice versa, since IIB allows only odd p, whereas IIA only allows even p.

26.3 T-duality on AdS space for scattering amplitudes

We now turn to the description of gluon scattering amplitudes using AdS/CFT. Gluon scattering amplitudes are not really observables, since they suffer from IR divergences, as gluons are massless states. But they are quantities that can be used for observables. For instance, in a QCD scattering experiment performed at accelerators, we can describe the scattering as a composition of the scattering of quarks and gluons, together with a model for hadronization of initial and final states (the nonperturbative process turning quark and gluon states into hadronic states). Also, we describe gluon amplitudes in $\mathcal{N} = 4$ SYM, which is a conformal theory, and in a conformal theory we cannot really define asymptotic states and therefore scattering, since all scales are equivalent, thus there is nothing special about "infinity" (the boundary, where we are supposed to separate states).

Nevertheless, we can formally define the gluon scattering amplitudes as the sum of Feynman diagrams, but we get IR divergent results. To get finite results, we need to regularize the theory. One useful way to do this is via dimensional regularization, since the dimensionally regularized theory is not conformal anymore, so we can define asymptotic states and scattering, but it is still relativistically invariant and supersymmetric. Another way would be a direct IR momentum cut-off, which as we will see is relevant to understanding scattering in the gravity dual.

To describe gluon scattering in the gravity dual, we consider the fact that gluons are open strings states, and open strings end on D-branes. Then in AdS/CFT, where the D-branes exist on the boundary at infinity, the scattering of gluons would correspond to thin open strings ending on the D-branes at infinity and the scattering is described by the worldsheet that extends into AdS and ends on the thin strings, as in Fig. 26.1a.

More precisely, in the AdS space in Poincaré coordinates,

$$ds^2 = R^2 \frac{dx^\mu dx_\mu + dz^2}{z^2}, \tag{26.25}$$

one can consider a D3-brane in the IR of AdS, at $z = z_{\text{IR}} \to \infty$, and strings ending on it. This provides an IR regularization for the scattering amplitudes.

We can now consider the T-dual 3+1-dimensional coordinates y^μ, defined by

$$\partial_\alpha y^\mu = i \frac{R^2}{z^2} \epsilon_{\alpha\beta} \partial_\beta x^\mu. \tag{26.26}$$

The factor of i is because the 3+1-dimensional space is Minkowski, but the string worldsheet has Euclidean signature, and we have added the factor R^2/z^2 in the transformation (compared to (26.23)). If we formally make the T-duality on the flat 3+1-dimensional coordinates, which means just using y^μ instead of x^μ and applying the Buscher rules on the

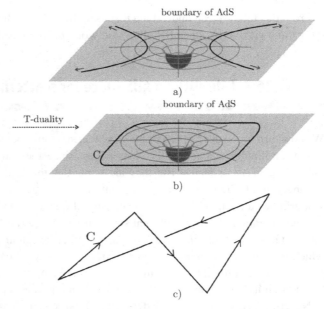

Figure 26.1 a) String worldsheet ending on the boundary on thin open strings going to infinity, giving the scattering amplitude of open strings; b) After T-duality, it is mapped to a string worldsheet ending on the boundary on a (Wilson loop type) contour; c) The contour is a null polygon in four flat dimensions of zigzag type.

supergravity background, namely the AdS space, i.e. $\tilde{g}_{00} = 1/g_{00}$ in the 3+1 directions, and redefine the radial coordinate, we obtain

$$ds^2 = R^2 \frac{dy^\mu dy_\mu + dr^2}{r^2}, \quad r = \frac{R^2}{z}, \tag{26.27}$$

which is again AdS space! But the T-duality exchanges the momentum integer $n = \Delta X^\mu/(2\pi R)$ with the winding number $m = p^\mu R$, which means it exchanges Δy^μ with $2\pi p^\mu R^2$. The R^2 factor has already been used in the definition of y^μ, so extending this definition of T-duality to the noncompact case means that, after extending it, the string extends in the 3+1 dimensions a distance

$$\Delta y^\mu = 2\pi p^\mu. \tag{26.28}$$

Since p^μ is null, and for a scattering amplitude $\sum_i p_i^\mu = 0$ for momentum conservation, after the T-duality, at the D-brane position we have a *closed polygon contour made up of null lines* (null polygon), and a string worldsheet ends on it, as in Fig. 26.1b and c.

After the T-duality, the regulator D-brane is at $r = R^2/z_{\rm IR} \to 0$, i.e. on the boundary of the T-dual AdS space, so we have a string worldsheet corresponding to a Wilson loop for a null polygonal contour.

26.4 Scattering amplitude prescription

The gluon scattering amplitude prescription of Alday and Maldacena is then the following. The amplitude can be written as the tree amplitude, that contains all the dependence on polarization tensors ϵ_i, and a scalar factor,

$$\mathcal{A}_n(p_i, \epsilon_i) = \mathcal{A}_{n,\text{tree}}(p_i, \epsilon_i)a(p_i), \tag{26.29}$$

and then the scalar factor $a(p_i)$ is given at strong coupling from AdS/CFT by

$$a(p_i) = e^{iS_{\text{string}}[C]}, \tag{26.30}$$

where C is the null polygonal contour described in the previous subsection, with sides equal to $2\pi p_i^\mu$.

One puzzle arises, since in the original AdS_5 space, before the duality, we should have quantities defined on the boundary, but they were defined at $z \to \infty$, which seems not to be there. But from the T-duality relation (26.26) we see that if y^μ is to be finite, and z is infinite, we need that the original coordinates also have $x^\mu \to \infty$. But this region, $x^\mu \to \infty, z \to \infty$ with all ratios fixed, is actually a part of the boundary, as one can check by analyzing the Penrose diagram, so we have no problem.

Dimensional regularization

As mentioned, by calculating the string action $S_{\text{string}}[C]$ we obtain an infinite result. Indeed, we have already calculated it in Chapter 14 for the Wilson loop, where we subtracted the infinity. But in the case of the gluon scattering amplitude, the result is supposed to be infinite, so it only needs to be regularized. For calculations, it is easier to use dimensional regularization.

We consider Dp-branes, where $p = D - 1$ and $D = 4 - 2\epsilon$. That means that we still have a theory with 16 supercharges, i.e. we have $\mathcal{N} = 1$ SYM in ten dimensions reduced to D dimensions.

The dimensionally regularized gravity dual is

$$ds^2 = f^{-1/2}dx_D^2 + f^{1/2}(dr^2 + r^2 d\Omega_{9-D}^2),$$
$$f = \frac{c_D \lambda_D}{r^{8-D}}, \quad c_D = 2^{4\epsilon}\pi^{3\epsilon}\Gamma(2 + \epsilon), \quad \lambda = g_D^2 N. \tag{26.31}$$

Since the 't Hooft coupling in D dimensions λ_D has dimensions, it is written in terms of the dimensionless 4-dimensional 't Hooft coupling λ and the IR cut-off scale μ (dimensional transmutation parameter) as

$$\lambda_D = \frac{\lambda\mu^{2\epsilon}}{(4\pi e^{-\gamma})^\epsilon}. \tag{26.32}$$

Performing the T-duality on x_D directions of the metric, we obtain the T-dual dimensionally regularized AdS metric,

$$ds^2 = f^{1/2}(dy_D^2 + dr^2) = \sqrt{c_D\lambda_D}\frac{dy_\mu dy^\mu + dr^2}{r^{2+\epsilon}}. \tag{26.33}$$

Note that the way we have calculated the gluon amplitude involved a string worldsheet ending on a contour at infinity, similarly to the Wilson loop, but in the T-dual AdS space. Thus, at least at strong coupling, there is a duality between Wilson loops and gluon scattering amplitudes. Such a duality has also been proven perturbatively, after the Alday and Maldacena prescription was put forward. We say more on it in the last section of this chapter.

26.5 4-point amplitude

We now turn to calculating the 4-point amplitude using the Alday–Maldacena prescription. We need to find a worldsheet that ends on the boundary on a null contour C of lengths proportional to the momenta k_i^μ, $i = 1, \ldots, 4$. We consider $2 \rightarrow 2$ scattering with the spatial momenta on a plane. That means that we can choose $y^3 = 0$, and the worldsheet is in the directions (y^0, y^1, y^2, r), and of Euclidean signature. We can choose any ordering of the external momenta in order to calculate the amplitude, but if we choose k_1, k_2 incoming and k_3, k_4 outgoing, the projection of C on the (y^1, y^2) plane is singular (two null lines going up, followed by two going down), so we choose instead k_1, k_3 incoming and k_2, k_4 outgoing, which means that we have a zigzag pattern in the (y^0, y^1, y^2) space (one line up, one down, one up, one down), as in Fig. 26.1c.

If we choose the "static gauge" for the string worldsheet, $y^1 = \sigma^1, y^2 = \sigma^2$ (since the worldsheet has Euclidean signature, we can choose the worldsheet coordinates $\sigma^{1,2}$ to be equal to two spatial coordinates), the string action is found to reduce to

$$S = \frac{R^2}{2\pi} \int dy^1 dy^2 \frac{\sqrt{1 + (\partial_i r)^2 - (\partial_i y^0)^2 - (\partial_1 r \partial_2 y^0 - \partial_2 r \partial_1 y^0)^2}}{r^2}. \tag{26.34}$$

Alternatively, if we choose the conformal gauge ($\gamma_{ab} = \delta_{ab}$ on the worldsheet), we obtain

$$iS = -\frac{R^2}{2\pi} \int d\sigma^1 d\sigma^2 \frac{1}{2} \frac{\partial_a r \partial_a r + \partial_a y^\mu \partial_a y_\mu}{r^2}. \tag{26.35}$$

We leave the details of this as an exercise.

One first finds the string worldsheet that ends on the Wilson contour that corresponds to equal Mandelstam variables $s = t$ ($s = -(k_1 + k_2)^2$ and $t = -(k_2 + k_3)^2$), whose projection in the (y^1, y^2) plane is a square, since $s = t$ means that all $|\vec{k}_i|$ are equal. The edges of the square are chosen to be $(y^1, y^2) = (\pm 1, \pm 1)$.

The boundary conditions in static gauge for ending on the null polygon with square projection in (y^1, y^2) are

$$r(\pm 1, y^2) = r(y^1, \pm 1) = 0; \quad y^0(\pm 1, y^2) = \pm y^2; \quad y^0(y^1, \pm y^1) = \pm y^1. \tag{26.36}$$

Then a solution that satisfies these boundary conditions and the equations of motion of the action (26.34) is

$$y^0(y^1, y^2) = y^1 y^2; \quad r(y^1, y^2) = \sqrt{(1 - (y^1)^2)(1 - (y^2)^2)}. \tag{26.37}$$

The same solution in conformal gauge, that satisfies the equations of motion of (26.35), is

$$y^1 = \tanh\sigma^1; \quad y^2 = \tanh\sigma^2; \quad y^0 = \tanh\sigma^1 \tanh\sigma^2; \quad r = \frac{1}{\cosh\sigma^1 \cosh\sigma^2}. \quad (26.38)$$

One can find the solution for $s \neq t$, when the contour is projected in the (y^1, y^2) plane onto a parallelogram, by boosting the solution in conformal gauge with $b = v\gamma$ in the embedding directions for AdS_5, that define it via $\sum_{i=1}^4 (Y_i)^2 - (Y^0)^2 - (Y^5)^2 = -R^2$. The result after the boost is

$$y^1 = \frac{a\tanh\sigma^1}{1 + b\tanh\sigma^1 \tanh\sigma^2}; \quad y^2 = \frac{a\tanh\sigma^2}{1 + b\tanh\sigma^1 \tanh\sigma^2},$$

$$y^0 = \frac{a\sqrt{1+b^2}\tanh\sigma^1 \tanh\sigma^2}{1 + b\tanh\sigma^1 \tanh\sigma^2}; \quad r = \frac{a}{\cosh\sigma^1 \cosh\sigma^2}\frac{1}{1 + b\tanh\sigma^1 \tanh\sigma^2}. \quad (26.39)$$

By considering the limit $\sigma^1 \to \infty$, we find $r = 0$ and that we span a line between the points

$$A: \ y^1 = y^2 = \frac{a}{1+b}; \quad B: \ y^1 = -y^2 = \frac{a}{1-b}, \quad (26.40)$$

reached for $\sigma^2 \to \pm\infty$. The other two vertices are obtained by taking $\sigma^1 \to -\infty$, again finding $r = 0$, and then $\sigma^2 \to \pm\infty$. Here s and t are given by the square of the diagonals of the parallelograms, since $-s = (k_1 + k_2)^2 = (2\pi)^2(y_1 - y_2 + y_2 - y_3)^2$ and $-t = (k_2 + k_3)^2 = (2\pi)^2(y_2 - y_3 + y_3 - y_4)^2$, and y_i are the four vertices. We obtain

$$s = \frac{-8a/(2\pi)^2}{(1-b)^2}; \quad t = \frac{-8a/(2\pi)^2}{(1+b)^2} \Rightarrow \frac{s}{t} = \frac{(1+b)^2}{(1-b)^2}. \quad (26.41)$$

However, in order to obtain a finite result for the action of the worldsheet, we have already seen that we need to regularize. We use dimensional regularization. In the dimensionally regularized metric (26.33), an approximate solution is

$$r_\epsilon \simeq \sqrt{1 + \epsilon/2}\, r_{\epsilon=0}; \quad y_\epsilon^\mu \simeq y_{\epsilon=0}^\mu. \quad (26.42)$$

Substituting this back in the dimensionally regularized action and using the fact that $(\partial_a r \partial_a r + \partial_a y^\mu \partial_a y_\mu)/(2r^2)|_{\epsilon=0} = 1$, we obtain

$$S = \frac{\sqrt{c_D \lambda_D}}{2\pi}\int \frac{\mathcal{L}_{\epsilon=0}}{r^\epsilon}$$

$$= i\frac{\sqrt{c_D \lambda_D}}{2\pi}\int_{-\infty}^{+\infty} d\sigma^1 d\sigma^2 (r_{\epsilon=0})^{-\epsilon}\left[1 + \frac{\epsilon}{2}\left(\frac{\partial_a r \partial_a r}{2r^2}\bigg|_{\epsilon=0} - 1\right)\right.$$

$$\left. -\frac{\epsilon^2}{4}\left(\frac{\partial_a r \partial_a r}{2r^2}\bigg|_{\epsilon=0} - 1\right) - \frac{\epsilon^2}{4}\right]. \quad (26.43)$$

One can calculate this integral and obtain for the amplitude (note that $iS_{\text{string}} = -\frac{R^2}{2\pi}A = -\frac{\sqrt{\lambda}}{2\pi}A$, with A being the area of the worldsheet):

$$\mathcal{A}/\mathcal{A}_{\text{tree}} = e^{iS_{\text{string}}} = \exp\left[iS_{\text{div}} + \frac{\sqrt{\lambda}}{8\pi}\left(\log\frac{s}{t}\right)^2 + \tilde{C}\right],$$

$$\tilde{C} = \frac{\sqrt{\lambda}}{4\pi}\left(\frac{\pi^2}{3} + 2\log 2 - (\log 2)^2\right),$$

$$iS_{\text{div}} = 2iS_{\text{div,s}} + 2iS_{\text{div,t}},$$

$$iS_{\text{div,s}} = -\frac{1}{\epsilon^2}\frac{1}{2\pi}\sqrt{\frac{\lambda\mu^{2\epsilon}}{(-s)^\epsilon}} - \frac{1}{\epsilon}\frac{1}{4\pi}(1-\log 2)\sqrt{\frac{\lambda\mu^{2\epsilon}}{(-s)^\epsilon}}, \quad S_{\text{div,t}} = S_{\text{div,s}}(s \to t). \quad (26.44)$$

The result was found later to be completely fixed by conformal invariance except for the constant part $\tilde{C}$, as is the result for the 5-point function, so possible freedom (reflecting dynamics, not just conformal invariance) starts to appear only from the 6-point function on.

26.6 IR divergences

The IR divergent structure of the 4-point amplitude is in general given by

$$\mathcal{A}_{\text{div,s}} = \exp\left[-\frac{f(\lambda)}{16}\left(\log\frac{\mu^2}{-s}\right)^2 - \frac{g(\lambda)}{4}\left(\log\frac{\mu^2}{-s}\right)\right], \quad (26.45)$$

where $f(\lambda)$ is called the cusp anomalous dimension, since it appears in other contexts, like the dimension of operators of high spin S, $\Delta - S \simeq f(\lambda)\log S$, or soft anomalous dimension or Wilson loop anomalous dimension (since it appears in the Wilson loop divergence), and $g(\lambda)$ is called the collinear anomalous dimension.

In dimensional regularization, where there is an ϵ defining the dimension ($D = 4 - 2\epsilon$), the divergence is written as

$$\mathcal{A}_{\text{div,s}} = \exp\left[-\frac{1}{8\epsilon^2}f^{(-2)}\left(\frac{\lambda\mu^{2\epsilon}}{s^\epsilon}\right) - \frac{1}{4\epsilon}g^{(-1)}\left(\frac{\lambda\mu^{2\epsilon}}{s^\epsilon}\right)\right], \quad (26.46)$$

and the functions $f^{(-2)}$ and $g^{(-1)}$ are defined through

$$\left(\lambda\frac{d}{d\lambda}\right)^2 f^{(-2)}(\lambda) = f(\lambda); \quad \lambda\frac{d}{d\lambda}g^{(-1)}(\lambda) = g(\lambda). \quad (26.47)$$

By comparing with the result in (26.44), we obtain

$$f = \frac{\sqrt{\lambda}}{\pi}, \quad g = \frac{\sqrt{\lambda}}{2\pi}(1 - \log 2). \quad (26.48)$$

In general, for an n-point amplitude, the IR divergence comes from the cusps joining neighboring momenta k_i^μ and k_{i+1}^μ, through $s_{i,i+1} = -(k_i + k_{i+1})^2$. In the exponent, we now get the sum of the same factors as before,

$$\log \mathcal{A}_n/\mathcal{A}_{n,\text{tree}}|_{\text{div.}} = \sum_{i=1}^{n}\log\mathcal{A}_{\text{div},s_{i,i+1}}$$

$$= -\frac{f(\lambda)}{16}\sum_{i=1}^{n}\log^2\left(\frac{\mu^2}{-s_{i,i+1}}\right) - \frac{g(\lambda)}{4}\sum_{i=1}^{n}\log\left(\frac{\mu^2}{-s_{i,i+1}}\right). \quad (26.49)$$

In the Alday–Maldacena prescription, the IR divergence is related to the cusps in the contour joining momenta k_i and k_{i+1}. We know that is the case for Wilson loops, and the calculation here is the same.

Therefore one has

$$\log \mathcal{A}_n/\mathcal{A}_{n,\text{tree}}|_{\text{div.}} = iS_{\text{div.}}(\epsilon) = \sum_{i=1}^{n} S_{i,i+1,\text{div.}}(\epsilon). \qquad (26.50)$$

Defining the variables $y^{\pm} = y^0 \pm y^1$ and choosing the static gauge $\sigma^1 = y^-, \sigma^2 = y^+$, the $\epsilon = 0$ (4-dimensional) action for the string is

$$S = \frac{R^2}{2\pi} \int dy^+ dy^- \frac{\sqrt{1 - 4\partial_- y^2 \partial_+ y^2 - 4\partial_- r\partial_+ r - 4(\partial_- y^2 \partial_+ r - \partial_- r\partial_+ y^2)^2}}{2r^2}, \qquad (26.51)$$

and if the 3-momenta $\vec{k}_i, \vec{k}_{i+1}$ are chosen (without loss of generality) to be

$$2\pi \vec{k}_i = z_1(1,0,0); \quad 2\pi \vec{k}_{i+1} = z_2(\alpha, 1, \sqrt{\alpha}) \Rightarrow (2\pi)^2 s = -\alpha z_1 z_2, \qquad (26.52)$$

the solution of the string action ending on the wedge formed by these lines is

$$r(y^-, y^+) = \sqrt{1 + \epsilon/2}\sqrt{2}\sqrt{y^- \left(y^+ - \frac{y^-}{\alpha}\right)}; \quad y^2(y^-, y^+) = \frac{1}{\sqrt{\alpha}} y^-. \qquad (26.53)$$

One can substitute this solution back into the string action and find

$$iS_{i,i+1}(\epsilon) = \frac{\sqrt{\lambda_D c_D}}{2\pi} \int \frac{\mathcal{L}_{\epsilon=0}}{r^\epsilon} = -\frac{1}{\epsilon^2} \frac{\sqrt{\lambda}}{2\pi} \sqrt{\frac{\mu^{2\epsilon}}{(-s_{i,i+1})^\epsilon}} C(\epsilon),$$

$$C(\epsilon) = \frac{\sqrt{c_D}}{2^{\epsilon/2}} \frac{(2\pi)^{-\epsilon}}{(4\pi e^{-\gamma})^{\epsilon/2}} \frac{\sqrt{1 + \epsilon}}{(1 + \epsilon/2)^{1+\epsilon/2}}, \qquad (26.54)$$

which is the same as $iS_{\text{div},s}(\epsilon)$ for $s \to s_{i,i+1}$. Therefore, indeed, the IR divergence in the 4-point amplitude comes from the cusps, and then the total IR divergence of the n-point amplitude comes from the sum of these IR divergences inside the exponential, as we described.

This is an example of a result which would have been very hard to calculate by perturbation theory: the full IR divergent structure, together with the exact form of the anomalous dimensions $f(\lambda)$ and $g(\lambda)$ at strong coupling $\lambda \to \infty$.

26.7 Fermionic T-duality

We mentioned that, because the gluon amplitudes are found from the same calculation as the Wilson loops, this implies that there is a duality symmetry that exchanges Wilson loops with gluon amplitudes, at least at strong coupling $\lambda \to \infty$. This duality was in fact also proven perturbatively to all orders in the coupling.

Another important implication related to the Alday and Maldacena calculation is that there is a new conformal symmetry, called "dual conformal symmetry," of $\mathcal{N} = 4$ SYM, which acts on coordinates y^μ defined by $\Delta y^\mu = 2\pi k^\mu$. This is reflected in the gravity dual by the fact that, after making the (bosonic) T-duality transformation, we obtain a *different* AdS space, with its own isometry, dual to a conformal invariance, unrelated to the isometry (mapped to a conformal invariance) of the original AdS space. So dual conformal symmetry is a symmetry of amplitudes, easily seen from the gravity dual. It has also been proven in the mean time, and it was shown that when commuting its generators with the generators of the usual conformal invariance, we generate an infinite symmetry (with an infinite number of generators) called the Yangian, which is a symmetry of $\mathcal{N} = 4$ SYM.

But more precisely, Berkovits and Maldacena showed that the symmetry that relates the original superstring in $AdS_5 \times S^5$ background to another superstring in another $AdS_5 \times S^5$ background is the bosonic T-duality that we have shown, together with a new type of T-duality called *fermionic T-duality*, which acts on eight of the 32 fermions of the superstring $\theta^{\alpha i}$. On a single fermion θ^1, it acts as (similarly to the action of bosonic T-duality on a single worldsheet boson = spacetime coordinate X)

$$\theta^1 \to \theta^1 + \rho, \quad X^\mu \to X^\mu, \quad \theta^{\tilde{A}} \to \theta^{\tilde{A}}, \quad \tilde{A} = A, \neq 1, \quad A = (\alpha i). \tag{26.55}$$

The resulting combined bosonic and fermionic T-duality was proven to be an exact symmetry of string theory, and to relate gluon amplitudes to Wilson loops and conformal symmetry to dual conformal symmetry.

Important concepts to remember

- T-duality is an exact string symmetry that exchanges the compact radius R with $\tilde{R} = \alpha'/R$, momentum and winding charges, $m \leftrightarrow n$, and X with X', where $\partial_\alpha X = \epsilon_{\alpha\beta} \partial_\beta X'$.
- For D-branes, T-duality parallel to the Dp-brane turns it into a D$(p-1)$-brane, and T-duality perpendicular to it into a $D(p+1)$-brane.
- A formal T-duality on AdS space turns it into a dual AdS space, with center and boundary interchanged, and gluon momenta into a boundary condition for the string worldsheet to end on a null segment.
- The prescription to calculate $\mathcal{A}_n/\mathcal{A}_{n,\text{tree}}$ is to calculate $e^{iS[C]}$ in the T-dual AdS space, as for the Wilson loop, ending on the boundary contour C, which is now a null polygonal contour, with sides $\Delta y^\mu = 2\pi k^\mu$.
- The result is infinite due to IR divergences, and one needs to dimensionally regularize the background and the string solution to get a finite result.
- The IR divergences are due to the cusps where the null lines meet, and can be calculated. They are given by two anomalous dimension functions $f(\lambda)$ and $g(\lambda)$, the cusp and collinear anomalous dimension functions.
- A bosonic plus a fermionic T-duality exchanges an AdS background with itself, conformal invariance with dual conformal invariance and gluon amplitudes with Wilson loops, and is an exact symmetry of superstring theory.

References and further reading

The prescription developed to calculate gluon scattering amplitudes at strong coupling using AdS/CFT was proposed and tested on the 4-point amplitude in [114]. The IR divergences for a general n-point amplitude were derived in [115]. The fermionic T-duality was proposed in [116].

Exercises

1. Consider T-duality of closed strings on both directions of a torus T^2. Write down the mass formula generalizing (26.12), and the Buscher rules for this duality.

2. Consider T-duality in the presence of two intersecting Dp-branes, with $p - k$ common spatial directions. What possibilities for the resulting intersecting branes are allowed after the T-duality? Is this consistent for any k, and why?

3. The gluon scattering amplitude in the conformal theory $\mathcal{N} = 4$ SYM has a "conformally anomalous" part, plus a part that can depend only on the conformally invariant "cross ratios" u_{ijkl}, that can be expressed in terms of the vertex positions for the null polygon on which the string worldsheet ends, specifically in terms of

$$x_{ij} \equiv x_i - x_j = 2\pi(k_i + \ldots + k_{j-1}), \tag{26.56}$$

as

$$u_{ijkl} = \frac{x_{ij}^2 x_{kl}^2}{x_{il}^2 x_{jk}^2}. \tag{26.57}$$

How many independent cross ratios are there for the n-point amplitude? Specialize for $n = 4, 5, 6$ and comment on the result.

4. Prove that the string action in AdS_5 in static gauge reduces to (26.34), and in conformal gauge reduces to (26.35).

5. Prove that the solution (26.37) satisfies the equations of motion of (26.34) and the solution (26.38) satisfies (26.35).

6. Verify that the solution in (26.53) solves the equations of motion of (26.51).

27

Holographic entanglement entropy: the Ryu–Takayanagi prescription

In this chapter we describe how to calculate holographically a type of entropy that depends on dividing a system into two subsystems, called entanglement entropy. We see that the result is given holographically by the same kind of minimal surface that was used for the calculation of the Wilson loop and the gluon scattering amplitudes, but with a different geometry and dimensionality for its boundary (a closed 2-dimensional surface bounding a spatial volume on the boundary on AdS, on which a 3-dimensional worldvolume ends).

27.1 Entanglement entropy

Consider a quantum mechanical system with many degrees of freedom, in various space-time dimensions. Spin chains, lattices, and quantum field theories are standard examples. Then consider the (total) system in the pure state $|\Psi\rangle$, with density matrix

$$\rho_{\text{tot}} = |\Psi\rangle\langle\Psi|. \qquad (27.1)$$

It has total von Neumann entropy, $S_{\text{tot}} = -\text{Tr}\,\rho_{\text{tot}} \log \rho_{\text{tot}} = 0$.

However, if we divide the system into two subsystems A and B (imaginary division, not any physical procedure), for instance in the case of a spin chain we consider in A only the sites left of site l and for B the sites right of l, we can define a notion of entropy called *entanglement entropy of the subsystem A* as follows. The total Hilbert space is a product of the Hilbert spaces of A and B, $\mathcal{H}_{\text{tot}} = \mathcal{H}_A \otimes \mathcal{H}_B$. For an observer who only has access to A, we need to trace the density matrix over the states of B, so he will experience the reduced density matrix

$$\rho_A = \text{Tr}\,_B \rho_{\text{tot}}. \qquad (27.2)$$

The entanglement entropy is then the von Neumann entropy of the reduced density matrix,

$$S_A = -\text{Tr}\,_A \rho_A \log \rho_A. \qquad (27.3)$$

This is a measure of how entangled the wave function $|\Psi\rangle\langle\Psi|$ is.

One can also define the entanglement entropy at finite temperature T, $S_A(\beta)$, obtained from using instead of the ρ_{tot} above, the thermal $\rho_{\text{tot,thermal}} = e^{-\beta\hat{H}}$. Of course, if A is the total system, the thermal entropy and the entanglement entropy are the same.

Properties

1. We find that if ρ_{tot} is pure (as for $T = 0$ above), the entanglement entropy of subsystem A is the same as of subsystem B,

$$S_A = S_B, \tag{27.4}$$

since the entropy is entirely due to the division of S into two subsystems. In particular, this means that the entanglement entropy is *non-extensive*.

2. For three subsystems A, B, C that do not intersect each other, we have

$$S_{A+B+C} + S_B \leq S_{A+B} + S_{B+C},$$
$$S_A + S_C \leq S_{A+B} + S_{B+C}, \tag{27.5}$$

called strong subadditivity.

3. In particular, one can put $B = 0$ in the above and find

$$S_{A+C} \leq S_A + S_C, \tag{27.6}$$

which allows the definition of *mutual information* $I(A, B)$ by

$$I(A, B) = S_A + S_B - S_{A+B} \geq 0. \tag{27.7}$$

27.2 Application for black holes

One of the original motivations of the definition of entanglement entropy was to describe the physics of black holes. Indeed, the black hole information paradox can be stated as follows. If we start with a system in a pure quantum state (non-entangled, with zero von Neumann entropy) $|\Psi\rangle$, and the system evolves unitarily according to the laws of quantum mechanics, i.e. with the unitary evolution operator U, but such that when describing the final state classically, a black hole forms, then the final state should be thermal, i.e. a mixed state, with $\rho_{\text{tot,thermal}} = e^{-\beta \hat{H}}$, since we have a Hawking radiation at temperature $T = 1/\beta$. But this is a paradoxical situation, since we were supposed to be in a pure state $|\Psi'\rangle = U|\Psi\rangle$.

A possible way to explain this is if we consider that in the presence of the black hole, the event horizon shields the interior of the black hole from causal relation with the outside, $\mathcal{H}_{\text{total}} = \mathcal{H}_{\text{inside}} \otimes \mathcal{H}_{\text{outside}}$. And since we can never have access to the states inside the horizon, we should be tracing over them, to have the density matrix

$$\rho_{\text{outside}} = \text{Tr }_{\text{inside}} |\Psi'\rangle\langle\Psi'|. \tag{27.8}$$

This allows for the possibility that ρ_{outside} is a thermal density matrix $e^{-\beta \hat{H}}$, as one needs. One immediate problem seems to be that in general the entanglement entropy is not extensive, whereas the thermal entropy should be. Another arises since, as we shortly see, the entanglement entropy in quantum field theories contains divergences, whereas the black hole entropy does not.

But there are ways to fix these problems, and in particular it was found that quantum corrections to the entropy of the black hole in the presence of matter fields equals the entanglement entropy. Nevertheless, as the apparent black hole information paradox is one of the most important unsolved problems of theoretical physics, it is clear that the details of any description of black hole entropy and radiation from entanglement are in general difficult to realize.

27.3 Entanglement entropy in quantum field theory

Consider a quantum field theory on a spatial manifold M, moving in time $\mathbb{R}_t$. Consider a submanifold A at time $t = t_0$, $A \in M$, bounded by ∂A, that separates the submanifolds A and B, the complement of A inside M. Then we can define the entanglement entropy of A as before. But since the entanglement entropy is always divergent in quantum field theory, we need to introduce an UV cut-off a (in a lattice regularization, it would be the lattice cut-off).

The divergence in $d + 1$ spacetime dimensions is proportional to the area of ∂A,

$$S_A = \gamma \frac{\text{Area}(\partial A)}{a^{d-1}} + \text{subleading,} \tag{27.9}$$

for $d > 1$, where γ is a constant that depends on the system. For a $d = 1$ conformal field theory (in 1+1 spacetime dimensions), and for an infinitely long total system and a subsystem A of length l, we have

$$S_A = \frac{c}{3} \log \frac{l}{a}, \tag{27.10}$$

where c is the central charge.

In higher dimensions, the general form for d even is

$$S_A = p_1(l/a)^{d-1} + p_3(l/a)^{d-3} + \ldots + p_{d-1}(l/a) + p_d, \tag{27.11}$$

and for odd d we have

$$S_A = p_1(l/a)^{d-1} + p_3(l/a)^{d-3} + \ldots + p_{d-2}(l/a)^2 + \tilde{c} \log(l/a), \tag{27.12}$$

which of course includes the case $d = 1$ for $\tilde{c} = c/3$.

Calculation of entanglement entropy in QFT using path integrals

The calculation of the entanglement entropy is done using the *replica trick*, of formal differentiation with respect to an integer n,

$$S_A \equiv -\text{Tr}_A \rho_A \log \rho_A = -\frac{\partial}{\partial n} \text{Tr}_A \rho_A^n \big|_{n=1}. \tag{27.13}$$

We calculate $\text{Tr}_A \rho_A^n$ for integer n and then analytically continue the result in n and calculate S_A from it.

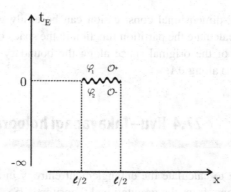

Figure 27.1 Contour for calculation of the reduced density matrix ρ_A.

We consider a $1 + 1$-dimensional quantum field theory to exemplify the main idea.

We can calculate the ground state density matrix $\rho = |\Psi\rangle\langle\Psi|$ from the path integral formalism. Consider flat Euclidean coordinates (t_E, x) and at time $t_E = 0$ take A to be the interval $x \in [-l/2, +l/2]$. The ground state wave functional Ψ depends on the value $\phi_0(x)$ of the field on the $t_E = 0$ line, and is found by path integrating from $t_E = -\infty$ to $t_E = 0$, such that $t_E = 0$ is the boundary of the region of integration,

$$\Psi[\phi_0(x)] = \int_{t_E=-\infty}^{t_E=0} \mathcal{D}\phi \, e^{-S}\Big|_{\phi(t_E=0,x)=\phi_0(x)}. \tag{27.14}$$

This wave functional corresponds to the ket state $|\Psi\rangle$. To find its complex conjugate $\bar{\Psi}[\phi_0(x)]$, corresponding to the bra state $\langle\Psi|$, we path integrate from $t_E = +\infty$ to t_E instead. Then the *total* density matrix ρ's matrix element between states defined by ϕ_0 and ϕ_0' is

$$(\rho)_{\phi_0\phi_0'} = \Psi[\phi_0(x)]\bar{\Psi}[\phi_0'(x)]. \tag{27.15}$$

The reduced density matrix ρ_A is obtained by taking $\phi_0(x) = \phi_0'(x)$ for $x \in B$ and integrating over it, giving for its matrix element between ϕ_1 and ϕ_2,

$$(\rho_A)_{\phi_1\phi_2} = (Z_1)^{-1} \int_{t_E=-\infty}^{t_E=+\infty} \mathcal{D}\phi e^{-S} \prod_{x\in A} \delta(\phi(0+\epsilon,x)-\phi_1(x))\delta(\phi(0-\epsilon,x)-\phi_2(x)). \tag{27.16}$$

Here Z_1 is a normalization factor such that $\mathrm{Tr}_A \rho_A = 1$. So one considers the plane $\mathbb{R}^2$ in (t_E, x) and in it a cut at $t_E = 0$ between $-l/2$ and $+l/2$, and the boundary condition ϕ_1 above the cut, and ϕ_2 below it, as in Fig. 27.1.

Finally, to calculate $\mathrm{Tr}\,\rho_A^n$, we need to calculate

$$(\rho_A)_{\phi_1\phi_2}(\rho_A)_{\phi_2\phi_3} \ldots (\rho_A)_{\phi_n\phi_{n+1}}. \tag{27.17}$$

We see that we need n copies of the above construction on $\mathbb{R}^2$, but with $\phi(0 + \epsilon, x)$ in the first identified with $\phi(0 - \epsilon, x)$ in the next, etc. We easily see that this gives a Riemann surface with n sheets $\mathcal{R}_n$, so we need to calculate the partition function

$$\mathrm{Tr}_A \rho_A^n = (Z_1)^{-n} \int_{(t_E,x)\in\mathcal{R}_n} \mathcal{D}\phi e^{-S[\phi]} \equiv \frac{Z_n}{Z_1^n}, \tag{27.18}$$

where Z_n is the partition function on $\mathcal{R}_n$.

This 2-dimensional construction can be easily generalized to higher dimensions. We need to calculate the partition function in the space obtained by gluing n copies ("Riemann sheets") of the original space along the boundary of A, ∂A, resulting in a deficit angle $2\pi(1-n)$ along ∂A.

27.4 Ryu–Takayanagi holographic prescription

We wish to calculate the entanglement entropy in a conformal field theory in $d+1$ flat dimensions, from the gravitational theory in AdS_{d+2} in Poincaré coordinates (t, x^i, z), $i = 1, \ldots, d$. Then $M = \mathbb{R}^d$ is the spatial manifold, existing on the UV-regularized surface $z = a$. Note that the UV cut-off $z_{\text{UV}} = \epsilon$ was identified with the UV regulator used in the definition of the entanglement entropy scaling. In the case of gluon amplitudes, the preferred regulator was dimensional regularization, but now, as in the case of the Wilson loop, the preferred regulator is $z = \epsilon$, also used in the (related) holographic renormalization method.

We consider a spatial volume $A \subset M$ (at $t = t_0$) bounded by a closed $d-1$-dimensional surface ∂A, and B its complement. In the bulk, we consider the same time slice $t = t_0$, and we consider a d-dimensional surface γ_A that ends on the closed surface ∂A, $\partial \gamma_A = \partial A$. Similarly to the 2-dimensional case of the string worldsheet giving the Wilson loop and also gluon amplitudes, in AdS there is a minimal surface for γ_A, meaning one of minimal area. Then the holographic formula for the entanglement entropy is

$$S_A = \frac{\text{Area}(\gamma_{A,\text{min.}})}{4G_N^{(d+2)}}. \tag{27.19}$$

Note that this formula is very similar to the Bekenstein–Hawking formula for the entropy of a black hole. In fact, it is a kind of generalization of that, since if we consider the AdS Schwarzschild black hole, the minimal surface tends to wrap the horizon, so we indeed obtain the Bekenstein–Hawking formula.

Heuristic derivation

To derive this formula using AdS/CFT, we must start by finding a solution to Einstein's equations with $\Lambda < 0$ (AdS background) that asymptotes to the surface $\mathcal{R}_n$ with n Riemann sheets and deficit angle $\delta = 2\pi(1-n)$ at the surface ∂A, at the boundary of AdS $z \to 0$.

This problem is difficult, but in AdS_3 (dual to CFT_2) we can find a solution. Indeed, gravity in three dimensions has no propagating degrees of freedom (the transverse traceless graviton $g_{\mu\nu}$ has $(D-1)(D-2)/2 - 1 = 0$ degrees of freedom), which implies that solutions are given by patches of AdS_3 glued together, with the only possible nontrivial fact being the existence of lines (codimension two surfaces in AdS_3) where deficit angle is localized, and where matter sources are localized. Such a solution, when lifted to four dimensions on a trivial fourth direction, is called a cosmic string in cosmological context. Then the Ricci scalar is a delta function,

$$R = -4\pi(1-n)\delta(\gamma_A) + \frac{d+2}{d}16\pi G_N^{(d+2)}\Lambda, \qquad (27.20)$$

where the last term is the constant Ricci scalar in AdS space ($\Lambda < 0$), and we have written the formula in AdS_{d+2}, since there is a natural generalization of this construction to γ_A being a codimension two surface of constant deficit angle (though in this case it is not obvious such a solution exists).

Minimizing the Einstein–Hilbert action with the above Ricci scalar implies that we need to find the γ_A of minimal area, $\gamma_{A,\min}$. Substituting in the supergravity action to obtain the on-shell action, we have

$$S_{\text{on-shell,sugra}}(\gamma_{A,\min}) = \int d^{d+2}x\sqrt{-g}\left(\frac{R}{16\pi G_N^{(d+2)}} - \Lambda\right) + \dots$$

$$= -\frac{(1-n)}{4G_N^{(d+2)}}\text{Area}(\gamma_{A,\min}) + \text{const.} \qquad (27.21)$$

But by AdS/CFT, the quantity we want to evaluate in order to find Tr (ρ_A^n) via (27.18), the CFT partition function on $\mathcal{R}_n$, Z_n, equals the supergravity partition function in the classical limit, $\exp[-S_{\text{on-shell,sugra}}]$. Since for $n = 1$, $S_{\text{on-shell,sugra}}(\gamma_{A,\min})$ gives zero, we obtain

$$S_A = -\frac{\partial}{\partial n}\text{Tr}\,[\rho_A^n]\Big|_{n=1} = -\frac{\partial}{\partial n}\left[\frac{(1-n)\text{Area}(\gamma_{A,\min})}{4G_N^{(d+2)}}\right]_{n=1} = \frac{\text{Area}(\gamma_{A,\min})}{4G_N^{(d+2)}}. \qquad (27.22)$$

q.e.d.

27.5 Holographic entanglement entropy in two dimensions

We now consider the application of the formalism above to field theories in two dimensions. For a conformal field theory in two dimensions, a holographic calculation of the central charge, as in the 4-dimensional case in (23.16), gives

$$c = \frac{3R}{2G_N^{(3)}}. \qquad (27.23)$$

We want then to reproduce holographically the formula (27.10) with the above central charge. For that, we need an infinitely long total system and a subsystem A of finite length l, chosen between $x = -l/2$ and $x = +l/2$. For the holographic calculation in Poincaré coordinates, γ_A is the geodesic line obeying the boundary condition at $z = a$ that the endpoints are $(x, z) = (-l/2, a)$ and $(x, z) = (+l/2, a)$.

The geodesic is found to be

$$(x, z) = \frac{l}{2}(\cos s, \sin s), \quad \epsilon \le s \le \pi - \epsilon, \qquad (27.24)$$

where $\epsilon = 2a/l$. One can check that this satisfies the geodesic equation, but the proof is left as an exercise. Then the total length ("Area" L) is

$$L(\gamma) = 2R \int_{\epsilon}^{\pi/2} \frac{ds}{\sin s} = 2R \log \frac{l}{a}. \tag{27.25}$$

Then the entanglement entropy is

$$S_A = \frac{L(\gamma_A)}{4G_N^{(3)}} = \frac{R}{2G_N^{(3)}} \log \frac{l}{a} = \frac{c}{3} \log \frac{l}{a}. \tag{27.26}$$

q.e.d.

27.6 Holographic entanglement as order parameter and confinement/deconfinement transition

The holographic entanglement entropy can act as an order parameter for the confinement/deconfinement phase in a confining gauge theory. Indeed, we can guess this fact from the analogous fact for the Wilson loop, calculated holographically from the string worldsheet of minimal area ending on a fixed contour on the boundary. We saw in that case that the Wilson loop gave an area law scaling at large distances in the confining phase, and a Coulomb scaling in the deconfined phase.

The simplest case is the case considered by Witten, of dimensional reduction on Euclidean time down to 3-dimensional pure Yang–Mills (since the fermions become massive, as they have antiperiodic boundary conditions on the compact direction), a confining theory. We can then Wick rotate back to Minkowski space on a different direction to obtain the *AdS soliton*, a double Wick rotated black hole,

$$ds^2 = \frac{r^2}{R^2}(-dt^2 + f(r)\chi^2 + dx_1^2 + dx_2^2) + R^2 \frac{dr^2}{r^2 f(r)}, \tag{27.27}$$

dual to pure Yang-Mills after reduction on the compact direction χ. Here $f(r) = 1 - r_0^4/r^4$ and the periodicity of χ is $L = \pi R^2/r_0$.

In the reduced theory (trivial χ dependence for everything), γ_A is 2-dimensional, like the string worldsheet giving a Wilson loop, so the calculation is very similar, but the contour C on which this "Euclidean worldsheet" ends is in the x_1, x_2 plane, where it bounds a closed spatial region, taken to be the same infinitely long rectangle as in the Wilson loop case, $-l/2 \leq x_1 \leq l/2, 0 \leq x_2 \leq V$, with $V \to \infty$ taking the role of T for the Wilson loop.

We can then parallel the calculation of the Wilson loop here and find we need to minimize

$$\frac{\text{Area}}{L} \simeq V \int_{-l/2}^{l/2} dx_1 \frac{r}{R} \sqrt{\left(\frac{dr}{dx_1}\right)^2 + \frac{r^4 f(r)}{R^4}}. \tag{27.28}$$

We then think of x_1 as "time," so the "Hamiltonian" $H = (dr/dx_1)\delta \text{Area}/\delta(dr/dx_1) - \text{Area}$ is conserved (x_1 independent), which gives

$$\frac{dr}{dx_1} = \frac{r^2}{R^2}\sqrt{f(r)\left(\frac{r^6 f(r)}{r_*^6 f(r_*)} - 1\right)}, \tag{27.29}$$

integrating to

$$\int dx_1 = \frac{l}{2} = \frac{Lr_0}{\pi}\int_{r_*}^{\Lambda} dr \frac{1}{r^2\sqrt{f(r)\left(\frac{r^6 f(r)}{r_*^6 f(r_*)} - 1\right)}}, \tag{27.30}$$

where we have used $R^2 = Lr_0/\pi$. Here r_* is defined as the minimal value of r, since we parameterized the constant "energy" such that $dr/dx_1 = 0$ (inflexion point of the curve) when $r = r_*$, and Λ is a UV cut-off. The relation above implies a maximum l for which we have a solution, obtained when $r_* \to r_0$, i.e. when the curve descends all the way to the horizon, touches it and goes back up,

$$l \leq l_{\max} \simeq 0.22L. \tag{27.31}$$

For $l > l_{\max}$, there is no *connected* minimal surface, and instead there are just two disconnected lines descending from the boundary to the horizon, so their lengths are independent of the separation l.

This means that the entanglement entropy, related to the area of the minimal "worldsheet," i.e. to the length Area/LV, depends on which is smaller. For $l \leq l_{\max}$, the minimal surface is the connected one, and we obtain a holographic entropy that is l dependent, which in fact has a finite part that scales as $l \to 0$ as

$$S_A(l)|_{\text{finite}} = -c_1 VN^2 l^{-(d-1)}, \tag{27.32}$$

and is a constant $(-c_2 VN^2)$ for $l > l_{\max}$, a behavior characteristic of a phase transition. One can make this more obvious by defining the derivative of S_A,

$$C(l) \equiv \frac{l^d}{V}\frac{dS_A(l)}{dl}, \tag{27.33}$$

which is discontinuous, jumping from a finite value at $l \leq l_{\max}$ to zero at $l > l_{\max}$. Therefore $C(l)$, which also does not depend on the constant divergence of S_A, acts as a good order parameter for the confinement/deconfinement phase transition.

The above example was for a confining background with constant dilaton $\phi = \phi_0$. But the Ryu–Takayanagi prescription has a proposed (conjectured) extension to confining backgrounds (e.g. Klebanov–Strassler),

$$S = -\frac{1}{4G_N}\int e^{-2\phi}\text{Area}, \tag{27.34}$$

where the $e^{-2\phi}$ factor is natural, because $1/G_N$ contains $1/g_s^2 = e^{-2\langle\phi\rangle}$. The entanglement entropy again acts as an order parameter for a phase transition. For a confining phase, there is an $l_{\max}$ as above, but it is found that sometimes there could be a confining $V_{q\bar{q}}$ from the Wilson loop, yet no $l_{\max}$ in the entropy, so the relation between these two order parameters is not clear.

Important concepts to remember

- The entanglement entropy is the von Neumann entropy of a pure total state of two entangled subsystems A and B, when we trace over the degrees of freedom of one of them, A.
- The entanglement entropy is non-extensive and $S_A = S_B$.
- The black hole information paradox can be intuitively understood as entanglement between the inside and outside of the event horizon. There is entanglement entropy for the outside observer even if the total state is a pure state.
- Entanglement entropy in a QFT is defined by separating space in two regions by a closed surface ∂A, and is UV divergent and the leading divergence is proportional to the area of ∂A.
- The S_A is calculated from Tr $[\rho_A^n]$ by the replica trick, and Tr $[\rho_A^n]$ is calculated from the partition function on a space $\mathcal{R}_n$ with n Riemann sheets at ∂A.
- The holographic entanglement entropy is found from the minimal area of a surface γ_A ending on ∂A at the boundary as $S_A = \text{Area}(\gamma_{A,\text{min}})/4G_N$.
- The derivative with respect to the size l of A of S_A acts as an order parameter for the confinement/deconfinement phase transition in gauge theories.

References and further reading

The prescription to holographically calculate entanglement entropy was proposed in [117]. A review of holographic entanglement entropy can be found in [118].

Exercises

1. Calculate the mutual information for the infinite 1-dimensional chain, with A the line between $x = -l/2$ and $x = +l/2$. Explore the limits $l \to 0$ and $l \to \infty$.
2. Use the holographic formula (27.19) to prove strong subadditivity, Eq. (27.5).
3. Generalize the construction of the entanglement entropy of QFT in 1+1 dimensions in Section 27.3 to 2+1 dimensions, for a boundary $\partial A=$ circle in the two spatial dimensions.
4. Prove that the solution in (27.24) is a geodesic, satisfying the geodesic equation

$$\frac{d^2 x^\mu}{d\tau} + \Gamma_{\nu\rho}^\mu \frac{dx^\nu}{d\tau} \frac{dx^\rho}{d\tau} = 0, \qquad (27.35)$$

where $d\tau$ is the length element along the trajectory.

5. Consider the gravity dual obtained from the near-horizon limit of (extremal) D2-branes. Calculate the holographic entanglement entropy for the boundary closed spatial contour C given by $-l/2 \le x_1 \le l/2, 0 \le x_2 \le V$, with $V \to \infty$.

6. Compare the holographic calculation in 2+1 dimensions of the Wilson loop to the calculation of the entanglement entropy (27.28). What are the similarities and differences? Keeping $f(r)$ general, can one find a result that shows a phase transition in the Wilson loop, but does not show a phase transition in the entanglement entropy?

References

[1] M. E. Peskin and D. V. Schroeder, *An Introduction to Quantum Field Theory*, Perseus Books, 1995.

[2] Steven Weinberg, *The Quantum Theory of Fields*, vols. 1,2,3, Cambridge University Press, 1995, 1996, and 2000.

[3] P. J. E. Peebles, *Principles of Physical Cosmology*, Princeton University Press, 1993.

[4] C. W. Misner, K. S. Thorne, and J. A. Wheeler, *Gravitation*, W. H. Freeman and Co., 1970.

[5] L. D. Landau and E. M. Lifshitz, *Mechanics*, Butterworth-Heinemann, 1982.

[6] R. M. Wald, *General Relativity*, University of Chicago Press, 1984.

[7] S. V. Ketov, *Fortsch. Phys.* **45**, 237 (1997) [hep-th/9611209].

[8] L. Alvarez-Gaume and S. F. Hassan, *Fortsch. Phys.* **45**, 159 (1997) [hep-th/9701069].

[9] P. West, *Introduction to Supersymmetry and Supergravity*, World Scientific, 1990.

[10] J. Wess and J. Bagger, *Supersymmetry and Supergravity*, Princeton University Press, 1992.

[11] S. J. Gates, M. T. Grisaru, M. Rocek, and W. Siegel, hep-th/0108200.

[12] M. Dine, *Supersymmetry and String Theory, Beyond the Standard Model*, Cambridge University Press, 2007.

[13] P. van Nieuwenhuizen, Les Houches 1983, Proceedings, "Relativity, groups and topology II," p. 823.

[14] P. Van Nieuwenhuizen, *Phys. Rept.* **68**, 189 (1981).

[15] D. Z. Freedman and A. van Proeyen, *Supergravity*, Cambridge University Press, 2012.

[16] M. J. Duff, B. E. W. Nilsson, and C. N. Pope, *Phys. Rept.* **130**, 1 (1986).

[17] S. W. Hawking and G. F. R. Ellis, *The Large Scale Structure of Space–Time*, Cambridge University Press, 1973.

[18] M. J. Duff, R. R. Khuri, and J. X. Lu, *Phys. Rept.* **259**, 213 (1995) [hep-th/9412184].

[19] A. A. Tseytlin, *Nucl. Phys.* B **475**, 149 (1996) [hep-th/9604035].

[20] M. Cvetic and A. A. Tseytlin, *Nucl. Phys. B* **478**, 181 (1996) [hep-th/9606033].

[21] B. Zwiebach, *A First Course in String Theory*, Cambridge University Press, 2004.

[22] M. B. Green, J. H. Schwarz, and E. Witten, *Superstring Theory*, vols. 1 and 2, Cambridge University Press, 1987.

[23] Joseph Polchinski, *String Theory*, vols. 1 and 2, Cambridge University Press, 1998.

[24] K. Becker, M. Becker, and J. H. Schwarz, *String Theory and M-theory*, Cambridge University Press, 2007.

[25] Clifford Johnson, *D-branes*, Cambridge University Press, 2003.

[26] O. Aharony, S. S. Gubser, J. M. Maldacena, H. Ooguri, and Y. Oz, *Phys. Rept.* **323**, 183 (2000) [hep-th/9905111].

[27] J. Dai, R. G. Leigh, and J. Polchinski, *Mod. Phys. Lett. A* **4**, 2073 (1989).

[28] J. Polchinski, *Phys. Rev. Lett.* **75**, 4724 (1995) [hep-th/9510017].

[29] E. D'Hoker and D. Z. Freedman, hep-th/0201253.

[30] J. M. Maldacena, *Adv. Theor. Math. Phys.* **2**, 231 (1998) [hep-th/9711200].

[31] S. S. Gubser, I. R. Klebanov, and A. M. Polyakov, *Phys. Lett.* B **428**, 105 (1998) [hep-th/9802109].

[32] E. Witten, *Adv. Theor. Math. Phys.* **2**, 253 (1998) [hep-th/9802150].

[33] H. J. Kim, L. J. Romans, and P. van Nieuwenhuizen, *Phys. Rev. D* **32**, 389 (1985).

[34] B. de Wit and H. Nicolai, *Nucl. Phys. B* **243**, 91 (1984); B. de Wit, H. Nicolai, and N. P. Warner, *Nucl. Phys. B* **255**, 29 (1985); B. de Wit and H. Nicolai, *Nucl. Phys. B* **274**, 363 (1986); B. de Wit and H. Nicolai, *Nucl. Phys. B* **281**, 211 (1987).

[35] H. Năstase, D. Vaman, and P. van Nieuwenhuizen, *Phys. Lett.* B **469**, 96 (1999) [hep-th/9905075].

[36] H. Năstase, D. Vaman, and P. van Nieuwenhuizen, *Nucl. Phys. B* **581** (2000) 179 [hep-th/9911238].

[37] W. Mueck and K. S. Viswanathan, *Phys. Rev. D* **58**, 041901 (1998) [hep-th/9804035].

[38] D. Z. Freedman, S. D. Mathur, A. Matusis, and L. Rastelli, *Nucl. Phys. B* **546**, 96 (1999) [hep-th/9804058].

[39] G. Chalmers, H. Năstase, K. Schalm, and R. Siebelink, *Nucl. Phys. B* **540**, 247 (1999) [hep-th/9805105].

[40] A. Smilga, *Lectures on Quantum Chromodynamics*, World Scientific, 2001.

[41] J. M. Maldacena, *Phys. Rev. Lett.* **80**, 4859 (1998) [hep-th/9803002].

[42] S. -J. Rey and J. -T. Yee, *Eur. Phys. J. C* **22**, 379 (2001) [hep-th/9803001].

[43] E. Witten, *Adv. Theor. Math. Phys.* **2**, 505 (1998) [hep-th/9803131].

[44] J. Polchinski and M. J. Strassler, *Phys. Rev. Lett.* **88**, 031601 (2002) [hep-th/0109174].

[45] D. E. Berenstein, J. M. Maldacena, and H. S. Năstase, *JHEP* **0204**, 013 (2002) [hep-th/0202021].

[46] J. C. Plefka, *Fortsch. Phys.* **52**, 264 (2004) [hep-th/0307101].

[47] J. Kowalski-Glikman, *Phys. Lett.* B **134**, 194 (1984).

[48] P. C. Aichelburg and R. U. Sexl, *Gen. Rel. Grav.* **2**, 303 (1971).

[49] G. T. Horowitz and A. R. Steif, *Phys. Rev. Lett.* **64**, 260 (1990).

[50] R. Penrose, Any spacetime has a plane wave as a limit, *Differential Geometry and Relativity*, Reidel, Dordrecht, 1976, pp. 271–275.

[51] M. Blau, J. M. Figueroa-O'Farrill, C. Hull, and G. Papadopoulos, *JHEP* **0201**, 047 (2002) [hep-th/0110242].

[52] M. Blau, J. M. Figueroa-O'Farrill, C. Hull, and G. Papadopoulos, *Class. Quant. Grav.* **19**, L87 (2002) [hep-th/0201081].

[53] J. A. Minahan and K. Zarembo, *JHEP* **0303**, 013 (2003) [hep-th/0212208].

[54] J. Plefka, *Living Rev. Rel.* **8**, 9 (2005) [hep-th/0507136].

[55] K. Zarembo, *Comptes Rendus Physique* **5**, 1081 (2004) [*Fortsch. Phys.* **53**, 647 (2005)] [hep-th/0411191].

[56] E. Witten, *Nucl. Phys. B* **443**, 85 (1995) [hep-th/9503124].

[57] S. B. Giddings, *Phys. Rev. D* **67**, 126001 (2003) [hep-th/0203004].

[58] K. Kang and H. Năstase, *Phys. Rev. D* **72**, 106003 (2005) [hep-th/0410173].

[59] K. Kang and H. Năstase, *Phys. Lett.* B **624**, 125 (2005) [hep-th/0501038].

[60] J. Polchinski and M. J. Strassler, hep-th/0003136.

[61] I. R. Klebanov and M. J. Strassler, *JHEP* **0008**, 052 (2000) [hep-th/0007191].

[62] J. M. Maldacena and C. Nunez, *Phys. Rev. Lett.* **86**, 588 (2001) [hep-th/0008001].

[63] J. M. Maldacena and H. S. Năstase, *JHEP* **0109**, 024 (2001) [hep-th/0105049].

[64] T. Sakai and S. Sugimoto, *Prog. Theor. Phys.* **113**, 843 (2005) [hep-th/0412141].

[65] V. Balasubramanian, P. Kraus, A. E. Lawrence, and S. P. Trivedi, *Phys. Rev. D* **59**, 104021 (1999) [hep-th/9808017].

[66] V. Balasubramanian, P. Kraus, and A. E. Lawrence, *Phys. Rev. D* **59**, 046003 (1999) [hep-th/9805171].

[67] S. B. Giddings, *Phys. Rev. D* **61**, 106008 (2000) [hep-th/9907129].

[68] E. Witten, *JHEP* **9807**, 006 (1998) [hep-th/9805112].

[69] S. A. Hartnoll and P. Kovtun, *Phys. Rev. D* **76**, 066001 (2007) [arXiv:0704.1160 [hep-th]].

[70] E. D'Hoker and P. Kraus, *JHEP* **0910**, 088 (2009) [arXiv:0908.3875 [hep-th]].

[71] S. A. Hartnoll, *Class. Quant. Grav.* **26**, 224002 (2009) [arXiv:0903.3246 [hep-th]].

[72] J. Casalderrey-Solana, H. Liu, D. Mateos, K. Rajagopal, and U. A. Wiedemann, arXiv:1101.0618 [hep-th].

[73] H. Liu, K. Rajagopal, and U. A. Wiedemann, *Phys. Rev. Lett.* **97**, 182301 (2006) [hep-ph/0605178].

[74] H. Năstase, hep-th/0501068.

[75] S. S. Gubser, S. S. Pufu, and A. Yarom, *Phys. Rev. D* **78**, 066014 (2008) [arXiv:0805.1551 [hep-th]].

[76] O. Aharony, A. Fayyazuddin, and J. M. Maldacena, *JHEP* **9807**, 013 (1998) [hep-th/9806159].

[77] D. E. Berenstein, E. Gava, J. M. Maldacena, K. S. Narain, and H. S. Năstase, hep-th/0203249.

[78] O. Aharony, O. Bergman, D. L. Jafferis, and J. Maldacena, *JHEP* **0810**, 091 (2008) [arXiv:0806.1218 [hep-th]].

[79] S. Mukhi and C. Papageorgakis, *JHEP* **0805**, 085 (2008) [arXiv:0803.3218 [hep-th]].

[80] H. Năstase, C. Papageorgakis, and S. Ramgoolam, *JHEP* **0905**, 123 (2009) [arXiv:0903.3966 [hep-th]].

[81] H. Năstase and C. Papageorgakis, *JHEP* **0912**, 049 (2009) [arXiv:0908.3263 [hep-th]].

[82] X. Chu, H. Năstase, B. E. W. Nilsson, and C. Papageorgakis, *JHEP* **1104**, 040 (2011) [arXiv:1012.5969 [hep-th]].

[83] J. Gomis, D. Rodriguez-Gomez, M. Van Raamsdonk, and H. Verlinde, *JHEP* **0809**, 113 (2008) [arXiv:0807.1074 [hep-th]].

[84] C. Csaki, H. Ooguri, Y. Oz, and J. Terning, *JHEP* **9901**, 017 (1999) [hep-th/9806021].

[85] A. Karch and E. Katz, *JHEP* **0206**, 043 (2002) [hep-th/0205236].

[86] K. Skenderis, *Class. Quant. Grav.* **19**, 5849 (2002) [hep-th/0209067].

[87] M. Henningson and K. Skenderis, *JHEP* **9807**, 023 (1998) [hep-th/9806087].

[88] D. Z. Freedman, S. S. Gubser, K. Pilch, and N. P. Warner, *Adv. Theor. Math. Phys.* **3**, 363 (1999) [hep-th/9904017].

[89] R. C. Myers and A. Sinha, *JHEP* **1101**, 125 (2011) [arXiv:1011.5819 [hep-th]].

[90] Z. Komargodski and A. Schwimmer, *JHEP* **1112**, 099 (2011) [arXiv:1107.3987 [hep-th]].

[91] D. Astefanesei, H. Năstase, H. Yavartanoo, and S. Yun, *JHEP* **0804**, 074 (2008) [arXiv:0711.0036 [hep-th]].

[92] J. Erlich, E. Katz, D. T. Son, and M. A. Stephanov, *Phys. Rev. Lett.* **95**, 261602 (2005) [hep-ph/0501128].

[93] A. Karch, E. Katz, D. T. Son, and M. A. Stephanov, *Phys. Rev. D* **74**, 015005 (2006) [hep-ph/0602229].

[94] U. Gursoy and E. Kiritsis, *JHEP* **0802**, 032 (2008) [arXiv:0707.1324 [hep-th]].

[95] S. Kachru, X. Liu, and M. Mulligan, *Phys. Rev. D* **78**, 106005 (2008) [arXiv:0808.1725 [hep-th]].

[96] D. T. Son, *Phys. Rev. D* **78**, 046003 (2008) [arXiv:0804.3972 [hep-th]].

[97] K. Balasubramanian and J. McGreevy, *Phys. Rev. Lett.* **101**, 061601 (2008) [arXiv:0804.4053 [hep-th]].

[98] T. Griffin, P. Horava, and C. M. Melby-Thompson, *Phys. Rev. Lett.* **110**, 081602 (2013) [arXiv:1211.4872 [hep-th]].

[99] C. P. Herzog, M. Rangamani, and S. F. Ross, *JHEP* **0811**, 080 (2008) [arXiv:0807.1099 [hep-th]].

[100] A. Adams, K. Balasubramanian, and J. McGreevy, *JHEP* **0811**, 059 (2008) [arXiv:0807.1111 [hep-th]].

[101] J. Maldacena, D. Martelli, and Y. Tachikawa, *JHEP* **0810**, 072 (2008) [arXiv:0807.1100 [hep-th]].

[102] S. A. Hartnoll, C. P. Herzog, and G. T. Horowitz, *Phys. Rev. Lett.* **101**, 031601 (2008) [arXiv:0803.3295 [hep-th]].

[103] P. Kovtun, *J. Phys. A* **45**, 473001 (2012) [arXiv:1205.5040 [hep-th]].

[104] D. T. Son and A. O. Starinets, *JHEP* **0209**, 042 (2002) [hep-th/0205051].

[105] G. Policastro, D. T. Son, and A. O. Starinets, *Phys. Rev. Lett.* **87**, 081601 (2001) [hep-th/0104066].

[106] P. Kovtun, D. T. Son, and A. O. Starinets, *Phys. Rev. Lett.* **94**, 111601 (2005) [hep-th/0405231].

[107] S. S. Gubser, *Phys. Rev. D* **78**, 065034 (2008) [arXiv:0801.2977 [hep-th]].

[108] R. C. Myers, S. Sachdev, and A. Singh, *Phys. Rev. D* **83**, 066017 (2011) [arXiv:1010.0443 [hep-th]].

[109] S. Sachdev, *Ann. Rev. Condensed Matter Phys.* **3**, 9 (2012) [arXiv:1108.1197 [cond-mat.str-el]].

[110] A. Mohammed, J. Murugan, and H. Năstase, *Phys. Rev. Lett.* **109**, 181601 (2012) [arXiv:1205.5833 [hep-th]].

[111] N. Iqbal, H. Liu, and M. Mezei, arXiv:1110.3814 [hep-th].

[112] C. P. Herzog, *J. Phys. A* **42**, 343001 (2009) [arXiv:0904.1975 [hep-th]].

[113] J. McGreevy, *Adv. High Energy Phys.* **2010**, 723105 (2010) [arXiv:0909.0518 [hep-th]].

[114] L. F. Alday and J. M. Maldacena, *JHEP* **0706**, 064 (2007) [arXiv:0705.0303 [hep-th]].

[115] E. I. Buchbinder, *Phys. Lett. B* **654**, 46 (2007) [arXiv:0706.2015 [hep-th]].

[116] N. Berkovits and J. Maldacena, *JHEP* **0809**, 062 (2008) [arXiv:0807.3196 [hep-th]].

[117] S. Ryu and T. Takayanagi, *Phys. Rev. Lett.* **96**, 181602 (2006) [hep-th/0603001].

[118] T. Nishioka, S. Ryu, and T. Takayanagi, *J. Phys. A* **42**, 504008 (2009) [arXiv:0905.0932 [hep-th]].

Index

Printed in the United States
by Baker & Taylor Publisher Services